D1765934

WITHDRAWN
FROM
UNIVERSITIES
AT
MEDWAY
LIBRARY

The Ecology of
Butterflies in Britain

The Ecology of Butterflies in Britain

Edited by
Roger L. H. Dennis

Figures prepared by
Derek A. A. Whiteley

Oxford New York Tokyo
OXFORD UNIVERSITY PRESS
1992

Oxford University Press, Walton Street, Oxford OX2 6DP

Oxford New York Toronto
Delhi Bombay Calcutta Madras Karachi
Petaling Jaya Singapore Hong Kong Tokyo
Nairobi Dar es Salaam Cape Town
Melbourne Auckland

and associated companies in
Berlin Ibadan

Oxford is a trade mark of Oxford University Press

Published in the United States
by Oxford University Press, New York

A catalogue record for this book is available from the British Library

Library of Congress Cataloging in Publication Data

The Ecology of Butterflies in Britain/edited by Roger L. H. Dennis;
figures prepared by Derek A. A. Whiteley.
1. Butterflies—Great Britain—Ecology. 2. Butterflies—Great Britain. I. Dennis, Roger L. H.
QL555.G7E26 1993 595.78'9045'0941—dc20 92–11406

ISBN 0 19 854025 6

Typeset by Joshua Associates Ltd, Oxford
Printed and bound in Great Britain by
BPCC Hazells Ltd, Aylesbury

To our families:

Margaret and Pamela; Alison; Bunty and Laurel; Rita, Hannah, and Oliver; Dee, Jennifer, and Richard; Titia, Lucille, and Timothy; Sarah, Emily, and Anna; David, Dan, and Jeremy.

Preface

There are now many books on British butterflies which provide a systematic treatment of species. These have become encyclopaedic in the detail they provide, and are often excellently illustrated. Yet very few books have adequately addressed biological themes, and made sense of, what must appear to a newcomer in the subject, the bewildering variation in life history and morphology presented by these organisms. The exception, of course, is E. B. Ford's landmark publication *Butterflies*. The present book has been stimulated by the extensive factual and conceptual advances in the subject since the third edition of that work, which was published in 1957. The book is the product of two meetings: one at Furzebrook, Dorset in March 1985, and the other at Princes Risborough, Buckinghamshire in June 1987; the venues were kindly provided by Jeremy Thomas and Tim Shreeve.

The dominant theme in the book, as in Ford's *Butterflies*, is evolution. The keyword 'ecology' in the title emphasizes that not only do most adaptations and evolutionary change arise from interactions between different organisms and between organisms and their environment, but that the conceptual framework and techniques responsible for most of the discoveries over the last two decades belong to ecology. The subject matter progresses from lower to higher levels of organization and complexity (from chapters 2 to 10), from individual behaviour and adaptations (chapters 2 and 3) to populations (chapters 4, 5, and 6) and communities (chapter 7), and finally to genetic and evolutionary theory (chapters 8, 9 and 10). The historical view, pioneered by Bryan Beirne and thoroughly approved by Ford, occupies the penultimate chapter. Differences in content between *Butterflies* and the current work largely reflect the present necessary preoccupation with the conservation of our fauna. Conservation primarily evokes spatial issues, and accounts for the biogeographical subject matter of the first chapter—statements and questions about the butterfly bank. As the successful conservation of our butterflies depends on the extent of our knowledge of them, as much as on our willingness to act on their behalf, the final chapter is an appropriate place for this topic.

The primary objective has been to produce a book that opens the subject to as wide an audience as possible, without devaluation of its scientific content, one that will entertain as well as inform. It should be of particular value for those who intend to undertake research and require an introduction to the subject and for others who will formulate policy that may influence land use and wildlife. The urgency of the need to encourage respect and consideration for the plight of butterflies and other organisms worldwide is regarded as paramount, as is the need for permanent and effective governmental commitment to Britain's excellent institutions that do crucial taxonomy, field research, mapping, and monitoring. Foremost amongst these are the Natural History Museum, the Institute of Terrestrial Ecology research stations, the Biological Records Centre, the Butterfly Monitoring Scheme, and derivative organizations of the Nature Conservancy Council.

The book includes several features to assist readership. English names for species are given at their first mention in each chapter, and there is an extensive glossary of terms. To establish some uniformity, the scientific nomenclature for species is that established in Emmet and Heath (1989). The Appendices include a check list of butterflies on mainland Britain, together with their larval hostplants and a detailed habitat classification. The main advantage of a regional structure for a thematic text is that it should enable the reader to understand how the various biological concepts interlock for a manageable number of species. Occasionally, it has been necessary to travel beyond the British islands to illustrate some ideas. An important feature, to illuminate concepts more fully, has been the inclusion of a large number and wide variety of figures. Among these, a small number of process–response models have been drawn to discourage 'tunnel' reasoning and to encourage an heuristic approach to biological systems. Unfortunately, not all my objectives could be attained. The production costs of this work precluded colour plates to illustrate habitats and behaviour and detailed distribution maps for the species. The latter can be found in Heath, Pollard, and Thomas (1984) and Emmet and Heath (1989).

The preparation of the book has absorbed my attention for some seven years and I am anxious to express my gratitude, as much as space will allow, to those friends and colleagues who have ensured its coming to fruition. I cannot eulogize sufficiently the enthusiastic collaboration, selflessness, and kindnesses of my fellow contributors: Tim Shreeve, Keith Porter, Martin Warren, Paul Brakefield, Jeremy Thomas, Caroline Steel, and Derek Whiteley. Each of the authors is a highly respected professional in the field of ecology and conservation. I am especially indebted to Tim Shreeve; he accumulated references for the bibliography and was instrumental in encouraging Derek Whiteley to apply his skill to preparing our figures, and also assisted him in this formidable task. I have benefited enormously from our conversations, his ceaseless optimism and encouragement, and have enjoyed immensely our collaboration over joint ventures that have arisen from preparing this book. Keith Porter also earns my deep gratitude for adopting chapter 7 amidst other demanding responsibilities in the Nature Conservancy Council. I would like to express my thanks to those acknowledged in the text and to those not mentioned there but who have contributed in their own unique manner, and apologize to those whose information could not be used ultimately due to limitations of space. The School of Biological and Molecular Sciences, Oxford Polytechnic, provided the facilities for preparation of the figures. I would like to thank the members of staff of the Oxford University Press who have been responsible for the rapid and efficient progress of the text towards publication during the past year; Pamela Gilbert for seeking out difficult references; Paul Harding and Geoff Radford for distribution data; Brian Gardiner and John Feltwell for their expert advice on *Pieris brassicae*, and Ove Høegh-Guldberg for his on *Aricia*; Ashley Morton for valuable information relating to chapter 7 and the long-term loan of his splendid Ph.D. thesis; Ian Morgan, Neville Birkett, Ian Rippey, and the late Russell Bretherton for distribution records; Keith Bennett for advice on Devensian and Holocene vegetation patterns; a number of friends for help and encouragement over the years; Steve Courtney—whose own superb research work on the Pieridae triggered my interest in butterfly ecology; Robert (Bob) Williams—for his mastery of mainframes and his readiness ever to assist on computer matters; Eric Classey, Tom Dunn, John Heath, the late Stanley Jacobs, Ian Lorimer, Hugh Michaelis, Ernie Pollard, Ian Rutherford, George Thomson, Gerry Tremewan, and Dick Vane-Wright. As this book includes reference to surveys at Brereton Heath and the Bollin valley in Cheshire during the 1980s, I would like to record my gratitude to the boys of the Manchester Grammar School Natural History Society who took part and added to my enjoyment and to the ranger service at the sites—to Malcolm Ainsworth, Ian Mottershead, and Philip Robinson—for their enthusiastic support. Finally, to Margaret, my wife, my deep appreciation for all her help and patience during this marathon.

Wilmslow, Cheshire R.L.H.D.
April 1992

Contents

Contributors

Professor P. M. Brakefield, Section of Evolutionary Biology, Department of Population Biology, University of Leiden, Schelpenkade 14a, 2313 ZT, Leiden, The Netherlands.

Dr. R. L. H. Dennis, The Manchester Grammar School, Manchester M13 0XT.

Dr. K. Porter, English Nature, Foxhold House, Thornford Road, Crookham Common, Newbury, Berkshire, RG15 8EL.

Dr. T. G. Shreeve, School of Biological and Molecular Sciences, Oxford Polytechnic, Gipsy Lane, Oxford OX3 0BP.

Mrs. C. A. Steel, Brasenose Farm, Eastern Bypass, Oxford OX4 2QZ.

Dr. J. A. Thomas, Institute of Terrestrial Ecology, Furzebrook Research Station, Wareham, Dorset, BH20 5AS.

Dr. M. S. Warren, 2 Fairside, Higher Ansty, near Dorchester, Dorset, DT2 7PS.

Mr. Derek Whiteley, School of Biological and Molecular Sciences, Oxford Polytechnic, Gipsy Lane, Oxford OX3 0BP.

1

Islands, regions, ranges, and gradients

Roger L. H. Dennis

Butterflies are conspicuous insects. Because of this and because there are fewer species of butterflies than in other groups of insects, they are usually easily recognized and their distributions for the most part well known. This book starts, then, by asking a basic but most important biogeographical question: why do different butterfly species occur where they do but not everywhere? The full answer is rarely simple. Essentially, distribution patterns depend on at least three factors: (i) the behavioural and ecological requirements of the species, (ii) the geographical extent of the appropriate environmental conditions and resources, and (iii) the historical opportunity for organisms to colonize new areas. The degree to which butterflies are influenced by different factors depends greatly on the scale at which observations are made, though scale interdependence ensures that what happens at one scale cannot be entirely separated, in terms of cause and effect, from another. Local, regional, and continental patterns are linked in numerous ways. In this chapter, an array of problems concerning distributions is investigated at a wide range of scales, from the British islands as a whole to issues amid local settings.

1.1 Butterflies on British islands

Islands typically have fewer species than mainland areas of a similar size. It has long been known that the numbers of organisms on islands relate to their area and their isolation from a continental source; the smaller the island and the further it is away from the continental source, the fewer the species (Fig. 1.1; Munroe 1948; Preston 1962; MacArthur and Wilson 1967; Bengtson and Enckell 1983). Both of these factors influence the butterfly fauna of British islands, but to date statistical analyses have produced conflicting results. Relatively greater significance is attributed by Hockin (1981) to distance from the nearest source and by Reed (1985) to island area—or more specifically habitat diversity as measured by the number of plant species—despite the similarity of simple correlations (Table 1.1). Several reasons emerge for the discrepancies. They can be resolved partly in the different numbers of islands investigated and the numbers of butterflies these authors associate with islands. There is naturally a bias towards having more information on the species of larger islands and greater certainty as to the breeding status of species occurring on them. As regards interpretation, it is less straightforward to apply area and isolation to British islands than it is to tropical and warm temperate counterparts (Scott 1972; Hockin 1980; Miller 1984). Although, the British islands do not cover a large area, they cut latitudinally across steep and distinctive climatic gradients (Dennis 1977). As a result, and in some measure to the geography of the islands, predictor variables selected to explain species richness tend to be highly correlated with one another, frustrating an island biogeography analysis. Numbers of butterflies on the islands relate closely to the number of potential colonizers at the nearest source. In turn, latitude, and thus environmental constraints, determine the number of species at this source (Table 1.1; Dennis 1977, 1985*a*; Dennis and Williams 1986). Thereafter, island area and isolation have a varying influence depending on how many of the British islands are considered. Most British islands are tiny and, as more of these are entered into the analysis, island area assumes an increasingly significant role.

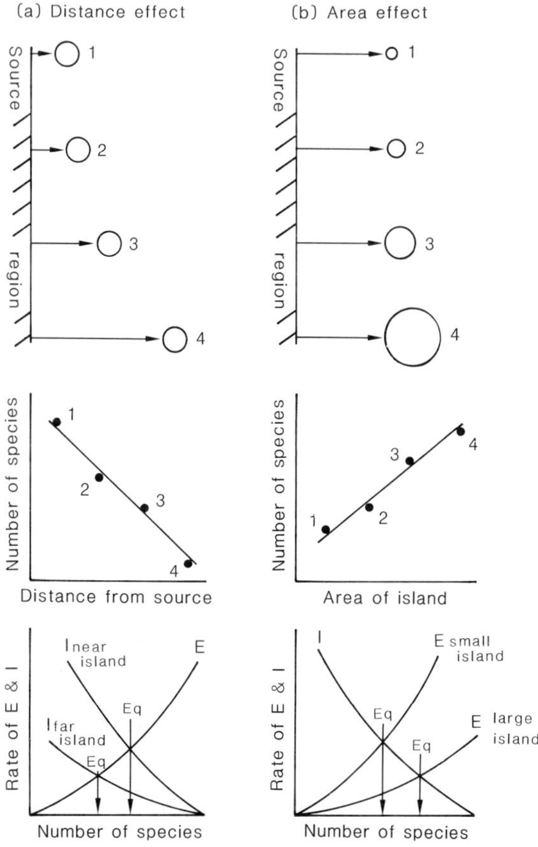

Fig. 1.1 A model for the equilibrium theory of island biogeography. The influence of (a) distance and (b) area on the number of species on islands. E, extinction; I, immigration; Eq, equilibrium number of species. (Adapted from MacArthur and Wilson 1967, and Dennis 1985*a*.)

Studies up to now have been biased toward islands exceeding 15 hectares.

The closest determinant of butterfly numbers, obtained in the most recent analysis by Reed (1985), is that of plant species (Table 1.1). Some caution is required in interpreting the high correlation obtained. Only a limited number of plants are larval hosts for British butterflies and these are wide ranging and occur on even the smallest of islands; thus the size of the island flora can be interpreted only as a measure of habitat richness or diversity. Also, plant species are subject to the same zonal influences (i.e. latitude, island area, and island isolation) as butterflies, and the correlation with latitude is much higher than reported by Reed (1985). It is also difficult to determine what part of the correlation between numbers of butterflies and plants on islands reflects similar standards of survey for the two groups of organisms. Studies of animals and plants on different islands contrast greatly in their thoroughness. Finally, there are significant discrepancies in the number of butterfly species attributed to specific islands (Dennis, unpublished data).

1.1.1 Island biogeography theory

Island biogeography theory has developed to explain the way numbers of species on islands relate to island area and isolation. It embraces two conflicting ideas: the equilibrium (Preston 1962; MacArthur and Wilson 1967) and non-equilibrium models (Williams 1964; Sauer 1969; Connor and McCoy 1979). The equilibrium model reasons that the number of species on islands is a dynamic balance between continual immigration and extinction; the number of species living on an island remains much the same but there is a continual turnover in the species composition of the island (Fig. 1.1). Non-equilibrium theory denies any fine balance between gain and loss; for instance, separate periods of colonization and extinction can be envisaged. Larger islands may have more species simply because greater habitat diversity is found on them (Williams 1964). 'Passive sampling' may also be responsible for more species being recorded on larger islands, which present larger targets for species emigrating from a source region (Connor and McCoy 1979). For Britain and its butterflies, the picture that emerges is not a simple and clear one. The British islands are continental (geologically a continuation of Europe) not oceanic (not specifically volcanic) in origin and different issues point to equilibrium and non-equilibrium forces.

Several factors may argue against the equilibrium model and favour an historical explanation for butterfly species on British islands. Firstly, many island populations differ physically and genetically from mainland populations (Ford 1975; Thomson 1987); indeed, regional populations of many species on the British mainland differ in many ways from each other and have been given different subspecies names

Table 1.1 Some simple correlation coefficients (Pearson *r*) between numbers of butterflies on British islands and five significant predictor variables.

	Hockin	Reed	Corrected data from Reed
Island area	0.22	0.36	0.19
Isolation	−0.67	−0.42	−0.55
Number of species at nearest source	0.77	0.46*	0.65
Plant species	0.72	0.73	0.80
Latitude	−0.74	−0.39	−0.69
Number of islands	29	52	45

The British mainland and Ireland have been removed from all analyses by Hockin and Reed. Note, in both analyses, substantial errors in species' numbers occur. All variables have been transformed to $\log_{10}$ by Hockin, but only area and distance by Reed, for which maximum values are quoted in the table.

* Calculated from available data. When the effect of latitude is removed, area accounts for more of the residual variation (0.49) than does isolation (−0. 23).

From Hockin (1981) and Reed (1985) with recalculations using data corrected from Reed.

to highlight the distinctions. This has been interpreted as the result of isolation and separate development of the island populations and would point away from continual admixture induced by gene-flow from the British mainland stock. Secondly, numbers of both butterfly and plant species decline northwards on the British mainland, emphasizing the relationship with climatic constraints. Disjunctions in the range of butterflies (e.g. small mountain ringlet *Erebia epiphron*) and habitat relicts (e.g. butterflies restricted to the Burren in western Ireland) reinforce this view of 'faunal entrenchment' as opposed to fluidity of movement. Thirdly, several butterfly species have not been recorded on some islands on which one would expect them to occur on the basis of resources, proximity to the British mainland, and chance introductions. For instance, the scotch argus *Erebia aethiops*, the small pearl-bordered fritillary *Boloria selene* and the grass-feeding hesperiids are absent from Ireland. Moreover, during the last 150 years possibly five species have become extinct on the British mainland but have not been replaced by other species dispersing from the European continent (see Appendix 1; Table 1.2). Even when introduced, as in the case of Weaver's fritillary *Boloria dia* on the North Downs (Cribb 1986), they do not necessarily survive for many years. Finally, some butterflies seem to have great difficulty in colonizing new areas without artificial introduction, as in the case of the silver-studded blue *Plebejus argus* from the Great Orme to the Dulas valley, separated by eight miles of farmland and town in North Wales (Dennis 1977; Thomas, C. 1985*a*).

Substantial evidence seems to favour an historical account. Many British butterflies have been shown to form compact and closed populations. Occasional vagrant French butterflies (e.g. swallowtail *Papilio machaon gorganus*; black-veined white *Aporia crataegi*; large wall brown *Lasiommata maera*) have failed to establish themselves in Britain. Even those species that clearly migrate in large numbers over the North Sea and Channel, such as the red admiral *Vanessa atalanta*, the painted lady *Cynthia cardui*, *Colias* species, and the Camberwell beauty *Nymphalis antiopa*, only attain temporary status. There are, however, important counter arguments. Several butterflies are known to colonize islands successfully; there are numerous records of breeding and over-wintering of the large white *Pieris brassicae* on Hebridean and northern islands, as there is of the peacock *Inachis io*, another vagrant in the north. Of greater significance are the records of allegedly non-migratory small heath *Coenonympha pamphilus* from the Isle of May and St Kilda, meadow brown *Maniola jurtina* from Fair Isle, Isle of May, and the Farne Islands, and common blue *Polyommatus icarus* from

Table 1.2 Butterfly species absent from Britain but which are found along the coastal fringe of France, Belgium, and Holland from Brest (France) to Gröningen (Holland)

HESPERIIDAE	NYMPHALIDAE
Pyrgus alveus	*Apatura ilia*
Pyrgus armoricanus	*Liminitis populi*
Spialia sertorius	*Nymphalis antiopa*
Carcharodus alceae	*Araschnia levana*
Heteropterus morpheus	*Argynnis pandora*
PAPILIONIDAE	*Argynnis niobe*
Papilio machaon gorganus	*Argynnis lathonia*
Iphiclides podalirius	*Brenthis ino*
	Boloria aquilonaris
PIERIDAE	*Clossiana dia*
Aporia crataegi	*Melitaea phobe*
Pontia daplidice	*Melitaea didyma*
Colias croceus	*Mellicta parthenoides*
Colias hyale	*Neohipparchia statilinus*
Colias alfacariensis	*Minois dryas*
	Arethusana arethusa
LYCAENIDAE	*Coenonympha arcania*
Nordmannia ilicis	*Coenonympha hero*
Lycaena tityrus	*Lasiommata maera*
Lycaena hippothöe	*Lopinga achine*
Everes argiades	
Glaucopsyche alexis	
Maculinea alcon	
Maculinea teleius	
Pseudophilotes baton	
Lycaeides idas	
Cyaniris semiargus	

From Higgins and Riley (1983); Higgins and Hargreaves (1983).

Heiskeir Isle and the Farne Islands (Heslop-Harrison 1953; Eggeling 1957; Robertson 1980, via Harding, personal communication; Dunn, personal communication). Such islands are little more than sea-washed rocks and lighthouses, and these observations demonstrate that even small butterflies are capable of colonizing distant shores if 'landing facilities' are available. Evidently, they are capable of colonizing islands in cooler northern Britain. Moreover, selection for distinct phenotypes in north and north-west Scotland (see chapter 10) does not preclude the build up of populations during warm sunny seasons and the dispersal of individuals to nearby islands. Observations on dispersal of such butterflies across 'bare' fields by Baker (1968*a*,*b* 1969) and of unusual records in urban gardens give this view increasing credibility (Dennis, personal observation; Owen, D. 1975; Owen, J. 1983; see chapter 6). It would appear that some butterflies must have survived a sea crossing, as the Hebridean islands have had no land bridge in Post Glacial times and the faunas were decimated by glaciation. Man, from the Mesolithic (10 000 years BP) to the present, may well have been responsible for introducing a number of species to British islands quite inadvertently. This process can be envisaged as continuing and part of the equilibrium model.

Many past and present factors have contributed and continue to contribute to our butterfly 'bank' (Fig. 1.2). Historical events post-dating the last major glacial advance 18 000 years BP have been most significant in influencing island butterfly faunas (see chapter 10). A series of initial regulators determined the movement of species to islands which then had connections with the continent; subsequent constraints, arising from vegetation succession and climatic change, have pruned away at these arrivals. Islands without land connections must have depended at the outset on dispersal from overseas. The role of humans has probably been important, by effecting introductions during their own colonization of islands and by modifying habitats. Throughout much of the Post Glacial period the influence was positive, as hunting and subsequently agricultural demands retained space for species requiring open habitats. However, during the last 200 years 'interference' has been decidedly detrimental (see chapter 11).

No statistical examination of butterfly data can separate non-equilibrium from equilibrium forces, as different influences can have similar effects, but the weight of evidence may point in favour of one solution rather than another. It is especially significant that the British mainland, which had a long post-glacial connection with the Continent (to *c*.7800 years BP), has far fewer species than the continental shoreline of France. Thirty-eight non-migratory species additional to those found in Britain occur along the coastline from Brittany to northern Holland (Table 1.2), but not one has successfully colonized Britain during the last 50 years. Even though Jersey is so close to France it has only one resident species which is not found in Britain, the

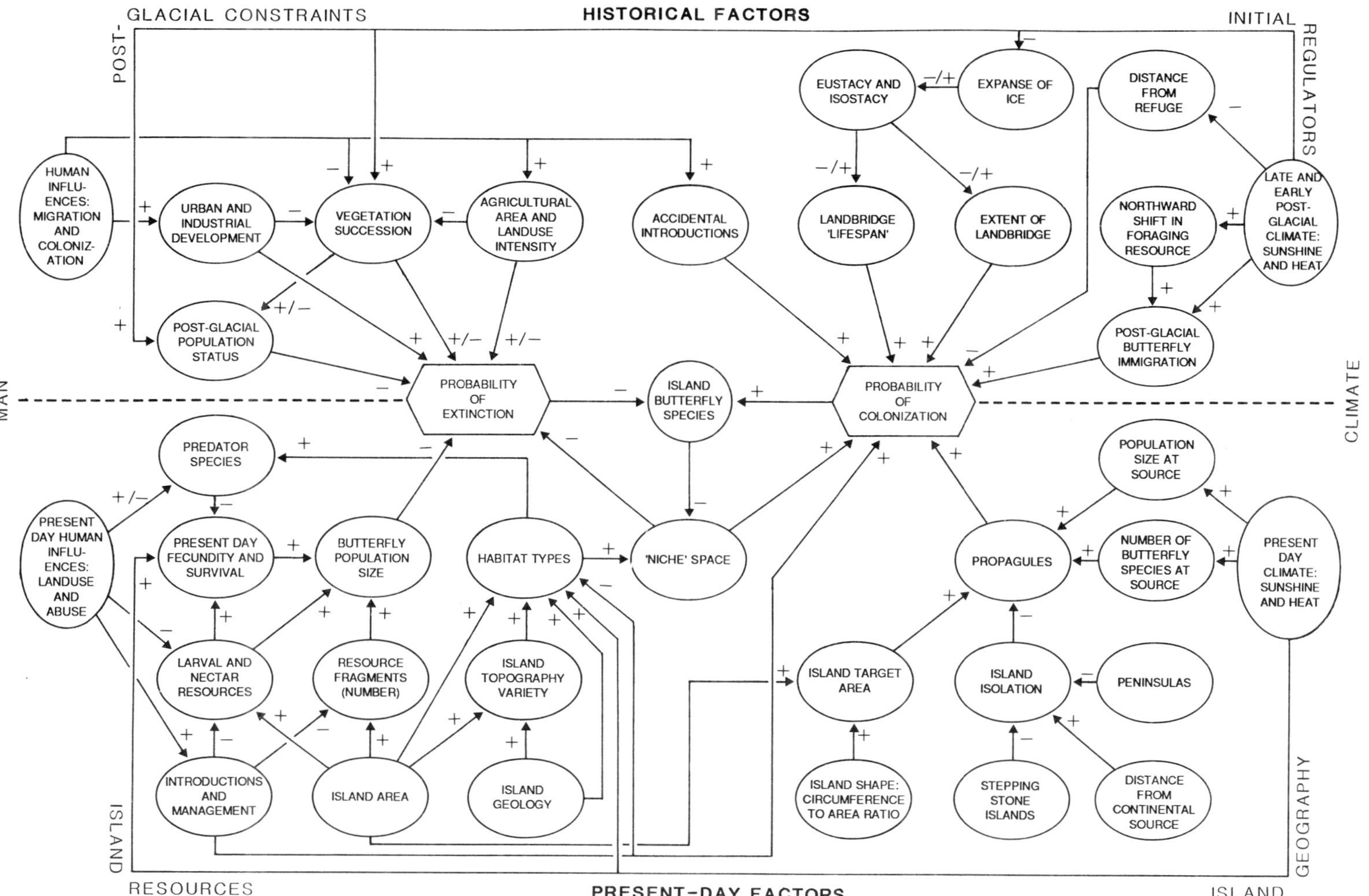

Fig. 1.2 Factors accounting for the numbers of butterfly species on British islands. For interpretation of signs, refer to **process-response models** in the glossary. A large number of factors may account for the number of species on islands by influencing extinction and colonization rates.

large chequered skipper *Heteropterus morpheus*, believed to have been introduced during the German occupation of 1940–45 (Long, 1970). Exotic butterflies that have found their way to the British mainland have not been able to establish a foothold.

The key to understanding this is to appreciate that British environments are distinctly marginal for continental butterflies (Dennis 1977). Only a fraction of species at the European coastal fringe can effectively colonize Britain; in practice, the source populations for most smaller British islands occur on the nearest landmass, often other small islands, which have similar species. Many species on small islands in the Hebridean, Orkney, and Shetland groups no doubt frequently become extinct, to be colonized again later on from larger neighbouring islands, which have more numerous, larger populations, and more varied microclimates and resources. Colonization and extinction probably operate in pulses in response to periodic climatic changes. The number of butterflies at the nearest source, equating with zonal factors such as latitude, are strongly influenced by summer warmth and sunshine (Dennis and Williams 1986; Turner 1986). Colonization tends to occur in hot, sunny summers like the ones of 1976 and 1984, and extinction, which particularly affects small islands, occurs during cool, cloudy seasons.

1.1.2 Species' combinations and 'missing species'

The main disadvantage with a statistical approach that gives consideration only to gross numbers is that it takes no account of which species occur on which islands. Yet, the pattern of presence and absence of particular butterflies may tell us much more, especially when their habitat and hostplant demands are also considered. There are two ways of investigating these faunal patterns: (i) to compare the assemblages of species found on different islands and (ii) to consider the distribution of each species in turn. These approaches also hint at under-recording, especially for islands that lack butterflies that are found elsewhere in more isolated and bleak surroundings.

A convenient way of comparing species' combinations for different islands is to apply a computer technique such as non-metric scaling. This involves moving points (representing islands) about until their relative positions equate as well as is possible with their faunal affinities (Fig. 1.3; Dennis 1977). The resulting plot can subsequently be matched against a map of Britain. Distortions in the relative positions of adjacent islands or regions then point to differences not just in numbers but also in types of species and in turn to the influence of such factors as area and isolation. The picture that emerges is very much one of a systematic trend reproducing the British Isles, but several notable irregularities occur. Each island has been dislocated northwards, pointing to 'losses' in the number of species. Those islands least affected (such as Ireland, Man, Anglesey, and Wight) are either large or close to mainland

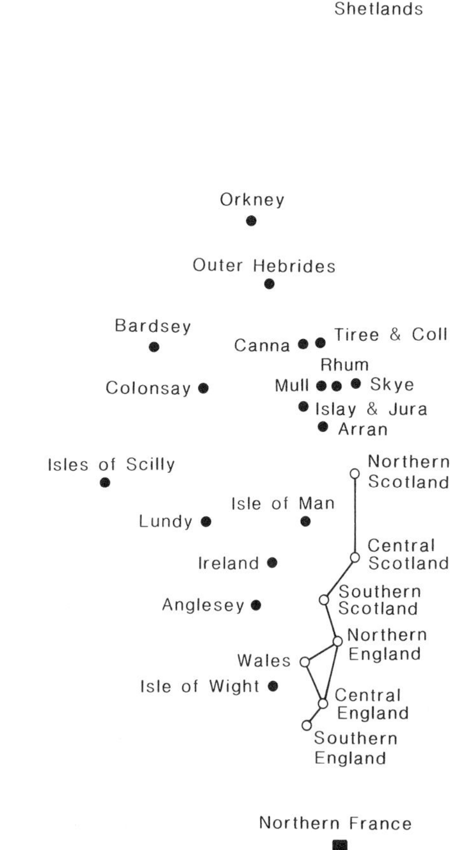

Fig. 1.3 A non-metric scaling computer map illustrating the relationships between a selection of British islands based on their associations for butterfly species. See text for technique. (After Dennis 1977.)

Britain and are thus most resilient to population extinction or are open to 'gains'. Perhaps the most interesting finding of this technique is the identification of a 'core' group of butterflies (*Maniola jurtina*, *Coenonympha pamphilus*, small tortoiseshell *Aglais urticae*, green-veined white *Pieris napi*, *Polyommatus icarus*, dark green fritillary *Argynnis aglaja*, and less often, the grayling *Hipparchia semele*). These occur virtually wherever there are a few hectares of rough ground on islands and disappear only in the Orkney and Shetland groups. Small, and in many cases, isolated southern islands (such as Isles of Scilly, Lundy, and Bardsey) are far out of alignment (Fig. 1.3) and reveal interesting and different attachments. Part of their fauna comprises the same robust species which have colonized the Hebrides and Orkneys, but a number of southern species are resident, often different butterflies on different islands, giving each island group a unique species assemblage. For instance, the Isles of Scilly do not have *H. semele*, the gatekeeper *Pyronia tithonus*, the ringlet *Aphantopus hyperantus*, nor *A. aglaja*, all of which can be found on Lundy, but do have the holly blue *Celastrina argiolus* and the speckled wood *Pararge aegeria* which are not found on Lundy. These differences are not always easily explained, but Lundy was more accessible by way of overland dispersal in the immediate Post Glacial period. Studies of faunal content are useful then for determining 'sins of omission and commission'. Colonsay in the Inner Hebrides is an interesting case, with its record of the purple hairstreak *Quercusia quercus* (Dunn 1965). It points to deforestation over much of the western highlands and islands and important human constraints imposed during the Post Glacial period (see chapter 10). The 'under-heather' ecology of the wood violet *Viola riviniana*, the larval host-plant for Hebridean *A. aglaja* (Dunn, personal communication), corroborates this observation.

Altogether the computer plot of island fauna confirms the previous island biogeography analysis, the influence of isolation and area superimposed on effects of latitude and dependent climatic controls. However, only indirectly does it match islands for species content. That a more rigorous comparative approach can be valuable is demonstrated by looking closely at the butterfly fauna of Ireland which presents a number of interesting problems (Fig. 1.4). Ireland is virtually a mini-continent with its own island satellites, many poorly surveyed, and too large to be expected to lose species from random population fluctuations. The consensus of opinion is that it has not had a land connection with the British mainland since the final ice retreat 10 000 years BP and is well isolated from parts of the mainland that have additional butterflies and which must have contributed originally to the Irish stocks. Two important observations can be made (Nash 1975); (i) generally, butterfly species which occur in Ireland as well as on the British mainland have more northerly distributions on the British mainland than those species that are absent from Ireland. This difference is statistically significant. (ii) Nevertheless, some 15 butterflies not found in Ireland extend as far north on the British mainland as many species that also occur in Ireland. This includes three of five species (*Erebia aethiops*, northern brown argus *Aricia artaxerxes*, and chequered skipper *Carterocephalus palaemon*) which have southern as well as northern limits to their ranges in Britain.

Why some butterfly species are not found in Ireland is something of a mystery. The answer is not a simple one and is likely to be different for each species. The existence of such anomalies points more towards unique historical events than to a continual turnover of species, as envisaged by equilibrium theory. Two solutions are feasible: (i) these butterflies may have entered Ireland but subsequently died out, perhaps due to climatic change, changes of forest cover, or land use, or (ii) they may never have entered, either the timing of entry, the route taken, or ecological factors preventing them. The latter may seem to be the more likely solution for ubiquitous, northern butterflies like *Boloria selene* and *Erebia aethiops*, which are tolerant of damp habitats and for which conditions seem to be suitable today. They may have entered Britain by a longer route from the east, once sea barriers were well-established. However, the subsequent expanse of lowland moss and dense wood on till may have caused the extinction of many butterfly species. It suppressed the distribution of *Helianthemum* spp. to all but small isolated limestone exposures in the west of Ireland and this may explain the absence of *Aricia artaxerxes*. The more southern species in Britain, such as grass-feeding skippers, may well have been squeezed out by forest development and climate change or in historical times as land use intensified. Such species

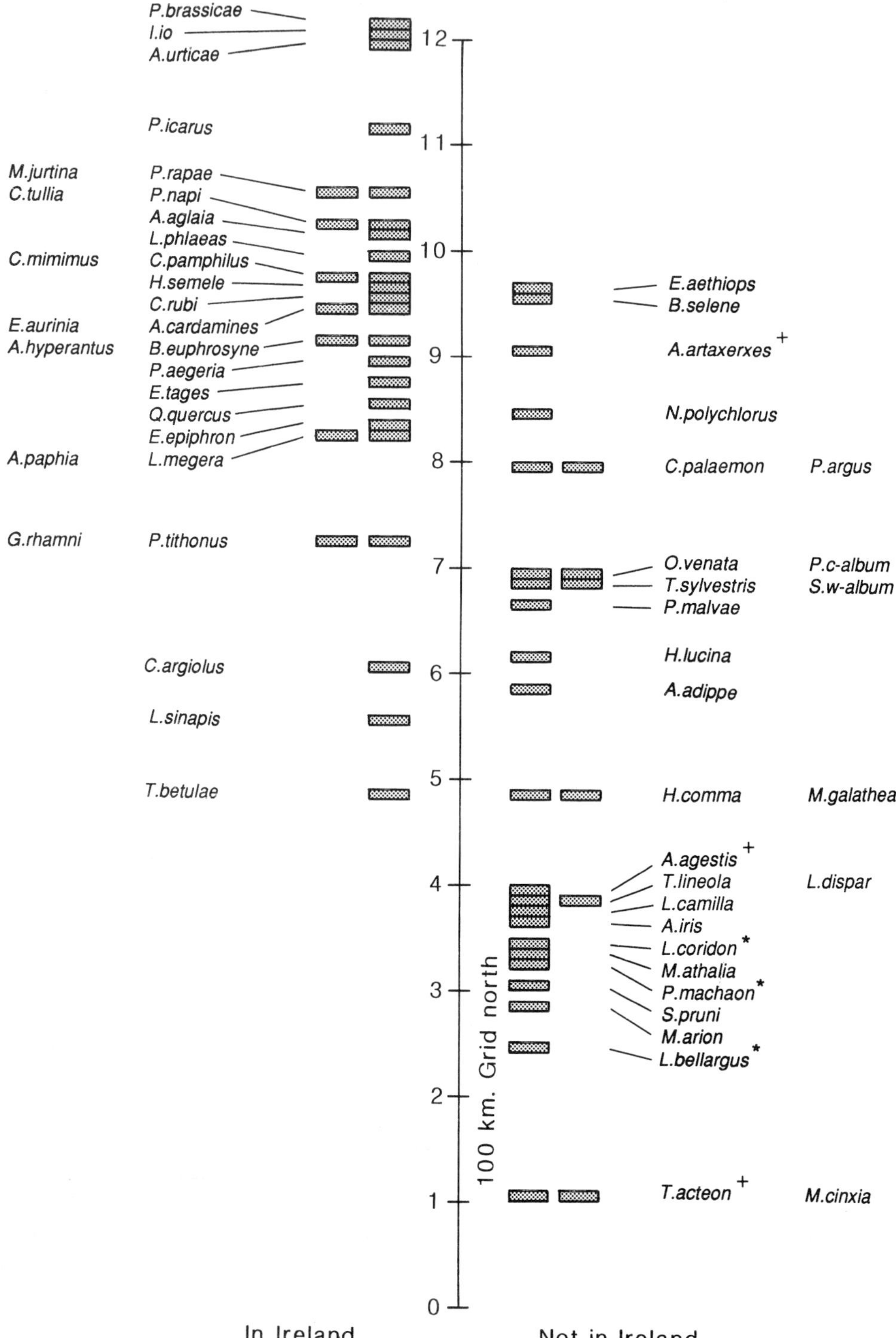

Fig. 1.4 A dispersion diagram comparing the British mainland ranges of butterflies found in Ireland and on the British mainland respectively. *, hostplant not found in Ireland; +, hostplant found in Ireland but with very restricted distribution.

have experienced substantial withdrawals of range on the British mainland during the last century (see chapter 11). Although the history of *Erebia epiphron* in Ireland is unclear, as indeed is its present status there (Redway 1981), Ireland has certainly lost species. Nevertheless, some care is required in interpreting past records. The heath fritillary *Mellicta athalia* was allegedly abundant in Killarney about 1866 (Nash 1975) but it is unlikely that it was ever recorded there (Lavery 1989). This historical viewpoint is taken up again in chapter 10. Before leaving Ireland, the reader may like to reflect on the fact that two species, the wood white *Leptidea sinapis* and the silver-washed fritillary *Argynnis paphia*, occur further north and seem generally to do better in Ireland than on mainland Britain.

1.2 Butterfly distributions on the British mainland

1.2.1 Patterns and trends in species' richness

Knowledge of British wildlife has greatly benefited from the detailed mapping of species by the Biological Records Centre (Heath *et al*. 1984); a data source unavailable when Ford (1945) wrote his book and only now being matched by similar schemes in Europe (see Geraedis 1986; Gonseth 1987; Jakšić 1988; Tax 1989). Amalgamating the distributions of butterflies produces a map of species abundance, full of evocative patterns and trends (Fig. 1.5). However,

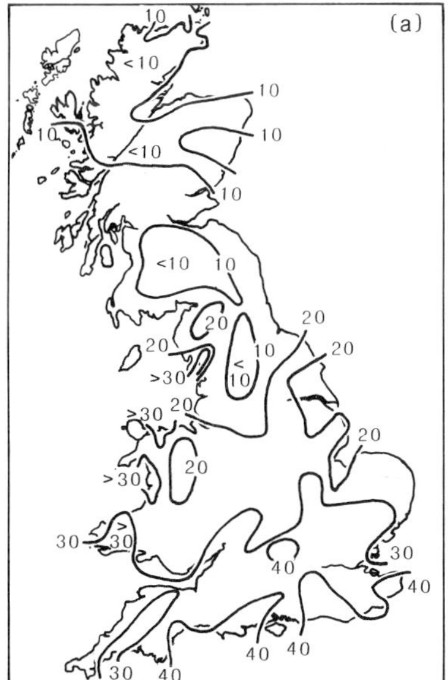

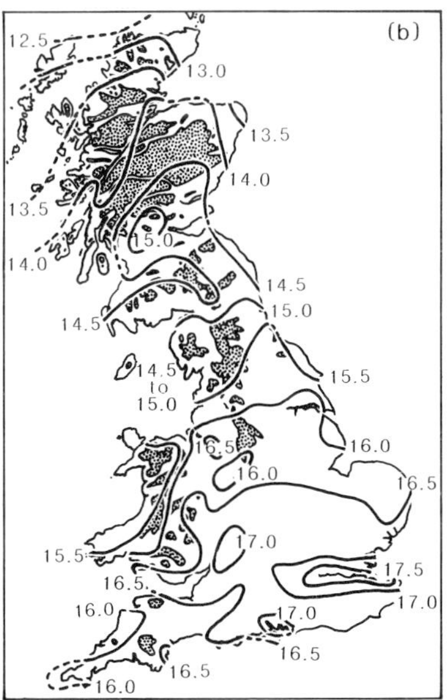

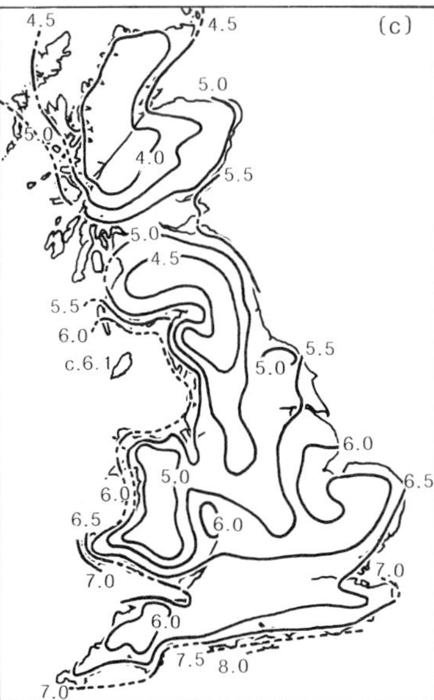

Fig. 1.5 Trends in numbers of butterfly species on the British mainland. (a) Generalized species' richness map for butterflies on the British mainland, based on tetrads for 50 km grid intersections. (Data courtesy of the Biological Records Centre, Institute of Terrestrial Ecology, and P. T. Harding.) (b) Isotherms for sea-level July mean temperatures (°C), 1941–71 (adapted from Meteorological Office 1975*a*, with permission of the Controller of HMSO), and the distribution of upland areas over 400 m in altitude (stippled areas). The lapse rate for January and July is 0.6°C for each 100 m. (c) The distribution of summer sunshine (average daily duration of bright sunshine hours, 1941–71). (Adapted from Meteorological Office 1975*b*, with permission of the Controller of HMSO). Summer sunshine and temperatures accounted for over 75 per cent of the variation in species richness in two separate studies using different sampling frames (Turner 1986; Turner *et al*. 1987; Dennis and Williams 1986).

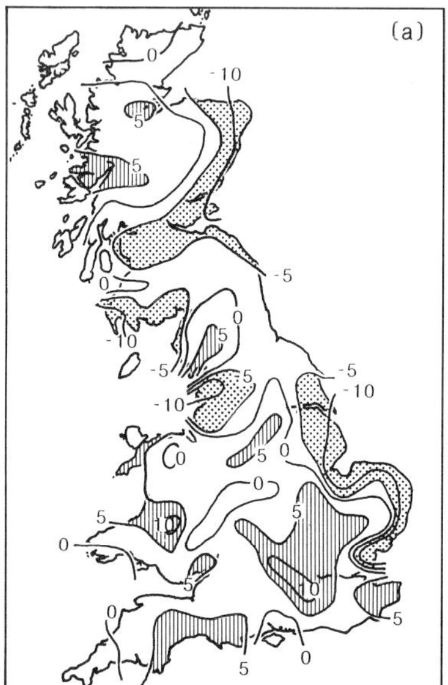

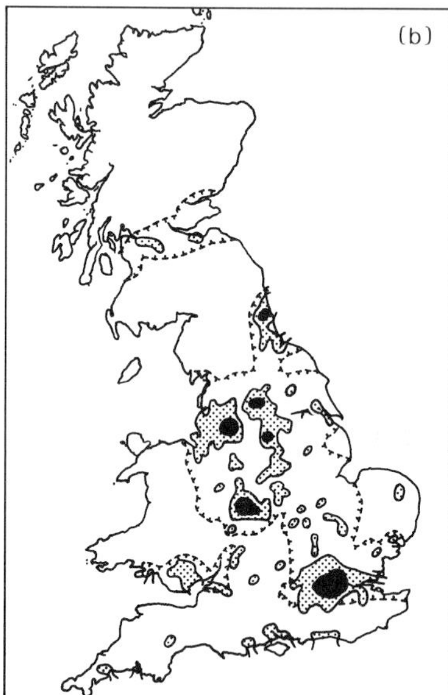

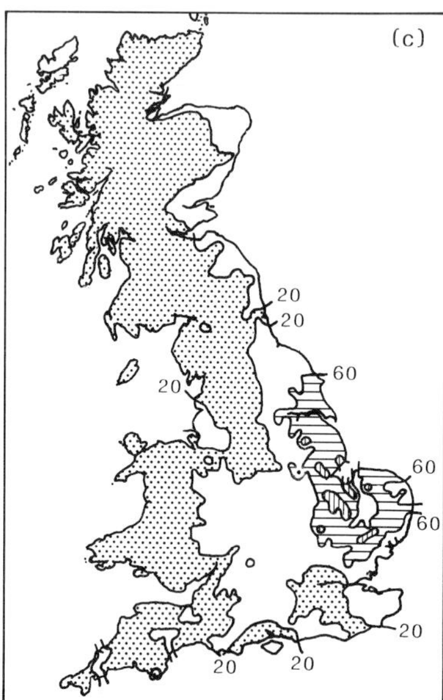

Fig. 1.6 Residual patterns in the abundance of butterfly species on the British mainland. (a) Areas having more or less butterfly species than expected on the basis of summer sunshine and temperatures. Isolines at intervals of five species. The plot is based on tetrad values at 50 km grid intersections. Vertical hatching, main zones of excess; stippled areas, main zones of deficit. (b) The distribution of urban cover (%) and of air pollution. (Data for air pollution from Hawksworth and Rose 1976; O'Hare 1974; Bell 1982; data for urban cover, courtesy of the Land Characteristic Data Bank, Bangor Institute of Terrestrial Ecology and G. L. Radford.) Black areas, >75 per cent urban cover; stippled areas, 25–75 per cent urban cover; pecked lines, lichen zones 5 and below which correspond to about mean winter SO_2 50 μg m $^{-3}$ and over. (c) The distribution of arable land (% cover); vertically hatched areas, >80 per cent arable cover; horizontally hatched areas, 60–80 per cent arable cover; stippled areas, <20 per cent arable cover. (Courtesy of the Land Characteristic Data Bank, Bangor Institute of Terrestrial Ecology and G. L. Radford.) (cf. Turner 1986; Turner *et al.* 1987.)

as recording is biased to low-lying areas in southern Britain, some caution is required when interpreting this map and any data derived from it (Fig. 1.6; Dennis and Williams 1986; Turner 1986; Turner *et al.* 1987).

The most distinctive trend is the decline in the number of species with latitude. Latitude is a zonal influence that controls climate, particularly sunshine and temperature, and there are a number of important ways in which climate could determine species abundance—some are direct in nature, others indirect. Both sunshine and temperature decline in northern latitudes. Butterflies, with their tropical origins, and lacking in ability to generate such internal heat, depend on both. All British species are known to build up their temperatures by using their wings as solar panels and basking in the sun (thermoregulation) (see section 2.2). Using sunshine alone, they can establish body temperatures 10 °C in excess of ambient air temperatures. This allows more efficient and longer periods of mate-location, and successful courtships, egg-maturation, egg-laying, and nectaring, all influencing reproductive success. Larvae, especially those which are black in colour, as in several nymphalids, do the same. Those particularly that live in clusters during the early stages, build up large temperature excesses by a bask–feed cycle (Porter 1982). This, in turn, speeds up feeding, digestion, and development, crucial for survival in northern climes where the growing season is short. Lower sunshine levels and temperatures to the north counter all these processes. An array of indirect effects also occur. Warmer and more active insects can avoid predators more easily and butterflies that have larvae which are able to absorb sunshine efficiently can avoid synchronized attacks at particular life stages by their specific parasitoids. Good examples are the marsh fritillary *Eurodryas aurinia* and *Aglais urticae* (Porter 1983; Pyörnilä 1976). Fungal and viral infections of early stages may also be related to lower temperatures and higher relative humidity. There are other ways in which climate may influence butterfly populations. Most plants are also influenced by temperatures, and butterflies decline to the north in relation to numbers of plant species, although most of our butterfly species do not occupy the full range of their larval hostplants (Dennis and Shreeve 1991). We have yet to investigate the way in which climate may alter the hostplant condition (chemistry, new growth production) or habitat diversity, and how this may ultimately affect butterfly populations. Many British butterflies are peculiar in that they use fewer larval hostplants than their counterparts in adjacent regions on the continent and this implies that they are subject to some degree of stress (section 10.4).

Much of northern Britain tends to be higher in altitude than it is in the south, making the influence of latitude less simple to determine. Higher areas are colder, cloudier, wetter, and windier. These are the conditions which accelerate nutrient depletion in northern soils and result in land use having relatively greater impact on natural habitats, which in turn affect butterflies. Northern soils, even those on calcareous substrate, are leached as a result of heavier rainfall and low evapotranspiration rates. Because of a lack of economic potential, northern uplands have also devolved to early seral habitat uniformity since Mesolithic and Neolithic deforestation. As a result of forest removal and lack of competitive land uses, British uplands are intensively grazed, without shelter, damper, windier, and thus colder still. Consequently, few butterfly species are found above 400 m in Britain (see section 10.4). One way or another, latitude has its greatest impact on butterflies through climate. All this makes for lower reproduction and survival for smaller, scattered northern butterfly populations and, as a result, smaller populations more easily become extinct as conditions fluctuate (see chapters 4 and 11).

That climate has a significant impact is evident in butterflies that occur in southern regions too. Surveys on the silver-spotted skipper *Hesperia comma* (Thomas *et al.* 1986), the orange-tip *Anthocharis cardamines* (Courtney 1980), the brown hairstreak *Thecla betulae* (Thomas, 1974), the small copper *Lycaena phlaeas* (Dempster 1971*b*), *Plebejus argus* (Dennis 1972*b*; Thomas, C. 1985*a*,*b*), the Adonis blue *Lysandra bellargus* (Thomas 1983*a*), the Glanville fritillary *Melitaea cinxia* (Simcox and Thomas 1979; Thomas and Simcox 1982) and the wall brown *Lasiommata megera* (Dennis and Bramley 1985) emphasize the importance of either aspect or grazing on soil-surface microclimates and thus on early stage development and survival. The British large blue *Maculinea arion*, for example, depended on a symbiotic relationship with an ant, *Myrmica sabuleti* which in turn requires short turf habitats and high daytime ground temperatures (Thomas 1976, 1977; see Fig. 4.3). However, not all the factors limiting butterfly populations are climatic (see next section). Detailed research on 11 species in addition to those mentioned (Lulworth skipper *Thymelicus acteon*, Thomas 1983*b*; *Papilio machaon*, Dempster *et al.* 1976; *Leptidea sinapis*, Warren 1984, 1985*a*; brimstone *Gonepteryx rhamni*, Pollard and Hall 1980; black hairstreak *Satyrium pruni*, Thomas 1974; large copper *Lycaena dispar*, Duffey 1968; small blue *Cupido minimus*, Morton 1985; *Plebejus argus*, Thomas, C. 1985*a*,*b*; white admiral *Ladoga camilla*, Pollard 1979*b*; *Eurodryas aurinia*, Porter, 1981; and *Mellicta athalia*, Warren 1987*a*,*b*,*c*) demonstrates that distributions are often more strictly determined by resources relating to edaphic conditions, colonizing ability, and vegetation succession than by climate (Turner 1986; Dennis and Williams 1986). Another five species absent from southern Britain (*Erebia epiphron*, *E. aethiops*, large heath *Coenonympha tullia*, *Aricia artaxerxes*, *Carterocephalus palaemon*) are almost certainly restricted by habitat and less by lack of sun and heat in Britain, although *A. artaxerxes* in Scotland is usually found on south-facing slopes.

The second pattern of variation in butterfly richness that needs explaining is that which statisticians call residual variation—regions which have rather more or fewer species than one would expect from some major influence, in this case based on latitude or summer temperatures and sunshine (Fig. 1.6). The most significant areas of deficit are the Fenland and adjoining areas in eastern England, the Lancashire plain, and the central lowlands of Scotland. These areas are low-lying, flat, and have rich soils and excellent agricultural potential. The most obvious explanation is that intensive cultivation practices, extractive industry and urbanization have destroyed essential butterfly habitats (e.g. *Lycaena dispar* through drainage of fens; Duffey 1968). Barbour (1986*a*) argues that another factor may be involved. A map of butterfly extinctions since 1970, which embraces some urban zones and a larger part of eastern England, coincides with moderate to severe depletion of lichens and thus correlates with sulphur dioxide pollution. There is certainly concern about the influence pollution has on butterfly populations at the margin of many European industrial cities (Heath 1981), but as the two maps, of residuals and extinctions, are not the same, several factors are probably implicated. On the whole, habitat destruction, through agriculture, urbanization, and industrialization, rather than pollution, seems to account for the residual patterns and extinctions (see section 11.2), but this can only be adequately tested through field surveys.

Urban areas often have more butterflies than we expect, generally because we are not considering population size but the presence of a species, where a few records suffice to give the impression of wide coverage. Urban environments are invariably warmer than the surrounding countryside (Barry and Chorley 1982). They also have greater habitat diversity than is usually appreciated; parkland and waste ground provide suitable habitats as do large, mature gardens. Thus woodland butterflies, such as *Pararge aegeria* and *Quercusia quercus*, have been observed in Greater London and *Celastrina argiolus* can be abundant in suburban settings (Brakefield, personal communication; Stubbs 1979; Heath *et al*. 1984; Plant 1987).

Areas of butterfly surplus—more species than expected on the basis of temperature and sunshine—points to rich habitat diversity. A large zone in central southern England stands out, in particular, and has several potential advantages. It includes the Cotswolds, Chilterns, and other downland on Jurassic limestone and chalk; steep scarp slopes and dry valleys with their floristically varied calcareous grasslands; also, extensive areas of forest occur as on the Chiltern scarp or on lower clays and sands (Rockingham and Bernwood forests). The map of residuals reveals other smaller zones, noteworthy butterfly hot spots, especially the North Wales coast and the southern edge of the Lake District. Here, much of the rich natural flora remains intact on Carboniferous limestone scars and screes, with their thin rendzina and brown earth soils supporting rough pasture and woodland and, in the case of the southern Lake District, on adjoining mosses. The butterfly fauna in the vicinity of Morecambe Bay is especially interesting including, as it does, the overlap of northern elements (*Aricia artaxerxes*, *Erebia aethiops*) and southern elements (*Hamearis lucina*, high brown fritillary *Argynnis adippe*, *Pyronia tithonus*, *Gonepteryx rhamni*, and formerly *Thecla betulae* and *Leptidea sinapis*), each virtually at their limit. The steep-sided south facing valleys along the southern flank of the North Yorkshire Moors have much the same status. They too offer sheltered suntraps and deter human interference. These areas and their immediate surrounds deserve far more attention using detailed mapping and ecological surveys.

1.2.2 Ranges, distribution patterns, and inference

Statistical summaries have considerable value in generating ideas and testing the relative importance of potential influences. However, to explain butterfly geography requires a detailed programme of work on each species. Perhaps the most basic feature of butterfly biogeography is the range. The range describes a boundary around the locations at which populations of the butterfly are found. In a crude way it marks the tolerance limit of a species. It is perhaps easy enough to define as long as vagrants can be distinguished from residents, but it is not a simple concept as the factors affecting survival at different points on the perimeter of the range may vary substantially. After all, only one factor need be critical at each location.

All but a handful of butterflies in Britain have ranges which fall short of the north Scottish coast (Emmet and Heath 1989). Each limit forms a critical zone worthy of study as we know too little about how aspects of climate, habitat, and hostplant combine to influence different butterflies, although some excellent work has been achieved on less common species (Thomas 1983*a*,*b*; Thomas *et al*. 1986; Warren 1984, 1987*a*,*b*,*c*; Warren *et al*. 1984; Warren *et al*. 1986; Morton 1985). The availability of hostplants may restrict *Pieris brassicae*, *Aglais urticae*, *Aricia artaxerxes* and *Gonepteryx rhamni*. The distribution of *A. artaxerxes* follows common rockrose

Helianthemum chamaecistus closely and it is interesting how they are both absent from western Scotland (Thomson 1980). Not all range limitations of British butterflies are to the north. Five species (*Erebia epiphron*, *E. aethiops*, *A. artaxerxes*, *Coenonympha tullia*, *Carterocephalus palaemon*) have southern limits in Britain. *C. tullia* for instance is probably restricted by the southern limit of *Eriophorum–Molinia–Erica tetralix* communities associated with ombrogenous peat and raised bogs. Many butterflies, especially fritillaries, are fast declining in eastern and central England and are becoming increasingly confined to the west. Analysis of these patterns requires a wider vantage. The limit to each butterfly in Britain is a fraction of its true range. In the Palaearctic, British species occupy eastern and northern environments with lower summer and very much lower winter temperatures (Dennis 1977). Compensation for the lower summer temperatures may result from extended hours of sunshine, whereas entomologists have long suspected that mild, wet winters, not cold, dry ones typical of western Europe, are detrimental to butterflies (Beirne 1955).

Within the range, distribution patterns hold other clues of history and adaptation. Distributions may differ in a number of ways which are open to measurement and comparison. They may differ in

- location
- size
- shape
- orientation
- gradient (for physiological, phenetic, and phenological attributes) and
- texture (density, spacing including disjunctions, contiguity).

Each of these patterns carries various inferences. To interpret a pattern, locational details are always important as these permit comparison with potential causative agents. Extensive cover by a butterfly has several implications:

(1) tolerance of varying conditions, and/or
(2) uniformity and ubiquity of required conditions, and/or
(3) wide dispersal capability.

Those of our butterflies which do have wide distributions (e.g. small white *Pieris rapae*, *P. brassicae*, *P. napi*, *Lycaena phlaeas*, *Aglais urticae*, *Polyommatus icarus*, *Inachis io*, *Maniola jurtina*, *Coenonympha pamphilus*) point to combinations of all three parameters. Localized distributions suggest some restriction. *Melitaea cinxia* and *Thymelicus acteon* are both limited to warm, south facing slopes on the south coast, but significantly to very different areas of Britain, namely the Isle of Wight and Dorset, respectively (Fig. 1.7). Despite similar climatic requirements they have very different habitats—chines (deep ravines) and unstable cliffs with early successional ribwort plantain *Plantago lanceolata* for *M. cinxia* (Simcox and Thomas 1979), and thriving expanses of tor grass *Brachypodium pinnatum* for *T. acteon* (Thomas 1983*b*). These two studies emphasize the dangers of making generalizations from coarse-scale mapping. Detailed studies revealed critical ecological demands by both species.

The shape of distributions is even more suggestive of limitations. The linear belt of *Satyrium pruni* has now been identified as belonging to the East Midland remnants of ancient forest on heavy clays. This region comprises large stands of species of blackthorn characterized by a practice of long-rotation coppice (18–20 years) (Thomas 1974). The absence of *S. pruni* from similar areas in Essex, Sussex, and Suffolk may point to past management practices, but certainly to poor dispersal capability in this butterfly as it readily colonized a locality in the Weald when artificially transferred there (Collier 1959). More evocative are distribution shapes that outline the fens and Norfolk Broads (extinct *Lycaena dispar dispar* and extant *Papilio machaon britannicus*) and the calcareous exposures of southern England (*Cupido minimus*, *Aricia agestis*, *Hamearis lucina*, but especially *Hesperia comma*, chalk hill blue *Lysandra coridon* and *L. bellargus*) (see Fig. 10.6) and the Burren Carboniferous limestone in Ireland (*Thecla betulae*, pearl-bordered fritillary *Boloria euphrosyne*). Implicated are the dry conditions due to the permeability of the rocks and shallow soils which warm up quickly and provide suitable microclimates for early stages, speeding up development. More significant for the lycaenids are the hostplants (calcicoles), which favour shallow limestone soils, and their symbiotic association with ant species (see section 7.3.3).

Orientation in a distribution pattern may also point to preference in environmental conditions. Examples can be found at the local scale in those

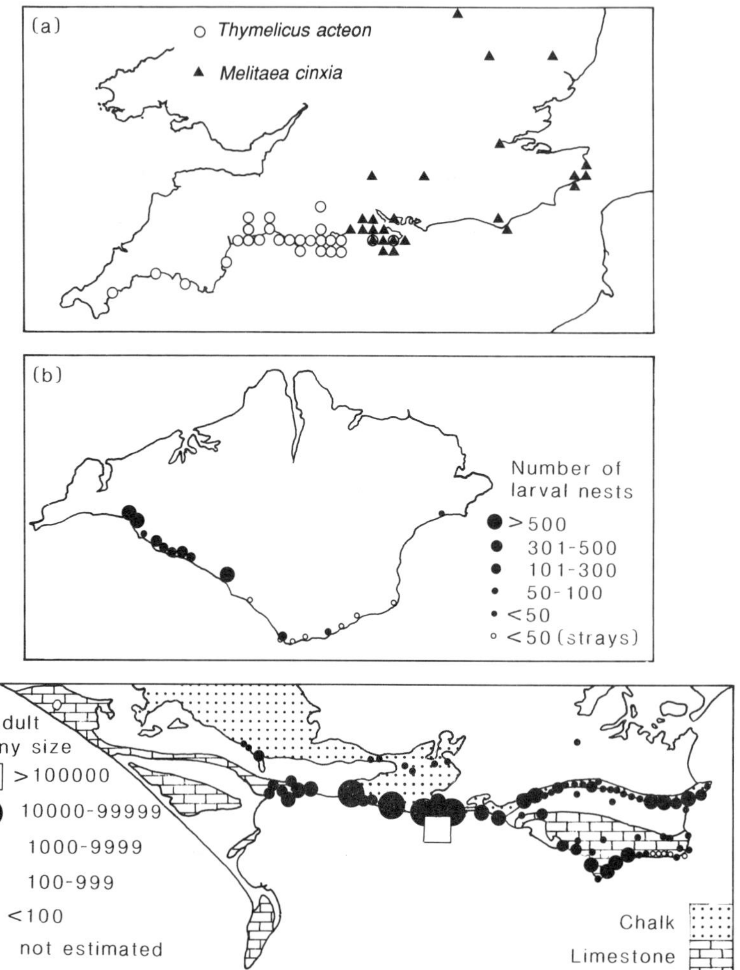

Fig. 1.7 The distributions of the Lulworth skipper *Thymelicus acteon* and the Glanville fritillary *Melitaea cinxia* in southern England. (a) The distribution of the two species in southern England (all records). (Courtesy of Biological Records Centre, Institute of Terrestrial Ecology and P. T. Harding.) (b) The population size (larval stage) of *M. cinxia* on the unstable cliffs of the Isle of Wight. (From Thomas and Simcox 1982; courtesy of Biological Conservation and Elsevier Applied Sciences.) (c) The distribution of colonies and the population size of *T. acteon* in Dorset. (Redrawn from Thomas 1983*b*; courtesy of The Royal Entomological Society.)

species which have south-westerly aspects to their distributions (see Fig. 10.8); but the concept may be less useful at the regional scale. Nevertheless, it might be used to describe the alignment of core and periphery in distributions. For example, the Essex skipper *Thymelicus lineola* seems to decline in status to the west and this may imply a dependence on a continental climate with drier conditions—one having greater extremes of winter cold and summer warmth, respectively.

An empty space on a map implies one of three things:

(1) inability to survive,
(2) inability to colonize, or
(3) under-recording.

For convenience we can distinguish between a disjunction (a clean break) and a void (a vacant space in an otherwise continuous distribution). Size, shape, and location are again important parameters for inference, and attention is focused on both the distributional segment as well as on the vacant space. Compactness of the distributional segment and frequency of breaks are additional attributes of such distributions. Lacunae are usually accredited significance on their size. The implication is that the larger the break, the older it is, but the speed of distributional change, gain or loss, can confound this relationship. Some test of these alternatives is provided by looking for old records. Disjunctions and voids usually proceed by way of local population extinction and plotting old records has, in some cases (e.g. *Antho-*

charis cardamines, Smith 1949; *Pararge aegeria*, Downes 1948*b*) reconstructed patchwork distributions which mark a stage in this process. However, if the disjunction occurred before 1800, there may be little evidence for it, and matching contemporary with historical environments will then be necessary in order to date it. Some notable disjunctions occur for *Carterocephalus palaemon*, the dingy skipper *Erynnis tages*, *A. cardamines*, the green hairstreak *Callophrys rubi*, *Hamearis lucina*, *Argynnis adippe*, *Eurodryas aurinia*, *P. aegeria*, *Erebia epiphron*, *E. aethiops*, and the marbled white *Melanargia galathea*.

A large number of species also have punctiform (dotted) distributions. These may occur because of (i) irregular colonization or (ii) population extinction. Expansion is perhaps suggested by the records for *Thymelicus lineola* in the West Country. Large numbers of old, unconfirmed records demonstrate that the majority of species, disturbingly, belong to the latter category (see section 11.2). Fragmentation of distributions is far advanced in some cases. *Cupido minimus* may always have been restricted by its localized hostplant, *Anthyllis vulneraria* (Heath *et al*. 1984), and *Plebejus argus* by its association with ants of the genus *Lasius* (Thomas, C. 1985*a*). *Nymphalis polychloros* has never been very abundant but has been a permanent resident in some areas, as along the North Wales coast. It is now extremely rare and this probably can be blamed largely on Dutch elm disease. Similar fears have surrounded the white-letter hairstreak *Satyrium w-album* although it survives well on sucker re-growth, which is abundant in some southern counties, and on more resistant wych elms; furthermore, conservation practice carries hopes of increasing the larval resource by planting less susceptible hybrid elms. The reasons for local population extinctions can be numerous, but extensive voids point to overwhelming odds against survival. These, including desolate British mountain terrain, the Lancashire plain, and the fenland, have been touched on. For *Erebia epiphron*, the reverse is true; it is lowland with warmer climes that forms the barrier. The larval hostplant is more widespread and occurs over more extensive areas of high ground above 700 metres as well as at lower altitude, but may attain the right condition—for instance sufficiently high density and the appropriate growth form—for the butterfly in restricted areas.

Most remarkable for voids is the hollowed-out coastal distribution of *Hipparchia semele* (Fig. 1.8). This butterfly usefully demonstrates the value of comparing life history detail with environmental data to interpret distribution patterns. A number of hypotheses can be put forward to explain *H. semele*'s distribution:

(1) a preference for marram grass *Ammophila arenaria*, from which the larvae have been swept (Shaw 1977; see Fig. 7.3);
(2) its relatively slow development and therefore its restriction to areas with large cumulative temperatures (Thomson 1980);
(3) a greater dependency on bright sunshine for thermoregulation to maintain its high rate of activity (see Fig. 1.5c; Findlay *et al*. 1983; Tinbergen 1942); and
(4) the need for bare, dry, warm soil conditions for rapid pupal development and which may also house fewer arthropod predators.

T. Shreeve (personal communication) finds that *Hipparchia semele* larvae use *Festuca rubra* on sand dune areas in South Wales but marram could act as an important secondary hostplant for older larvae in May and June. However, it is not found everywhere where there are dunes and marram. The butterfly certainly inhabits zones with bright sunshine and an extended growing season, which follows the coastal fringe, but it is also absent from areas where both are available. *H. semele* is noteworthy as being our only butterfly which constructs a genuine subterranean cocoon (Ford 1945). High soil moisture conditions may lead to increased disease and thus mortality amongst pupae. It may also make for problems at eclosion (Brakefield, personal communication). Coastal areas have higher sunshine levels and windspeeds which in turn lead to greater evapotranspiration and drier conditions. The driest conditions are provided by coastal sand-dunes for which marram provides a useful indicator. Where the butterfly extends inland it generally does so on calcareous or sandy bedrock (e.g. East Anglian Breckland), both of which are permeable and dry. Its occurrence along the Welsh border, from Llangollen to the Malvern Hills, coincides with the rain shadow of the Cambrian mountains and steep rocky hillsides. When it has been found on impermeable bedrock such as shales, as for instance on Coed Hyrddyn near Llangollen, North Wales,

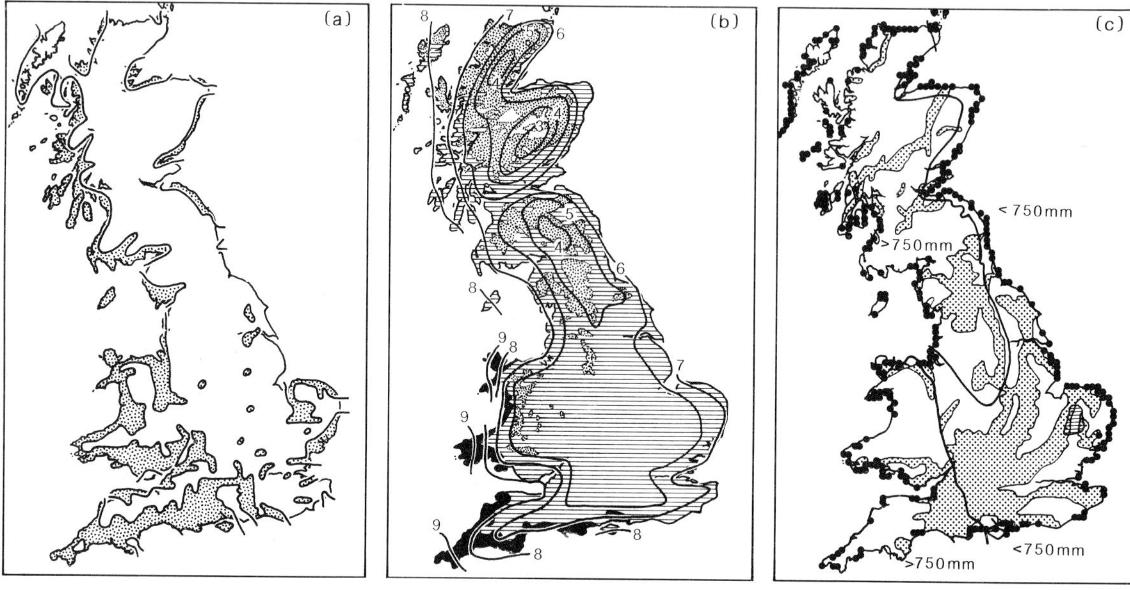

Fig. 1.8 The distribution of the grayling *Hipparchia semele* in relation to selected climatic and geological factors. (a) The current generalized distribution of *H. semele*. (Courtesy of the Biological Records Centre, Institute of Terrestrial Ecology and P. T. Harding.) (b) The length of the growing season (number of months with a mean temperature above 6°C) and the mean length of the frost-free period (in months). Growing season: black, 9 to 12; shading, 7 and 8; stippled, 5 and 6; white, less than 4. Frost-free period shown as isolines. (After Gregory 1954; courtesy of the Institute of British Geographers.) (c) The distribution of two kinds of dry habitat for *H. semele*. Marram grass *Ammophila arenaria* covered coastal dunes (circles) and limestones (stippled) probably act as habitat reservoirs for the butterfly which has spread inland, intermittently, onto other suitable substrates. Among the most important alternatives are dry heaths on sand and gravel, for example, Breckland (shading), moorland with thin soils and steep hillsides (see text), and regions of low rainfall, here illustrated as areas with <750 mm in 30 per cent of years (continuous line).

and Brechfa, east of Carmarthen in South Wales, it selects steep hillsides with very thin soils facing south, which shed rain water rapidly and dry out quickly. Soils which dry out quickly because they are excessively freely-drained or subject to rapid surface run-off also warm up quickly. The butterfly may then be able to develop faster as well as avoid damp-associated diseases. Any one, a combination or perhaps none of these hypotheses may be correct. They nevertheless illustrate something important, that the comparison of mapped data can lead to the formulation of useful hypotheses which, it is hoped, will then be explored.

Comparing mapped data can be useful even when distributions are extensive and apparently unremarkable. A comparison of the distribution of *Callophrys rubi* with its main northern hostplant bilberry *Vaccinium myrtillus*, demonstrates that other hostplants (such as gorse *Ulex europaeus*, dogwood *Thelycrania sanguinea*, common rockrose *Helianthemum chamaecistus*, birdsfoot trefoil *Lotus corniculatus* and dyer's greenweed *Genista tinctoria*) must be used in southern and eastern England, if not in the north and west. It would be interesting to know if this affects the phenology of the insect and other aspects of its physiology. Similarly, on a local scale, the distribution of *Aricia agestis* in Warwickshire off the limestone suggests the use of a hostplant other than *Helianthemum* spp., probably common storksbill *Erodium cicutarium* (Smith and Brown 1979). It is known to use dovesfoot cranesbill *Geranium molle* as well (Warren 1986; Blab and Kudrna 1982).

1.3 Focusing on regional and local issues

Investigating distributions at different scales reveals different issues and, for at least two reasons, there is a need for biogeographical studies of butterflies at more local scales. First, for the British Isles as a whole, latitude dominates other potential influences but is irrelevant at a local scale. Secondly, much variation in geology, soils, and land use occurs within a 10 km square, the impact of which can only properly be accounted by local mapping projects. Local mapping schemes fall into two basic groups: (i) county or district data banks intent on itemizing their entire flora and fauna, and (ii) autecological surveys that investigate one organism in relation to its requirements. Both may have the same objectives, which are generally to learn as much as possible about each insect's preferences in respect of conservation policy. For both, too, the concepts of habitat and population become more important, especially knowledge of the local scale concentration of resources for adult feeding, larval foraging, adult roosting, hibernation of early stages, mate-location, and courtship, on which individual survival depends.

1.3.1 County mapping schemes

Many county or district schemes are now underway, using a mapping resolution of one kilometre squares or tetrads (2 km squares). Several have published reports (Emmet and Heath 1989). The varied and unique pressures within towns and cities, their unusual habitats, makes such surveys as important for urban as for rural counties. These maps provide an invaluable inferential tool for future study. This is nicely illustrated by Garland (1981) in his booklet on butterflies of the Sheffield district which includes a transparent environmental overlay. Figure 1.9 reproduces maps from this work contrasting three butterflies with very different distributions. Smith and Brown's (1979) atlas for Warwickshire records data for different decades. Supreme among the regional publications, many of which are now appearing, are the excellent books of Thomson (1980), Thomas and Webb (1984), and Mendel and Piotrowski (1986), Thomson's work establishing the model for regional texts.

1.3.2 Factors in local distribution patterns

Few single-species mapping surveys have been undertaken at a local scale. These, more than county schemes, directly attempt to determine the structure of populations and to search for causal inference. Much of this work has been on endangered species and is unavailable to the public (ITE and JCCBI confidental reports, see Heath *et al*. 1984). However, Rafe and Jefferson's (1983) survey of *Melanargia galathea* in the Yorkshire wolds is a useful example of this approach. Mapping records of the butterfly for 1 km squares revealed three main extant and two extinct populations. These were found to occupy the few extensive areas of rough grassland now restricted to steep-sided dry valley systems amid the intensive cereal farming on the chalk. Forty-three per cent of unimproved grassland has been lost to other land uses between 1960 and 1982, restricting potential habitats for the butterfly in the wolds. The distribution also implicated climatic controls, as would be expected for a butterfly at the edge of its range, since it was most numerous on slopes with a southerly aspect.

More detailed mapping still, at resolutions of 100 metres or less, is required to sift out properly the reasons for most local distributions. The distribution of *Erebia aethiops* on Arnside Knott in south Cumbria has been shown to coincide with areas of dense growth of an unusual hostplant there, *Sesleria caerulea*, which is limited to thinner rendzina soils on steeper slopes (Dennis 1982*c*). More remarkable are the parallel distributions of the two races of *Plebejus argus* and *Hipparchia semele* occurring on the Great Ormes Head, in North Wales (see Fig. 10.8). Their distributions depend mainly on local climate, on the dry, sunny south and western slopes, and the influence these factors have on microclimates, habitats, and hostplants (Dennis 1972*a*,*b*; 1977). At a finer scale too the distribution of male *Lasiommata megera* on Brereton Heath, Cheshire, depends on climatic constraints (Fig. 1.10; Dennis and Bramley 1985). Such detailed mapping begins to account for differences in population density and should form an integral part of ecological survey work (see Douwes 1975, 1976). Nectar sources, hostplants, microclimates, and other resources can all be mapped and

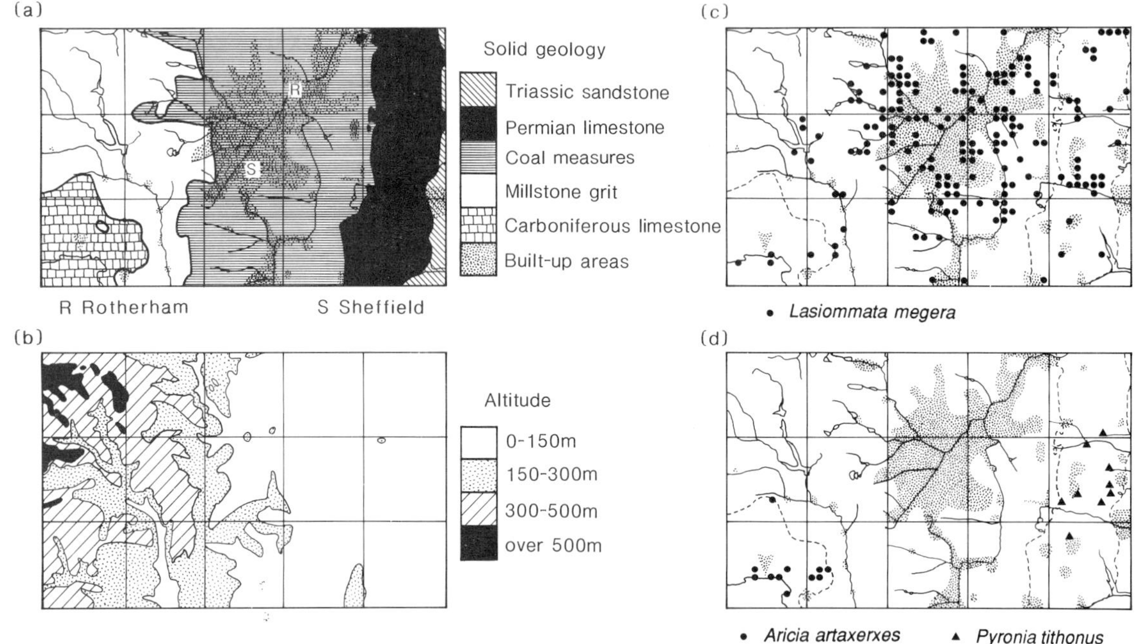

Fig. 1.9 The distribution of selected butterflies in the vicinity of Sheffield's urban landscape. (a) Geology and urban development; (b) altitude; (c) distribution of the wall brown *Lasiommata megera*; (d) distribution of the gatekeeper *Pyronia tithonus* and the northern brown argus *Aricia artaxerxes*.

As *L. megera* frequents bare ground it is able to colonize derelict sites in cities. *P. tithonus* and *A. artaxerxes* are restricted to limestones in this area but at very different altitudes and associated with different habitats. (Redrawn from Garland, 1981; courtesy of Sorby Natural History Society, Sheffield.)

compared to similar details for the insect, such as its distribution, density, and individual movement. Ordnance Survey maps at a variety of scales (1:2500; 1:10 000; 1:25 000) provide ideal frameworks for such studies, but it is not too difficult to construct one's own maps for smaller areas applying a variety of simple techniques such as compass survey, line and offset, and quadrats (Debenham 1962; Bennett and Humphries 1974; Miller 1977).

1.3.3 Habitats and species' diversity

Within a small area, managing the resources required for one butterfly species may inadvertently cause the loss of another butterfly or some other organism (see section 7.5). In view of the desperate need now to maintain the variety of wildlife in the British countryside, it is surprising that very few studies have been undertaken to map butterfly diversity at the local scale. Species diversity is by no means a simple concept and many measures have been devised to express it (see Peet 1974; Usher 1983). Fundamentally, species diversity is a function of the number of species present (species abundance) and the evenness with which individuals are distributed among these species (species equitability) (Hurlbert 1971; Pielou 1975). The number of species alone is an inadequate measure of the fauna as two habitats may share the same butterfly species, but in one they may be abundant while in the other all but one could be rare. Thus, some population assessment is essential to determine the status of species in diversity studies. Simultaneously, an assessment should also be made of the type of species as the two habitats may have similar diversity indices but contain a very different butterfly fauna (see section 7.5).

Repeated transects, as used by the Institute of Terrestrial Ecology (ITE) at Monks Wood Experimental Station in Huntingdon, to monitor butterfly

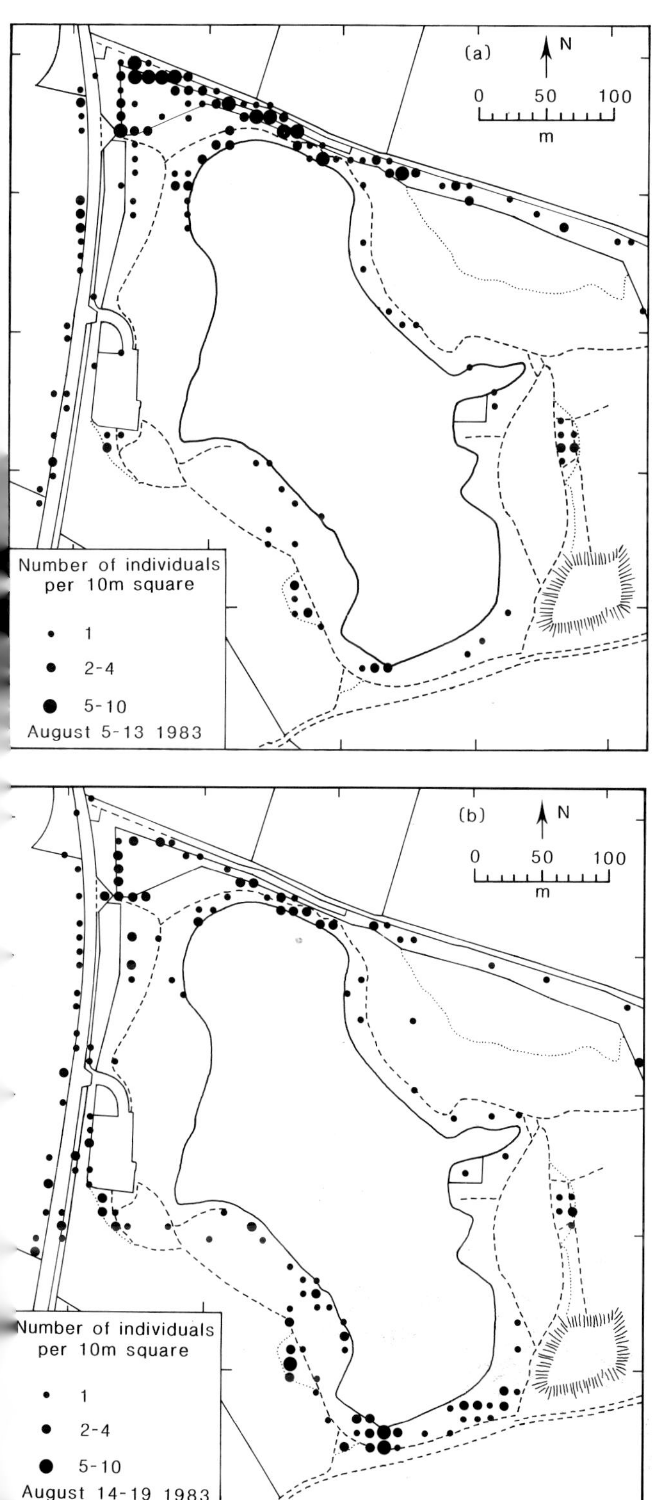

Fig. 1.10 The distribution of male wall brown *Lasiommata megera* on Brereton Heath, Cheshire, for the date periods: (a) 5–13 August, 1983 (b) 14–19 August, 1983.

The significant southerly shift in distribution between the two date periods was caused by a later emergence of butterflies from cooler north facing slopes on the southern borders of the lake and not from a north–south movement of individuals. In cool seasons, such as in 1985, the southern population is probably unsuccessful or comprises few individuals. (From Dennis and Bramley 1985; courtesy of Societas Europaea Lepidopterologica.)

populations at some 80 British sites, provides the most suitable technique for diversity studies (Pollard 1979*a*; Pollard *et al*. 1986), especially if linked to some method for assessing absolute population size (Thomas 1983*c*). However, this important long-term project deals mainly with nature reserves that have habitats atypical of those found in the countryside at large. It could, nevertheless, easily be extended to farmland and to urban landscapes. The short-term nationwide WATCH census, carried out in 1981–82 during 'Butterfly Year', provides a valuable precedent for future more intensive work at county level or more local scales (Thomas 1983*d*). Adult butterfly numbers were recorded along standardized transects on 670 sites, covering most commonplace biotopes. Apart from five conspicuous species, butterflies were necessarily identified at family level to simplify recording. Ten biotopes were sufficiently well sampled for general conclusions to be made about their relative value as butterfly habitats.

The findings were not altogether unexpected but revealed some striking differences in adult densities and they confirmed earlier subjective impressions. The figures were particularly disappointing. Although high counts of individuals and species, particularly rare ones, were obtained for improved pasture, very little natural grassland now survives in southern Britain (Smith 1980; Mellanby 1981; section 11.2). Abandoned railway tracks and wasteland were also shown to support large numbers of the more common sedentary species, but these habitats similarly occupy a small area. The importance of hedgerows has often rightly been expressed as a wildlife reserve, but perhaps the most disappointing feature of the survey was the low densities of butterflies found in this habitat; they were often no greater than in the arable or improved pasture fields they bordered. However, better figures emerged for roadside verges (see Way 1977). Mature gardens recorded similar species to those in modern housing estates, but densities were generally twice as high; these comprised mobile nymphalids and pierids including *Gonepteryx rhamni*, as well as the lycaenid *Celastrina argiolus*. Woods were inadequately sampled, particularly conifer plantations.

At the local scale, multiple species' studies (synecological surveys) additionally require a careful assessment of resources—their quality and size. Habitats for butterflies in Britain have never been continuous. At one time, 5000 years ago, open areas in our landscape were extremely limited (see section 10.4). But now that semi-natural habitats are severely curtailed, it is not difficult to envisage how the island biogeography model discussed earlier in the chapter can have relevance in a terrestrial context. Shreeve and Mason (1980) have carried out such an investigation of the butterfly fauna of 22 woods in north-east Essex and south Suffolk. They found that both the number of vagile and less vagile species correlated significantly with the size of the woodlands. In the case of more sedentary species, woodland size may actually measure the expanse of particular resources although they could not be ascertained from any correlation with the area of glades and rides. For more sedentary butterflies found in woodlands, there is no 'continental' stock providing source populations, and colonization of habitats depends on dispersal from nearby 'islands' as suggested earlier when colonization of islands in the Outer Hebrides and Orkneys was discussed. For vagile species, the woods may be considered as different sized samples from a relatively uniform distribution of species.

Much theoretical work (Higgs and Usher 1980; Game 1980; Higgs 1981; Margules *et al*. 1982; Janzen 1983; Simberloff and Abele 1984; Boecklen and Gotelli 1984) and many empirical studies (Middleton and Merriam 1983; Woolhouse 1983; Peterken and Game 1984; Game and Peterken 1984) have now been carried out on the implications of terrestrial island biogeography, mainly in respect of acquiring and designing nature reserves for conservation (see section 11.5). The spatial organization of such habitat islands is clearly important. Island size and spacing, potential stepping stones, 'flyways' such as roadside verges, railway embankments, and canal towpaths, the ratio of perimeter to area as a measure of potential interference from agriculture, emerging and sinking islands, as new habitats appear and old ones are developed—all have their relevance. But these studies are of little use if no account is made of the precise resources required by the species investigated. Terrestrial island biogeography requires two tiers of research: (i) of semi-natural habitats in a sea of farmland and urban landscape, and (ii) of resource islands (e.g. nectar sites, larval hostplant sites, roost sites, hibernation sites, mate location sites) within each semi-natural habitat—

pattern of islands within islands. In the survey of Essex and Suffolk woods, glade and ride area evidently did not coincide either with the size or quality of resources used by butterfly species requiring open conditions in those woodlands.

Two surveys on species diversity within woodland rides begin to address the need for more detailed surveys of terrestrial habitat islands (Peachey 1980; Warren 1981; see section 7.5.2). Butterflies differ greatly in their requirements. This is clear from anecdotal information on habitat preferences (Appendix 2) and, increasingly, from detailed research (autecological surveys) on individual species (see Heath *et al*. 1984; BUTT 1985). Species' diversity depends to a large extent on a mix of resources which, in turn, depend on a variety of environmental and geographical conditions, not just of light and shade and of related climatic variables such as temperature and humidity, but of soil conditions (i.e. soil moisture and soil nutrient levels), and land use (see Reichholf 1973; Bierhals 1976; Gerlach 1976; Erhardt 1985). Analysis of the Bernwood forest data reveals groups of species within rides which differ according to the light environment, but also on the basis of tall and short plant communities, and the proximity of specific resources in different habitats outside the woodland (see Fig. 7.9). Although much valuable work on the ecology of single species has now been undertaken, studies of species-associations become essential for managing butterfly habitats and creating new ones (see section 11.5). However, butterflies vary substantially in behaviour and ecology, and a review of these topics in the following chapters provides an insight into the difficulties faced by conservation practices.

2

Adult behaviour

Tim G. Shreeve

2.1 The significance of behaviour patterns

The success of an individual adult butterfly is measured by its reproductive output. The most successful males are those that gain access to the greatest number of females, that is, are able to maximize the number of matings. Female reproductive success corresponds with the number of eggs deposited and the survival of offspring. Although the two sexes may have different requirements, they both need to locate adult resources (i.e. mates, nectar, and/or roost sites) in suitable areas. They must be able to use these resources efficiently and be able to avoid predators. All these activities are interdependent; they are governed by physiological processes (i.e. body temperature) and sensory ability, and are also influenced by the structure of the habitat in which the butterfly occurs. These factors are examined in this chapter.

2.2 Regulating body temperature

2.2.1 Body temperature and microclimates

Flight is essential for finding mates, food, new habitats, to escape from predators and in the case of females for locating egg-laying sites. Flight is dependent on the contraction of the flight muscles of the thorax which, like all insect muscles, have the capacity to function over a wide temperature range. When muscles are warm, less energy is used for a given amount of movement than when they are cool, and warm muscle is also capable of the most rapid contraction. Because different species have different wingbeat frequencies in normal flight, the temperature at which flight is initiated may differ between species; it is perhaps greater in those species with rapid wingbeats than in butterflies that adopt gliding flight.

Thoracic temperatures of between 30 and 40°C at the time of voluntary flight initiation have been recorded for various pierid butterflies (Watt 1968; Roland 1982; Kingsolver 1985*a*), of between 32 and 34.5°C for the speckled wood *Pararge aegeria* (Shreeve 1984) and within the range 28–30°C for the swallowtail *Papilio machaon* (Wasserthal 1975). All these species have moderate wingbeat frequencies. By contrast, overwintering monarchs *Danaus plexippus* of North America and the small apollo *Parnassius phoebus* of alpine areas in North America and Europe both engage in gliding flights with infrequent wingbeats when their thoracic temperatures are at 10°C and 17–18°C, respectively (Chaplin and Wells 1982; Guppy 1986). Not only is body temperature important to flight but it is also suggested that it can affect the rate of production and maturation of eggs within the female abdomen (Stern and Smith 1960; see section 3.1). This may be of the greatest importance to those species (e.g. meadow brown, *Maniola jurtina*, Essex skipper, *Thymelicus lineola*) which do not emerge from the pupa with a full complement of mature eggs.

Insects do not maintain a constant body temperature; some groups (endotherms) are capable of generating body heat by muscular activity and by circulating body fluids, but the remainder (ectotherms) are dependent on the external environment for temperature regulation. Unlike many moths, flies and wasps, few butterflies regularly engage in the practice of wing vibration to elevate their body temperature before engaging in flight. However, the role of this mechanism should not be overlooked; shivering by the small tortoiseshell *Aglais urticae*, *Pararge aegeria*, the wall brown *Lasiommata megera*, and the grayling *Hipparchia semele* has been noted to occur in cool, dull weather. This may serve as an energetically expensive mechanism of heat generation (Heinrich 1972*a*) which is used only in extreme conditions. This behaviour may well occur in other species, but butterflies are primarily ectotherms that bask in sunshine and rely on the warmth of the air immediately surrounding their bodies for heat.

Air temperatures in temperate regions rarely reach those required for efficient flight muscle (28–38 °C) of non-gliding species and the need to keep warm is often a major constraint on the activity of butterflies in these regions. Being mobile and of small size, butterflies are able to take advantage of particular microclimates at vegetation and soil surfaces, sites with temperatures that can be very different from those of the surrounding air. In sunlit areas under a deciduous woodland canopy the temperature of the boundary layer of air at the surface of vegetation can reach 35 °C when air temperature is at 25 °C (Rackham 1975), and on sandy soils and in areas with highly reflective or absorptive surfaces, such as bare chalk, dry earth, lichens, or leaf-litter, surface temperature can exceed 45 °C when air temperature is in the region of 20 °C. By using such sites in cool weather the difference between the optimal body temperature for flight and that of the air around the butterfly body is minimized, increasing the opportunity to engage in activities that are dependent on flight. In hot weather species can suffer from heat stress, and activity may then need to be restricted to cool microhabitats.

The distribution of warming sites can have a profound effect on where a species is located and how it performs its activities. For instance, *Pararge aegeria* settles on dry stones and soil on woodland rides early in the morning, but as air and vegetation temperatures rise during the day individuals move from rides to cooler sunlit areas under woodland canopy (Shreeve 1984). These daily changes of habitat use are matched by seasonal changes. In spring and late summer the majority of individuals are located along rides but in midsummer when ground and air temperatures are greater most individuals are located in woodland. In this species, habitat type and the distribution of warming sites has an influence on mate-locating behaviour (see section 2.4.3).

2.2.2 Basking and morphology

As well as using warm microclimates, butterflies are recognized as using three methods of basking, usually described as: lateral, dorsal (Clench 1966), and reflectance (Kingsolver 1985*a*). In each, the wings of the butterfly are used as devices to collect incoming solar radiation and pass this energy to the body, thereby increasing body temperature (Fig. 2.1). In lateral baskers it is the ventral wing surfaces that are involved and in both dorsal and reflectance baskers the dorsal wing surfaces. Because of morphological (physical) similarities between related species, which affect how a butterfly basks, closely related species frequently adopt the same method. Lateral baskers (including the Theclinae, Coliadinae, and some Satyrinae; see Table 2.1) are those that bask with closed wings oriented to sunlight, relying on the heating of the ventral wing surfaces and the conduction of this heat to the thorax and abdomen. Dorsal and reflectance baskers both angle open wings to sunlight but are distinguished by the method in which heat is transferred to the body. Dorsal baskers (most Nymphalidae, Hesperiidae, and some Lycaenidae; see Table 2.1) rely on the heating of the wings and the conduction of this heat to the flight muscles and body organs, but reflectance baskers (Pierinae and perhaps some Lycaenidae) primarily rely on the reflection of solar radiation by the wings on to the thorax and abdomen.

All basking butterflies make continuous adjustments of their posture in relation to solar radiation intensity and thoracic and basking-site temperatures (Fig. 2.1). When thoracic temperatures are low both dorsal and lateral basking species will press their bodies and wing-tips to the ground, reducing convection currents around the body. With rising thoracic temperature convective heat loss is not so important and the wings are steadily raised, decreasing the wing area exposed to solar radiation. When

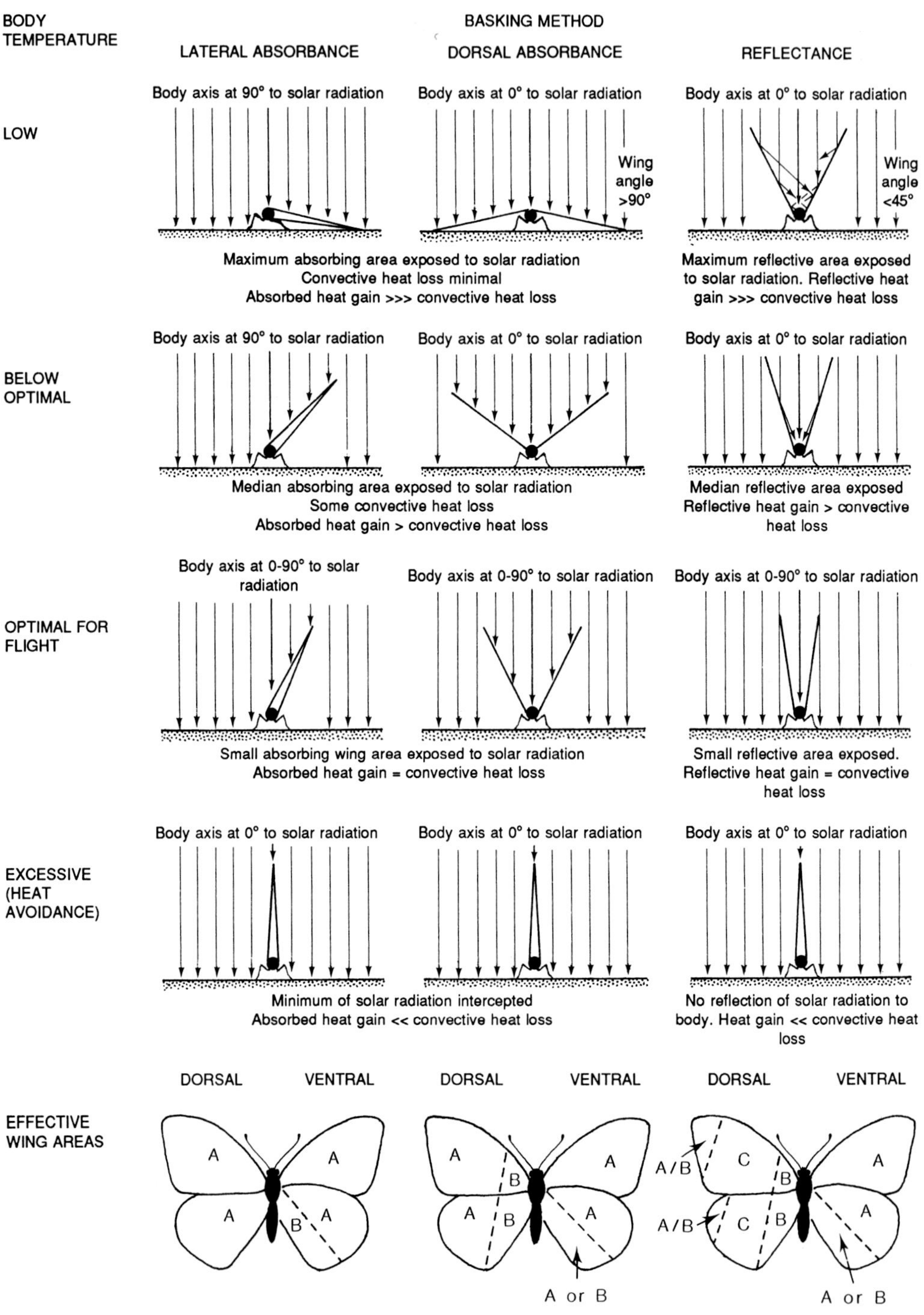

Fig. 2.1 The basking postures adopted by dorsal absorbance, lateral absorbance, and dorsal reflectance basking butterflies, and the main wing areas of thermal importance to these basking methods. Species with different basking methods adopt different wing positions in relation to body and air temperature, making them more or less conspicuous to conspecifics and potential predators.

Table 2.1 The major basking methods of British butterfly species

Species	Absorption		Reflection
	lateral basking	dorsal basking	
HESPERIIDAE			
Carterocephalus palaemon		+	
Thymelicus sylvestris		+	
Thymelicus lineola		+	
Thymelicus acteon		+	
Hesperia comma		+	
Ochlodes venata		+	
Erynnis tages		+	
Pyrgus malvae		+	
PAPILIONIDAE			
Papilio machaon		+	
PIERIDAE			
Leptidea sinapis	+		
Colias croceus	+		
Gonepteryx rhamni	+		
Pieris brassicae		+	+
Pieris rapae		+	+
Pieris napi		+	+
Anthocharis cardamines		+	+
LYCAENIDAE			
Callophrys rubi	+		
Thecla betulae	+		
Quercusia quercus	+		?
Satyrium w-album	+		
Satyrium pruni	+		
Lycaena phlaeas			?
Lycaena dispar			?
Cupido minimus		+	?
Plebejus argus		+	+
Aricia agestis		+	
Aricia artaxerxes		+	
*Polyommatus icarus**		+	+
*Lysandra coridon**		+	+
Lysandra bellargus		+	+
Celastrina argiolus	?		?
Hamearis lucina		+	
NYMPHALIDAE			
Ladoga camilla		+	
Apatura iris		+	
Vanessa atalanta		+	
Cynthia cardui		+	
Aglais urticae		+	

Table 2.1 (*cont.*)

Species	Absorption		Reflection
	lateral basking	dorsal basking	
Nymphalis polychloros		+	
Inachis io		+	
Polygonia c-album		+	
Boloria selene		+	
Boloria euphrosyne		+	
Argynnis adippe		+	
Argynnis aglaja		+	
Argynnis paphia		+	
Eurodryas aurinia		+	
Melitaea cinxia			+
Mellicta athalia		+	
Pararge aegeria		+	
Lasiommata megera		+	
Erebia epiphron	?	+	
Erebia aethiops	?	+	
Melanargia galathea		+	
Hipparchia semele	+		
Pyronia tithonus		+	
Maniola jurtina		+	
Aphantopus hyperantus	+	+	
Coenonympha pamphilus	+		
Coenonympha tullia	+		

* Differences may occur between the sexes; the males are brighter than the females and are probably more dependent on reflectance basking.

body temperature reaches that at which flight can occur the butterfly may take to the wing or it may remain settled. Without flight, and when body temperature exceeds an optimal level, the body and wings may angle away from the sun to cast a minimum shadow, minimizing the area of the wing exposed to sunlight. In extreme cases the butterfly may move to shade or raise its body on extended legs (stilting) out of the warmest layer of air on the substrate surface. Because reflectance baskers rely on a different method of body warming than dorsal and lateral baskers the postural changes associated with body temperature are somewhat different. In any species of this group there is a particular wing angle which intercepts and reflects the most light on to the body; this angle is dependent on wing size and pigmentation (Kingsolver 1985*a*,*b*). Wing angles greater than or smaller than that which intercepts the most light reduce the rate of body warming. Thus in these species, closing or fully opening the wings can occur in response to excessive body temperature. Whichever occurs may be closely related to the benefits and costs of conspicuousness and concealment (see chapter 5).

Because butterflies rely on absorption, conduction, and reflection while basking, differences of morphological characters which affect these physical processes can have significant effects on adult performance. For example, a reduction of reflective wing area, of a reflectance basker, will slow the rate of heating, and a decrease in the amount and area of radiation-absorbing pigment on the wing of a dorsal or lateral basker will also decrease the rate of warming. Likewise, the necessity to maintain an adequate heat balance in a particular environment can be a selective agent in the determination of these characters, in particular wing colour and pattern, size, and the amount of hair-scales (Table 2.2; see section 10.5). Variation and change in these characters are also influenced by developmental constraints, by intraspecific communication and by predation pressure (Dennis and Shreeve 1989), so any detailed consideration of the thermoregulatory significance of particular characters and behaviours must also include a consideration of these other factors.

Wing colour is of undoubted significance to thermoregulation, but no generalizations can be made without reference to the properties of wing pigments, to the part of the wing important to thermoregulation and to the method of basking. Pigments vary in their capacity to absorb and reflect solar radiation. Melanin derived blacks and browns, which are commonly found in the Nymphalidae and Pieridae are effective absorbers of infra-red radiation, (Watt 1968) whilst pteridine derived whites and yellows, which are commonest in the Pierinae, are highly reflective of infra-red radiation (Kingsolver 1985*b*). It cannot be assumed, however, that all black pigments are melanin derived and good absorbers of infra-red. Heinrich (1972*b*) found that some tropical black butterflies of Papua New Guinea reflect solar radiation. The function of melanin pigments in thermoregulation is given in detail for lateral basking *Colias* butterflies by Watt (1968), Roland (1982), Kingsolver (1983), and Kingsolver and Watt (1983*a*,*b*), in some dorsal basking Nymphalinae and Papilionidae by Wasserthal (1975) and in reflectance basking Pieridae by Kingsolver (1985*a*,*b*). Kingsolver also details the role of reflective pteridine derived white pigments.

In dorsal- and lateral-basking butterflies the most important area of the wing involved in thermoregulation is that nearest the body, this being the region which conducts heat to the body most rapidly, either directly via the cuticle or via the body fluid (haemolymph) which fills the interior of the wing-veins and the body (Wasserthal 1983). In reflectance baskers the whole wing surface is of importance (Kingsolver 1985*a*). Melanization is most effective on the wing

Table 2.2 Morphological factors affecting the thermal balance of dorsal, lateral, and reflectance basking adult butterflies

Character	Effect of warming rate and thermal stability Basking method		
	dorsal	lateral	reflectance
Dorsal wing surface			
Basal melanization	+	Neutral	+
Melanization of main wing area	Neutral	Neutral	−
Melanization of wing-tips	Neutral	Neutral	−
Ventral wing surface			
Basal melanization	Neutral or +*	+	Neutral or +*
Melanization of main wing areas	Neutral	Neutral	−
Body size	Increase in thermal stability and increase in warming time		
Hair-scales	Reduction in convective heat loss and increase in thermal stability		

* Basal melanization of the ventral wing surface is of significance to those species which use both reflectance and lateral basking.

\+ Increase in character leads to more rapid warming in cool cloudy conditions.

− Increase in character leads to slower warming in cool cloudy conditions.

surface exposed to solar radiation; in lateral baskers this is the underside and in dorsal and reflectance baskers the upper surface. With the exception of reflectance baskers, darker individuals warm more rapidly than paler individuals of the same species, and in the cooler parts of the ranges of dorsal and lateral basking species there is a tendency for individuals to be more heavily melanized than in the warmer parts of their ranges. In reflectance baskers this pattern is reversed (Table 2.3).

2.2.3 Thermoregulation and activity

Variation within populations of those characteristics which affect body-heating and cooling is of significance to both individual and population perform-

Table 2.3 Geographic and seasonal variation in morphological characters in relation to climate of selected pierid and satyrine butterflies within Britain

Species	Main basking method	Variable morphological characters of greatest significance to thermoregulation	Differences in relation to climate and season and functional significance
PIERIDAE			
Leptidea sinapis	Lateral	Melanization of ventral hindwing	Darker in cool, cloudy regions, increasing warming rates
Pieris brassicae } *Pieris rapae* }	Reflectance	Melanization of dorsal wing-tips	Darker in summer broods, decreasing reflective wing area
Pieris napi	Reflectance and lateral	Melanization of dorsal wing-tips and ventral hindwing coloration	Dorsal wing-tips darker in summer broods, with decreasing reflective area. Ventral melanization in spring increases warming rates. In northern areas all broods are yellower on the ventral surface and darker on the dorsal surface
SATYRINAE			
Pararge aegeria	Dorsal	Melanization of basal area of dorsal wing surface and size	Lighter in summer broods, decreasing absorptive wing area. Spring broods larger. North-western subspecies (*oblita*) also larger, increasing thermal stability in cooler variable conditions
Lasiommata megera	Dorsal	Melanization of basal area of dorsal wing-surface	Little seasonal difference but darker in northern areas, increasing warming rates in cool areas
Aphantopus hyperantus	Dorsal and lateral	Size	Smaller in northern and cool and cloudy areas, decreasing warming time in cool climates but also decreasing thermal stability. Greying of underside in northern areas probably of no thermal significance

ance. In cool, dull weather dark *Colias* butterflies fly and feed more often than do paler members of the same population (Roland 1982). Because dark specimens are able to warm more rapidly than pale individuals they maintain a higher average body temperature throughout their activity (basking and flight). Thus they are always nearer their optimum temperature than pale specimens which have to stop flying more frequently to warm, and which also have a lower average body temperature. In hot, sunny weather the performance of dark and light butterflies is reversed, as dark ones can suffer from heat stress and their activity can be severely curtailed because of this overheating. Variation in populations is probably maintained because no individual can consistently outperform others when weather is variable. Reflectance baskers differ from both dorsal and lateral baskers because the significance of colour, particularly melanization is reversed. Absorbent pigments on the upper wing surface reduce the reflective area of the wing and in cool conditions specimens with more reflective wing surfaces are the most active because they can warm up rapidly.

Size also affects activity because small individuals of dorsal and lateral basking species warm more rapidly than large ones because of their reduced mass, but are subject to greater internal temperature fluctuation. In cool, dull weather large specimens are more active than small ones. Once warmed, they are best able to maintain a stable body temperature. In cool but bright weather small individuals are more active because of their ability to reach optimal body temperature more rapidly. Irrespective of basking posture, hair-scales on the body reduce convective cooling and the only studies of the relationship between hair-scales and warming rates (Kingsolver 1981; Kingsolver and Moffat 1983) demonstrate that *Colias* butterflies in cool, windy environments have more hair than in warmer less exposed localities.

Flight activity produces metabolic heat; almost 80 per cent of energy expended by insect flight-muscle is converted to heat (Weis-Fogh 1972). However, the total amount of heat produced is related to effort, dependent on wing-stroke amplitude and wingbeat frequency (Tsuji *et al*. 1986). Most heat is dissipated by convection and the rate of heat loss is inversely proportional to body size (Digby 1955). The efficiency of flight and the capacity of individuals to sustain flight activity is thus related to the initial thoracic temperature on take-off, and rates of heat production and loss, factors which will vary between species, and which are also dependent on local meteorological conditions. Tsuji *et al*. (1986) have found that *Colias* butterflies can maintain a steady body temperature during flight but Heinrich (1985) studying the inornate ringlet *Coenonympha inornata* of North America, Shreeve (1984) studying *Pararge aegeria*, and Shelley and Ludwig (1985) working with the tropical satyrine *Calisto nubila*, have all demonstrated that flight can cause body cooling. In these three satyrine species at least, flight durations and hence general performance is correlated with air temperature (Fig. 2.2). Cool conditions reduce flight durations and hence potentially reduce effective mate-locating behaviour and egg-laying.

Although no measurements of body temperatures were made, Dennis (1982*d*) found that the flight activity of *Lasiommata megera* was also correlated with air temperature. This may be because body-cooling could be occurring during flight. In contrast to species having fairly rapid wingbeats, which disturbs the layer of air around the body, the body temperature of *Parnassius phoebus* is maintained

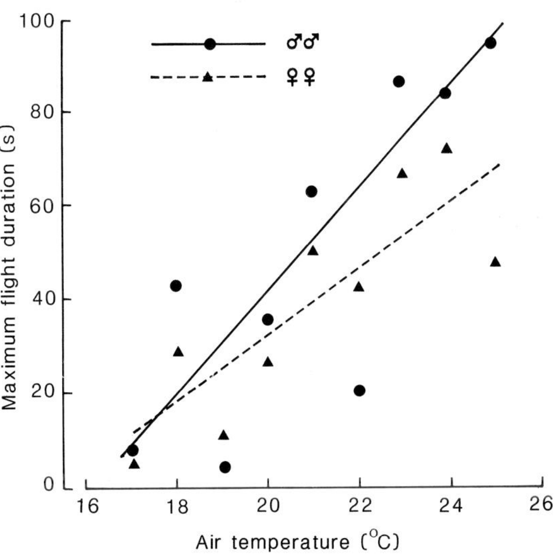

Fig. 2.2 Flight durations of speckled wood *Pararge aegeria* in relation to air temperature. Flight durations increase with increasing air temperature, body cooling being slower at high temperature when the difference between optimal body temperature and air temperature is less. (From Shreeve 1984; courtesy of *Oikos*.)

during gliding flight because the boundary layer of air around the flying individual is not disturbed (Guppy 1986). Thus convective cooling of flying individuals of this species is no greater than of settled individuals; solar radiation, rather than metabolic heat, is the main source of heat during flight activity.

Additional to the effects of interactions between morphology and behaviour on flight activity are differences in enzyme systems. Watt *et al.* (1983) document variation in phosphoglucose isomerase (PGI) enzyme systems, which are involved in glycolysis and the producton of fuel for flight in *Colias* butterflies. Different PGI genotypes facilitate activity in particular climatic and weather conditions, and it is argued that this polymorphic enzyme system, in this group of butterflies, enables activity in a wide range of altitudes with dissimilar environmental conditions.

2.3 Finding nutrients

2.3.1 Nutritional requirements

Some species of butterflies are frequent visitors to flowers; others feed on honeydew, and particularly males, on dung, urine, dead animals, and even wood ash, dry earth, and stones. Clearly there is a diversity of 'foods' providing an array of very different nutritional resources and different species of butterfly, by their various feeding behaviours, appear to have dissimilar requirements.

Because of the dependence of butterflies on sunshine and their common habit of basking in full sun in warm areas, there is a potential for adults to be at risk of desiccation. Consequently most butterflies require water. Muscular activity uses energy and prolonged activity can be increased if individuals have access to carbohydrates and sugars. Egg-production is dependent on the availability of certain nutrients and trace elements (see section 3.1), which can be obtained by the adult butterfly. In order to mate successfully the male of most species must have an adequate supply of pheromones to elicit the response of courted females as well as have sufficient energy to acquire females. Although female pheremones may also be involved in mating (see p. 43), in short-lived species they are most likely to be formed from substances obtained during the larval stage since females mate soon after emergence. The nutritional requirements of males may, therefore, differ from those of females. In addition, species that overwinter as adults (e.g. the brimstone *Gonepteryx rhamni*, and *Aglais urticae*) probably have different requirements before and after hibernation. In autumn they feed in order to build up a supply of lipids to be used during hibernation. In spring their requirements are probably for those nutrients which enhance their immediate reproductive output.

Unpalatable butterflies may also enhance their distastefulness to predators by obtaining certain chemicals during the adult stage. This has been recognized for a variety of American and African butterflies, which feed on nectar containing pyrrolizidine alkaloids, substances which act either with other noxious chemicals or possibly alone to deter predators (reviewed by Boppré 1986; see chapter 5). Although little is known about the chemical defences in adults of species resident in Britain, it is worth observing that some species, particularly those which feed on ragwort (*Senecio* spp.) nectar which contains alkaloids, might enhance their probability of escaping bird predation, although there is no direct evidence to support this hypothesis.

When a butterfly emerges from the pupa it has a reserve of energy stored in the fat-body of the abdomen, a limited reserve of nitrogen, and some trace elements, although the actual amounts are dependent on how favourable conditions were in the larval stage. These reserves are of importance to activity and reproductive output but in some species adult feeding may increase egg-production (see chapter 3) and contribute to longevity (Watt *et al.* 1974; Murphy *et al.* 1983; Pivnick and McNeil 1985). This may be most important for females of species which emerge with no or few developed eggs (e.g. brown hairstreak *Thecla betulae*, *Thymelicus lineola*, *Maniola jurtina*.)

It is often assumed that adult butterflies have small energy requirements because insect muscle used in flight, is highly efficient at converting food-energy to mechanical movement. However, such an assumption should be treated with caution; 56 per cent of the total adult energy budget of the small white *Pieris rapae*, comes from adult feeding (Labine and Turk, in Gilbert and Singer 1975). Feeding activity can also comprise a considerable proportion of an adult's time. Dennis (1982*d*, 1983*c*) gives values of 14 and 11.6 per cent, respectively, for the proportion of the total amount of time that male and female *Lasiommata megera* spend feeding, and Wiklund and Åhrberg (1978) estimate that in the orange-tip *Anthocharis cardamines* 18 per cent of male and 16 per cent of female time is spent feeding. Both of these estimates exclude time spent searching for food so these figures underestimate the amount of time spent by these butterflies in feeding-related activities.

2.3.2 The effects of food quality on feeding

Floral nectar is probably the most common food of adult butterflies in northern Europe but, because different species of flowers have different nectar compositions, nectar is a variable resource. Also, within any species the amount, concentration, and composition of nectar varies with flower age, time of day, weather, and the activities of nectar feeding insects (Percival 1961; Corbet 1978). Consequently the 'quality' of nectar in any individual flower can be unpredictable. Within nectar there is a variety of sugars—mono- and oligosaccharides—and also amino acids, and occasionally lipids and alkaloids, all dissolved in water (Percival 1961; Watt *et al*. 1974; Baker and Baker 1975). All may be important to foraging butterflies, the various constituents fulfilling different requirements. Marshall (1982) has shown that amino acids and lipids are immediately used in the production of sperm and eggs, whilst sugars are stored in the fat-body for later use.

Nectar is most dilute and abundant in the early morning, and by mid-afternoon it is most concentrated and scarce, a consequence of depletion by insect visitors and decreasing air humidity. With these daily changes in nectar quality and quantity, feeding at different times of day can produce different rewards for a given time spent foraging. However, it is not known whether feeding is influenced more by nectar volume or by concentration. *Colias* butterflies (Watt *et al*. 1974) visit plants with dilute nectar (25 per cent by volume), but those of *Thymelicus lineola* use nectar that is more concentrated (30–40 per cent by volume; Pivnick and McNeil 1985), a concentration which gives maximal energy gain per unit time spent imbibing nectar. However, considerations of selectivity for particular nectar concentrations must take into account the energetic cost of finding sources which yield optimal nectar. If, for example, the best nectar sources are scarce or difficult to locate, then those individuals which specialize on these sources may gain less energy per unit time spent foraging than those individuals which feed from less rewarding sources, particularly if the less rewarding sources are more common or more easily located. Individuals feeding on the latter will frequently encounter small energy sources and so gain more energy per unit time spent foraging than those individuals which search for scarce resources.

Males and females may have different nutritional requirements because of the demands of egg and sperm production, and therefore they will use different food resources and spend different amounts of time feeding. It has been suggested that males may directly contribute to egg production by passing a complex mixture of proteins, hydrocarbons, glycerols, sterols and phospholipids in water as well as sperm during mating (Marshall 1980; see chapter 3). If males can obtain these substances before they mate and if females can assess the likely contribution of males to their reproductive output, for which there is some evidence, (Rutowski 1979; Marshall 1980; Silberglied 1984), then it will be to their advantage to feed on those substances that yield the greatest amounts of these materials. Proteins, lipids, and hydrocarbons can be synthesized from floral nectar but other carbons, particularly protein rich organic debris, may be the best sources of nitrogeneous compounds. In addition, it is suspected (Porter, personal communication) that sodium and potassium salts also play a role in the male investment in female reproductive effort. Thus the propensity of males of many species to feed from mud puddles and even dry stones (Downes, 1984*a*) may be related to nutritional aspects of reproduction, though the more usual explanation for this behaviour is that males are gathering the precursors of pheremones which are used in courtship (Arms *et al*.

1974; Gilbert and Singer 1975). That feeding from such substances may be related to immediate reproductive effort is suggested by occasional observations of old females of species which usually mate once, feeding from stones and litter (e.g. *Pararge aegeria*, *Hipparchia semele*). In such cases females may be replenishing supplies of salts that are normally supplied by males.

Because butterflies are liquid feeders, the water content of foods is of importance. The range of concentrations of liquids that a butterfly can imbibe is related to the diameter and length of the proboscis, the viscosity of the liquid, and the suction pressure that the butterfly can exert (Kingsolver and Daniel 1979; May 1985). A short, wide-diameter proboscis has a smaller surface area than a long, narrow proboscis, the former facilitating uptake of more viscous liquids than the latter. Proboscis morphology may be important in determining both the types of food and the timing of feeding activity of individual species, and probably plays an important part in interactions between nectar feeding species (see section 7.2). To date there is little information on these aspects, but some species are believed to dilute concentrated nectars (Kevan and Baker 1983) while others (e.g. purple emperor *Apatura iris* and the white admiral *Ladoga camilla*) are known to do this when feeding on solids, by exuding water from the proboscis. Other species may exude watery fluid from the anus (e.g. large skipper *Ochlodes venata*; Dennis, personal observation) before feeding on substances such as ash.

Although the relationship between nutritional requirements and food type is poorly understood it is recognized that the distribution of food resources and feeding behaviour can influence other activities. Where food resources occur in different localities to mate-locating sites and egg-laying sites, butterflies tend to serialize their behaviour, engaging in periods of feeding to the exclusion of all other activities (Fig. 2.3). Where food and other resources are in close proximity, feeding behaviour can be interspersed with other activities such as mate-location or egg-laying. This behaviour is often shown by *Maniola jurtina*, the ringlet *Aphantopus hyperantus*, and chalk hill blue *Lysandra coridon*, which used habitats where adult foodplants, larval hostplants, and mate-locating sites are in close proximity. However, the behaviour of individuals in such habitats can be as serialized as those in segregated habitats (e.g. *Ochlodes venata*; Fig. 2.3). On the Great Orme, in North Wales, sites for egg-laying, mate-locating, and feeding for *Hipparchia semele* all occur in close proximity and both sexes engage in feeding throughout the day. In other locations, especially on dunes (e.g. Cefn Sidan, South Wales) where food plants are spatially segregated from other resources, these behaviours are periodic. At Cefn Sidan most feeding, particularly that of males, occurs in the morning before individuals of either sex engage in other activities. Species with widespread egg-laying sites and adult food resources (e.g. Pieridae) tend to engage in all activities throughout the day. Thus male and female *Anthocharis cardamines* feed on those nectar plants that they encounter while searching for mates and egg-laying sites, respectively.

In dry environments such as sand-dunes, or chalk downland in midsummer, the control of water balance is more difficult than in moister environments such as woodland. It is interesting to speculate whether species which are nectar feeders engage in more early morning feeding in dry than in wet environments, to take advantage of dilute nectar in the morning. However, no conclusion can be drawn without other details of the activities of these species and a knowledge of the behaviour of other nectar-feeding species (e.g. bees) which share the same resources (see Fig. 7.7).

Butterfly sensory acuity extends to vision (in both the human visible range and the ultraviolet), scent, touch, and taste. All are used in the location of food, but touch and taste are most important in eliciting proboscis extension after food sources have been detected from long distances with the other senses. Floral colour and pattern are important determinants of butterfly visits to flowers (Faegri and van der Pijl 1979); most species are attracted to flowers which are either blue-red or yellow in colour (Watt *et al*. 1974; Jennersten 1984). Visited flowers also tend to correspond to particular ultra-violet patterns. Feeding is usually opportunistic; different species of butterfly do not usually specialize on particular nectar sources but tend to use those sources which are abundant and accessible in any habitat. For example, the principal nectar sources of *Maniola jurtina* were found to be thistle *Cirsium* and knapweed *Centaurea* species at Hightown, near Liverpool (Brakefield

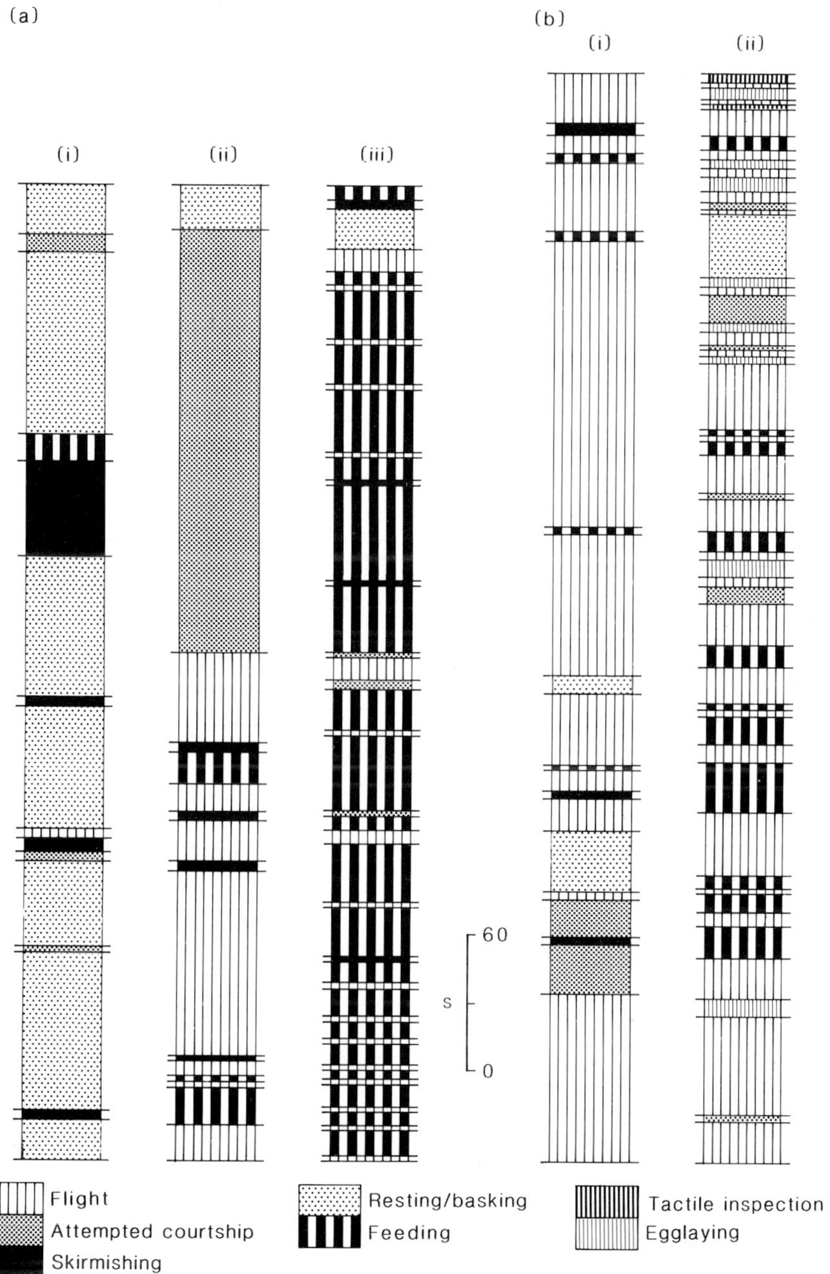

Fig. 2.3 The activities of male large skipper *Ochlodes venata* and male and female orange-tip *Anthocharis cardamines*. (After Dennis 1986*b*; courtesy of the Amateur Entomologists' Society; Dennis and Williams 1987; courtesy of The Lepidopterist's Society, USA.) (a) *O. venata*: (i) territorial activity; (ii) patrolling; (iii) feeding. (b) *A. cardamines*: (i) male patrolling; (ii) female egg-laying and nectaring. Male *O. venata* locate mates by alternating between establishing territories (see Fig. 2.4) and patrolling, but also opportunistically attempt to mate with passing females during long bouts of nectaring. Individuals of both sexes of *A. cardamines* use habitats where mates, egg-laying sites, and nectar sources are interspersed, engaging in short periods of feeding between other activities; males patrol in search of females and do not establish territories.

1982*a*) but in Bernwood Forest, Buckinghamshire it is bramble *Rubus fruticosus*, and on chalk downland in the same county, majoram *Origanum vulgare*. Wiklund (1977*a*) and Wiklund and Åhrberg (1978) demonstrated that the wood white *Leptidea sinapis* and *Anthocharis cardamines* can learn the location and type of flowers which are used as nectar sources, presumably forming a 'search image' for a particular plant species. Throughout a single flight period individuals switched between different species of flower in response to the seasonality of flowering times, tending to use those nectar sources which gave the best reward at any one time. This learning and switching, it is believed, can reduce searching time. Learning the location of plants is perhaps most refined in tropical *Heliconius* species in which particular butterflies feed from the same flowers in a regular sequence on successive days (Gilbert and Singer 1975).

Seasonal changes of nutrient types, and their abundance at any time, can affect activities not normally associated with feeding. For example, *Pararge aegeria* forms aggregations at nectar sources in early autumn when feeding switches from honeydew to floral nectar. When forming such assemblies there are no interactions between males or between the two sexes, but earlier in the year, when individuals feed alone, interactions between individuals are common. The apparent curtailment of sexual behaviour is also common in species which form male aggregations at puddle margins (e.g. green-veined white *Pieris napi*, small copper *Lycaena phlaeas*), but in mixed aggregations of those species (e.g. *Pieris rapae*) mate-rejection postures of females may be maintained (Shreeve, personal observation).

Research into nectar feeding is more advanced than into other types of feeding. A rare resource that is used by at least two species is the initial stage of the infection of grass flowers by an ergot-type fungi. In this stage of infection grass flowers are covered by a sweet sickly secretion with a powerful musty smell and it is probable that scent is the primary attractant. This is best illustrated by the observation that one area of float-grass *Glyceria plicata* on a ride in Bernwood Forest in midsummer 1982 was so attractive that between 20 and 30 *Ladoga camilla* and 40 adult *Pararge aegeria* were feeding on this resource, in addition to large numbers of flies and beetles. In the vicinity of this resource, adults of these two butterfly species were abundant but away from this area, scarce.

2.4 Mate-locating behaviour

2.4.1 Sex ratios and mating frequencies

In Britain most species of butterfly are expected to have equal numbers of males and females per generation, though Porter (1984) reports that this ratio can be distorted in the marsh fritillary *Eurodryas aurinia* by the combination of weather and parasitoids (see chapter 4), and all female broods of *Pieris napi* have also been reported (Bowden 1987). In all species the individual adults that comprise a single generation do not all emerge from the pupae at the same time. In some species the emergence of a single generation can extend over a period as long as four months (e.g. *Maniola jurtina*), in others it may be as short as two weeks (e.g. the small blue, *Cupido minimus*). Because adults emerge over a period of time and also because the two sexes tend to have different average development rates the precise ratio of the sexes changes during the flight period of a single generation.

In the majority of species females mate once, but males have the capacity to mate more than once. In some species (e.g. *Hipparchia semele*, *Euphydryas editha*) female monogamy is maintained by males secreting a plug during copulation, which prevents further matings. In others it is maintained by female behaviour in which mated females move to areas where males are scarce, or they adopt characteristic 'mate-rejection' postures when encountered by males. Even in species in which females regularly mate more than once (e.g. the Adonis blue *Lysandra bellargus*) females are not continuously available, repeat matings occurring only when sperm reserves or nutrients are depleted. Because of female monogamy or unreceptiveness, the ratio of males to receptive females is rarely in a 1:1 ratio; for most of the time there are more males than receptive females, and in such circumstances males are in competition to gain access to scarce receptive females.

The commonest strategy of males, which enhances their probability of mating, is developmental (Scott 1977; Wiklund and Fagerström 1977)—males emerging before females to maximize their access to emerging virgin females. However, there are a variety of behaviours which are also used, differing between and within species.

2.4.2 The nature of mate-locating behaviour

Scott (1974) defined two main methods of butterfly mate-location behaviour: perching and patrolling. Perching males settle at particular locations and rise to intercept passing females in an attempt to elicit courtship, and patrolling males fly in search for settled and flying females. Within these two broad behavioural definitions there is a variety of different behaviours. Patrolling is a behaviour that involves active searching for females. In some species (e.g. *Pieris napi* and *P. rapae*) males may move over a large area in search of receptive females, infrequently returning to the same area (Baker 1978; Dennis 1982*b*), but males of other species (e.g. *Leptidea sinapis* and *Anthocharis cardamines*) may patrol within a small area, frequently returning to the same locality (Wiklund 1977*a*; Warren 1984; Wiklund and Åhrberg 1978; Dennis 1982*b*, 1986*b*). Perching behaviour is equally variable. Males of the mobile *Aglais urticae* and peacock *Inachis io* (see chapter 6), move between perching sites throughout their adult lives (Baker 1972*a*). By contrast certain males of *Pararge aegeria* and *Cupido minimus* may use a set of perching sites within a small area for a period of days, or even their whole adult life (Wickman and Wiklund 1983; Morton 1985; see Fig. 2.4).

In addition to the perching and patrolling mate-locating behaviours are assembling and hilltopping (Table 2.4). In assembling, males form aggregations (leks) at specific sites, such as tree tops in wait for

Table 2.4 Variation of mate-locating behaviour of some British butterfly species in relation to habitat features and individual apparency

Species	Mate-locating behaviour				Seral stage	Habitat features		Apparency
	lek assembly	perch	patrol	territorial behaviour		1	2	
Ochlodes venata		+	+	+	3	P	S	Low
Erynnis tages		+	+	+	2	P	S	Low
Papilio machaon		+	+	+	3	P	S	High
Leptidea sinapis			+		3/5	U	S	High
Colias croceus			+		2–3	U	S	High
Gonepteryx rhamni			+		4	U	S	High
Pieris brassicae			+		2–3	U	S	High
Pieris napi			+		2–3	U	S	High
Pieris rapae			+		2–3	U	S	High
Anthocharis cardamines			+		3	U	S	High
Callophrys rubi		+	+	+	3–4	P	S	Low
Thecla betulae	+	+	?		5	P	C	Low
Quercusia quercus	+	+	?		6	P	C	Low
Satyrium w-album	?	+	?		6	P	C	Low
Satyrium pruni	?	+	?		5	P	C	Low
Lycaena phlaeas		+	?	+	2–3	P	S	Low
Cupido minimus	+	+	?		2	U	C	Low
Plebejus argus	+	+	+		2	U	C	High

Table 2.4 (*cont.*)

Species	Mate-locating behaviour				Seral stage	Habitat features		Apparency
	lek assembly	perch	patrol	territorial behaviour		1	2	
Lysandra coridon			+		2	U	C	High
Lysandra bellargus			+		2	U	C	High
Hamearis lucina		+	?	+	3–4	P	C	Low
Ladoga camilla		+	?	+	5–6	P	S	Low
Apatura iris		+	?	+	6	P	C	Low
Vanessa atalanta		+	?	+	3	P	C	Low
Aglais urticae		+	?	+	3	P	C	Low
Inachis io		+	?	+	3	P	C	Low
Polygonia c-album		+	?	+	3–4	P	C	Low
Boloria selene			+		2/5	U	S/C	High
Boloria euphrosyne			+		2/5	U	S/C	High
Argynnis aglaja			+		2–3	U	S	High
Argynnis paphia			+		5–6	U	S	High
Eurodryas aurinia			+		3	U	S	High
Melitaea cinxia			+		1	U	S/C	High
Mellicta athalia			+		3/5	U	S/C	High
Pararge aegeria		+	+	+	4–6	P	C	Low
Lasiommata megera		+	+	+	1–2	P	C	Low
Erebia epiphron			+		2	U	S/C	Low
Erebia aethiops		+	+		3	U	S	Low
Melanargia galathea			+		3	U	C	High
Hipparchia semele		+	?	+	1–2	P	C	Low
Pyronia tithonus			+		3	U	C	Low
Maniola jurtina		+	+		3	U	C	Low
Aphantopus hyperantus			+		3	U	C	Low
Coenonympha pamphilus		+	+	+	2	P	C	Low
Coenonympha tullia			+		3	U	C	Low

Species with unknown behaviours are omitted.

Seral stage
- 1: Bare ground, screes, and slopes with scarce pioneer herbs and grasses
- 2: Short herbs and grasses
- 3: Tall herbs, grasses, and scattered shrubs
- 4: Shrubs and invading trees
- 5: Pre-climax forest
- 6: Climax forest with regeneration patches
- /: Indicates occupation of alternative seral stages

Habitat features
- 1 Hostplant-habitat features
  - P: Predictable
  - U: Unpredictable or too numerous to use as recognizable features
- 2 Hostplant distribution
  - S: Scattered
  - C: Clumped
  - S/C: Occupies habitats with either scattered or clumped hostplants

Apparency (see Glossary)

Low: Individuals hard to detect from a distance.

High: Individuals highly conspicuous to conspecifics. This definition is only appropriate to apparency from a long distance. At short distances factors such as chemical and behavioural communications intervene to increase intraspecific apparency.

The behaviour of some species in relation to the habitat and hostplant may seem inappropriate, but was developed in their ancestral habitats and retained, other constraints being more important (see Dennis and Shreeve 1988).

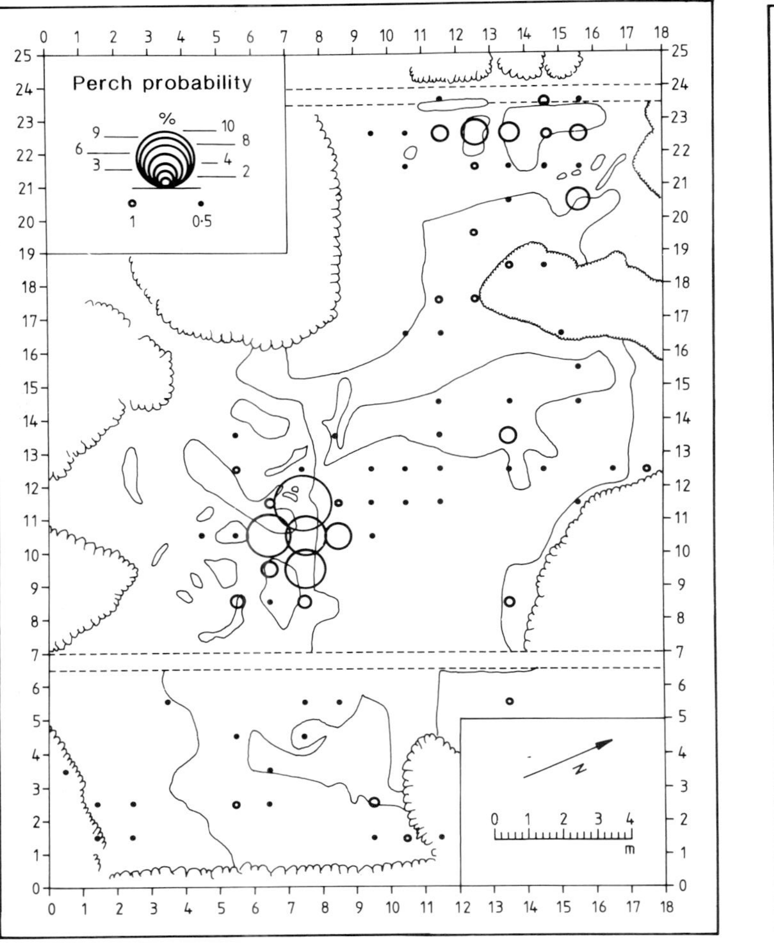

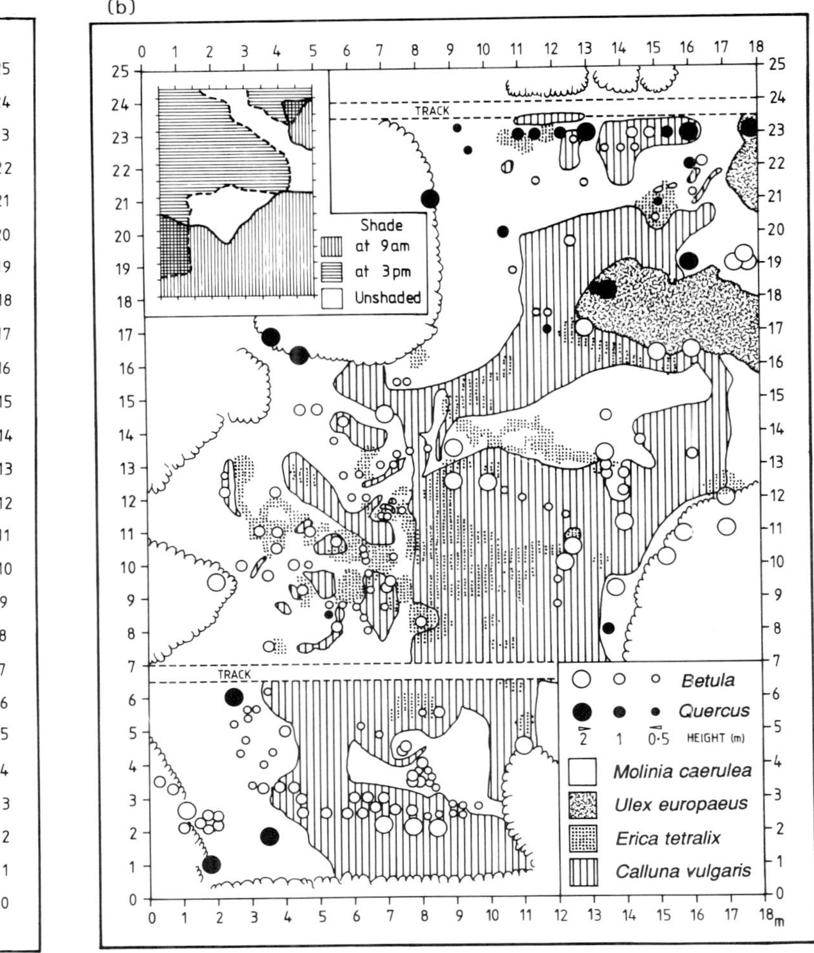

Fig. 2.4 Territorial perches of the male large skipper *Ochlodes venata* in heathland on Lindow Common, Wilmslow, Cheshire. (a) The location of territorial perches (per cent observations based on 162 transects) of male *O. venata*. The majority of perches are situated at 'edge' sites in sunshine, typically at the junction of woodland margins with shorter vegetation. Two well established territories occur around co-ordinates 07 10 and 14 22. (b) Vegetation structure of the heathland area illustrated in (a). Tiny birch seedlings under 15 cm tall form typical perches and are used repeatedly by different individuals. (After Dennis and Williams 1987; courtesy of the Lepidopterists' Society, USA.)

females (e.g. white-letter hairstreak *Satyrium w-album*, purple hairstreak *Quercusia quercus*, *Thecla betulae*). While hilltopping, males move to distinct topographic landmarks ('hilltops') in order to obtain mates, adopting either perching or patrolling (Shields 1967) and occasionally forming aggregations (Dennis and Williams 1987; Wickman, 1987).

A variety of factors which influence the ability of the two sexes to locate each other are important in determining the mate-locating behaviours adopted by any species. Most important are those which influence the predictability and apparency of receptive females. If receptive females are abundant or easy to locate then those males using patrolling mate-locating behaviour may be the most successful. When receptive females are scarce or difficult to detect, male perching may be more effective because patrolling males would expend considerable time and energy in unsuccessful searching (Scott 1974; Ehrlich 1984). Perching is dependent on habitat structure (Dennis 1982*d*; Dennis and Shreeve 1988). Where there are no distinct landmarks in the habitat which can be recognized and used by both sexes (i.e. hills, food plant clumps, bare ground) perching sites will be indistinguishable from other parts of the habitat. In such circumstances male patrolling is probably the only effective mate-locating behaviour. Where recognizable topographic features are available perching is facilitated (Fig. 2.5).

Because receptive females can be scarce, competition between males for females may be intense. If a species has specific mating areas which are occupied by perching males and if these sites are scarce, then those males which occupy these sites are at an advantage over other males. In such circumstances males may adopt territorial behaviour, defending their perching sites from other males. Defence can be by ritual display flights, as used by *Inachis io* (Baker 1972*a*), or can involve physical fighting in which contesting males collide with each other (Fig. 2.6). This has been described for *Pararge aegeria* (Wickman and Wiklund 1983) and for *Ochlodes venata* (Dennis and Williams 1987). In other species which adopt perching mate-locating behaviour (e.g. *Cupido minimus*, *Maniola jurtina*) males may never engage in territorial defence (Morton 1985; Dennis personal observation). This may be because the cost of engaging in territorial defence, either in time or in energy, is greater than any potential gain in increasing the chance of exclusive access to mates. This may occur when males concentrate at particular locations and are therefore very abundant, forming a lek. Those individuals which engage in defence will spend more time and energy interacting with other males than with potential mates. Territoriality may also be absent when perching sites are abundant. Any male engaged in defending such a site gains no advantage over other males since the latter can perch elsewhere with an equal probability of encountering a receptive mate but without the cost of defence.

Although any mate-locating behaviour can be classified as either some form of sit-and-wait tactic or a more active searching behaviour, most species adopt more than one type of behaviour (Table 2.4). Species which adopt perching behaviour also adopt patrolling behaviour (Fig. 2.3), but males of some species (e.g. *Anthocharis cardamines*, *Leptidea sinapis*) only patrol.

2.4.3 Factors affecting mate-locating behaviours

Most butterflies are opportunistic and males of most species will adopt supposedly 'untypical' mate-locating behaviours. Thus *Pieris rapae*, which spends most of its time patrolling for females will adopt, when feeding, a behaviour akin to perching, rising to inspect passing females. Recent studies of other species, most notably of *Pararge aegeria* (Davies 1978; Wickman and Wiklund 1983; Shreeve 1984, 1987), *Lasiommata megera* (Dennis 1982*d*, 1987; Wickman 1988), small heath *Coenonympha pamphilus* (Wickman 1985*a*,*b*), and *Ochlodes venata* (Dennis and Williams 1987) demonstrate how variation in weather, population structure, and habitat type influence mate-location. Davies (1978) found that some males of *P. aegeria* adopted territorial behaviour, perching in sunlit areas under tree canopies and defending these sites from other males. Contests between males were either spiral flights over the sunlit area or horizontal chases away from the site. In all contests it was the resident male which returned to the territory. Although territorial males gained the most matings not all males were territorial; some patrolled in the tree canopy in search of females. Wickman and Wiklund (1983) found that territoriality in a Swedish population, measured by the duration of contests between males, declined with season. This correlated with a rise in air

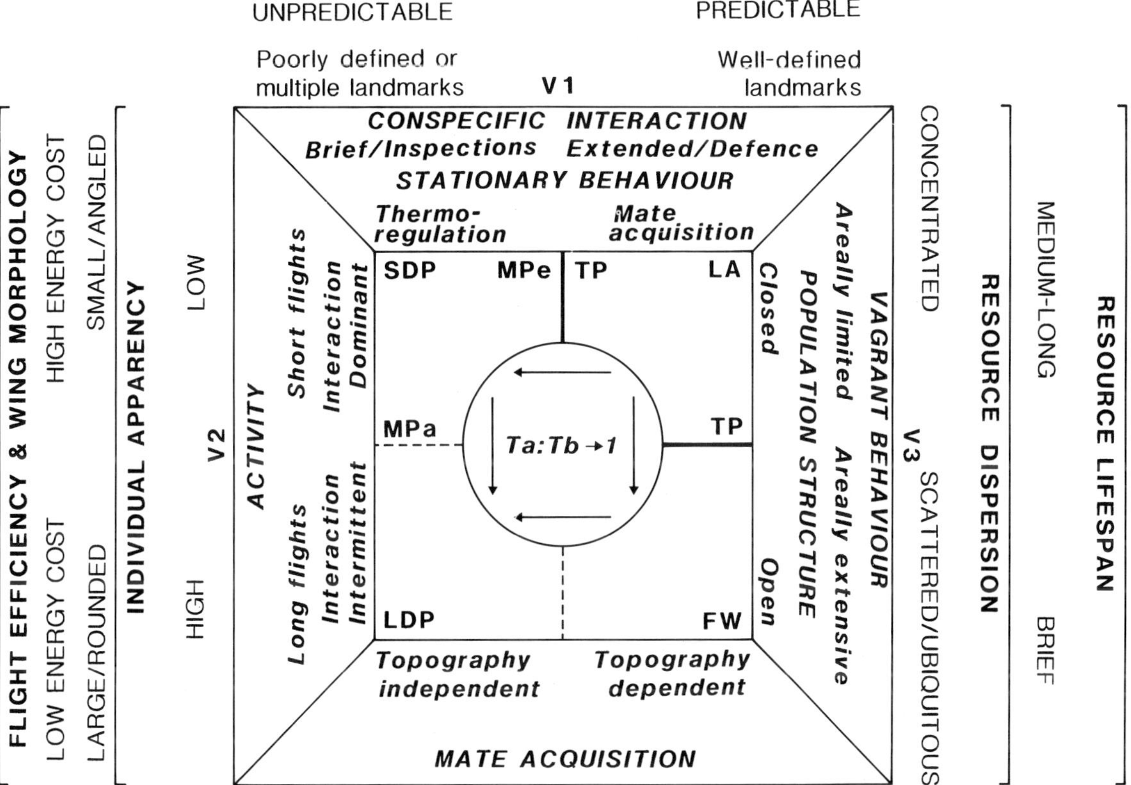

Fig. 2.5 The major factors influencing the evolution of species-specific mate-locating behaviours. Ta, ambient temperature; Tb, body temperatures; LA, lek assembly; TP, territorial perching; MPe, multiple perching with intermittent patrolling; MPa, patrolling dominant with intermittent perching; SDP, short distance patrolling; LDP, long distance patrolling; FW, use of flyways. Mate-locating behaviours are variable and those adopted by any species are influenced by temperature. Changes with lower temperatures occur in the reversed direction to the arrows in the centre of the diagram. V1, V2, and V3 are the main habitat factors and morphological characters which control the range of variation in the behaviour of individual species. (After Dennis and Shreeve 1988; courtesy of The Linnean Society and Academic Press.)

temperature, which reduced the necessity to use these areas as basking sites, and a decrease in the number of receptive females. Certain males would always win contests irrespective of ownership; this they ascribed to individuals learning the location of the best sites.

Shreeve (1984) found that the duration of residence of male *Pararge aegeria* in sunlit areas is closely related to air temperature and solar radiation intensity, males using these sites for basking before engaging in patrolling behaviour, a strategy which in open sunny woodland maximizes female encounter rates. Continuous patrolling does not occur because flight durations are limited by body-cooling (Fig. 2.1; section 2.2.3). The outcome of contests between males in open woodland could not be predicted.

Further studies of *Pararge aegeria* in open and shaded woodland and rides (Shreeve 1987) revealed that the behaviour of individuals is related to their hindwing spot pattern (Fig. 2.7). Males with three upper hindwing spots tend to engage in patrolling behaviour, using sunlit areas as warming sites which are not defended and are located in open habitats where patrolling is the best mate-locating strategy.

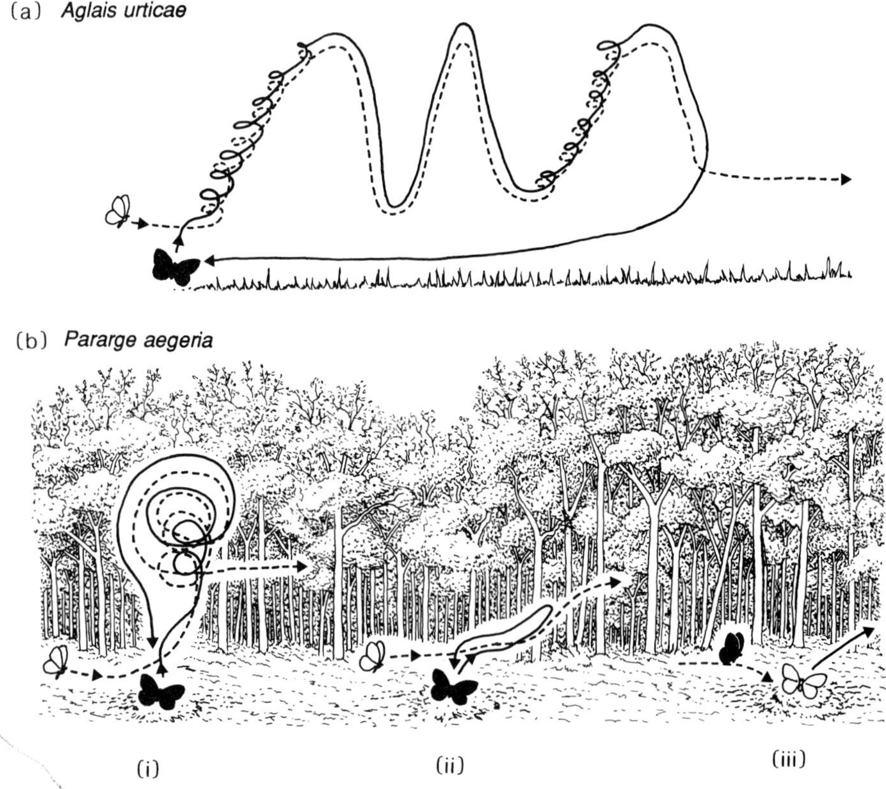

Fig. 2.6 The flight patterns of territorial male small tortoiseshell *Aglais urticae* in interactions with other males, and typical male–male interactions of the speckled wood *Pararge aegeria*. (a) *A. urticae*; a settled male chases away an incoming male in a series of spiralling and soaring flights. Either type of interaction may occur in succession and the sequence of events is flexible. (After Baker 1972*a*; courtesy of *Journal of Animal Ecology*, Blackwell Scientific Publications.) (b) *P. aegeria* in woodlands; the nature of the contest is dependent on the identity of the males, the dominant males being illustrated in black. (i) Two males which dispute ownership of the territory may engage in a 'spinning-wheel' flight which may last up to eight minutes. In such contests males clash into each other. The outcome of such contests is not predictable. (ii) A territorial male may chase an intruding male away with a horizontal chase. Most usually the intruding male has probably no territorial intent. (iii) A non-territorial male is displaced from a sunlit area by a territorial male without any pursuit flight.

Males with four hindwing spots were found to be territorial, defending sunlit areas from other males. They also tended to comprise the majority of the population in areas where perching behaviour was the most successful—in dense woodland. The proportion of individuals which engaged in these two behaviours changed seasonally, but this was found to correlate with seasonal changes in the relative abundance of the two male phenotypes.

In *Pararge aegeria* the behaviour of males was, therefore, found to be variable but related to weather, habitat structure, and to phenotype, the last being an example of polyphenism (see chapter 8). Wickman (1985*a*) and Dennis and Williams (1987) clearly demonstrate that males of *Coenonympha pamphilus* and *Ochlodes venata* switch between perching and patrolling behaviour during a single day. In the former species, individual males switch from territorial perching behaviour in the morning to patrolling in the afternoon. As air temperature increases during the morning, flight capacity is enhanced and males switch to patrolling because this increases their probability of encountering a receptive female. Perching may also be ineffective at

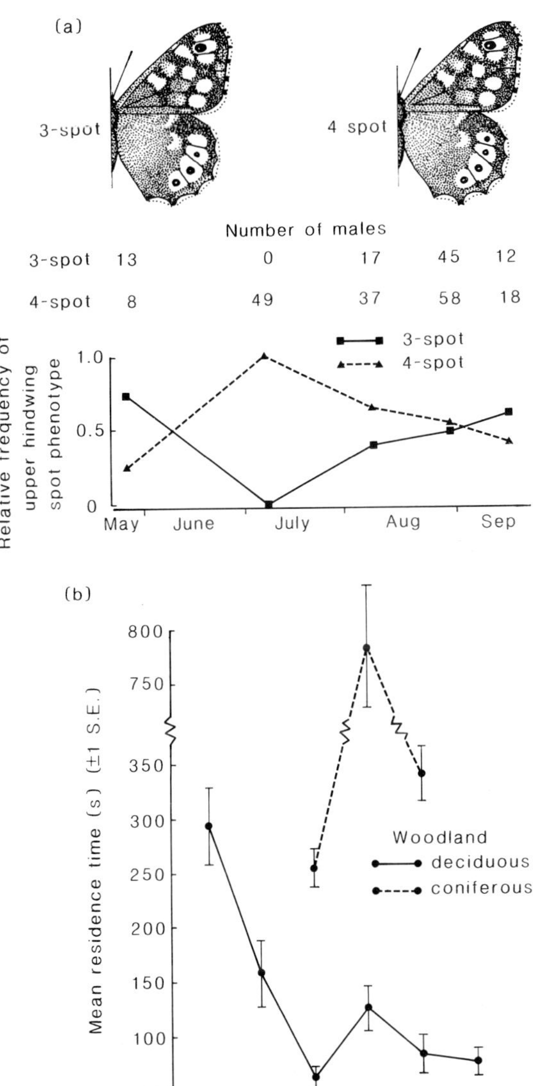

Fig. 2.7 Seasonal activities of the male speckled wood *Pararge aegeria* in Bernwood Forest, Buckinghamshire during 1984. (a) Males of the 3-spot hindwing phenotype tend to patrol in search of females whereas 4-spot males tend to adopt perching territorial perching behaviour, the latter occupying sunlit areas for longer than the former. Seasonal changes in the duration of residence in the woodland areas are related to changes in the relative frequencies of the two phenotypes. (b) Seasonal changes in the time spent by male *P. aegeria* in sunlit areas in deciduous and coniferous woodland. S.E., standard error. (After Shreeve 1987; courtesy of *Animal Behaviour* and Academic Press.)

high temperatures becuse flying males will encounter and mate females before they reach sites where males perch. By contrast, Dennis and Williams (1987) demonstrated that male *O. venata* switch from morning patrolling to afternoon perching (Fig. 2.8). This, it is suggested, is not related to thermal constraints on activity but could be related to the time of day when females are receptive. If perching occurs when receptive females are scarce, then the frequency of perchers may increase in the afternoon because the majority of females have been mated in the morning by patrolling males. It is interesting that after cool, cloudy mornings when the insects are inactive, male *O. venata* patrol in the afternoon. This work raises the question of how males detect females when the former are in flight. The butterfly is dull in colour and individuals are probably not easily detected using visual cues. Observations of males searching the vicinity where females emerge, strongly suggest that the primary attractant of females for males is scent or female pheromones. Over what distance males of this species can detect these pheromones is not known.

Hilltopping may arise under two different circumstances: (i) when hilltops are used as landmarks for assembly points, and (ii) when their use is linked to other aspects of behaviour and habitat use. In hot Mediterranean regions especially, but also in cool temperate northern Europe, *Lasiommata megera* engages in hilltopping (Dennis 1987; Wickman 1988), a behaviour linked to thermal constraints on activity, individuals moving to cooler hilltops in response to high temperatures and probable heat stress at lower altitudes. When on hilltops individuals engage in both perching and patrolling.

Most butterflies that primarily use patrolling mate-location behaviour are brightly coloured. In such species (e.g. *Anthocharis cardamines*, marbled white *Melanargia galathea*) the primary cue used by males to detect females is probably visual, females contrasting with their resting background. This brightness may also be enhanced by strong reflection in the ultraviolet wavelengths (Silberglied 1984), particularly by pteridine pigments in the wings of Pieridae. However, not all patrolling species are brightly coloured. Both *Maniola jurtina* and *Aphantopus hyperantus* are dull; yet males primarily engage in patrolling. This may be related in part to population density but possibly also to the lack of predictable landmarks in

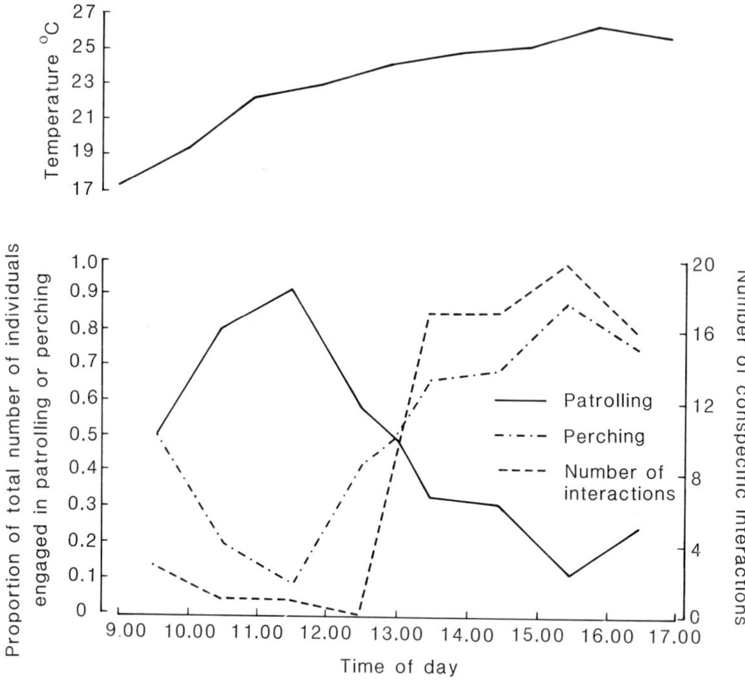

Fig. 2.8 The frequency of perching and patrolling mate-locating behaviour of the male large skipper *Ochlodes venata* during 7 July 1984 on Lindow common, Wilmslow, Cheshire. During the morning, patrolling is the dominant behaviour, but during the afternoon the majority of males establish territorial perches. (After Dennis and Williams 1987; courtesy of The Lepidopterist's Society, USA.)

their ancestral habitats that could be used as perching sites (Dennis and Shreeve 1988). When a dull butterfly occurs in dense populations, patrolling can be effective because males rapidly encounter receptive females. In *A. hyperantus*, at least, mate-location by patrolling males is enhanced by the behaviour of receptive females (Wiklund 1982), which actively solicit courtship by settling on prominent vegetation with open wings and flying up in front of males when they are detected. By contrast mated females avoid courtship by remaining within the vegetation, flying only to feed and lay eggs, and spending most of their time settled with closed wings. In perching species, males probably also rely on vision, detecting a moving object against the sky or far background and then inspecting it to determine whether or not it is a receptive female. Most perching species are known to rise and inspect numerous different types of flying objects including falling leaves, birds, and other species of insect.

Many species (such as *Pararge aegeria* or *Lasiommata megera*) have a broad repertoire of mate-locating behaviours but other species, especially among the Pieridae and Argynninae are more restricted (Table 2.4). Those with the most limited behavioural repertoire are species that only patrol. These differences may be related to the predictability of habitats and hostplants (Dennis and Shreeve 1988). Species that occupy habitats which are variable in structure or contain an excessive number of potential perch sites, often in association with short-lived hostplants, usually patrol. In such cases perching behaviour is not able to develop because of the unpredictability of the nature of areas which could be used for perching sites. Species in more permanently structured habitats may adopt perching. Dennis and Shreeve also argue that patrolling is an ancestral behaviour—perching having evolved when species or ancestral species groups became specialists on particular hostplants. As the hostplant range becomes restricted then there is a greater probability that species become habitat specialists, occurring in vegetation types with predictable structures where perching sites may occur.

2.5 Communication and courtship

In mate-acquisition, any individual which encounters another butterfly must determine if that individual is of the same species, of the same or different sex, and also whether it is a suitable partner with which to mate. Communication is by a variety of senses; vision, including colour and reflectance patterns, scent, touch, and in some cases, sound (Kane 1982). All may be used in courtship displays. Of these, the role of scents is the least understood (see Boppré 1984; Silberglied 1984), particularly that of female scents, or attractants.

Visual cues can be used in the detection and discrimination of conspecific individuals by both sexes. An insect's vision extends from the human visible range to include ultraviolet wavelengths, being most sensitive to the latter. Therefore, bright or iridescent wings enhance communication, especially if the level of brightness extends into the ultraviolet range. However, not all butterflies are brightly coloured; other factors such as predator avoidance (see chapter 5) influence coloration. In those species with bright wing areas it is usually the dorsal wing surfaces, concealed when the insect is at rest, that are the most colourful. Males also tend to be brighter than females. Vision is often used for long-distance communication and the brightest species tend either to be those which occur in habitats with few visual 'peaks', such as the iridescent lycaenid species of short grassland, or at low density, such as *Apatura iris* of the woodland canopy. Vision may be used by females to locate males at perching sites, by patrolling males to locate settled females, or in male–male communication. For example, patrolling pierids are attracted to a range of objects and species that visually resemble females. But visual stimuli may also act as a deterrent. Males of pierid butterflies which are quiescent with closed wings or which are mating may suddenly expose the dorsal wing surfaces if approached by another male. This sudden flash, which may be enhanced by ultraviolet reflection, serves to repel males that were initially attracted by visual stimuli (Silberglied and Taylor 1978).

Males of some species can detect females by scent alone. This is suspected for *Ochlodes venata* (section 2.4.3) and known to operate in some *Heliconius* species of South and Central America, in which males are able to detect female pupae (Brown 1981). More usually both scent and vision are involved. Wiklund (1977*b*) found that male *Leptidea sinapis* are able to recognize females by their scent which also serves as a cue to indicate their readiness to mate. In the same species females rely on visual cues to recognize males.

Effective communication is essential for recognizing suitable mates and different species may respond differently to the same stimuli. Thus visual cues alone are not sufficient to prevent mistakes in mate-recognition by the North American orange sulphur *Colias eurytheme* and the closely related yellow sulphur *C. philodice* (Silberglied and Taylor 1978). Males of *C. eurytheme* are orange and ultraviolet reflecting and *C. philodice* males are yellow and ultraviolet absorbing. Both will approach females of either species. Although female *C. eurytheme* are able to discriminate between males by their ultraviolet reflecting or absorbing properties, *C. philodice* females are unable to do so. The probability of interspecific mating is minimized by females of either species being able to distinguish males by their highly specific scents (Taylor 1973).

Male butterflies have androconial scales which are most frequently located on the upper forewings. These organs comprise modified wing scales with glandular bases which produce pheromones. Scent producing organs are not restricted to the forewing surfaces, being located on the hindwings and associated with the genitalia in some groups (e.g. Danainae). Scents are used by males in the latter stages of courtship in the Nymphalidae at least to elicit female acceptance. Tinbergen (1941, 1972) describes their use by *Hipparchia semele* (Fig. 2.9), and Magnus (1950) their role in the silver-washed fritillary *Argynnis paphia*. In the latter species, female wing colour plays an important role in their attractiveness to males (see Fig. 8.4). In both species once a male has located a female and caused it to land, it will repeatedly stroke or clash the female's antenna against the androconial patches on the wing and in doing so may prompt the female to accept the male. In these and other species this courtship ritual does not always elicit the female's acceptance of the male. In *Lasiommata megera* and *Maniola jurtina* repeated clashing of the androconial scales against

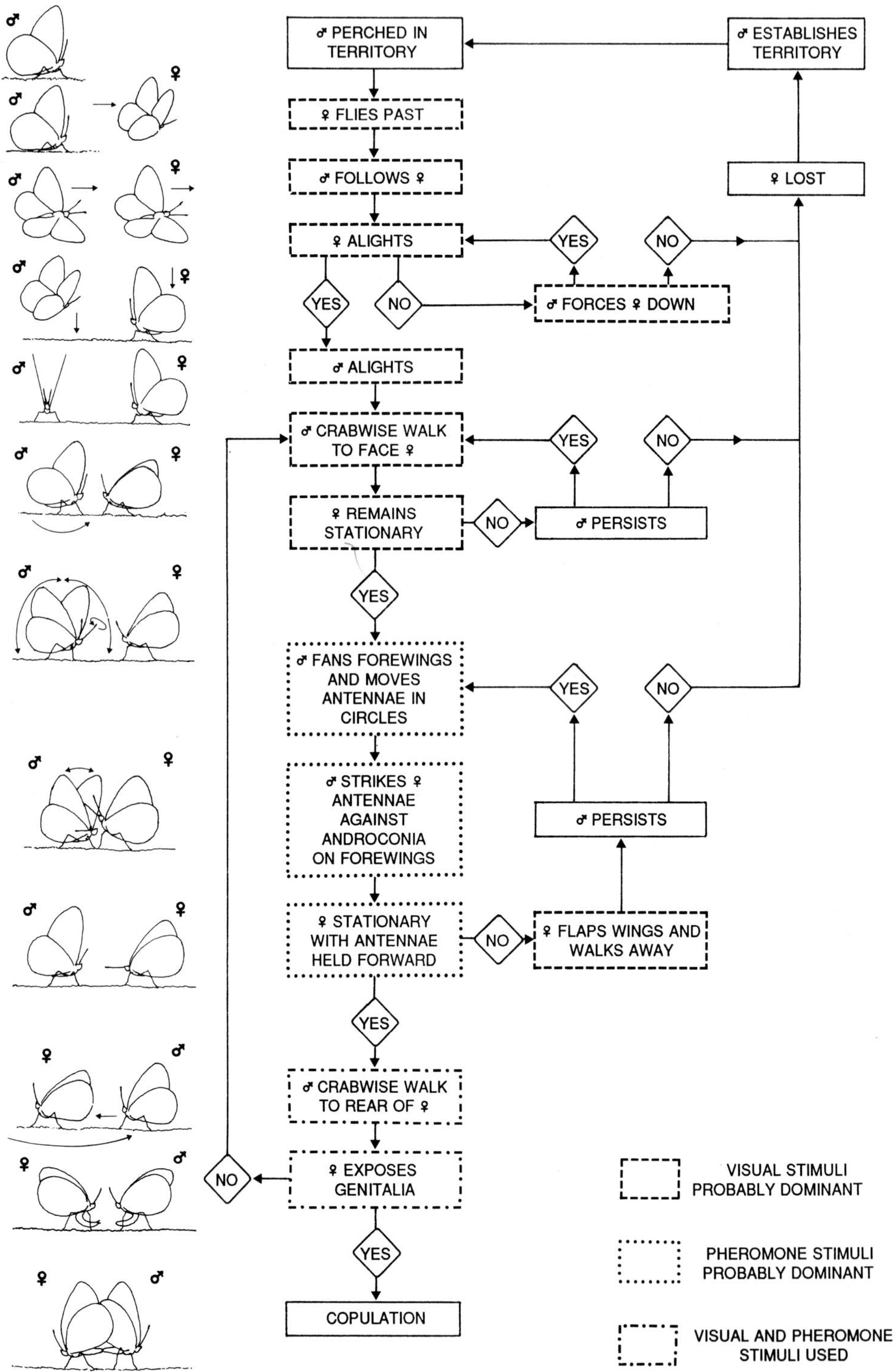

Fig. 2.9 The courtship behaviour of the grayling *Hipparchia semele*. There are visual and chemical signals between the sexes during mate-recognition and courtship, though the primary cue is visual when males first detect females. (Adapted from Tinbergen 1941, 1972; Tinbergen *et al*. 1942.)

the antennae of non-receptive females is ineffective: in both cases the female does not open its wings to facilitate the male's access to its genitalia. In other species an open-winged posture and raised abdomen may prevent access to males. Whether such an attitude is solely visual or additionally involves chemical communication is not known.

Courtship displays differ between species, although closely related species may have certain common sequences in these rituals. Whether the display may be as apparently simple as that of *Aphantopus hyperantus*, in which an unmated female may rise up in front of a flying male and immediately be mated on alighting (Wiklund 1982), or as complex as those of *Hipparchia semele*, their function is probably twofold. Because of their specificity they may serve to maintain reproductive isolation between closely related species (Silberglied 1977). They may also permit females to assess the quality of a potential mate (Rutowski 1982), since male scents and certain aspects of male performance may be indicative of the likely contribution that any particular male may make to a female's reproductive output.

2.6 Behaviour and butterfly biology

In this chapter, attention has been focused on three aspects of adult behaviour—thermoregulation, feeding, and mate-location. An emphasis has been placed on the requirements of the individual and the influence of extrinsic and intrinsic factors on behaviour. Whilst it is possible to describe the influence of single factors on behaviour, such as the effects of wing melanization on thermoregulation and flight activity, few factors operate in isolation. For example wing coloration in pierids may act as a warning to predators (Marsh and Rothschild 1974), brighter individuals providing a stronger visual stimuli. Wing colour may also be important in species and mate-recognition (Wiernasz, in Kingsolver and Wiernasz 1987). Thus variation in wing colour and posture, which may be expected to vary in response to climate and weather, may also be influenced by predation and mate-selection. The ultimate colour and pattern may then be the product of selection for different attributes (Kingsolver 1987; Kingsolver and Wiernasz 1987), variation in response to any single factor being constrained by different requirements. Similar arguments will apply to other aspects of morphology and behaviour, such as the influence of habitat structure, apparency, and predation on mate-locating behaviour, and the effect of weather and egg-load on the flights and hostplant searching ability of egg-laying females.

3

Eggs and egg-laying

Keith Porter

For success, a female butterfly must develop eggs, mate, and then place those eggs in sites where there is the greatest likelihood of larval success. This involves complex physiological and behavioural processes. Eggs have their origin in undifferentiated cell masses, or germinal buds, which are formed in the development of the embryo. These cells remain dormant throughout the life of the larva and begin to proliferate during the pupal stage. Upon emergence, many species of butterfly will already have a major part of their egg load fully developed and ready to be fertilized. In other species the process of egg growth will not begin until the female has had the opportunity to feed.

After mating (see chapter 2) and when mature eggs are ready for laying, the female butterfly is ready to find acceptable hostplants on which to oviposit—a process which involves location and recognition of habitats as well as of hostplants. These activities and the physiological processes involved are the subject of this chapter.

3.1 Egg production

3.1.1 The structure of eggs

The variety of shapes and sizes in the eggs of British butterflies can readily be seen using a hand lens or by reference to any good standard work on butterflies. The complexity of a butterfly egg was not realized until the advent of the scanning electron microscope, when highly magnified three-dimensional images of eggs were seen for the first time (Hinton 1970, 1981). The features revealed how the eggshell protects and services the developing embryo within the egg.

The main function of the butterfly egg is to provide a receptacle in which the embryo can develop without the threat of desiccation or mechanical damage. The unfertilized egg contains a tiny 'female' cell and large amounts of food material which is used by the embryo as it grows. The egg needs only a single male sperm and externally derived oxygen to develop normally but the need to obtain gaseous oxygen from the air creates a dilemma for the egg. Any hole large enough to allow an oxygen molecule into the egg will easily be large enough to enable water molecules to escape and ultimately to desiccate the egg. This is not a problem for insects that lay their eggs in water, but desiccation is a major threat to the eggs of terrestrial insects such as butterflies. It has been calculated that a typical insect egg has about 50 cm^2 of surface area for every 1 cm^3 of volume; obviously there is a great potential for water to evaporate (Hinton 1970). For an egg of the marbled white *Melanargia galathea* this would mean that it has an effective surface area of approximately 21 mm^2 as opposed to 3.14 mm^2, that is, if the egg had a simple smooth surface shell.

The problem of drying out can be lessened if the eggs are laid in a humid environment, but this is not always possible. As a general rule all butterfly eggs have a thin waterproof waxy layer on the outside of the shell to prevent water loss. The surface of the shell contains microscopic holes (aeropyles), often grouped in patches or along the summit of ribs or spines, and passing through the waxy layer and outer shell to a honeycomb of air-filled spaces within the shell itself (Fig. 3.1). The honeycomb links

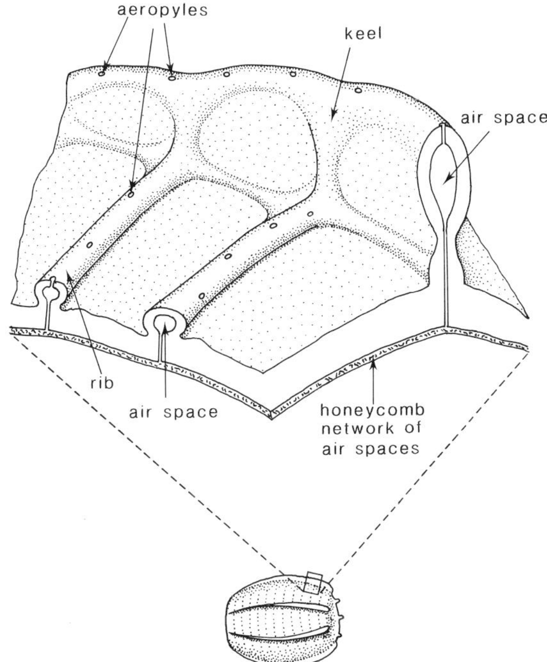

Fig. 3.1 A diagram of a butterfly eggshell. Unlike the egg of a bird, the 'shell' of a butterfly egg is impermeable to both water and gases. The water requirements of the developing embryo are met by supplies provided when the egg was formed; the shell prevents this valuable nutrient from evaporating into the atmosphere. The embryo requires access to oxygen and a means of expelling carbon dioxide. In a butterfly egg this gaseous exchange is facilitated by a series of air-filled canals that permeate the ribs. These are connected to the atmosphere via aeropyles, and to the membranes of the embryo via a layer of honeycombed cuticle. These structures permit sufficient movement of oxygen inwards and carbon dioxide outwards, but avoid any significant loss of water.

through to a very thin, air-permeable layer on the inner face of the eggshell. In butterfly eggs with prominent ribs or spines these structures have an extensively developed honeycomb within and they greatly facilitate oxygen uptake.

Within a family or subfamily, butterfly eggs are fairly uniform in shape and texture, having characteristic patterns of sculpture (Table 3.1). The pattern of sculpture is fairly constant within a species and can be used as an aid to identification. However, some care should be taken over the numerical aspects of sculpture. In the eggs of the small tortoiseshell *Aglais urticae* the number of keels, or ribs, was found to be either eight or nine, with both types being found in the same batch of eggs (Dennis and Richman 1985). Egg shape can be generally related to habitat, especially where the practical needs of attachment are considered, thus eggs that require firm attachment on hairy leaves (e.g. small blue *Cupido minimus*, green hairstreak *Callophrys rubi*) or for long periods (other hairstreaks) have a broad, flat base. At the other extreme are species which do not attach their eggs (e.g. *Melanargia galathea* or the ringlet, *Aphantopus hyperantus*); these lack any base.

Any comparison of egg size between species must take into consideration the size of the female (Table 3.2), but within given limits egg size is a variable feature. This has been demonstrated for several species of satyrine butterflies by Karlsson and Wiklund (1985). This variation in size can be due to one of several reasons, varying from differences in reproductive strategy (see chapter 4; Labine 1968; Courtney 1986) to ageing in individual butterflies, as shown for example for the small white *Pieris rapae* by Jones *et al*. (1982) and for the speckled wood *Pararge aegeria* (Karlsson 1987). The most commonly observed variation is when egg size diminishes through the lifespan of the female butterfly. This has been correlated with the amount of material available for egg production (Jones *et al*. 1982; Murphy *et al*. 1983; Wiklund and Persson 1983; Karlsson 1987; Wickman and Karlsson 1987). Reduced egg size has been suggested as an influence on viability (Murphy, in Chew and Robbins 1984). In *P. aegeria* (Wiklund and Persson, 1983), and the wall *Lasiommata megera* (Karlsson and Wiklund 1985) the size of the egg had no perceptible influence on the survival of offspring. In the natural state it is likely that few females ever survive long enough for diminishing egg size to be a problem; most females die long before their total egg production potential has been exhausted (see Wiklund *et al*. 1987 and chapter 4).

Egg colour changes with age in a predictable manner as summarized in Table 3.3. However, a polyphenism does occur in the small heath *Coenonympha pamphilus* (Wickman and Karlsson 1987)—the first eggs laid are green, and later eggs yellow. Virtually all butterfly eggs are cream or pale yellow when first laid and all change within a matter of a few hours or a few days to a stable colour. At this stage most eggs are monochromatic, but a few British species have a mottled pattern, as in the small mountain ringlet *Erebia epiphron*, or distinct colour zoning as in the purple emperor *Apatura iris*. Most of these colour changes occur in the inner layers of the eggshell, or chorion, often through the chemical deposition of melanin which darkens the base colour. The colour can be modified by changes in the embryo colour. As it completes its development and becomes a pharate first instar larva, then the colour of the egg often changes dramatically, the larva becoming visible through the transparent eggshell. Egg colour has obvious implications in camouflage and defence and these are discussed in chapter 5. Long-lived eggs, such as those of the purple hairstreak *Quercusia quercus* and brown hairstreak *Thecla betulae* are often tinged with green at the end of the winter. This colour results from the growth of algae on the outer surface of the egg and may contribute to the camouflage of these eggs.

3.1.2 Egg development in butterflies

The reproductive systems of males and females are fairly constant in form within all British butterflies and follow the general pattern shown in Fig. 3.2. The female system is essentially a paired structure with two ovaries, each ovary comprising four ovarioles. Eggs begin their development at the tips of the ovarioles and move along the increasingly wider tube as they grow in size. Each egg is surrounded by seven nurse cells which absorb nutrients from the female's haemolymph and pass it to the growing egg. The nurse cells eventually form a membrane (termed the follicle) which surrounds the mature egg and this is shed as the eggs pass into the common oviduct. The follicle cells secrete the materials that build the eggshell and these same cells form a 'mould' which determines the ultimate sculpture pattern of the eggshell (Hinton 1970).

During copulation the male's sperm is transferred inside the membranous sac, or spermatophore, and stored in the female's spermatheca. Fertilization takes place in the common oviduct as the eggs pass the duct leading to the spermatheca. Sperm pass along this duct and enter the egg through holes (micropyles) located in the shallow depression at the 'pole' of the egg. The females of many species can mate several times (e.g. *Pieris rapae*) while at least two other British species (the marsh fritillary *Eurodryas aurinia* and grayling *Hipparchia semele*), have a mechanism which prevents multiple matings. Males plug the female bursal duct with a hard secretion (sphragis) which prevents further matings. This strategy was first noted in the North American checkerspot *Euphydryas editha* (Labine 1964).

Table 3.1 Egg shape, size and duration of the egg stage in British butterflies

Species	Egg outline × 10 mag.	Egg size (mm)		Duration of stage* (days)			
		height	width	4–10	11–20	21–30	31+
HESPERIIDAE							
Carterocephalus palaemon		0.5	0.6	•			
Thymelicus sylvestris		0.4	0.9			•	
Thymelicus lineola		0.3	0.8				• (8 months)
Thymelicus acteon		0.6	1.1			•	

Species	Egg outline × 10 mag.	Egg size (mm)		Duration of stage* (days)			
		height	width	4–10	11–20	21–30	31+
Hesperia comma		1.0	0.8				• (7–8 months)
Ochlodes venata		0.6	0.8		•		
Erynnis tages		0.6	0.6	•			
Pyrgus malvae		0.6	0.5	•			
PAPILIONIDAE							
Papilio machaon		0.9	1.0	•			
PIERIDAE							
Leptidea sinapis		1.3	0.4		•		
Colias croceus		1.1	0.4	•			
Gonepteryx rhamni		1.3	0.4	•			
Pieris brassicae		1.2	0.5	•			
Pieris rapae		1.0	0.5	•			
Pieris napi		1.0	0.5	•			
Anthocharis cardamines		1.2	0.5	•			
LYCAENIDAE							
Callophrys rubi		0.5	0.7	•			
Thecla betulae		0.5	0.7				• (7 months)

Table 3.1 (*cont.*)

Species	Egg outline × 10 mag.	Egg size (mm)		Duration of stage* (days)			
		height	width	4–10	11–20	21–30	31+
Quercusia quercus		0.5	0.8				• (8 months)
Satyrium w-album		0.4	0.8				• (9 months)
Satyrium pruni		0.4	0.8				• (9 months)
Lycaena phlaeas		0.3	0.6	•			
Cupido minimus		0.2	0.4	•			
Plebejus argus		0.3	0.6				• (8 months)
Aricia agestis		0.3	0.5	•			
Aricia artaxerxes		0.3	0.5	•			
Polyommatus icarus		0.3	0.6	•			
Lysandra coridon		0.3	0.5				• (8 months)
Lysandra bellargus		0.3	0.5			•	
Celastrina argiolus		0.3	0.6	•			
Hamearis lucina		0.6	0.6		•		
NYMPHALIDAE							
Ladoga camilla		0.9	0.9	•			

Species	Egg outline × 10 mag.	Egg size (mm)		Duration of stage* (days)			
		height	width	4–10	11–20	21–30	31+
Apatura iris	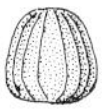	1.0	1.0		•		
Vanessa atalanta	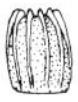	0.8	0.7	•			
Cynthia cardui		0.7	0.5	•			
Aglais urticae		0.8	0.7	•			
Nymphalis polychloros	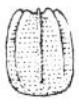	0.8	0.7			•	
Inachis io		0.8	0.6		•		
Polygonia c-album		0.8	0.7		•		
Boloria selene		0.6	0.5	•			
Boloria euphrosyne		0.8	0.6		•		
Argynnis adippe	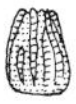	0.8	0.7				• (8 months)
Argynnis aglaja	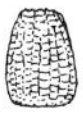	1.0	0.8		•		
Argynnis paphia	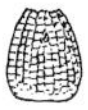	1.0	0.8		•		
Eurodryas aurinia		0.8	0.7				• (35 days)
Melitaea cinxia		0.5	0.5			•	
Mellicta athalia		0.5	0.4			•	

Table 3.1 (*cont.*)

Species	Egg outline × 10 mag.	Egg size (mm)		Duration of stage* (days)			
		height	width	4–10	11–20	21–30	31+
Pararge aegeria		0.8	0.7	•			
Lasiommata megera		0.9	0.8	•			
Erebia epiphron		1.0	0.7		•	•	
Erebia aethiops		1.3	1.0		•		
Melanargia galathea		1.0	1.0		•	•	
Hipparchia semele		0.8	0.7		•		
Pyronia tithonus		0.7	0.7		•	•	
Maniola jurtina		0.5	0.5		•	•	
Aphantopus hyperantus		0.8	0.9		•		
Coenonympha pamphilus		0.7	0.6		•		
Coenonympha tullia		0.8	0.6		•		

* This table gives approximate values for the size of egg and duration of the egg stage as derived from K. Porter, (unpublished data) and by reference to standard works including Frohawk (1924); Heath *et al*. (1984); and Brooks and Knight (1982). Egg size is variable according to the age of the female, her nutritional state, and genetic factors; the quoted sizes are approximate. The duration of egg stage is often quoted as a given number of days but this practice has been avoided in the above table due to the direct relationship between egg duration and ambient temperature. This is demonstrated by egg durations for *P. aegeria* at the following constant temperatures (Shreeve 1985):

°C	Egg duration (days)
25	6
20	7–8
12	23

Table 3.2 The size of eggs relative to the size of females in British butterflies

HESPERIIDAE (skippers)	largest
Hesperia comma	
Ochlodes venata	
Carterocephalus palaemon	
Erynnis tages	
Pyrgus malvae	
Thymelicus acteon	
Thymelicus lineola	
Thymelicus sylvestris	

PIERIDAE (whites)	largest
Anthocharis cardamines	
Pieris napi	
Pieris rapae	
Colias croceus	
Leptidea sinapis	
Gonepteryx rhamni	
Pieris brassicae	

LYCAENIDAE (blues, etc.)	largest
Quercusia quercus	
Callophrys rubi	
Satyrium pruni	
Plebejus argus	
Celastrina argiolus	
Thecla betulae	
Satyrium w-album	
Lycaena phlaeas	
Polyommatus icarus	
Lysandra coridon	
Aricia agestis	
Aricia artaxerxes	
Cupido minimus	
Lysandra bellargus	

NYMPHALIDAE (fritillaries, etc.)	largest
Apatura iris	
Argynnis aglaja	
Argynnis paphia	
Ladoga camilla	
Boloria euphrosyne	
Eurodryas aurinia	
Aglais urticae	
Polygonia c-album	
Vanessa atalanta	
Inachis io	
Nymphalis polychloros	
Argynnis adippe	
Boloria selene	
Cynthia cardui	
Mellicta athalia	
Melitaea cinxia	

SATYRINAE (browns)	largest
Erebia aethiops	
Erebia epiphron	
Lasiommata megera	
Melanargia galathea	
Coenonympha tullia	
Coenonympha pamphilus	
Pararge aegeria	
Aphantopus hyperantus	
Hipparchia semele	
Pyronia tithonus	
Maniola jurtina	

This table shows how, within a family, some species lay comparatively larger eggs than their relatives. The order in the table is derived from estimates of egg volume calculated empirically using plasticine models. Egg sizes and references for female wing-span were obtained from the plates and text of Frohawk (1924). Species at the top of each group have the largest eggs per wingspan.

Table 3.3 Egg-colour changes in British butterflies

	Colour		
	when laid	mid-stage colour	prior to hatching
Carterocephalus palaemon	White →	→	→
Thymelicus sylvestris	Pale cream	Off-white →	→
Thymelicus lineola	Pale yellow	Dull yellow	White/grey head
Thymelicus acteon	Pale cream	Pale yellow	Dull cream/grey head
Hesperia comma	Pale yellow	Apricot yellow	Opaque white
Ochlodes venata	White	Orange-yellow	Grey
Erynnis tages	Cream	Orange	Dusky orange
Pyrgus malvae	Pale green-white	Cream	Grey
Papilio machaon	Pale yellow	Blotched brown	Purple-grey
Leptidea sinapis	Pale cream →	→	→
Colias croceus	Pale cream		Pink-brown
Gonepteryx rhamni	Pale green-white	Yellow	Grey
Pieris brassicae	Pale yellow	Yellow	Grey
Pieris rapae	Pale yellow	Primrose yellow	Grey
Pieris napi	Pale yellow	Pale yellow-green	Grey
Anthocharis cardamines	Pale cream	Red-orange	Grey
Callophrys rubi	Pale green →	→	Grey-green
Thecla betulae	White-green	Opaque white →	→
Quercusia quercus	Pale blue-white	Grey-white →	→
Satyrium w-album	Blue-green	Purple-brown/white rim →	→
Satyrium pruni	Cream-brown	Pale brown →	→
Lycaena phlaeas	Pale green-white →	→	→
Cupido minimus	Pale blue-white →	→	→
Plebejus argus	White →	→	→
Aricia agestis	Pale green-white →	→	→
Aricia artaxerxes	White →	→	→
Polyommatus icarus	Pale green-grey →	→	→
Lysandra coridon	White →	→	→
Lysandra bellargus	Pale green-white →	→	→
Celastrina argiolus	Pale blue-white →	→	→
Hamearis lucina	Pale yellow	Cream-white	Grey
Ladoga camilla	Pale green-white	Off-white	Grey
Apatura iris	Pale green	Green/purple band	Grey-green
Vanessa atalanta	Pale green	Yellow-green	Grey
Cynthia cardui	Pale green	Dull green	Grey
Aglais urticae	Pale green	Dull green	Grey
Nymphalis polychloros	Dull yellow	Apricot yellow	Purple
Inachis io	Pale yellow-green	Blotched dark green	Grey
Polygonia c-album	Pale green	Yellow-green	Grey-green
Boloria selene	Pale orange	Pale brown	Grey
Boloria euphrosyne	Cream	Pale yellow	Grey
Argynnis adippe	Pale green	Apricot	Pink-grey
Argynnis aglaja	Pale cream	Banded purple/cream	Brown/grey
Argynnis paphia	Pale yellow	Off-white	Grey
Eurodryas aurinia	Pale cream	Red-brown	Grey

	Colour		
	when laid	mid-stage colour	prior to hatching
Melitaea cinxia	Pale cream	Dull white	Grey
Mellicta athalia	Pale green-white	Cream	Grey
Pararge aegeria	Pale green-yellow ———	———————→	Grey
Lasiommata megera	Pale green-yellow ———	———————→	Grey
Erebia epiphron	Yellow	Cream/brown blotched	Grey
Erebia aethiops	Pale yellow	Yellow/purple blotched	Grey
Melanargia galathea	Pale green-white	White	Grey
Hipparchia semele	Opaque white ———	———————→	Lilac-grey
Pyronia tithonus	Pale yellow	White/brown blotches	Grey
Maniola jurtina	Pale cream	Cream/red blotches	Grey
Aphantopus hyperantus	Pale yellow	Off-white	Grey
Coenonympha pamphilus	Pale green or pale yellow	Dull green/red blotches	Grey
Coenonympha tullia	Pale green	Cream/brown blotches	Grey

The colour changes given above are largely derived from Frohawk (1924) but contain additional information on *Coenonympha pamphilus* from Wickman and Karlsson (1987). The colours are merely approximations of the colour of the egg at various points in its development and will be useful as a guide. All species show some darkening of the egg a day or so before hatching as a result of the eggshell becoming semi-transparent and the first instar larva being visible.

3.1.3 The timing of egg production

Butterflies differ in the way they develop their eggs and this can be related to how they distribute the eggs on foodplants. The most frequent differences between species is the state of egg production on emergence of the adult female. Most species fall into one of three categories (Table 3.4) which are based upon the number of mature eggs present on emergence. At one extreme are the colonial, cluster-laying species (e.g. *Eurodryas aurinia*; heath fritillary *Mellicta athalia*) which emerge with the majority of their eggs fully developed, and at the other extreme are species which do not begin to develop eggs until a few days after emergence (e.g. the meadow brown *Maniola jurtina* and *Thecla betulae*). In the intermediate category are species which emerge with some eggs that are well developed but with the potential to develop the main egg load after feeding (e.g. Pieridae).

The female *Eurodryas aurinia* emerges with approximately 36 mature eggs in each of the eight ovarioles, in addition to a large number of developing eggs graduated in size. These mature eggs are usually deposited together in one large batch within 24 hours of emergence. Species with a few mature eggs on emergence (e.g. pierids) go on to develop more eggs over the lifetime of the female (Stern and Smith 1960; Chew and Robbins 1984; Courtney 1986), often reaching a peak after a few days, then declining.

Butterflies that emerge with no mature eggs do so for one of several reasons. Included in this category are species which hibernate as adults (*Aglais urticae*, peacock *Inachis io*, comma *Polygonia c-album*, brimstone *Gonepteryx rhamni*) or which invariably migrate (e.g. some red admirals *Vanessa atalanta* and clouded yellows *Colias* species). These species are obviously flexible in their approach to egg development, as some broods do not hibernate and develop eggs directly. Such processes are doubtlessly controlled by hormone levels, which in turn are triggered by daylength or food quality (Wigglesworth 1970). A few species emerge with egg development at a very rudimentary stage, with maturation of the ovaries occurring after emergence (e.g. *Maniola jurtina*). This strategy may be the result of selection for a shorter pupal stage at the expense of the maturation of eggs. Alternatively, increasing

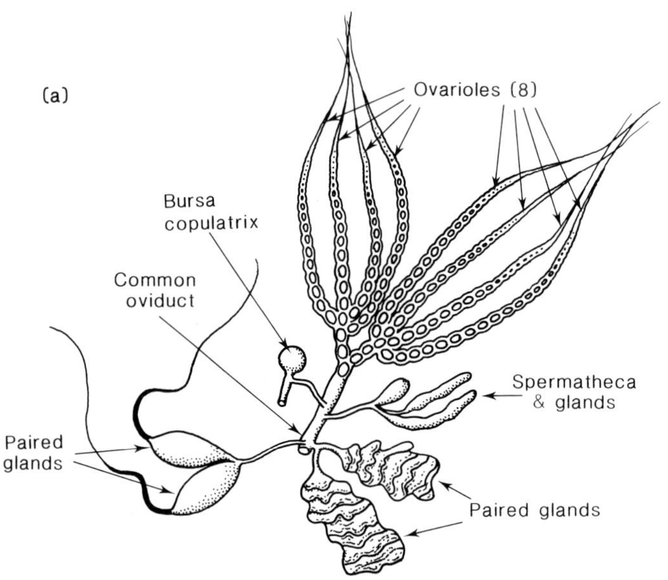

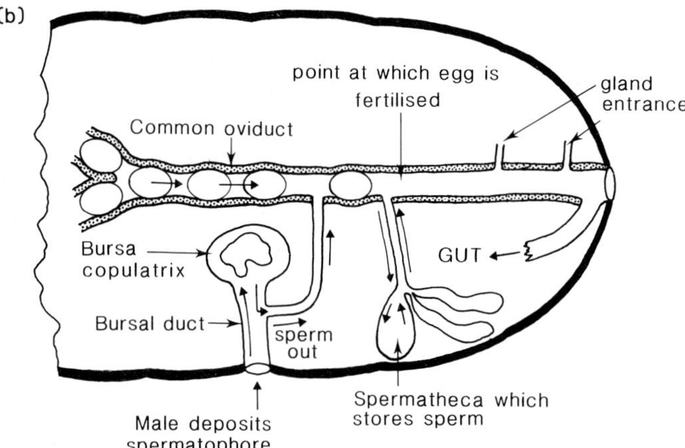

Fig. 3.2 The layout of the female butterfly reproductive system as exemplified by the marsh fritillary *Eurodryas aurinia* (see text). The bursa may contain the remnants of several spermatophores. The glands shown venting into the common oviduct are of largely unknown function, but it is thought that the large convoluted pair of glands provide the adhesive which cements eggs to leaves. (a) Reproductive organs. (b) Cross-sectional view.

fecundity by relying on adult resources and using larval energy to produce a larger adult body size could also be beneficial. In the first instance the vulnerable pupal stage is shortened and the risk of predation lessened. The larger adult body size would enable the adult to have a greater body reserve to migrate or to live longer.

3.1.4 Adult feeding and egg production

The majority of egg development in British butterflies depends on food reserves accumulated in the larval stages. These reserves are often visible in freshly emerged females, and late instar larvae, as extensive yellowish deposits of fat-body. This is a loosely connected mass of cells which accumulate reserves of food for future use. Thus, to a great extent the adult female has a given amount of

Table 3.4 The level of egg-development on emergence of the female butterfly

No eggs mature on emergence. Ovary usually immature: *Gonepteryx rhamni* (?); *Satyrium pruni*; *Thecla betulae*; *Inachis io* (?); *Aglais urticae*; *Maniola jurtina*; *Pyronia tithonus*
Few eggs mature on emergence: *Pararge aegeria*; *Melanargia galathea*; *Anthocharis cardamines*; *Pieris rapae*
Large number of eggs mature on emergence *Eurodryas aurinia*; *Erebia epiphron* (?)

This table gives examples of species known to fit into the above categories. We have very little information on the reproductive state of most of our butterflies on emergence. From K. Porter, (unpublished data) and Courtney (1980).

material available for egg production and her own energy needs at the start of her life. It is a logical extension of this to find that the largest, or heaviest females produce the most eggs (Courtney 1986).

A few butterflies, such as *Euphydryas editha*, are capable of completing the adult stage without feeding (Labine 1966), since males can mate after emergence and females emerge with a complement of mature eggs. By contrast adult feeding is almost essential for egg production in those butterflies (e.g. *Thecla betulae*, Essex skipper *Thymelicus lineola*, *Maniola jurtina*) which emerge from the pupa with no developed eggs. Adult feeding by *T. lineola* can increase egg production by 27 times (Pivnick and McNeil 1985) and *Colias* butterflies, which also emerge without mature eggs, realize only three to five per cent of their potential reproductive output in the absence of adult feeding (Watt *et al*. 1974). Even in those species that do not feed as adults, adult nutrients, when available, contribute to egg production and longevity.

Water is perhaps the most essential and easily obtained 'food' that adults need, and is necessary for continued egg production. Other nutrients can be obtained from nectar, tree sap, rotting fruit, or mineral pans (see chapter 2). The amount of egg-building nutrients in nectar is small (Wykes 1952) with the most useful constituents being the energy giving sugars. The most common sugars in nectar are sucrose, glucose, and fructose, with smaller amounts of maltose, melibiose, and ribose. The main nutrients that can directly contribute to egg production are the amino acids, the building blocks of proteins. One of the richest sources of these desirable foods is the honeydew secretions of aphids. The composition varies between aphid and plant species (Dixon 1973) but it contains one unique sugar; melizotose. Honeydew feeding is common, particularly among woodland butterflies (e.g. black hairstreak *Satyrium pruni*, *Apatura iris*), but whether melizotose has a significant role in the nutrition of these species is not known.

In some experiments with *Euphydryas editha* Murphy *et al*. (1983) showed that although sugars do indirectly contribute something towards egg building, their main role is in prolonging the female's life and hence giving more time for egg development. Small amounts of amino acids did help to prolong the production of large eggs, but were not as important in overall egg production as sugars. Similar effects have been found in *Colias* butterflies (Stern and Smith 1960). In the South American heliconid butterflies a major source of food for egg production is obtained by digesting pollen (Dunlap-Pianka *et al*. 1977). This is not a behaviour yet recorded from British species but it is known that many butterflies do pick up pollen from flowers and may act as pollinators (see Chapter 7).

Males may contribute towards egg production by supplying the female with minerals and other compounds which are enclosed in the spermatophore (Marshall 1980; Boggs and Watt 1981). The amount of material involved is very small and may not contribute significantly towards egg-laying, although the passing across of minerals such as sodium may be essential to the female in maintaining egg production. Sodium is present in very low levels in plant tissue, therefore butterfly larvae can only accumulate small amounts. All insects require sodium to maintain correct function of muscle and nerves, and the normal body levels can be reduced by up to 75 per cent when the female invests this mineral in her eggs (Adler and Pearson 1982). This deficit can be made up by sodium passed across to the female in the male's spermatophore. In this way the male can contribute indirectly to egg production (see also chapter 2).

In terms of other compounds in the spermatophore (i.e. proteins, fats, and sugars) there is evidence to show that these food materials are absorbed by the female (Boggs and Gilbert 1979). It is argued by Jones *et al*. (1986) that this does not infer that these materials constitute a significant investment towards egg production. This has been shown to be the case in two species of North American checkerspot butterflies, although the same may not be true of species maturing eggs over a long period. It is probable that with the exception of sodium, most of the nutrients required for egg production derive from larval feeding with supplements from the feeding of adult females in some species.

3.2 Finding and recognizing larval hostplants

Butterflies are highly selective in their choice of larval hostplants. It is also true that early instar larvae have little choice as to what they eat and depend on the judgement of the female butterfly to place eggs on or near the correct plant. Later instars are often surprisingly mobile and can seek out suitable hostplants. Indeed, many species have to move after totally denuding their hostplant (e.g. large white *Pieris brassicae*, Glanville fritillary *Melitaea cinxia*, *Eurodryas aurinia*, *Aglais urticae*).

In searching for an egg-laying site the female is responding to several demands which will influence the survival of her offspring. A prerequisite is that she chooses a species of plant or site acceptable to the larva, and ideally, this plant should be placed to provide optimum microclimatic conditions for the egg and larval development; it should be large enough to support the larva to a stage when it can move to other plants should the need arise; the plant should be at the appropriate stage of growth to meet the needs of the larva; and finally the plant should not attract undue attention from predators or parasitoids. Some of these needs can be catered for by the searching female, whereas others are the subject of chance events.

3.2.1 Searching patterns and cues

In many butterflies the initial stage of search involves locating the correct habitat. For example female *Aphantopus hyperantus* search for patches of damp grassland and the pearl-bordered fritillary *Boloria euphrosyne* seeks out open glades with bare ground in woodland. These broad search patterns place the female in an area where the hostplants, grasses and violets, respectively, are likely to be encountered and where environmental conditions are suitable. A larva with a requirement to bask, such as that of *B. euphrosyne*, will not thrive on violets growing beneath shrubs, despite the apparent high quality of the violets growing there. Many other instances of habitat selection, as a prelude to locating hostplants can be found among butterflies (Chew and Robbins 1984).

Female butterflies intent upon oviposition can often be seen to have a fluttering flight, most uncharacteristic of their determined spurts between flowers or during dispersal. In these searching flights the female may be using visual cues to locate possible hostplants, but scent may also be important to some species (Courtney 1988). *Pieris brassicae* has been shown to use the colour of plant leaves in the initial stages of search (Ilse, 1937), while orange-tips *Anthocharis cardamines* use a combination of colour and flower shape (Courtney 1980). Several species, including *Colias*, have been shown to use leaf-shape as a guide to hostplant identification (Rausher 1978; Stanton 1982). The precise cues used by females are poorly known, but what is certain is that visual searching plays a major role in many species.

A common phenomenon is that females choose plants that are large and obvious in the general vegetation; such plants are referred to as 'apparent'. *Eurodryas aurinia* and the Duke of Burgundy *Hamearis lucina* show a definite tendency to lay eggs on large, obvious, or apparent plants and this suggests that visual attraction and assessment of plant size is involved. The small skipper *Thymelicus sylvestris* is most obviously attracted to flowering heads of Yorkshire fog *Holcus lanatus* when seeking egg-laying sites (Porter, personal observation). A similar pattern of searching for apparent plants has also been shown in species such as the white admiral *Ladoga camilla* (McKay, personal communication), *Anthocharis cardamines* (Courtney 1982*a*; Dennis 1983a) and wood white *Leptidea sinapis* (Wiklund 1977*a*; Warren 1984). The use of large, apparent hostplants is not a general rule among British butterflies as some observations suggest that small, inconspicuous plants are selected by egg-laying females. This has been shown for the green-veined white *Pieris napi* by Dennis (1985*b*) in observations of females laying on turnip *Brassica rapa*.

In general, butterflies that lay on apparent hosts rarely alight on any other plants during their search for egg sites. This type of behaviour has been contrasted with species such as the small copper *Lycaena phlaeas*, high brown fritillary *Argynnis adippe*, dark green fritillary *Argynnis aglaja*, *Pieris rapae* and *P. napi* which often use foodplants hidden among vegetation (Wiklund 1984; Courtney 1986). Their technique is to land frequently on several species of plant which may or may not be suitable hosts. On encountering a suitable hostplant they will oviposit on or near that plant. In *A. aglaja* and some other fritillaries this may involve laying eggs on adjacent dead or living vegetation. In the silver-washed fritillary *Argynnis paphia* the female flies a short distance after locating suitable violets and lays her eggs in crevices on the bark of trees (Frohawk 1934). This behaviour does not appear to need a specific type of tree, as females in Sweden have been seen to lay on beech, oak, and birch (Wiklund 1984). The behaviour of *Boloria euphrosyne* is intriguing in that the female usually lays her eggs away from violets, the hostplant for the larvae, yet does not land on the violet plant prior to selecting an egg-laying site (Shreeve, personal communication). In comparison, the small pearl-bordered fritillary *Boloria selene* has been observed to drop eggs into the base of quite dense, tall, red fescue grass *Festuca rubra* where the hostplant, the marsh violet *Viola palustris* was growing (Dennis, personal communication). *B. selene* will also lay eggs on the hostplant.

In species such as *Maniola jurtina* and *Melanargia galathea* which habitually 'scatter' their eggs, the initial choice of habitat is obviously important and *Maniola jurtina* in particular, will seek warm hollows with short grass or stubble. These species and *Aphantopus hyperantus* will alight before laying their eggs, not spraying them in flight as often stated. *M. jurtina* even goes as far as bending the abdomen as if to place eggs before ejecting the egg towards the ground, or rejecting the site and moving on. The larvae of these butterflies are generally non-specialist feeders and have no difficulty in finding food in areas of finer-leaved grasses. There is evidence to suggest (Wilson 1985) that *Melanargia galathea* may depend upon the fine-leaved *Festuca rubra* as a source of pigment for the adult coloration.

3.2.2 Hostplant specificity and decisions in egg-laying

Once a female butterfly has landed on a plant she has to make a decision whether to lay eggs or fly off again. This decision involves certain stages of assessment on the part of the female (Fig. 3.3). The butterfly has first to confirm that the plant is a species that will be eaten by the larvae. The cues used at this stage are probably chemical and perhaps even textural. All plants contain certain organic compounds which are characteristic of the plant family, genus, or species and are referred to as secondary compounds (Ehrlich and Raven 1965, 1967; Chew and Robbins 1984). A well known example are the compounds termed mustard oils (sinigrin and its derivatives) which occur in cabbage *Brassica oleracea* and many other Cruciferae. These compounds are known specifically to attract egg-laying *Pieris brassicae* (Mitchell 1978) and no doubt other pierids (see section 7.4). In many instances females may be responding to a mixture of secondary compounds which act in a synergistic manner, rather than a single compound. We have much to learn about how butterflies recognize their hostplants. Female butterflies detect these chemicals through 'taste' sensors in their feet, antennae (Odendaal, *et al*. 1985), or proboscis, and this physiological contact with a plant allows the female to take her first decision on whether to stay or fly off. The next stage of the decision process may involve the presence of eggs, the presence of other herbivores, or the quality of the leaf. *P. brassicae* has been

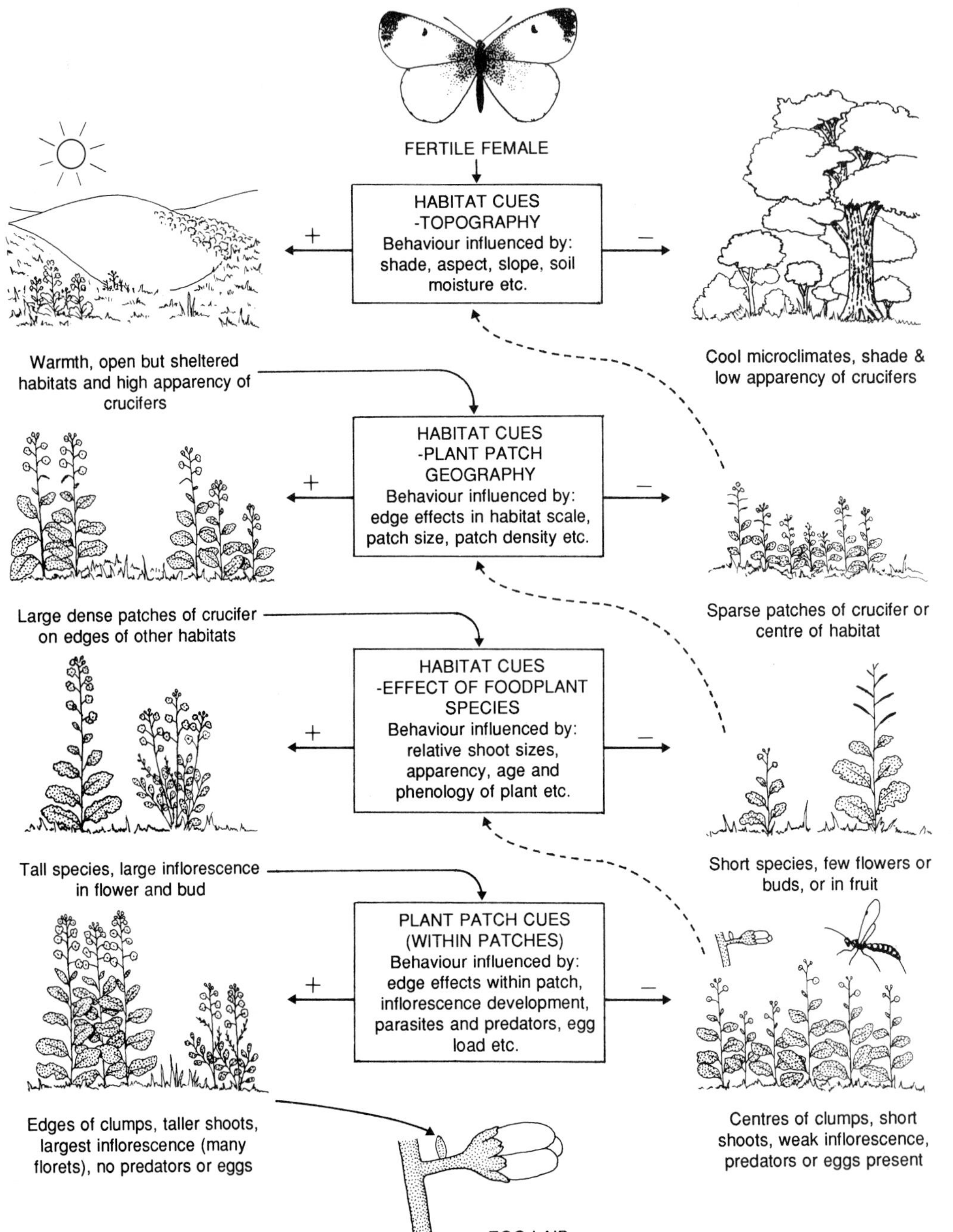

Fig. 3.3 Pathways of 'decision' for a fertile female orange-tip *Anthocharis cardamines*. The decisions taken by her before laying an egg can be related to many variables (given in the central boxes). This diagram leads through the stages of: within habitat cues, between patch cues, cues between different hostplants, and finally, cues within a single hostplant patch. This model provides a basic plan to describe how the butterfly may go about egg-site choice. The hostplant species chosen and the details of selection will be influenced by phenology of potential hostplants and their relative physiological state. (Adapted from Dennis 1983*a*.)

shown to avoid plants that already hold egg batches (Rothschild and Schoonhoven 1977), while leaf water content or quality may deter other pierid species (Scriber and Slansky 1981). An interesting association exists between ants and many lycaenid butterflies (Atsatt 1981*b*; Pierce 1984). Several species of Australian and South African lycaenids oviposit on hostplants frequented by ants and in so doing gain protection from predators (see sections 5.4 and 7.3). These butterflies show a preference for ovipositing in areas where ants are active and this begs the question as to whether British lycaenids preferentially lay on plants where ants occur.

Assessment of plant size is an important requisite of cluster-laying butterflies; *Eurodryas aurinia* and *Hamearis lucina* each lay more eggs on larger plants (Fig. 3.4), presumably because these larger plants will support more larvae past the vulnerable early instars than will small plants. Other species may respond to large plants by laying more than one egg on the plant. The process used by butterflies in deciding whether or not to lay on a plant, and if so how many eggs to lay, are poorly understood and must certainly involve other features of the female's oviposition history such as movement pattern, egg-load, weather conditions and age. These factors are covered in detail elsewhere in the chapter (p. 67).

Hostplant quality is important in many species. The eggs of *Gonepteryx rhamni* are always laid on young leaves or buds on or towards the tip of purging buckthorn *Rhamnus catharticus* or alder buckthorn *Frangula alnus* bushes; *Leptidea sinapis* will seek out the young shoots of meadow vetch *Lathyrus pratensis* in preference to other parts (Warren 1984); the holly blue *Celastrina argiolus* and *Callophrys rubi* will selectively lay on shoots and buds of all their hostplants. Egg-laying on plants of a certain quality reflects the needs of the larvae as demonstrated by the preference of female *Inachis io* to lay eggs on young plants of nettle *Urtica dioica* (Dennis 1984*c*; Pullin 1986*a*). The nitrogen level of young growth of nettle is higher than in older plants and as a consequence larvae grow more rapidly on young plants. The choice of hostplant quality is often linked to other factors such as apparency, as in *Anthocharis cardamines* (Courtney 1981, 1982*a*), or

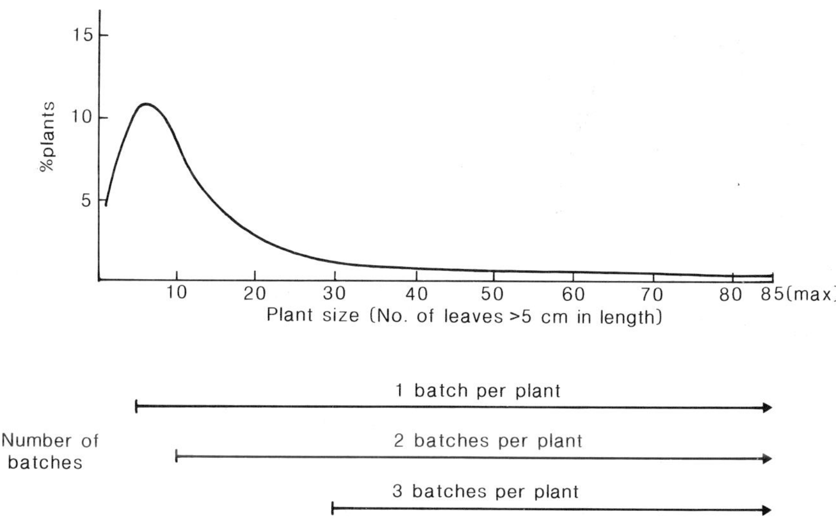

Fig. 3.4 The relationship between hostplant size and egg-laying in the marsh fritillary *Eurodryas aurinia*. The range of plants available to ovipositing females invariably follows a characteristic pattern, approximating crudely to a skewed bell-shaped curve. Plant size is described as the number of 'mature' leaves which could be used by ovipositing females. The range of batch numbers per plant of devil's-bit scabious *Succisa pratensis* demonstrates that ovipositing *E. aurinia* females select larger than average plants. Each egg batch laid represents a single choice by an individual female and each batch may contain more than 300 eggs.

microclimate as in the silver studded blue *Plebejus argus* (Thomas, 1985*a*).

3.2.3 Microclimate of egg sites

The microclimate of plant and ground surfaces is an intriguing topic. The humidity, wind speed, and temperature of leaf surfaces and stems are fairly well known due to the researches of plant physiologists and agriculturalists. Butterfly eggs are small enough to fit into the 'space' or boundary layer that envelopes leaves, stems, and other surfaces with a depth of a few millimetres or so. Temperatures within the boundary layer on the sunny side of a leaf may be 20°C higher than air temperature, yet at night, boundary layer temperature may fall well below that of the air (Edwards and Wratten 1980). These micro-effects are unnoticed by any life-form greater than 5 mm or so in size, yet can have a profound effect upon butterfly eggs. Relative humidity is also significantly different within the boundary layer. As a general rule, the boundary layer is much more humid than air 10 mm above the surface. This rule does not hold for surfaces that are in direct sunlight when the elevation of boundary layer temperature may create a zone of very low relative humidity. Related to temperature and humidity is the rate of mixing through wind turbulence in the boundary zone. The boundary exists because of a lack of air movement close to surfaces; leaf surfaces and hairs create friction which considerably reduces mixing. This results in the boundary layer being thickest adjacent to ribs and this may explain why many butterflies such as the large skipper *Ochlodes venata* place their eggs next to ribs, or in the crevices on plants.

The microclimate within the ground vegetation has several features in common with the boundary layer described above (Fig. 3.5). In general the windspeed falls, temperature and light levels decline and relative humidity rises as one moves down into the vegetation (Cox *et al.* 1973). Thus tall grasses have a moist, calm, dark, cool microclimate at ground level while shorter grasses in lowland Britain are warmer during the day and cooler at night, and generally drier and lighter at ground level. The effects of this difference in microclimate have been demonstrated in the Adonis blue *Lysandra bellargus* by Thomas (1983*a*) (Fig. 3.6). A similar need for short grass/herb vegetation also exists for the chalk hill blue *Lysandra coridon*, *Lasiommata megera* (Dennis 1983*c*; Heath *et al.* 1984), *Maniola jurtina* and *Boloria selene* (Willmott, in Heath *et al.* 1984), the silver-spotted skipper *Hesperia comma* (Thomas *et al.* 1986) and possibly *Coenonympha pamphilus*. Other species show a reverse preference, choosing taller grass (chequered skipper *Carterocephalus palaemon*, Lulworth skipper *Thymelicus acteon*, *Aphantopus hyperantus*, *Ochlodes venata*, and others). A similar conclusion may be drawn for species that prefer to

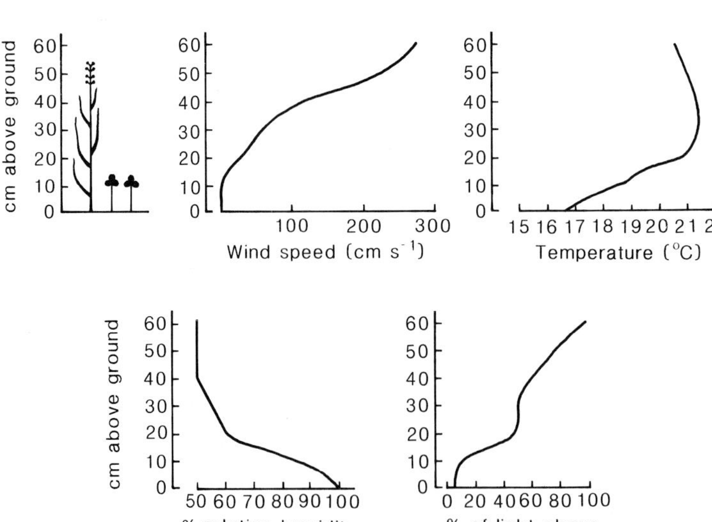

Fig. 3.5 The influence of vegetation structure on microclimate. The microclimate of vegetation layers can differ greatly from that of the air above. In general, layers near the ground tend to be cooler, moister, and darker than higher layers. (After Cox *et al*. 1973; courtesy of Blackwell Scientific Publications.)

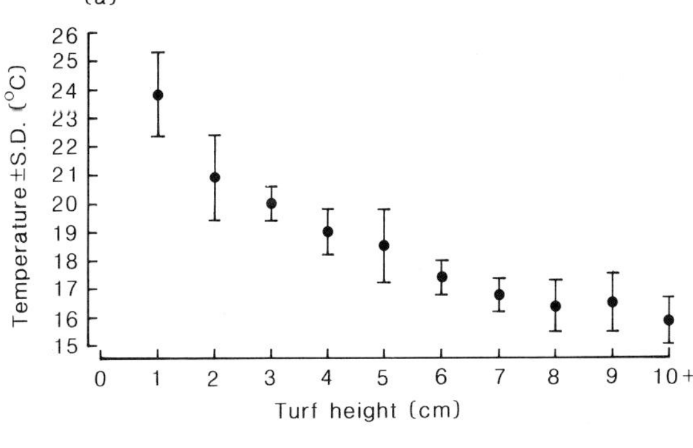

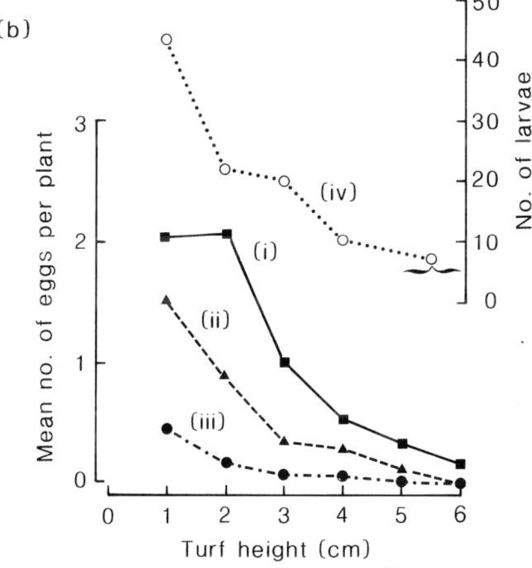

Fig. 3.6 The microclimate and choice of egg-sites in the Adonis blue *Lysandra bellargus*. Breeding sites of *L. bellargus* show a strong association with vegetation structure, such as grass height and microclimate. The combination of short turf and shelter at the edges of scrapes or hollows provides ideal egg-sites for this butterfly. (a) The shortest turf provides the highest temperatures beneath the hostplant, horseshoe vetch *Hippocrepis comosa*. This is demonstrated by the plot of mean ground level temperatures against turf height. Measured on 18 April 1981 near Swanage, 12.00–12.15 hours, during calm, hazy-bright conditions. For S.D., see 'standard error' in the Glossary. (b) The warmest sites under the shortest turf and in sheltered situations receive the greatest numbers of eggs, and subsequently hold the highest larval densities. This plot illustrates the relationship between numbers of eggs and larvae, turf height, and degree of shelter. Mean density of eggs in (i) completely sheltered situation, (ii) limited shelter and (iii) no shelter; (iv) total number of larvae on 80 plants in each turf category. (After Thomas 1982*a*; courtesy of *the Journal of Applied Ecology* and the British Ecological Society.)

lay eggs on the undersides of leaves (e.g. *Pieris napi, P. rapae, Hamearis lucina, Aglais urticae*). The preference of *P. napi* and *P. rapae* have been shown to differ in Italy (Petersen 1954), where *P. rapae* prefers to lay eggs on plants in sunny situations while *P. napi* prefers shaded plants; the latter has a lower temperature preference (28°C) than open ground *Pieris* species. Some butterflies may also use light levels as a cue to egg-site as light receptors have recently been described from the female genitalia of some species (Arikawa *et al*. 1980).

Argynnis paphia, which lays in crevices in bark, may be subject to temperature requirements that correspond to the type of bark. A preference is shown for the north side of oaks (Heath *et al*. 1984) which may infer a need for cooler climates. Coarse bark will have shaded crevices with relatively lower temperatures and higher humidity. Pale coloured bark can reflect much of the heat (Nicolai 1986), while the outer surface of bark in sunshine can reach 50°C (Edwards and Wratten 1980).

3.2.4 Plant topography and egg position

Sites for eggs vary in a characteristic manner between species, with position related to the needs of eggs and larvae. This can be shown by comparing species which hatch within a week or so of laying with those which are intended to pass through the winter (see section 5.2). Virtually all eggs which hatch quickly are laid on their respective hostplants (or in close proximity) in positions which provide optimal conditions for egg development. Many butterflies, such as the pierids, satyrines and some nymphalids, may place their eggs in a relatively exposed position with little protection from mechanical damage (e.g. *Pieris napi, Polygonia c-album, Pararge aegeria, Apatura iris, Ladoga camilla*). The lycaenids which have fast hatching eggs often oviposit in relatively concealed sites such as on flower stems (e.g. *Celastrina argiolus*), on leaf axils or stems (e.g. *Callophrys rubi*), or in flower clusters (e.g. *Cupido minimus*). Species with eggs that overwinter, place their eggs in mechanically protected woody parts of a plant. The egg-sites of some of the skippers demonstrate a precise adaptation to site. *Thymelicus sylvestris* and *T. acteon* lay within the relatively loose leaf sheaths of *Holcus lanatus* and tor grass *Brachypodium pinnatum* respectively. *Thymelicus lineola* usually lays its eggs within the tight leaf sheaths of cock's-foot *Dactylis glomerata* and creeping soft-grass *Holcus mollis*. *T. sylvestris* and *T. acteon* overwinter as tiny larvae inside a silken cocoon which is spun immediately after hatching, while *T. lineola* hibernates as a tiny caterpillar still within the eggshell. These differences are probably related to the relative security of the overwintering stage, as pointed out by Frohawk (1934). The silken cocooned larvae are firmly contained within their leaf sheath, while the tightly furled sheaths used by *T. lineola* do not demand this additional protection.

The site chosen by a single female butterfly will represent the end of a series of cues and decisions on plant suitability, as shown in Fig. 3.3 for *Anthocharis cardamines*. When several females choose the same site independently of one another, then this suggests that this site presents all the correct cues for that species of butterfly. Such multiple depositions of eggs can consist of each female adding to a previous egg-cluster (*Eurodryas aurinia*, *Hamearis lucina*, *Aglais urticae*, *Thymelicus sylvestris*, *T. lineola*) or may even extend to two different species laying adjacent to one another, as reported for *A. urticae* and *Inachis io* (Dennis 1984*c*; Dennis and Richman 1985). The use of one plant, and often the same leaf on a plant, either can lead to increased survival by defence through reinforcement of numbers as in the gregarious species (see Fig. 5.7), or may lead to a lowering of survival chance for eggs of single layers. This is taken to an extreme in *Anthocharis cardamines* and *Callophrys rubi* as the larvae are cannibalistic and will eat their smaller relatives. Some species, notably *Anthocharis cardamines*, will avoid sites, however suitable, if eggs of another female are already present (Wiklund and Åhrberg 1978; Courtney 1980; Thomas, C. 1984), thus evading competition.

3.3 Distribution of egg load

Patterns of egg distribution vary between species. Females should maximize the proportion of the egg load which is deposited in sites where larval success is the greatest. Different species have evolved physiological and behavioural adaptations in response to hostplant abundance and distribution patterns, habitat stability, and the activity of competitors and enemies. These adaptations can be classified into three main types, those which: lead to eggs being laid singly; lead to eggs being laid in small batches; result in egg clustering.

3.3.1 Single egg layers

Most species which lay single eggs (Table 3.5) have a strategy similar to that of *Maniola jurtina*, that is, their egg production and rate of laying reaches a peak a few days after emergence. In optimal conditions, species such as *Pieris rapae* can be expected to reach a maximum rate of 35 eggs laid per day after a day or so, and then declining rapidly as they age (Gossard and Jones 1977). The advantage of laying single eggs include benefits to female butterfly and larvae (Stamp 1980). An egg-laying female only needs to spend a few seconds actually placing her single egg and thus is less at risk from attracting attention to the plant than is a cluster-laying female. However, in doing so the single-laying female will be more noticeable when in flight and this increases its risk of predation. The single-laying female distributes her eggs over a large number of hostplants, sometimes of different species (e.g. *Anthocharis cardamines*, *Pieris rapae*, *P. napi*) and this will spread the risks of predation and reduce competition from other larvae or cannibalism through the use of the same hostplant (*A. cardamines*). Other advantages related to laying eggs singly are associated with the effectiveness of crypsis for the eggs and the resultant larvae (see chapter 5).

In temperate countries butterflies will experience dull, cool, overcast days when, at most, only a few eggs are laid. In such circumstances the butterfly will compensate for this shortfall by laying more eggs on the next sunny day. This may result in 'mistakes', when a single-laying species may deposit two or three eggs on one twig (e.g. *Thecla betulae*) or leaf (e.g. *Apatura iris*). In *Pieris rapae* intensive episodes of egg deposition occur when sunny weather follows dull, cloudy, or rainy conditions (Gossard and Jones 1977) and this may result in the otherwise abnormal deposition of small batches. Gossard and Jones suggest that search behaviour for hostplants may

Table 3.5 Egg-laying strategy of British butterflies

Species which lay single eggs:

Carterocephalus palaemon; Hesperia comma; Ochlodes venata; Erynnis tages; Pyrgus malvae

Papilio machaon

Leptidea sinapis; Colias croceus; Gonepteryx rhamni; Pieris rapae; Pieris napi; Anthocharis cardamines

Callophrys rubi; Thecla betulae; Quercusia quercus; Satyrium w-album; Satyrium pruni; Lycaena phlaeas; Cupido minimus; Plebejus argus; Aricia agestis; Aricia artaxerxes; Polyommatus icarus; Lysandra coridon; Lysandra bellargus; Celastrina argiolus

Ladoga camilla; Apatura iris; Vanessa atalanta; Cynthia cardui; Polygonia c-album; Boloria selene; Boloria euphrosyne; Argynnis adippe; Argynnis aglaja; Argynnis paphia

Pararge aegeria; Lasiommata megera; Erebia epiphron; Erebia aethiops; Melanargia galathea; Hipparchia semele; Pyronia tithonus; Maniola jurtina; Coenonympha pamphilus; Coenonympha tullia; Aphantopus hyperantus

Species which lay up to 15 eggs in a batch (observed maximum in parentheses):

Thymelicus sylvestris (10); *Thymelicus lineola* (14); *Thymelicus acteon* (15)

Hamearis lucina (8)

Cluster-laying species (approximate maximum in parentheses):

Pieris brassicae (120)

Aglais urticae (200); *Inachis io* (500); *Nymphalis polychloros* (220)
Eurodryas aurinia (350); *Melitaea cinxia* (300); *Mellicta athalia* (140)

also be affected by the number of mature eggs present in *P. rapae*.

3.3.2 Small egg-batch layers

Four species of butterfly resident in Britain lay small batches of eggs (Table 3.5). This number is often extended to include *Pieris rapae* which sometimes lays two to five eggs on the same leaf, but in the strictest sense it is not a habitual batch-layer. One advantage of laying small batches is presumably that the butterfly can afford to spend more time searching for suitable sites before laying a moderate investment of eggs. Alternatively, they do not depend upon continuous favourable flight conditions and can lay most of their egg-load in the few favourable 'windows' in the weather. *Thymelicus sylvestris*, *T. lineola*, and *T. acteon*, insert their smooth, oval eggs into the leaf sheaths of the appropriate grasses. *T. sylvestris*, which normally lays three to six eggs in a batch, is known to lay up to at least eight eggs at a time, but many grass sheaths contain many more eggs (up to 33 in one instance) and these large numbers of eggs are produced by different females independently choosing the same grass stem (Porter, unpublished data). A similar situation probably holds for the other two skippers; Frohawk (1934) recorded as many as 14 eggs in one stem for *T. lineola* and 15 for *T. acteon*. The oviposition strategy for these related skippers should provide an interesting study.

Hamearis lucina lays batches of up to eight eggs, although some reports do give higher figures which are most often attributable to two or more females adding eggs to a previous batch. The number of eggs in this species can be related to the size and apparency of the cowslip *Primula veris* or primrose *Primula vulgaris* used as a hostplant (Fig. 3.7), which suggests that females of this species can assess plant size (Porter, unpublished data). Each plant may be isolated by several metres from adjacent plants, and adjusting egg-batch size to plant size is one way of avoiding starvation or predation losses if larvae are forced to travel far to reach another plant.

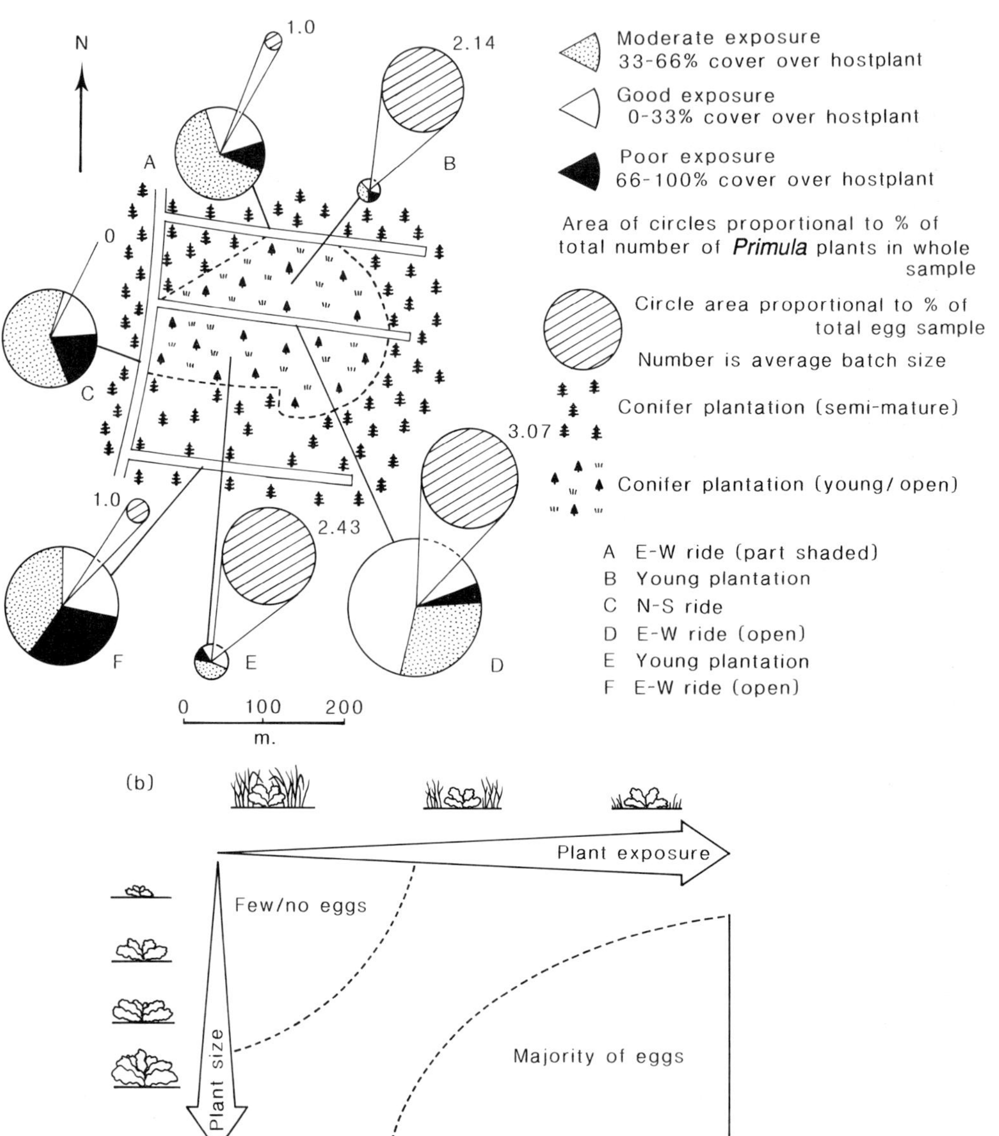

Fig. 3.7 The distribution of Duke of Burgundy *Hamearis lucina* eggs in a woodland site in Bernwood Forest, Buckinghamshire. Data based on observations from a single season. (a) The distribution of *H. lucina* eggs on primrose *Primula vulgaris* in a woodland ride/plantation as related to plant exposure and ride orientation. For each sample location of the site (A–F), the pie diagram shows the proportion of poor, moderate and well-exposed plants, the area of each circle representing the proportion of the total hostplant population sampled at that site. This area is a true reflection of the density of hostplants at each location. The area of each satellite circle represents the proportional contribution of that site towards the total egg sample (e.g. none was present at site C). The eggs were concentrated in a few sites, notably the open, young conifer plantation and on open east–west rides. (b) Egg distribution of *H. lucina* on primrose *Primula vulgaris*, relative to plant size and apparency. A distinct trend is evident towards the majority of eggs being laid on the largest, most apparent plants. A similar trend is seen for the average batch size; an increase in plant size and apparency leads to an increase in average batch size.

Laying small batches is a behaviour often present in ant-associated lycaenids in Australia and South Africa (Atsatt 1981*a*; Kitching 1981). Such behaviour also exists in some South American lycaenids and riodinines including *Hamearis susanae*, a relative of *H. lucina* (Bourquin 1953; Robbins and Aiello 1982; see also chapters 5 and 7).

3.3.3 Egg-cluster layers

Seven species of resident British butterfly lay clusters of at least 100 eggs (Table 3.5); the Camberwell beauty *Nymphalis antiopa* and black-veined white *Aporia crataegi* also show similar behaviour. The habit of cluster laying seems a risky way of distributing an egg-load especially if the cluster is large (e.g. *Eurodryas aurinia*). The reasons put forward to explain cluster-laying include benefits to the adult female, eggs, and larvae. Cluster laying has probably evolved in response to the clumping and wide spacing of patches of hostplants and the need for the female to lay as many eggs as possible (Courtney 1984). This is thought to be the case for the cluster-laying nymphalids, such as *Aglais urticae*, which use *Urtica dioica* as their hostplant (Dennis 1984*c*) and is no doubt equally likely for the cluster-laying fritillaries. There are several factors which consolidate cluster-laying behaviour including those that must originally have 'encouraged' this behaviour and those which today act as feedback-mechanisms to maintain it. Among the first group of factors are the constraints of weather, which place restrictions upon the time available for oviposition (Stamp 1980). Given a patchy resource and a limited amount of flight-time the female butterfly can achieve a higher realized fecundity (see Chapter 4) by clumping her eggs in large egg-batches. The short period that females are exposed when moving from one patch to the next may also reduce the risk of adult predation, while the longer time spent actually laying an egg-batch (up to one hour in *Eurodryas aurinia*) also means that the female is concealed from predators (Courtney 1984). Other factors belonging to this first group include pressures associated with low adult population density when females may need to spend potential oviposition time seeking a mate (Stamp 1980). The second group of factors which reinforce cluster-laying involve the use of egg toxins and aposematic coloration as a defence against predators (see Fig. 5.7) and the benefits of clustering eggs and larvae to reinforce chemical and behavioural deterrents to parasitoids and predators. These types of factor will act as feedback-mechanisms to maintain cluster-laying as a strategy (Dennis 1984*c*).

Cluster-layers have one main prerequisite; they need to lay on relatively large plants or closely grouped plants in order that their offspring may attain a size sufficient to allow them mobility to seek out new sources of food. This is apparent in the behaviour of the nettle-feeding nymphalids and the colonial fritillaries. In *Eurodryas aurinia*, the female selects larger than average sized plants on which to lay eggs (Fig. 3.4) and unusually large plants often hold two or more batches. Other species, such as *Pieris brassicae*, can adjust the number of eggs per batch to suit plant size (Rothschild and Schoonhoven 1977).

Altogether cluster-layers follow this strategy for a number of reasons, and these are inextricably linked to other features of the life history such as gregarious larval behaviour and the possession of toxins. In general, it has been suggested (Courtney 1984) that cluster-layers are very successful for achieving a high realized fecundity and thus their populations can increase rapidly. It may be that this ability to lay large numbers of eggs is balanced by the susceptibility of the resultant larvae to parasitoids (chapter 4).

3.4 Patterns in egg-laying

The British butterflies can be roughly divided into three groups with regard to how they use hostplants (Table 3.6), and each group can be related to a particular strategy. At one extreme are the monophagous species which use only one hostplant (although others may be acceptable in captivity) as exemplified by *Thecla betulae* and *Satyrium pruni*. At the other extreme are polyphagous species such as

Table 3.6 The natural use of hostplants by British butterflies

Monophagous species—butterfly species with only one main hostplant:
Thymelicus sylvestris; Thymelicus acteon; Hesperia comma; Papilio machaon; Thecla betulae; Quercusia quercus; Satyrium pruni; Cupido minimus; Lysandra coridon; Lysandra bellargus; Ladoga camilla; Inachis io; Argynnis adippe; Argynnis paphia; Eurodryas aurinia; Melitaea cinxia; E. epiphron

Oligophagous species
(a) butterfly species with two main hostplants within the same plant family:
Carterocephalus palaemon; Ochlodes venata; Gonepteryx rhamni; Satyrium w-album; Aricia artaxerxes; Lycaena phlaeas; Hamearis lucina; Apatura iris; Aglais urticae; Boloria selene; Boloria euphrosyne; Erebia aethiops; Coenonympha tullia
(b) butterfly species with more than two main hostplants, but within one plant family or closely related families:
Thymelicus lineola; Erynnis tages; Pyrgus malvae; Leptidea sinapis; Colias croceus; Pieris brassicae; Pieris rapae; Pieris napi; Anthocharis cardamines; Aricia agestis; Polyommatus icarus; Vanessa atalanta; Polygonia c-album; Argynnis aglaja; Pararge aegeria; Lasiommata megera; Melanargia galathea; Hipparchia semele; Pyronia tithonus; Maniola jurtina; Coenonympha pamphilus; Aphantopus hyperantus

Polyphagous species—butterfly species which have several hostplants in several plant families:
Callophrys rubi; Plebejus argus; Celastrina argiolus; Cynthia cardui; Nymphalis polychloros; Mellicta athalia

Many of the species listed above will accept a wide variety of plants as larval hostplants in captivity but the table has been compiled with reference to the main hostplants used in natural situations throughout the British Isles. The hostplants are listed in Appendix 1.

the painted lady *Cynthia cardui* and *Celastrina argiolus* which use a wide range of hostplants from several families (Heath *et al*. 1984). Species which lie between these extremes can be called oligophagous; they use hostplants within one family or from a few closely related families. Such species (e.g. *Thymelicus lineola, Lycaena phlaeas, Hamearis lucina*) can have one or two main hostplants; others (e.g. *Anthocharis cardamines, Pieris rapae, Pararge aegeria*) have several alternative hostplants which appear to be equally acceptable (Courtney and Duggan 1983; Shreeve 1986*a*).

3.4.1 Geographical patterns in egg-laying

The main hostplant chosen as an egg site often depends on the availability of alternatives. This is clearly seen in *Lycaena phlaeas* which will use common sorrel *Rumex acetosa* on base-rich soils and sheep's sorrel *R. acetosella* on acidic or sandy soils. The same is also true, though for less obvious reasons, in *Celastrina argiolus* which uses gorse *Ulex europaeus* in some localities in preference to the range of shrubs used elsewhere (Heath *et al*. 1984). The apparent non-selectivity of some satyrines (e.g. *Maniola jurtina*) can perhaps be related to a dependency on microclimate rather than any particular hostplant. Although *Pararge aegeria* shows some selection, Shreeve (1986*a*) found that this species chose sites that were at least 21 °C in summer. Other reasons for grass choice by satyrines may involve hostplant quality and seasonal needs of larvae (Bink 1985). Species which use unrelated hostplants as egg sites appear to do so for two main reasons. The first is the opportunist strategy of species such as *Cynthia cardui*, whose larvae are adapted to use many hostplants over their pan-continental distribution. The other reason for such polyphagous behaviour is seen in *Celastrina argiolus* and *Callophrys rubi*, where larvae are specialist feeders on buds or other soft tissues of the hostplant, and thus a range of shrubs are acceptable providing they have suitable vegetation at the right time of year (Heath *et al*. 1984; Pollard 1985).

Multivoltine species often have seasonal changes of hostplants as in *Celastrina argiolus*, which relies heavily upon developing buds of holly *Ilex aquifolium* and other shrubs in spring and buds and berries of ivy *Hedera helix* in late summer. A similar situation has been shown in the common blue *Polyommatus icarus* at a site in Cheshire (Dennis 1985*c*) where females of the first brood had laid the most eggs on lesser trefoil *Trifolium dubium* while the summer generation had laid their eggs mainly on common bird's-foot trefoil *Lotus corniculatus* or marsh bird's-foot trefoil *L. uliginosus*. These differences were attributed to the abundance and state of the hostplant and their importance as cues to egg-laying females.

The way in which species use hostplants can differ between populations (White and Singer 1974; Ehrlich *et al*. 1975) or may differ seasonally between broods as described above. Within a given population the pattern of hostplant use is generally fixed (Wiklund 1974), individuals showing a graded preference for a series of hostplants. This gradation of preference may reflect the suitability of the hostplant as larval food as in Swedish populations of the swallowtail *Papilio machaon* (Wiklund 1981) or may reflect the apparency of the hostplant, as in *Anthocharis cardamines* (Courtney 1982*a*). Wiklund (1981) has shown that in one Swedish population of *P. machaon* there are two broad types of egg-laying behaviour among females. In the first, females are specialists and only lay eggs on the preferred hostplant, provided it was available, whilst generalist females of the second type laid eggs on several hostplant species even in the presence of the preferred hostplant. Their patterns of egg-laying can be considered as extremes; specialist females needed to spend more time searching for their preferred hostplant and thus laid fewer eggs, but these had a greater chance of survival. Conversely, the generalist females laid more eggs on a series of hostplants, but their offspring had a lower survival rate as some eggs were placed on plants which are only marginally acceptable to larvae. The specialist females would be selected for in stable habitats with a continuity of preferred hostplants (i.e. milk parsley *Peucedanum palustre* in fens) while generalist females would be more successful at utilizing fringe habitats with alternative hostplants (in Sweden) such as cowbane *Cicuta virosa* on streambanks or caraway *Carum carvi* in meadows (Wiklund 1981; see section 10.4.2).

3.4.2 Distribution of eggs within hostplant patches

Observations on the distribution of butterfly eggs in natural situations show that eggs have a biased distribution despite the abundance and distribution of their hostplants. In some instances it is possible to explain these egg distributions in terms of special microclimate needs as in *Lysandra bellargus* (Thomas 1983*a*) or size and apparency of the hostplant as in *Hamearis lucina* (Porter, unpublished data). In other cases factors such as proximity of nectar sources (Murphy *et al*. 1984), presence or absence of male territories, or good basking sites may influence an egg distribution pattern.

A frequently reported observation is for eggs to be distributed towards the edge of hostplants or clumps of plants. This phenomenon has broadly been described as the edge effect (see Courtney and Courtney 1982). The term has been further refined by Dennis (1984*b*) and can be considered to cover three distinct types of egg distribution:

(1) density effects, when eggs are laid on separate outlying patches or shoots which are at low density within an overall larger zone of hostplant.
(2) edge effects, where eggs are placed on the edges of hostplant clumps; and
(3) recess effects, when eggs are laid in clefts, recesses, or hollows, typically where the hostplant abuts onto a hedge, wall, or bare ground.

The density effect is found in pierids such as *Pieris rapae*, *P. napi* and *Anthocharis cardamines* (Dennis 1983*a*,*b*; Courtney 1986; Fig. 3.8) and is, in its simplest form, a term describing the fact that the densest patches of foodplant will have the fewest eggs. Explanations for this phenomenon are generally related to plant distribution patterns and mobility patterns. Isolated plants are encountered more often than those within clumps (Jones 1977) and plants which outlie clumps are the first and last to be encountered by egg-laying females (Courtney and Courtney 1982) and often receive most eggs.

The true edge-effect has been described for *Anthocharis cardamines* (Courtney and Courtney 1982; Dennis 1983*a*, 1985*e*), and for *Aglais urticae* (Dennis 1984*c*; Fig. 3.9). Many reasons have been suggested as to why egg-laying is concentrated on the edges of hostplant patches. In *Anthocharis cardamines* it is

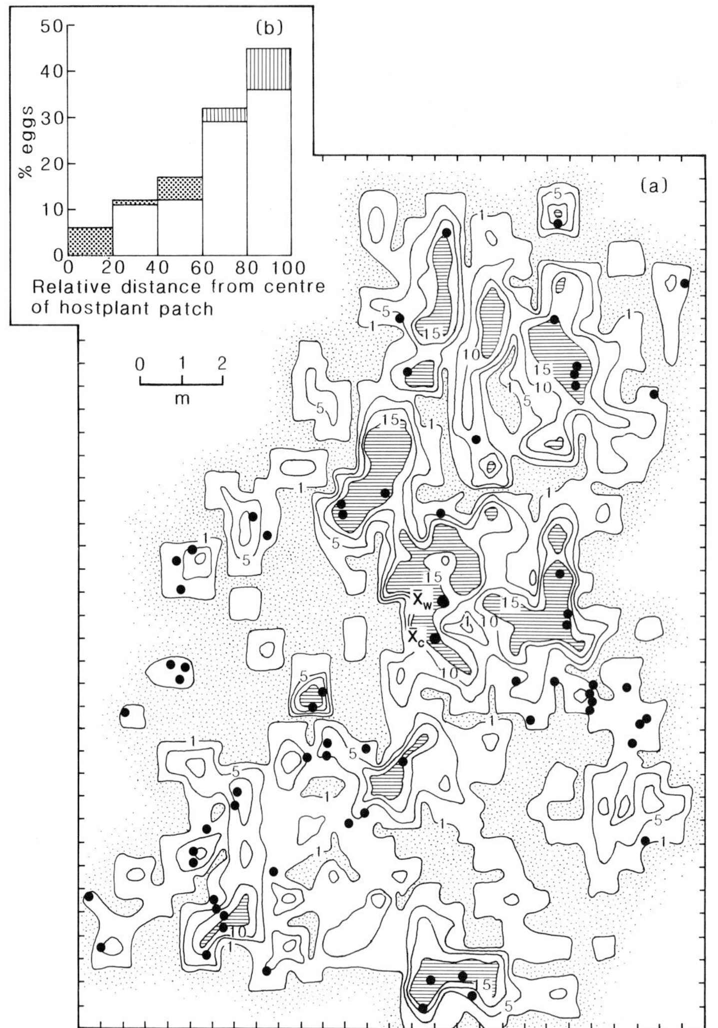

Fig. 3.8 The density-effect in egg-laying in the orange-tip *Anthocharis cardamines* on cuckoo flower *Cardamine pratensis* in the Bollin valley, north Cheshire. (a) Eggs tend to have a peripheral distribution around a large hostplant patch. $\bar{x}_c$ and $\bar{x}_w$ give the centre (unweighted, and weighted according to plant density) for the hostplant patch. Eggs and larvae are shown as black circles. Isolines give inflorescence density at intervals of 1, 5, 10, and 15. (b) A significant bias in distribution of eggs to the periphery of the hostplant patch is illustrated in a plot of the proportion of eggs located in five annuli extending from the centre of the hostplant patch. The numbers of eggs expected in each sector is given by the height of the bars and is based on the number of flowerheads in each sector; vertical hatched areas, the number of eggs exceeds that expected; stippled areas, a deficit of eggs occurs from that expected by the model. (After Dennis 1983*a*; courtesy of the Northern Naturalists' Union.)

suggested that the plant edges are used because females tend to lay eggs after flying long distances, but having done so lay successively fewer eggs; egg-laying females fly towards the nearest plant on their flight path, thus reaching the edges first (Courtney and Courtney 1982; Dennis 1983*b*). Alternatively, hostplant quality may be better at the edges of dense patches (Dennis 1984*c*, 1985*e*). In *A. cardamines* this behaviour may possibly be interpreted as an adaptation to searching for hostplants that are thinly and irregularly distributed, and is believed to be the best strategy for an opportunist species such as *A. cardamines*. However, *A. cardamines* may not use the edges of clumps for ovipositing when the quality of hostplant shoots is better in the centre of patches (Dennis 1985*e*).

In a study of *Aglais urticae* over two-thirds of all egg batches were found on the edge of nettle patches (Dennis 1984*c*; Fig. 3.9). The reasons for laying on the edges may be related to the quality of leaf and ease of access by females. *A. urticae* appear to choose large nettle clumps, especially those which are isolated in discrete clumps rather than extensive contiguous nettle-beds. The female may be able to judge the clump-size by flying around the circumference and responding appropriately with a batch of

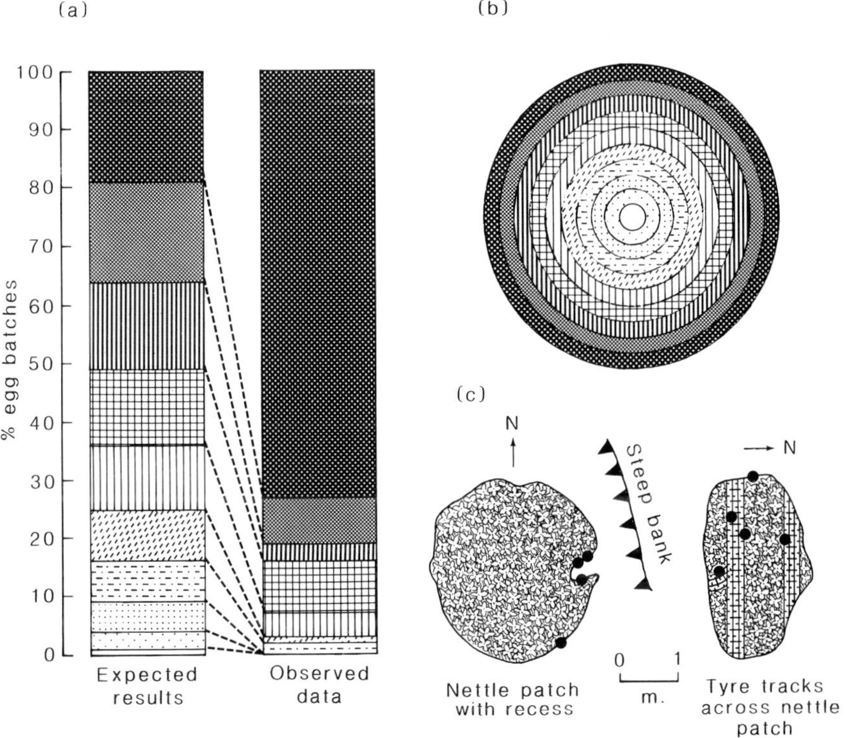

Fig. 3.9 The edge-effect and recess-effect in egg-laying in the small tortoiseshell *Aglais urticae* on nettle *Urtica dioica* in the Bollin valley, north Cheshire. (a) The relative frequency of egg-batches in 10 annuli of a circular nettle patch compared to that expected from the proportion of hostplant shoots available in the annuli; significantly, 73 per cent of the egg batches are laid in the outer ring compared to 19 per cent expected from the concentric model. (b) The concentric model for egg-laying on hostplant patches. Outer annuli offer a wider area for egg-laying despite being of equal width to inner annuli. (c) The use of recesses (natural enclaves, tyre tracks) for egg-laying by *A. urticae*, despite the more usual bias for edge plants. (After Dennis 1984*c*; courtesy of *The Entomologist's Gazette*.)

given size. This behaviour is not always rigidly followed as the butterfly shows a definite preference for young growth of nettles, and may prefer to lay in the centre of a patch of young nettles rather than on the edge of an old clump (Dennis 1985*e*).

As opposed to simple edges, other species, such as *Lasiommata megera*, *Polyommatus icarus*, and *Plebejus argus* show some preference for the hollows, clefts, and recesses within the outline of the plant (Dennis 1983*c*, 1984*b*; Thomas, C. 1985*a*). In these instances the female butterfly lays eggs in positions where they may be better concealed from predators and protected from herbivores (Dennis 1983*c*). Such positions may also provide particular microclimates which enable eggs to hatch rapidly and hence further reduce risks of predation. In *L. megera* females have been noted to show a particular preference for warm sheltered depressions along paths or slopes (Dennis 1983*c*, Heath *et al*. 1984). Similar observations also exist for *Lysandra bellargus*, *L. coridon* (Thomas, J. 1983*a*) and *Hesperia comma* (Thomas *et al*. 1986).

The egg-load deposited by females represents a population's total investment for the next generation. Once deposited, the eggs are abandoned and are subject to an array of selection agents. However, in South Leyte (Philippines) females of *Hypolimnas antelope* have been observed to defend their egg-batches (Johnston and Medicielo 1979). The distribution of the eggs reflects a compromise between the need to maximize fecundity and to ensure survival of the eggs and larvae. In this chapter it has been shown that where possible, females choose appropriate sites for the survival of their offspring. However, habitats rarely, if ever, present unlimited optimal sites for egg-laying, and in order to deposit their egg-loads, females are necessarily opportunists. The pressures on the survival of eggs and later stages are dealt with in the following two chapters.

4

Butterfly populations

Martin S. Warren

4.1 What is a population?

In an ecological context, the word 'population' has a more precise meaning than it has acquired in common usage, and needs to be defined to avoid confusion. Throughout this book, the term 'population' refers to a group of individuals of a particular species that are separated to some extent, either in space or in time, from other groups of the same species. Each group, or population, usually occupies a discrete site or region and is separated from other groups by an area which is generally uninhabited. Thus, a species of butterfly occurring on chalk downland may be separated into two distinct populations by an intervening belt of scrub, or by some other unsuitable type of vegetation. Of course, a population is rarely totally isolated from its neighbours and there may be some movement, or dispersal, of individuals from one population to another (see Chapter 6). Such a system of interacting local populations may together form a metapopulation which covers a much wider region (Levins 1970; Gilpin and Hanski 1991). Each population has its own measurable characteristics such as density, birth-rate, death-rate, genetic composition, and breeding area. The term 'colony' can be used instead of 'population', but the former is often applied more loosely, as for example when referring to a particular site.

4.2 Measuring and monitoring butterfly populations

The study of population dynamics is concerned primarily with identifying the factors influencing the annual fluctuations in abundance, long-term trends, and the overall levels of abundance. A prerequisite of these studies is to gather accurate quantitative data on a range of population parameters, using some of the methods summarized in this section. The most widely used methods of estimating the absolute size of the adult population is to mark, release, and recapture sequential samples of adult butterflies. Detailed descriptions of the various procedures, their underlying assumptions, and analysis are given by Begon (1979) and Blower *et al.* (1981). Estimates of adult numbers from a single day's sampling can be obtained by using the Petersen Estimate (also known as the Lincoln Index). If sampling is repeated on three or more days, a more detailed analysis is possible by using methods described in Fisher and Ford (1947), Jolly (1965), and Manley and Parr (1968). These enable the calculations of a range of parameters, including the survival rate of adults, the daily recruitment rate, and daily population estimates. The accuracy of these and other techniques when applied to butterfly populations, and the validity of the underlying assumptions, are discussed by Gall (1984) and Roff (1973*a*,*b*). There is some evidence that the marking procedure may effect the subsequent behaviour of certain species following release, particularly the more mobile species (Morton 1982). However, the majority of British species so far studied seem remarkably unaffected by marking, at least over intervals of 24 hours or more, and the procedure can yield invaluable information on longevity, population size, and also movement (see Chapter 6).

For many purposes, absolute estimates of

population size are not necessary, and relative estimates are sufficient to compare numbers from year to year, or from site to site. The latter are based on systematic counts, following standardized sampling procedures. The simplest, and most popular, is the Butterfly Monitoring Transect Method which involves counting butterflies, once a week between April and September, along a fixed route under prescribed weather conditions. Full details of the method are given by Pollard (1977) and in a small booklet available from the Butterfly Monitoring Scheme, Institute of Terrestrial Ecology, Monks Wood (address in Appendix 4). The results of the scheme indicate national and regional trends in the abundance of each species, the timing of emergence periods, and the effects of changes in habitat management (see Pollard 1982; Pollard *et al.* 1986).

By using an adaptation of the transect method a relative estimate of adult members can be obtained from a single sample. Here, butterflies are counted along a random, but measured, route which zigzags through a site. Adult density can then be calculated per unit length of transect (Thomas 1983*c*) or per unit time of search (method 2 in Warren *et al.* 1984). Despite the considerable sampling errors involved, both methods have been shown to provide sufficiently accurate results on national surveys of several endangered species (Thomas 1983*c*; Warren *et al.* 1984).

Some butterfly species have reasonably conspicuous egg, larval, or pupal stages, which can be counted with equal, if not greater, accuracy than the adult stage. Standardized counts can be carried out either by searching a fixed area or within randomly placed quadrats (see methods in Southwood, 1978). The most common use of such counts is the construction of life tables (see section 4.2.2), but they can also be used to compare population size on different sites. For example, Thomas and Simcox (1982) estimated the absolute size of Glanville fritillary *Melitaea cinxia* populations by counting the number of larval nests during the spring (see Fig. 1.7).

4.3 The structure of butterfly populations

4.3.1 Closed versus open populations

Many adult butterflies spend a substantial proportion of their active lives in flight, seeking out the location of various essential resources (see Chapter 6). Despite this considerable flight activity, the adults of most species tend to stay within certain well-defined areas, usually where resources such as nectar and foodplants are concentrated. For a butterfly like the Adonis blue *Lysandra bellargus* that 'home area' might be the slope of chalk downland, while for the wood white *Leptidea sinapis* it might be a series of rides or clearings within a wood. Species that form well-defined colonies within discrete areas are referred to as having 'closed' populations, because the majority of individuals hatch and die within the same 'home area'. Species that range more widely, or regularly migrate from one breeding area to another are referred to as having 'open' populations.

The population structure of British butterflies (determined primarily from mark–recapture studies) is shown in Table 4.1. About 80 per cent of our species are thought to form closed populations, and the great majority have been known to survive in very small areas of less than two hectares. The actual breeding habitat may be even smaller than this as butterflies are extremely selective when egg-laying. However, for a species to survive in such a small area, the habitat has to be highly suitable and it is more usual for colonies to be spread over a larger area of less suitable habitat. Only a small proportion (approximately 20 per cent) of British butterflies have closed populations with minimum breeding areas of more than two hectares. A few species, like the brown hairstreak *Thecla betulae* and purple emperor *Apatura iris*, breed at low density over wide areas, but congregate to mate at one or more prominent points; often tall trees or groups of trees (see section 2.4). The boundaries of their breeding areas are still usually distinct (although some colonies may overlap) and the populations are usually considered to be essentially closed. Where several closed populations exist in close proximity, there may be regular exchange of individuals leading

Table 4.1 The population structure of British butterflies and approximate minimum area from which a viable colony has been recorded

Minimum breeding area (ha)	Closed populations					Open or migratory populations
0.5–1	1–2	2–5	5–10	10–50	>50	
Thymelicus sylvestris	*Carterocephalus palaemon*	*Argynnis aglaja*	*Argynnis adippe*	*Papilio machaon*	*Apatura iris*	*Colias croceus*
Thymelicus lineola	*Thymelicus acteon*	*Argynnis paphia*		*Thecla betulae*		*Gonepteryx rhamni*
Hesperia comma	*Erynnis tages*	*Eurodryas aurinia*		*Lycaena dispar*		*Pieris brassicae*
Ochlodes venata	*Leptidea sinapis*	*Melitaea cinxia*		*Ladoga camilla*		*Pieris rapae*
Quercusia quercus	*Callophyrs rubi*	*Erebia epiphron*				*Pieris napi**
Satyrium w-album	*Lycaena phlaeas*	*Erebia aethiops*				*Anthocharis cardamines**
Satyrium pruni	*Aricia agestis*					*Celastrina argiolus*
Cupido minimus	*Aricia artaxerxes*					*Vanessa atalanta*
Plebejus argus	*Lysandra bellargus*					*Cynthia cardui*
Polyommatus icarus	*Maculinea arion*					*Aglais urticae*
Lysandra coridon	*Boloria selene*					*Nymphalis polychloros*
Hamearis lucina	*Boloria euphrosyne*					*Inachis io*
Mellicta athalia	*Pararge aegeria*					*Polygonia c-album*
Melanargia galathea	*Lasiommata megera*					
Maniola jurtina	*Hipparchia semele*					
Coenonympha pamphilus	*Pyronia tithonus*					
	Coenonympha tullia					
	Aphantopus hyperantus					
Total						
16	18	6	1	4	1	13

* Species which tend to have a closed population structure in northern and upland Britain.
Updated from Thomas (1984).

to the formation of a metapopulation. The latter may also be formed when there is a complex of habitat patches which are not all occupied at the same time, and within which there is periodic extinction and recolonization. The dynamics of metapopulations are crucial to the long-term survival of a species and are particularly relevant now that many formerly continuous populations have been split by habitat fragmentation (Hanski and Gilpin 1991). However, the concept of the metapopulation is relatively recent and has been examined in few British species. Perhaps the most obvious examples are amongst sedentary species which occupy ephemeral habitats, such as the heath fritillary *Mellicta athalia* (see section 11.4.4; Warren 1987*b*, 1991), and the marsh fritillary *Eurodryas aurinia*, which in some regions seems to have more or less permanent 'reservoir' populations and temporary 'satellite' ones that may shift over a period of several years (Warren 1990). The population structure of certain other species (e.g. green-veined white *Pieris napi* and orange-tip *Anthocharis cardamines*) appears to vary across the country. Towards the edge of their ranges in northern Britain, and at high altitudes, they tend to form closed populations in restricted areas (Courtney 1980), but further south (and elsewhere in Europe) they seem to be more mobile and form predominantly open populations (Dennis 1982*a*; Dempster 1989).

Of the British butterflies 13 species form open populations, having highly mobile adults that regularly move between breeding areas (see Table 4.1). Among these are some of our more familiar garden butterflies, including nettle-feeders like the small tortoiseshell *Aglais urticae* and peacock *Inachis io*, and the two 'pest' species, large white *Pieris brassicae*, and small white *Pieris rapae*. Also included in this group are the truly migratory species (red admiral *Vanessa atalanta*, painted lady *Cynthia cardui*, and clouded yellow *Colias croceus*) which regularly fly across to Britain from mainland Europe. Although they may breed here successfully during the summer, normally unable to survive our winters, they rely on continual replenishment from overseas. A much rarer butterfly, the large tortoiseshell *Nymphalis polychloros*, now appears to fall into this category as reported sightings are very sporadic. Established colonies have been recorded in Britain in the past, but in recent years there have been very few reports of regular breeding on the same site year after year.

4.3.2 Adult residence time

In the wild few adult butterflies live to their maximum possible age. Mark–recapture data has revealed that the average residence-time for most species ranges between five and 10 days, and is sometimes as little as two or three days (Table 4.2). Butterflies therefore have to be highly efficient in order to produce sufficient progeny for the next generation. The estimation of butterfly residence times is fraught with difficulty. Several statistical methods are available for estimating survival rates from mark–recapture data (see Begon 1979). Another major problem with mark–recapture data is the difficulty of separating losses of adults due to death from those due to emigration (see chapter 6). The term residence time has therefore been preferred to lifespan. For many species the difference may be small (i.e. if there is little emigration), but this cannot usually be inferred from the data.

A useful method for calculating the residence (survival) rate is the construction of recapture-duration plots described by Watt and his co-workers (1977). Examples of plots are given in Fig. 4.1 for *Leptidea sinapis* and the meadow brown *Maniola jurtina*. They show that, in common with most species, the loss rate of adults (e.g. from predation or migration) is fairly constant until they reach a certain age (8–10 days for *L. sinapis*). The loss rate then increases, suggesting the onset of senescence when adults begin to die faster through 'old age'.

Despite the short residence times of most adults in the population, a few individuals survive for much longer than average. The maximum residence times recorded for the majority of non-overwintering British species is between 20 and 30 days in the wild (Table 4.2). This figure probably indicates their maximum possible lifespan, as they rarely live longer in captivity. The precise reasons why so few adults survive to their maximum lifespan are largely unknown, but predation probably accounts for a large proportion of premature deaths (see section 4.5.3). Poor adult survival may also be caused by increased flight activity which may lead to faster wing deterioration and a more rapid exhaustion of energy reserves (see Brown and Chippendale, 1974;

Table 4.2 The average residence times of adult British butterflies calculated from mark-recapture data

Species	Sex	Average residence time (days)	Maximum residence time (days)	Reference
Thymelicus acteon	Both	6–7	—	Thomas (1983*b*)
Hesperia comma	Both	8	—	Thomas *et al*. (1986)
Leptidea sinapis	Male	8.2–9.8	27	Warren *et al*. (1986)
	Female	8.0–10.6	22	Warren *et al*. (1986)
Anthocharis cardamines	Male	4.5–8.3	—	Courtney and Duggan (1983)
Satyrium w-album	Male	5.4	21	Davies (1985)
	Female	8.4	13	Davies (1985)
Cupido minimus	Both	4.5	13	Coulthard (1982)
	Male	5.2–13.6	—	Morton (1985)
	Female	4.3–16.4	—	Morton (1985)
Plebejus argus	Male	3.0	25	Read (1985)
	Female	3.4	17	Read (1985)
Polyommatus icarus	Both	5.4	12	Dowdeswell *et al*. (1940)
	Both	3.4	—	Parr *et al*. (1968)
Lysandra coridon	Male	6.6	20	Davies *et al*. (1958)
	Female	4.7	13	Davies *et al*. (1958)
Lysandra bellargus	Male	4.2–9.5	15	Davies *et al*. (1958)
	Female	10.6–12.0	13	Davies *et al*. (1958)
Maculinea arion	Both	3–4	—	Thomas, J. A. (unpublished data)
Eurodryas aurinia	Male	4.1–5.1	14	Porter (1981)
	Female	3.1–3.8	10	Porter (1981)
	Male	8.9	20	Warren, M. S. (unpublished data)
	Female	5.8	14	Warren, M. S. (unpublished data)
Mellicta athalia	Male	2.7–6.0	19	Warren (1987*b*)
	Female	2.3–10.8	14	Warren (1987*b*)
Pararge aegeria	Male	7–11	23	Shreeve, T. G. (unpublished data)
	Female	5–8	27	Shreeve, T. G. (unpublished data)
Lasiommata megera	Both	4.2	—	Parr *et al*. (1968)
	Male	3.7	14	Dennis and Bramley (1985)
Erebia epiphron	Both	1.5	4	Porter, K. (unpublished data)
Maniola jurtina	Both	3.5–8.6	17	Dowdeswell *et al*. (1949)
	Male	8.6	28	Pollard (1981)
	Female	12.0	28	Pollard (1981)
	Both	6.2	—	Tudor and Parkin (1979)
	Both	3.3–12.0	22	Brakefield (1982*b*)
	Male	9.7	26	Dennis, R. L. H. (unpublished data)
	Female	7.2	24	Dennis, R. L. H. (unpublished data)
Coenonympha tullia	Both	3.3	9	Turner (1963)

For explanation of residence time, see Glossary.

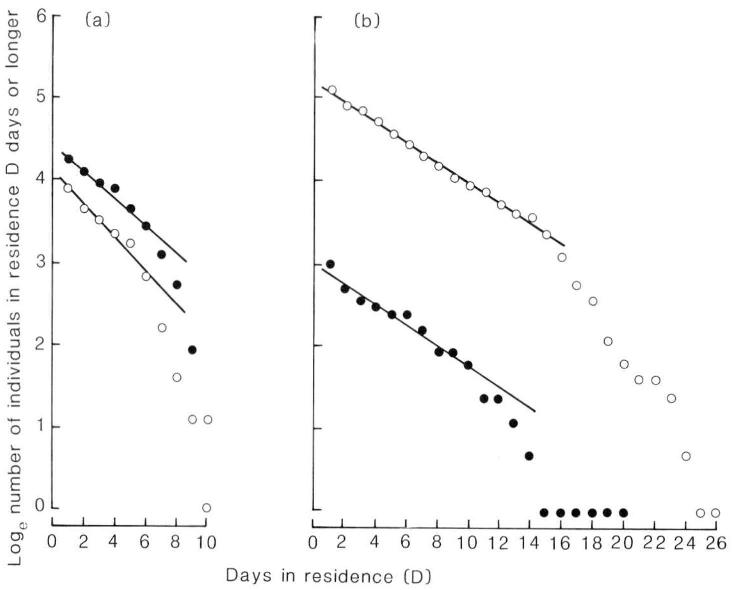

Fig. 4.1 Recapture-duration decay plots for the meadow brown *Maniola jurtina* and the wood white *Leptidea sinapis*. (a) *M. jurtina* at Hightown, Liverpool, 1976. (After Brakefield 1982*b*.) (b) *L. sinapis* at Yardley Chase, Northamptonshire, 1979. (After Warren *et al*. 1986.) Open circles, males; closed circles, females. Lines indicate initial decay rate which is used to calculate mean residence rate. (Courtesy of *Journal of Animal Ecology*, Blackwell Scientific Publications.)

Watt *et al*. 1974). Brakefield (1982*b*) suggests that this may have been the cause of lower survival rates of *Maniola jurtina* adults in the hot summer of 1976, compared to the following two cool summers.

A few of our butterflies have a completely different life cycle and overwinter as adults. Consequently, they must survive for far longer periods than those species with brief periods of adult emergence. Species such as *Aglais urticae*, *Inachis io* and the comma *Polygonia c-album* all hibernate as adults and regularly live for periods of eight to nine months. The record for average lifespans in the adult stage is probably held by the brimstone *Gonepteryx rhamni* which can survive for periods of 10 months from August one year until June or even July in the next. However, the proportion of the autumn population persisting this long is unknown.

4.3.3 Phenology and life history strategies

Butterflies have evolved numerous different life history strategies in order to maximize their reproductive potential (Table 4.3). Firstly, it is crucial that the appearance of each stage coincides with a limited part of the season when the correct resources are available. Adults must fly when the weather is warm enough for them to be active, and when sufficient nectar resources and egg-laying sites are available. Results from the Butterfly Monitoring Scheme show that, for many species, the timing of the flight period can vary by about two to three weeks between years, and that there is often a similar gap between the appearance of populations in the south of the country, compared with those in the north (Pollard *et al*. 1986). The main cause of this variation appears to be climatic and, in general, a warmer than average spring will lead to earlier emergence. Brakefield (1987*d*) has shown that variation in the emergence between years of *Maniola jurtina* and the gatekeeper *Pyronia tithonus* is related to the temperature in the summer months, particularly June. In contrast, the flight periods of a few species are fairly well synchronized across Britain so that they all emerge over the same period. This phenomenon occurs in the large skipper *Ochlodes venata*, small skipper *Thymelicus sylvestris*, dingy skipper *Erynnis tages* and to a lesser extent *P. tithonus*. For these species, it seems likely that the timing of emergence is related to day-length which shows far less variation at specific locations throughout the country than does temperature. Larvae must also appear when there are sufficiently nutritious food sources (high leaf water content and nitrogen), and the correct stage in the life cycle must be reached to be able to overwinter successfully. Most species can become dormant (diapause) in only one stage, which is often

Table 4.3 Life history strategies of butterflies in the British Isles

Species	Overwintering stage (E: egg; L: larva; P: pupa; A: adult)	Number of complete generations per year (P: partial)	Geographic variation in number of generations (P: partial)
HESPERIIDAE			
Carterocephalus palaemon	L	1	
Thymelicus sylvestris	L	1	
Thymelicus lineola	E	1	
Thymelicus acteon	L	1	
Hesperia comma	E	1	
Ochlodes venata	L	1	
Erynnis tages	L	1 (P2)	P2 in south only
Pyrgus malvae	P	1 (P2)	P2 in south only
PAPILIONIDAE			
Papilio machaon	P	1 (P2)	
PERIDAE			
Leptidea sinapis	P	1 (P2)	
Colias croceus	L (do not survive)	2+	Depends on migration
Gonepteryx rhamni	A	1	
Aporia crataegi†	L	1	
Pieris brassicae	P	2 (P3)	P3 in south only
Pieris rapae	P	2 (P3)	P3 in south only
Pieris napi	P	1–2 (P3)	1 generally in northern uplands
LYCAENIDAE			
Callophrys rubi	P	1	
Thecla betulae	E	1	
Quercusia quercus	E	1	
Satyrium pruni	E	1	
Satyrium w-album	E	1	
Lycaena phlaeas	L	2–3 (P4)	2 in north and cool years
Lycaena dispar	L	1	
Cupido minimus	L	1 (P2)	P2 in south
Plebejus argus	E	1	
Aricia agestis	L	2	
Aricia artaxerxes	L	1	
Polyommatus icarus	L	1–2 (P3)	1 in north and Ireland P3 in south
Lysandra coridon	E	1	
Lysandra bellargus	L	2	
Cyaniris semiargus†	L	1 (P2?)	
Celastrina argiolus	L	1–2 (P3)	1 in Ireland usually P3 in south
Maculinea arion	L	1	
Hamearis lucina	P	1	

Table 4.3 (*cont.*)

Species	Overwintering stage (E: egg; L: larva; P: pupa; A: adult)	Number of complete generations per year (P: partial)	Geographic variation in number of generations
NYMPHALIDAE			
Ladoga camilla	L	1	
Apatura iris	L	1	
Vanessa atalanta	A (do not survive)	1–2+	Depends on migration
Cynthia cardui	— (do not survive)	1–2+	Depends on migration
Aglais urticae	A	1–2	1 in north
Nymphalis polychloros	A	1	
Inachis io	A	1 (P2)	P2 in south
Polygonia c-album	A	1 (P2)	not known
Boloria selene	L	1 (P2)	P2 in south-west
Boloria euphrosyne	L	1 (P2)	P2 in south-west
Argynnis adippe	E	1	
Argynnis aglaja	L	1	
Argynnis paphia	L	1	
Eurodryas aurinia	L	1	
Melitaea cinxia	L	1	
Mellicta athalia	L	1	
Pararge aegeria	L/P	2–3 (P4)	variable throughout
Lasiommata megera	L	2 (P3)	P3 in south
Erebia epiphron	L	1*	
Erebia aethiops	L	1	
Melanargia galathea	L	1	
Hipparchia semele	L	1	
Pyronia tithonus	L	1	
Maniola jurtina	L	1	extended on chalk
Aphantopus hyperantus	L	1	
Coenonympha pamphilus	L	1–2	1 in north
Coenonympha tullia	L	1*	

* Has the ability to take two years to complete a single generation under simulated natural conditions in captivity.
† No longer in Britain.
Note: the overwinter survival of migratory species probably depends on the severity of weather conditions (see Table 6.1).

consistent within each subfamily or even family group. For example the majority of Satyrinae (browns) overwinter as larvae, the Theclinae (hairstreaks) as eggs, and the Lycaenidae (blues and coppers) and Argynnini (larger fritillaries) as larvae, or more rarely, eggs. Among British species, only the speckled wood *Pararge aegeria* has the ability to overwinter in two different stages, either as larvae or pupae. The mechanisms controlling this and other life history variations are discussed in section 8.6.

Secondly, butterflies have developed different strategies with respect to the number of generations per year (Table 4.3). Some species are univoltine (one generation a year), some are bivoltine (two generations a year), and others are multivoltine (several consecutive generations a year). As with the overwintering stage, voltinism tends to be consistent within related groups. For example, the Hesperiinae (skippers) and Theclinae are both always single brooded throughout Britain and the rest of Europe. However, the number of generations is variable in many species, and tends to increase further south. Thus, the small copper *Lycaena phlaeas* has two generations in northern England and three in southern England, except in very cool years. Many other species have a partial second or third generation in the more southerly or south-westerly parts of their range. In most cases the proportion of the first generation offspring that develops into the second is genetically determined, although it is usually triggered by factors such as climate. These species usually have two or more complete generations further south in Europe, but this varies with altitude as well as latitude. In Britain, the partial second generation does not always have time to develop successfully (see Warren 1984), especially in bad years, and so does not confer an advantage on those populations. A few species, such as *Lysandra bellargus*, always have two full generations throughout Britain and Europe. It is possible that a few northern species may have the ability to take two years to complete their life cycle, but this has never been confirmed in the wild in Britain. *Erebia epiphron* and *Coenonympha tullia* can be reared on a two-year cycle in captivity (Wheeler 1982; Melling 1987), and it would seem a useful adaptation in extreme climatic zones.

Shapiro (1975*a*) has observed that multivoltinism in butterflies is usually correlated with a high intrinsic rate of increase (an *r*-selected strategy), high mobility, association with disturbed habitats, wide-ranging geographical distribution, and overall good colonizing ability. In contrast univoltinism tends to be correlated with a comparatively low intrinsic rate of increase (a *K*-selected strategy), low mobility, association with constant habitats, more restricted geographical distribution, and poor colonizing ability. He suggests that a strategy of univoltinism may therefore have arisen from selection in a short growing-season at high latitudes or altitudes, or as part of an overall *K*-strategy evolved in environments with a long growing-season. As with most generalizations, a few species do not seem to fit (such as the highly mobile, batch laying, but single brooded *Nymphalis polychloros*), but it does seem to hold reasonably well for most British species.

A third aspect of life history strategy is the duration of each developmental stage. The duration of the adult flight period is highly variable in some species, while more or less fixed in others. For example, those of *Aricia artaxerxes*, *Maniola jurtina* and the heath fritillary *Mellicta athalia* are highly variable geographically (Thomson 1980; Brakefield 1987*d*; Warren 1987*a*). The length of the flight period of *M. jurtina* is also dependent on habitat, being far longer in chalk downland than in other grasslands (Pollard *et al*. 1986). The duration of the larval period can be highly variable, as in *Pararge aegeria* which can have four or five larval instars (Shreeve 1986*b*). The mechanisms that control these and other life history variations are discussed further in chapter 8.

Each of the different strategies confers advantages and disadvantages, but, as these vary from species to species, it is difficult to draw general conclusions. Each strategy has presumably been evolved to minimize the selective pressure caused by factors such as resource availability, competition, predation and parasitism. However, many of these factors can be overcome without affecting the life cycle, for example by the use of poisonous chemicals to avoid predators or larval webs to reduce parasite attacks (see Chapter 5). The benefits of each strategy are consequently highly complex, and a full understanding of their significance and evolution requires far more research.

4.4 Natural population fluctuations

4.4.1 Fluctuations, trends, and carrying capacity

Butterflies have long been noted for their enormous fluctuations in abundance from year to year. 1976, 1983, 1984, and 1989 were widely claimed to have been good for butterflies, while 1985 and 1986 were bad. Beirne (1955) has even attempted to classify good and bad years for butterflies going back to the nineteenth century. However, accurate data on the population fluctuations of a large number of British species has only been available since the start of the Butterfly Monitoring Scheme in 1976. The national trends for the first 10 years of the scheme confirm these general observations, but indicate that the fluctuations are not consistent from species to species. Pollard *et al.* (1986) have related the fluctuations of many species to weather, but they stress that the conclusions drawn from only 10 to 14 years' data must be tentative. The most striking association is between increased numbers and warm, dry summers. There is also a frequent association between wet conditions early in the previous year and increased numbers of butterflies in the current year (Pollard 1988, 1991). The 1976 drought provided a good example of the effects of weather since the populations of many species 'crashed' the following year—probably due to desiccation of their food plants. Detailed research on individual populations has also shown that weather can be a major cause of annual fluctuations. There are important exceptions though, some of which are discussed in section 4.4.2.

The upper limit of fluctuations in the size of a population indicates the carrying capacity of the site which it inhabits. The carrying capacity is determined by the quantity of certain critical resources, such as the number of suitable egg-laying sites or the abundance of food plants. Alternatively, populations may be limited by factors such as predators and parasites, or by inter- or intraspecific competition (see Dempster 1983). However, the carrying capacity of a site is rarely constant and, when it changes, can produce long-term trends in population size. These trends are quite separate from the rapid, annual fluctuations, and are usually caused by quite different factors. They are frequently brought about by changes in habitat quality and, can be far more significant to conservation than the short-term fluctuations. Another important feature of long-term trends is that they are often influenced by the human management of a site and can therefore be manipulated, unlike the annual fluctuations that are beyond our control. Examples of long-term trends are described later in section 4.4.3 and Chapter 11.

4.4.2 Life-tables

Life-tables provide one of the most valuable methods of examining the population dynamics of an insect population. They are constructed from estimates of the mortality (death-rate) at various stages of the life cycle. Such estimates can be obtained only from laborious and painstaking observations, usually over several years, but give a unique insight into the factors affecting population size (see Dempster 1975, and Southwood 1978, for details of the methods). Most studies on butterflies have used age-specific life tables, that are constructed by recording deaths in each stage from egg to adult within successive generations. To complete the analysis, an estimate is also needed of the number of individuals entering each generation, which for butterflies can be taken from the number of either adults or eggs. The usual way of compiling a life table is to mark the position of a large number of eggs laid in the wild, and regularly visit each location (usually every few days) to determine when, and what type of, mortality occurs. Species that lay eggs singly, and have larvae that rarely wander far, are the easiest species to study in this way as it is possible to determine individual mortalities. To date no complete life tables have been constructed for batch layers. An example of a typical life table is shown in Table 4.4.

Life-tables reveal two important characteristics of butterfly populations, a high birth-rate and a high death-rate. For *Leptidea sinapis*, the mortality within each developmental stage (including each larval instar) varied from 18–57 per cent (Table 4.4). The total mortality from egg to adult emergence during the study was 93–97 per cent. Similar large losses occur in every butterfly species, and can be withstood because of the large number of eggs laid. However, the average number of eggs laid by females (fecundity) can also fluctuate from year to year and has to be included in the life table analysis. Most butterflies have a maximum potential fecundity of several hundred eggs, but this is rarely achieved in the wild (see section 4.5.1). The failure of females to lay their full complement of eggs (the egg shortfall) is therefore treated as a mortality in the analysis. The precise figure used for maximum fecundity is unimportant in itself, as it is the changes in egg shortfall from year to year that are of interest.

Life-table data can be analysed by several different methods, but the most widely used is the 'key-factor' method. This enables the recognition of the effects of variation in fecundity and mortality, and the key-

Table 4.4 Life-table data for *Leptidea sinapis* (wood white) at Yardley Chase, Northamptonshire, 1977–83

Stage	1977–78	1978–79	1979–80	1980–81	1981–82	1982–83	1983–84
Adult index	1453	1550	658	875	623	1282	903
Number of eggs	185	201	155	77	171	181	215
	(31.4)	(44.7)	(36.0)	(54.5)	(38.6)	(59.1)	
First instar	127	111	114	35	105	74	
	(43.3)	(50.5)	(33.3)	(37.1)	(30.5)	(31.1)	
Second instar	72	55	76	22	73	51	
	(27.8)	(34.5)	(18.4)	(45.4)	(21.9)	(21.6)	
Third instar	52	36	62	12	57	40	
	(51.9)	(50.0)	(56.5)	(41.7)	(45.6)	(22.5)	
Fourth instar	25	18	27	7	31	31	
	(36.0)	(27.8)	(29.6)				
Fourth instar (wandering phase)	16	13	19				
Number of pupa observed	14	16	12	———	Not assessed	———	———
	(57.1)	(56.3)	(41.7)				
Number of adults emerged	6	7	7				
Total generation mortality (per cent)	96.3	97.2	92.8				

After Warren *et al.* (1986). Percentage mortality in each developmental stage is shown in parentheses.

factor (or factors) that determines the annual fluctuations in population size. The method is only applicable to species, like *Leptidea sinapis* illustrated in Fig. 4.2, with one generation per year. The logarithm of the annual mortality (death-rate) in each developmental stage must be plotted for successive years alongside the total generation mortality (K), which is the sum of each constituent mortality (e.g. k_1 = egg mortality, k_2 = first instar mortality, etc.). For convenience, total generation mortality is usually measured from the adult stage, but the egg stage may be used if accurate estimates of adults cannot be obtained (as in the case of studies on the *Thecla betulae* and white admiral *Ladoga camilla* by Thomas 1974, and Pollard 1979*b*, respectively). Total generation mortality can alternatively be expressed by the logarithm of the changes in adult abundance from one year to the next. This is particularly useful when the life table data are incomplete; as with the study of *L. sinapis* in Fig. 4.2, where pupal mortalities were not estimated every year and egg shortfall (k_0) was initially estimated by an indirect method.

The identification of key factors in a life table analysis is achieved simply by a visual examination of the plots of the k-values for each successive generation. Any correlation between the fluctuations in a series of k-values and the total generation mortality (K) suggests that this is one of the key factors. A more precise method is to calculate correlation coefficients for each variable (see Southwood 1978). In the example of *Leptidea sinapis*, the key factors that appear to be responsible for the annual fluctuations in adult numbers are the shortfall of egg production (k_0), and the mortality of eggs (k_1) and young larvae (k_2 and k_3). It is interesting to note that pupal mortality, although very high (42–57 per cent) does not seem to be responsible for the annual changes since it varied little during the three years when it was estimated.

The results of the life table studies carried out in Britain are summarized in Table 4.5. For three of the seven species where key factors were determined, annual fluctuations (in either egg or adult numbers) were thought to be caused by the shortfall in egg production. In the remaining species, the key-factors

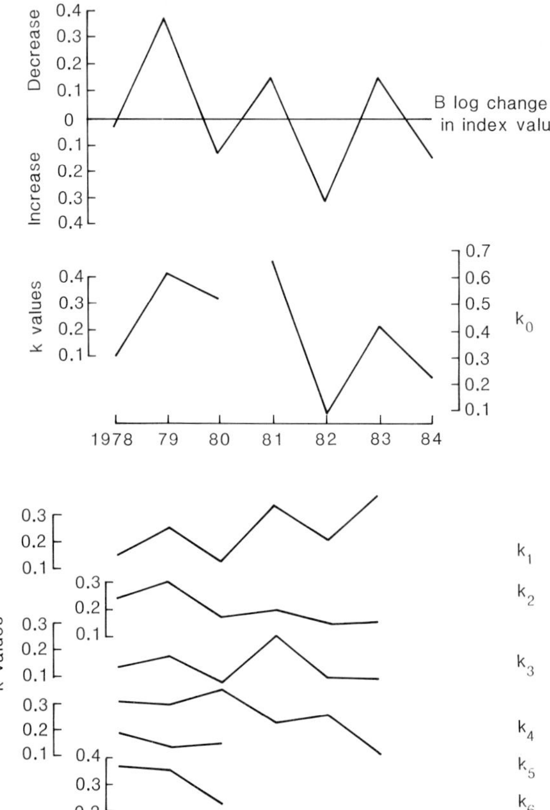

Fig. 4.2 Partial key-factor analysis of life table data for the wood white *Leptidea sinapis*. (Data from Warren *et al*. 1986.) Shortfall of eggs k_0 estimated by two distinct methods and therefore shown as two separated lines; k_1, egg mortality; k_2 to k_5, larval mortality; k_6 pupal mortality. Fourth instar larval mortality and pupal mortality not assessed from 1981–84. *B* is the logarithm of the annual change in the index of adult abundance, which is equivalent to a total *K* of a conventional key-factor analysis (see text). (After Warren *et al*. 1986; courtesy of *Journal of Animal Ecology*, Blackwell Scientific Publications.)

causing fluctuations were mortalities at various stages in the life cycle. The factors that are actually responsible for mortality and egg shortfall are discussed in section 4.5.

The analysis of life tables can sometimes reveal the mode of action of each mortality factor. It can, for instance, indicate whether a particular cause of mortality acts in a density-dependent manner or not. Such a phenomenon might occur with predators or parasites which cause greater mortalities when their larval prey becomes more numerous. Two other causes of density-dependent mortality have been identified in a detailed study of the large blue *Maculinea arion* (Thomas 1977, 1980). When eggs become crowded on the flower-heads of thyme (where the larvae subsequently feed on developing seeds), large mortalities are caused by the cannibalistic behaviour of the young larvae. A similar phenomenon occurs with *Anthocharis cardamines* (Courtney 1981). Also, when *M. arion* larvae enter their parasitic phase within red ants' nests, mortalities due to starvation increase if more than one larva enters each nest (Fig. 4.3). If five or more enter a single nest, they usually all die as there are rarely enough ant grubs to support them. Consequently, in years of abundance, most ants' nests become overcrowded with larvae and few will survive to become adults. Exceptionally high population densities can also lead to mass starvation in certain other butterflies. Populations of the marsh fritillary *Eurodryas aurinia*, for example, have been known to reach such numbers that the larvae eat all available food plants (Ford 1945).

Other mortality factors may demonstrate delayed or inverse density-dependence (i.e. the proportion killed decreases at high density), or may operate quite independently of population density. The latter applies to many of the key-factors so far identified

Table 4.5 Summary of life table studies on butterflies in Britain showing average percentage mortality per stage (range in parentheses)

Species	Locality	Number of generations studied	Average percentage mortality per stage						Pupa	Total	Key factor stage	Reference
			Egg	I	II	III (larval instars)	IV	V				
Papilio machaon	Hickling, Norfolk	2	10 (7–14)	31 (28–34)	24 (24)	25 (15–35)	47 (40–54)	—	92 (1 gen)	99	—	Dempster *et al* (1976)
Leptidea sinapis	Yardley Chase, Northants	6	44 (31–59)	30 (30–50)	24 (18–45)	45 (22–56)	31 (28–36)	— —	52 (42–57)	95 (93–97)	Adult fecundity (young larvae)	Warren *et al*. (1986)
Pieris rapae	Monks Wood, Cambs.	3	13 (11–16)	49 (38–61)	21 (15–23)	20 (12–24)	39 (30–52)	42 (35–55)	—	>91 (>88–92)	—	Dempster (1967)
Anthocharis cardamines	Durham	3	10	50	16	39	7	0	45 (31–68)	80	Adult fecundity?	Courtney and Duggan (1983)
Thecla betulae	Cranleigh, Surrey	4	33 (25–45)	31 (19–40)	25 (21–28)	36 (32–39)	32 (25–38)	—	63 (50–77)	94 (88–97)	Pupa + adult fecundity	Thomas (1974)
Satyrium pruni	Monks Wood, Cambs.	2	59 (54–64)	48 (47–48)	34 (19–50)	39 (22–57)	58 (44–71)	—	68 (50–86)	99 (98–99)	?Late larvae + pupae	Thomas (1974)
Lycaena dispar batavus	Woodwalton Fen, Cambs.	6	Egg . . . . spring emergence: 96 (88–98)			spring emergence . . . . pupation: 85 (62–100)				>99 (98–100)	Young larvae	Duffey (1968)
Lycaena phlaeas	Holme Fen, Cambs.	2	Egg to hibernation as I–III instar larvae: 82				—	—	—	—	—	Dempster (1971*b*)
Maculinea arion	Site 'x'	6	—	—	—	—	—	—	—	—	Late larvae in ants' nests	Thomas (1976)
Ladoga camilla	Monks Wood, Cambs.	6	30 (20–50)	43 (31–78)	28 (9–58)	56 (49–69)	17 (0–27)	69 (63–78)	45 (29–67)	96 (94–99)	Late larvae + pupae	Pollard (1979*b*)

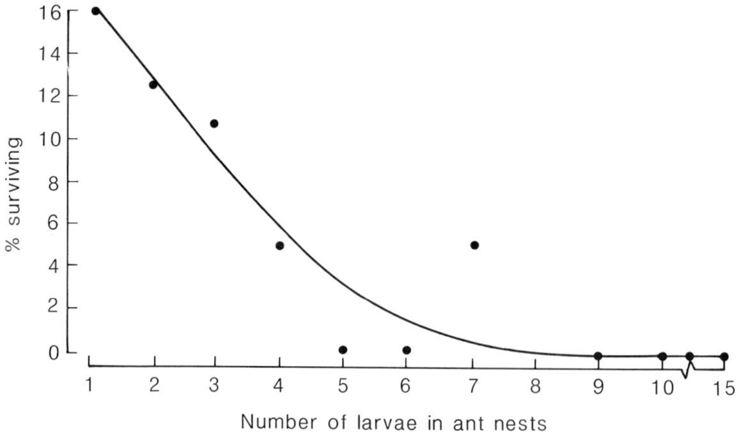

Fig. 4.3 Density-dependent mortality of the large blue *Maculinea arion* larvae due to starvation in red ants' nests. (J. A. Thomas, unpublished data.)

for butterflies, notably those affected in some way by the weather. The identification of these different density relationships can be complex and requires life table data for numerous successive generations (for details see Dempster 1975, and Southwood 1978).

4.4.3 Long-term population trends

The analysis of life tables by the key-factor method is concerned solely with the causes of annual fluctuations in population size and does not usually reveal anything about longer term trends in a population. These long-term trends are of the utmost importance to butterfly conservation as they can reflect changes in habitat suitability, which may in due course lead to extinction. The identification and analysis of long-term trends, therefore, requires a different approach.

One of the biggest problems in identifying any underlying trends in population size is that they are usually obscured by large, annual fluctuations. However, data from the Butterfly Monitoring Scheme can be used to separate national fluctuations (i.e. expected fluctuations), from local variations. If data from a site begin to depart significantly from the national, or regional patterns, changes in the carrying capacity of the site may be indicated. Examples are shown in Fig. 4.4 for three grassland species with rather different habitat requirements. On a small nature reserve in the Gomm Valley in Buckinghamshire, the small heath *Coenonympha pamphilus* and *Erynnis tages*, species which require fairly short turf, have suffered from the general lack of management which has resulted in tall vegetation. In contrast the marbled white *Melanargia galathea*, which prefers taller, unmanaged grassland, has benefited (Fig. 4.4a). On the National Nature Reserve at Old Winchester Hill in Hampshire, all three butterflies seem to have profited from the introduction of a system of rotational grazing on part of the reserve (section 7 of the transect, Fig. 4.4b). This type of management produces a range of turf heights suitable for most downland butterflies (see section 11.5). An interesting feature of both reserves is that the yearly fluctuations in abundance still reflect the national average, but slowly veer away from it.

A slightly different approach to identifying long-term trends was used in the study of *Leptidea sinapis* (Warren *et al.* 1986). No significant trends were detected in the total size of the population during the eight year study, but the distribution of the population within the site varied considerably. This was linked to the suitability of the breeding habitat within each woodland ride, which became more shaded during the study (see Fig. 4.5). The most open rides (<20 per cent shade) which initially contained low densities of *L. sinapis*, became gradually more suitable as their shade levels increased towards the predicted 'optimum' for the species (20–50 per cent shade). The rides that originally contained optimum shade levels slowly deteriorated, and one ride which was just past the

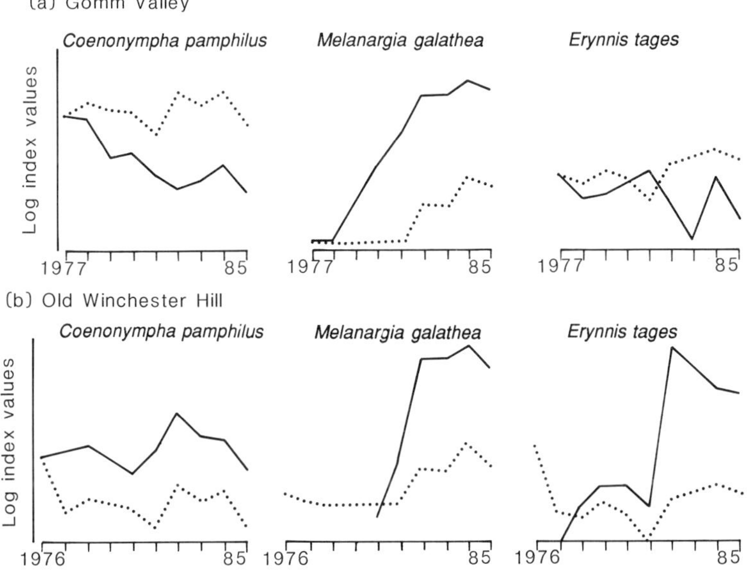

Fig. 4.4 Comparison of index values for the adult abundance of the small heath *Coenonympha pamphilus*, the marbled white *Melanargia galathea* and the dingy skipper *Erynnis tages*, at two sites with the regional or national collated index. (a) Gomm Valley, Buckinghamshire. (b) Old Winchester Hill (section 7), Hampshire. Solid lines, local index values; dotted lines, national collated index. (After Pollard *et al.* 1986; courtesy of Nature Conservancy Council.)

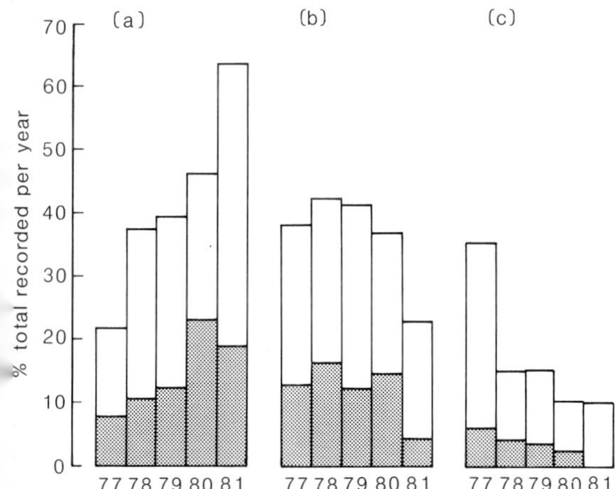

Fig. 4.5 Annual changes in the proportion of wood white *Leptidea sinapis* adults recorded in woodland rides at Yardley Chase, Northamptonshire between 1977 and 1981. Stippled portions, females; open portions, males. (After Warren 1985*a*; courtesy of *Biological Conservation* and Elsevier Applied Sciences.) (a) Five rides with less than 20 per cent shade (pre-optimum). (b) Two rides with 20–50 per cent shade (optimum). (c) One ride with 53 per cent shade (post-optimum).

optimum at the start of the study deteriorated very rapidly. When some of the shadier rides were opened up through clearance of their scrub margins, or thinning of adjacent conifer crops, these downward trends were reversed and *L. sinapis* numbers increased. Thus, in contrast to the annual fluctuations, long-term trends in *L. sinapis* populations are more likely to be caused by changes in the levels of shade in its ride habitats.

The effects of management on butterfly populations can occasionally be very large and rapid. In the case of woodland colonies of *Mellicta athalia*, where the butterfly breeds only in newly cut areas, these effects can completely dominate the annual fluctuations in abundance (Warren 1987*c*). Between 1980 and 1984, a total of 38 populations were monitored in coppiced woodland (the Blean Woods) in Kent. Over this period 16 died out when their habitat became too shaded, and 10 new ones were established naturally in fresh clearings. On average, population size reached a peak between two and three years after a stand of woodland was coppiced, and then declined rapidly as the coppice re-grew (see Fig. 4.6). In vigorous chestnut coppice, all the colonies became

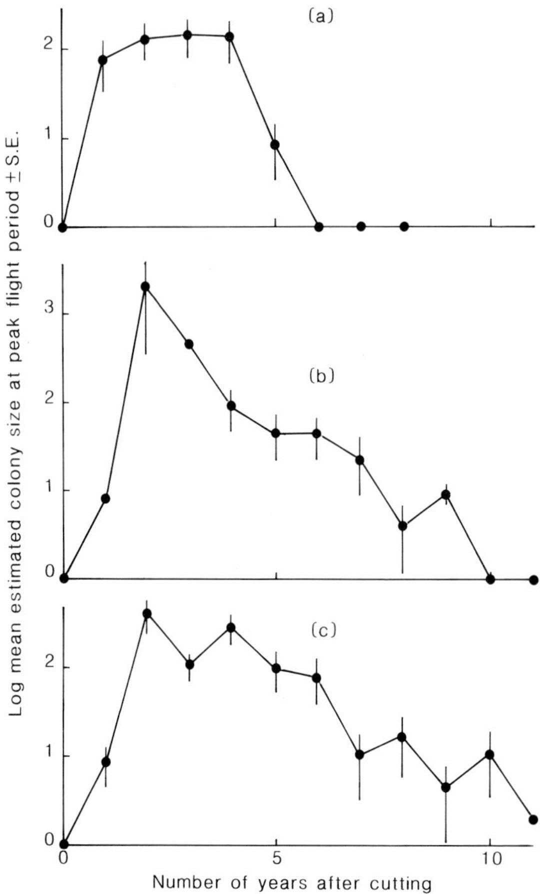

Fig. 4.6 Annual changes in the average size of heath fritillary *Mellicta athalia* colonies following cutting or clearance in three different types of woodland habitat in Kent, 1980–84. Note use of logarithmic scale on vertical axis. (S.E., standard error). (After Warren 1987*c*; courtesy of *Journal of Applied Ecology* and the British Ecological Society.) (a) Vigorous coppice. (b) Sparse or slow growing coppice. (c) Conifers with high forest edge.

extinct six years after cutting, by which time the tree canopy had completely closed, shading the larval foodplant (in this habitat, common cow-wheat *Melampyrum pratense*). In less vigorous coppice (often a mixture of chestnut and other slower growing trees) colonies survived a little longer, but all became extinct within 10 years. In young conifer plantations (on deciduous woodland sites) the butterfly thrived in the early stages when deciduous woodland was felled for replanting. However, the larval foodplant quickly dies out within plantations and becomes restricted to adjacent deciduous margins, where small populations survived for as long as 11 years. These rapid responses to habitat changes have serious, long-term implications for the conservation of this rare species which are discussed in Chapter 11.

4.5 Mortality factors affecting population size

4.5.1 Fecundity

In the wild, butterflies seldom (if ever) lay their full complement of eggs, and the egg shortfall is considered to be an important mortality factor when studying population dynamics. The variation between individuals has already been discussed in chapter 3, and this section is concerned only with average egg-laying performance of females within a population in different years. Estimates of fecundity in British butterflies, shown in Table 4.6 indicate tha

Table 4.6 Fecundity estimates for wild populations of British butterflies (Note: all species shown here lay their eggs singly)

Species	Period of study	Estimated fecundity (Eggs/female)	Maximum fecundity (Eggs/female)	Reference
Leptidea sinapis	1977–79	30–63	*c*.200?	Warren (1981)
Anthocharis cardamines	1977–80	6–23	*c*.150?	Courtney and Duggan (1983)
Lycaena dispar batavus	1960–63 1969–73	5–100 (av. 56)	*c*.600	Duffey (1968, 1977)
Thecla betulae	1970–76	12–64 (av. 28)	145	Thomas (1974) (and unpublished data)
Maculinea arion	1972–77	20–73 (av. 42)	378	Thomas, J. A. (unpublished data)
Maniola jurtina	1976	66	—	Brakefield (1982*b*)

there can be fourfold differences in the average number of eggs laid per female. A major factor determining the rate of egg-laying is the weather, as adults are active only during warm, sunny conditions (see section 2.2). In the study of *Leptidea sinapis*, observations on a sample length of woodland ride indicated that the number of eggs laid each day was related to the daily maximum temperature (Warren 1981). No eggs were laid when the temperature was below 16°C, and egg-laying increased steadily within the observed range 16–24°C. An independent analysis showed that the fluctuations in the size of the population were inversely correlated with the mean daily rainfall during the egg-laying period (Warren *et al*. 1986). Nevertheless, a number of weather factors probably combine to influence the fecundity of this species.

Poor weather has been shown to reduce fecundity in several other British butterflies, notably *Anthocharis cardamines* and *Thecla betulae* (Courtney and Duggan 1983; Thomas 1974 respectively). It is probably no coincidence that all these species lay their eggs singly, and therefore females spend much of their lives looking for suitable egg-laying sites. The fecundity of batch-laying species, like the *Euryas aurinia* and *Mellicta athalia*, seems to be far less affected by poor weather (unless this is prolonged) as each female needs only a short spell of good weather to lay a hundred or more eggs (Warren 1987*c*). Butterflies that lay eggs singly rarely achieve this figure in a whole lifetime (Table 4.6). Fecundity estimates are needed for many more species to test this hypothesis properly. Another subject requiring further research is the susceptibility of butterfly populations to weather variations near the edges of their geographic distributions. Many species reach the northern limit of their range in Britain (see chapter 1) and weather conditions may well be marginal for their survival unless habitat conditions are ideal. The Butterfly Monitoring Scheme has provided some evidence that yearly fluctuations in abundance tend to be greater in northern Britain (Pollard *et al*. 1986), and it would be fascinating if this were proved true for the majority of species (Fig. 4.7). If so, it may explain why so many butterflies appear to have such demanding and specific habitat requirements in Britain compared to the rest of Europe (see section 10.4).

Fecundity in butterflies may also be affected by the density of suitable egg-laying sites, although this is difficult (or impossible) to detect in population studies. When suitable egg-laying sites are scarce (e.g. if the carrying capacity of a site becomes reduced), females (and possibly males) tend to respond by flying off in search of better areas. This can usually be inferred if mark–recapture studies reveal that the average residence time on a site is unpredictably short. In the study of *Mellicta athalia*, the residence times of adults in an overgrown coppiced site, where conditions had almost become unsuitable, were far shorter than on a highly suitable site (Warren 1987*b*). Unfortunately, most other

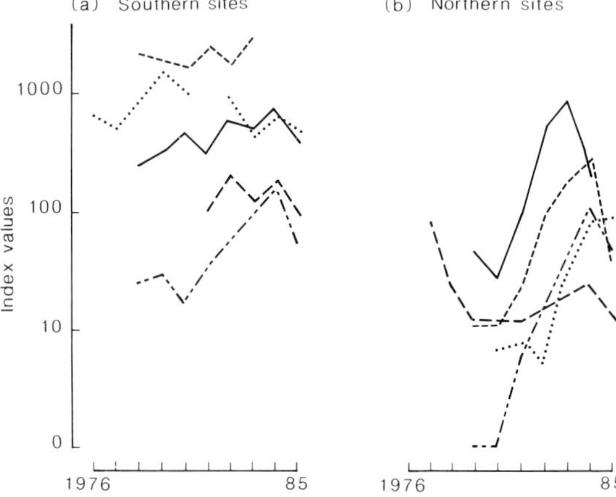

Fig. 4.7 Fluctuations in index values of the meadow brown *Maniola jurtina* at five sites in southern England (a) and five sites in Scotland (b), selected at random. In general, fluctuations have been much greater in the north of Britain. (After Pollard *et al*. 1986; courtesy of Nature Conservancy Council.)

population studies have been conducted in highly suitable habitats which may not be representative of the majority of sites where conditions are far from ideal.

4.5.2 The pattern of survival in the developmental stages

Eggs

Studies of two cluster-laying species, *Eurodryas aurinia* and *Mellicta athalia*, have revealed extremely low mortalities, often due to infertility (Porter 1981; Warren 1987*a*). In the multivoltine cluster-laying *Aglais urticae*, Pullin (1986*a*) found that egg mortality can differ between seasons, and suggests that this may have contributed to its development of a variable life history strategy. Over a six year period, Pollard (1979*b*) recorded egg losses of 20–50 per cent in *Ladoga camilla*, a species that lays eggs singly. Poor survival has been found in other single-egg layers including *Thecla betulae*, and black hairstreak *Satyrium pruni* which overwinter in the egg stage (Thomas 1974) and *Leptidea sinapis* (Warren *et al*. 1986; see Table 4.5). Egg mortality was also variable in a study of the large copper *Lycaena dispar batavus* in Holland, ranging from 0–79 per cent (Bink 1972). In *Pieris rapae*, egg mortality is usually low, particularly in the absence of birds. Richards (1940) recorded approximately five per cent mortality in open fields where bird activity was low and Dempster (1967) recorded 11–16 per cent mortality. Equally low mortality rates of 7–14 per cent have been found in eggs of the swallowtail *Papilio machaon* (Dempster *et al*. 1976).

Larvae

In the majority of butterflies studied, the heaviest mortalities occurred in the larval stage (see Table 4.5). On average, overall larval mortalities have been estimated to be 88 per cent in *Papilio machaon* (Dempster *et al*. 1976), 80 per cent in *Leptidea sinapis*, 89 per cent in *Pieris rapae* (Dempster 1967), 24 per cent in *Anthocharis cardamines* (Courtney and Duggan 1983), 76 per cent in *Thecla betulae* (Thomas 1974), 91 per cent in *Satyrium pruni*, 84 per cent in *Lycaena dispar batavus* (Bink 1972) and 95 per cent in *Ladoga camilla* (Pollard 1979*b*). These mortalities are usually spread fairly evenly between the larval instars, with the exception of *L. camilla*, which suffers particularly heavy predation by birds in later instars (Pollard 1979*b*). However, the average figures quoted in the above studies should only be taken as a guide as mortalities are expected to vary from site to site. In particular, the level of mortality may be related to the density of larvae, leading to substantial variation between years and between sites.

Pupae

Quantitative data on the survival rates of pupae are scarce, principally because for most species this stage is the hardest to locate in the field. The evidence available suggests that pupal mortality is usually very high, although the figures have often been obtained by placing caged-reared pupae in 'natural' situations in the wild. *Papilio machaon*, *Leptidea sinapis* and *Anthocharis cardamines* pupae are suspended from reed or grass stems 0.5–2 m above the ground and experience 92, 42–57 and 31–68 per cent mortality respectively (Dempster *et al*. 1976; Courtney and Duggan 1983; Warren *et al*. 1986). *Ladoga camilla* pupae, which usually hang from twigs and branches slightly higher above the ground, suffer 29–67 per cent mortality (Pollard 1979*b*). The pupae of the two lycaenids *Thecla betulae* and *Satyrium pruni*, which are attached to their food plant, blackthorn *Prunus spinosa* have similar mortalities ranging from 50–77 and 50–86 per cent, respectively (Thomas 1974).

4.5.3 Natural enemies of butterflies

The natural enemies of butterflies fall into three main categories: predators, parasites, and disease. At all stages of their development, butterflies have developed defence mechanisms against these other organisms (see chapter 5). A large proportion of deaths in the young stages of most butterflies are caused by predators; for example, in *Leptidea sinapis* they accounted for 70–80 per cent of all deaths (Warren *et al*. 1986). It is often extremely difficult to determine which species of predator is involved and predation can usually only be inferred when an individual egg or larva under study disappears between successive visits. This may be inaccurate since a few individuals may drop to the ground or wander off the foodplant, although these are rare occurrences in the few species studied to date.

Some predators conveniently leave remains of their prey, and occasionally these are so characteristic that the type of predator can be identified with reasonable accuracy. Egg predators often leave a jagged edge of the base still attached, quite unlike the egg remains of a successfully hatched larva which either makes just a small exit hole in the egg case, or consumes it completely. Pupal remains are often characteristic of particular predators (Frank 1967; Thomas 1974). The identification of prey remains usually relies on observation of various predators in captivity, and cannot be taken as absolute proof.

A test for identifying the predators of insects has been developed by Dempster (1960). This method, known as the precipitin test, requires taking a sample of predators operating in the vicinity of the study prey, and testing their gut contents with blood anti-serum obtained from a rabbit previously inoculated with an extract of prey material. A positive reaction indicates that the predator has consumed one of the prey being examined. In his study of *Pieris rapae*, Dempster (1967) used the precipitin test to discover that its eggs and younger larvae were eaten by a variety of arthropod predators, the most important being nocturnal beetles and spiders. These predators rarely seemed to attack the larger larvae, which were taken mainly by birds (established in a separate experiment where larvae were protected from birds by cages). A similar pattern of predation has been found in population studies of *Thecla betulae*, *Satyrium pruni* (Thomas 1974), *Ladoga camilla* (Pollard 1979*b*), and other Lepidoptera (Dempster 1971*a*, 1983).

The main predators of butterfly pupae vary from species to species, but, in most cases, birds or small mammals are thought to be responsible. In general terms, arthropod predators are responsible for predation in the early and smaller developmental stages of butterflies while birds and small mammals take the later and larger stages.

Comparatively little is known concerning the predators of adult butterflies, although there have been numerous casual observations of predation by birds, occasionally in large numbers. The adults of low flying butterfly species are sometimes caught in spiders' webs, but the proportion of the population killed in this way is likely to be fairly small. It is possible that most losses occur at night, when roosting butterflies must be at considerable risk from nocturnal predators. This has not yet been studied.

One final form of predation that needs to be mentioned is cannibalism—when a larva consumes an egg or larva of the same species. The phenomenon is usually confined to very young larvae and has been reported in a number of species including *Anthocharis cardamines* (Courtney and Duggan 1983), *Maculinea arion* (Thomas 1977), small blue *Cupido minimus* (Morton 1985), and green hairstreak *Callophrys rubi*. Most of these butterflies lay their eggs on flower-heads, and there is often only sufficient food for one larva to develop fully per flower-head. The larvae of most British butterflies, including all the gregarious species, are not cannibalistic even when conditions become severely crowded.

Insect parasitoids are the other main cause of mortality in the developing butterfly. They can affect the egg, larva, or pupa. Adults too can be parasitized externally by mites. Some parasitoids are host specific while others are generalists, attacking any potential host within a given size range. The specific parasitoids of many butterfly species can be far rarer than their hosts and may be more seriously threatened with extinction when their hosts become rare. Parasitoids can also serve a very important function in the regulation of population size, exemplified by *Eurodryas aurinia* with larvae that can reach such densities that they consume all available foodplants, leading to mass starvation. This catastrophic occurrence is prevented by the parasitoid *Apanteles bignellii*, which can kill as many as 86 per cent of larvae and has been found to be a major cause of the population cycles characteristic of *E. aurinia* colonies (Porter 1983).

The rate of parasitism observed in populations of other British butterflies is generally much lower than that recorded in *Eurodryas aurinia*. In the study of *Leptidea sinapis*, parasitoids were detected in eggs and third-instar larvae, but accounted for only 0–22 per cent of all deaths. The incidence of parasitism was also found to be fairly low in most life table studies referred to in Table 4.5.

A major cause of death in some non-British butterfly populations is disease, but there seems little evidence that it is a significant mortality factor in British species. A granulosis virus caused heavy larval mortalities in a Canadian population of *Pieris rapae* (Harcourt 1966), but a similar study of the species in Britain showed that it caused only a few deaths (Dempster 1967). In general, diseases affect butterfly populations only when they attain abnormally high densities, such as when species reach pest proportions on commercial crops (e.g. *P. rapae*). However, disease may be seriously underestimated in conventional life table studies as diseased individuals are undoubtedly more susceptible to predation. The cause of death may therefore be attributable to predators, even though the individual may be dying already from disease. Clearly more research on the effects of disease in butterflies is required before more general conclusions can be reached.

4.5.4 Weather as a mortality factor

Although relatively little is known about the function of natural enemies, the mortalities they cause may also be influenced by weather factors. In the life table study of *Ladoga camilla*, the key factor causing annual fluctuations was identified as predation of older larvae and pupae, probably by birds (Pollard 1979*b*). During poor weather, these critical stages take longer to develop and tend to suffer greater predation. Thus, the annual fluctuations in population size were found to be related, indirectly, to the temperature in June and July when older larvae and pupae were developing. Similarly, poor summer weather was also suspected to have had an adverse effect in *Leptidea sinapis*, by increasing mortalities of eggs and young larvae (Warren *et al*. 1986). Porter (1983) has also shown a major effect of weather in the parasitism of *Eurodryas aurinia* larvae by *Apanteles bignellii* (Fig. 7.4). The parasite regularly has three full generations per year to the host's one, with adult parasites emerging in late August, March, and June. The crucial stage occurs in the final parasite generation which can be delayed in cool weather until after most larvae have pupated, resulting in lower host mortalities. When spring weather is warm, the parasitoid generations are synchronized more closely with the host larvae, and causes greater mortality.

Climatic factors can affect butterfly populations at almost any stage in their life cycle, although the most susceptible stage varies from species to species. In addition to indirect effects on the mortality of eggs, larvae, and pupae, weather can influence adult survival and fecundity directly, as discussed in sections 4.3.2 and 4.5.1. It is therefore not surprising that weather is so often implicated in the fluctuations observed in adult population size (see section 4.4). Spooner (1963) has, for example, mentioned that good years for *Maculinea arion* often coincided with periods of warm weather, and poor years with periods of cool weather. Rainfall changes have been suggested as the cause of fluctuations in the black-veined white *Aporia crataegi*, as the timing of several local extinctions coincided with high September rainfall (Pratt 1983). However, in order to demonstrate the role of climate with any certainty detailed information is required about the function and the relative importance of each mortality factor. Currently, such data are available for only a few species. Nevertheless, the role of climate does appear to be important to most butterflies and will undoubtedly be a fruitful subject for further study.

5

Avoidance, concealment, and defence

Paul M. Brakefield and Tim G. Shreeve with Jeremy A. Thomas

Fluctuations in abundance are a regular feature of butterfly populations. Most of these changes can be related either to the performance of egg-laying females (chapter 3) or to survival at each stage of the life cycle (chapter 4). Every stage differs in size, form, location, mobility, duration, and usually in its season of occurrence. Because of these differences each offers a different reward to a specific type of natural enemy, and presents it with particular problems of finding and handling. The different life stages, therefore, have different enemies. Those of the earliest stages tend to be arthropod predators and parasitoids, whereas later, larger and often more mobile stages tend to be difficult for these small enemies to handle. In contrast, vertebrate predators are not usually the major enemies of eggs and small larvae because they offer small reward to these consumers compared to the energetic cost of finding them. They tend to be the most important enemies of the larger stages. Predators and parasitoids can be divided into two major groups: generalists, which use many types of food, and specialists, which attack particular species or groups of related species. All life stages of butterflies are vulnerable to both types of enemy.

To escape from natural enemies each stage has its own suite of defences (Fig. 5.7). These can be divided into primary and secondary defence mechanisms. The function of the former is to avoid detection. They may involve four main categories:

(1) adjustments of the timing of appearance;
(2) changes in habitat use or location;
(3) behavioural or morphological adaptations; and
(4) mimicry or resemblance to an inedible food item.

Secondary defence mechanisms, on the other hand, enhance the ability to escape from an enemy once it has been detected. These commonly include mechanisms that:

(1) make it difficult for the enemy to handle the prey;
(2) confuse or startle the predator or parasitoid; and
(3) cause distastefulness or toxicity.

This chapter discusses the effectiveness of these mechanisms in each of the various stages of butterfly life cycles.

5.1 Adult defences

5.1.1 The predators of adults

Of all the stages in the life cycle, the adult butterfly is the most active, whilst at the same time comprising the smallest number of individuals. Because of their generally conspicuous feeding, mate-location and egg-laying activities, adult butterflies have a unique set of problems in evading predators. Not only must they avoid predators when active but also when inactive, either during unsuitable weather conditions or when roosting. Predators may be more numerous in favourable habitats which support high densities of adults than in less favourable habitats where the reward for any particular predator is low. The most common predators of adults are vertebrates, most of which hunt in specific sites at particular times. Therefore, the risks for an adult differ with location and time of day.

Weather affects the activity of adults and predators in different ways. Cool weather inhibits butterfly activity while increasing the energetic requirements of warm blooded predators, but warm weather can act inversely. There is some evidence that survival rates of butterflies, such as the meadow brown *Maniola jurtina* (Brakefield 1982*b*), are lower when the flight seasons are hot and dry. Work on the monarch *Danaus plexippus* and sulphur butterflies (*Colias* spp.) in North America suggests that longer and more intensive periods of adult activity, likely to be associated with such conditions, may lead to more rapid wing wear and exhaustion of energy reserves (Watt *et al.* 1974; Brown and Chippendale 1974).

Although invertebrates such as spiders and dragonflies can take substantial numbers of butterflies (see Moore 1987), birds and lizards are recognized as their principal predators. There are, however, remarkably few observations of actual predation taking place (e.g. Ford 1945; Brockie 1972; Bengtson 1981). In studies of selective predation on the black and pale forms of the peppered moth *Biston betularia*, Kettlewell (1973) released the moths and directly observed their predation by birds. However, in these experiments there are problems of interpretation, as there was a higher density of released moths than would occur in the wild, and the resting positions on trees of many of the released moths were unnatural (Brakefield 1987*a*). Direct observations of predation on other species are rare. However, Chai (1986) has made careful observations of jacamar birds taking a variety of butterflies in flight in Costa Rica. Brower and Calvert (1985) estimated that orioles and grosbeaks took over two million butterflies or nine per cent of an overwintering colony of *Danaus plexippus*. Melling (personal communication) showed that 90 per cent of meadow pipit *Anthus pratensis* droppings on a Northumberland moor contained wing-scales, most probably of the large heath *Coenonympha tullia*, and perhaps of the small heath *C. pamphilus* and *Maniola jurtina*.

Another approach to collecting information about predation by birds and lizards is by examining wing damage. Sometimes butterflies are captured with clear impressions of beak-marks on the wings (Fig. 5.1). Usually the evidence is less clear cut. Many workers have discussed the analysis of different patterns of wing damage; some of which are probably consistent with predator attacks (see Bowers and Wiernasz 1979; Robbins 1980, 1981). However, the interpretation of data describing such damage is fraught with difficulties, especially because damage is indicative of a successful escape from a predator and not of mortality due to a predator. It is impossible to relate the data directly to the ratio of successful to unsuccessful attacks. Chai's (1986) observations of jacamars suggest that successful attacks may frequently outnumber, or even far outweigh, unsuccessful ones which result in wing damage. It is also difficult to use wing damage data to compare the levels of predation in different populations or on different morphs. The incidence of wing damage will depend on the length of time that individuals are exposed to predators as well as on the rates of predator attacks for each of the activity states of the butterfly. Edmunds (1974*a*) emphasizes similar problems when comparisons are made between species. Nevertheless, the almost universal incidence of wing damage (of a form consistent with vertebrate predator attacks) in large samples of butterflies from a wide variety of species is convincing evidence for the general importance of mortality due to vertebrates, especially birds.

5.1.2 Camouflage as a defence against vertebrates

Studies of butterfly mating and courtship show that the general colour of the wings, including ultraviolet reflecting areas, are important in communication between the sexes and the discrimination of mates (see section 2.4), but there is little evidence that the *details* of pattern play any substantial role in mate-location or mate-choice (Silberglied 1984). The general distribution and amount of dark melanin pigments and pale pigmentation on the wings of many butterflies is involved in thermoregulation (see chapter 2). Thus sex and climate play a role in wing pattern evolution. However, when the details of wing pattern elements and of their arrangement are considered, then undoubtedly the influences of visual predators are the most important factors shaping their evolution and general diversity.

It has recently been demonstrated by Bowers *et al.* (1985) for a North American fritillary, *Euphydryas chalcedona*, that bird predation can bring about rather subtle differences in colour pattern in a natural population of the butterfly. They demonstrated differences in colour pattern between 309 butterflies that had been attacked and eaten by birds (leaving detached wings), and a representative sample of 296 live butterflies from the same popula-

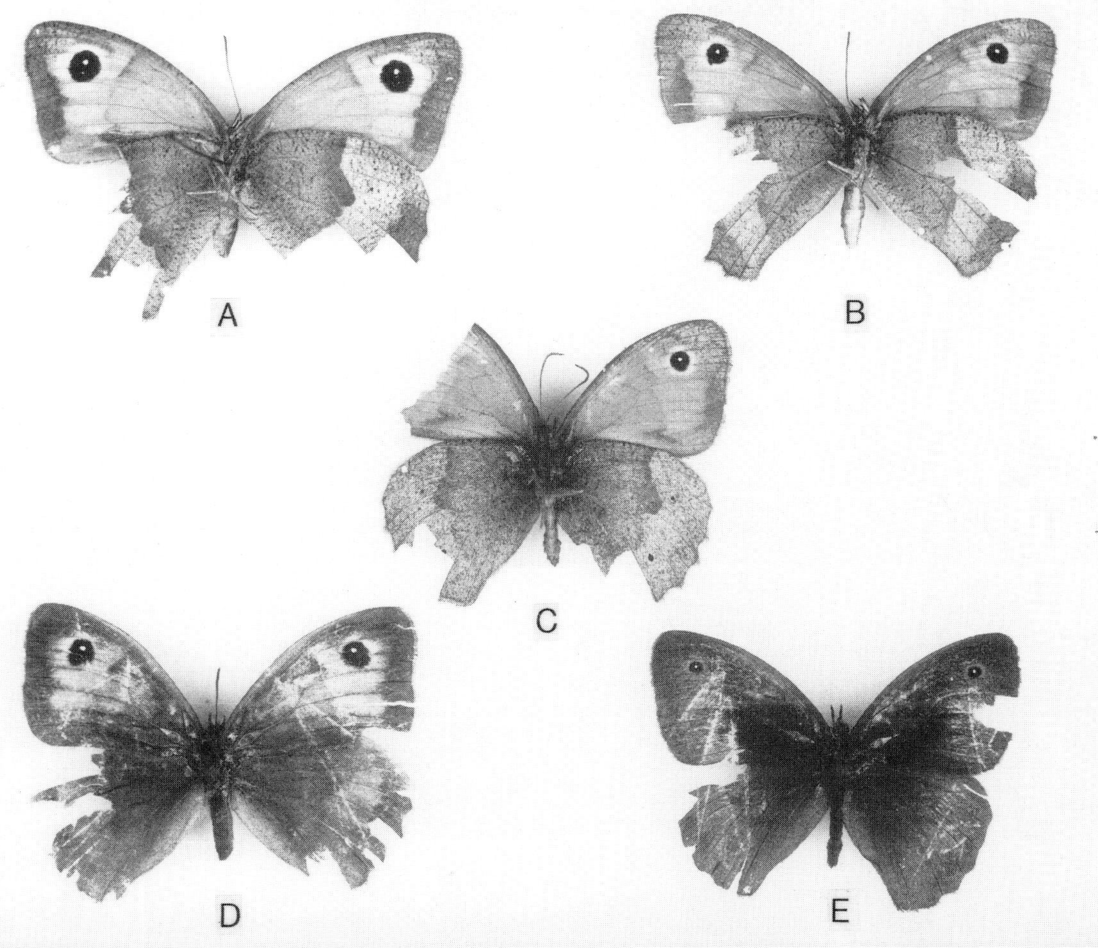

Fig. 5.1 Examples of wing damage found in wild-caught butterflies. Part of a sample of 29 meadow brown *Maniola jurtina* collected along a line of poplar trees on a coastal dyke on Goeree Overflakee in The Netherlands (A, B, and C show the ventral surface; C and E are males). Some of the specimens show clear evidence of partial evasion of attacks by vertebrate predators. Many insectivorous birds were feeding along the dyke. D and E have clear imprints of the bills of birds. C has had the one forewing tip cleanly cut off by a bird. The symmetrical damage to the hindwings of A and B is consistent with attacks by birds on butterflies at rest with their wings closed.

tion. In particular, birds attacked the palest individuals, independent of their age, and they also took proportionately fewer males.

The colour and pattern of the wings of many species can be described as camouflaged, thus enhancing the ability of the species visually to resemble its resting background making it difficult to be detected by a predator. Where species with such cryptic wing surfaces have been tested or bioassayed for palatability to vertebrate predators there has been little or no evidence of any chemical defence or distastefulness. Such studies have involved a wide variety of butterflies, usually in comparisons with warningly-coloured or aposematic species protected by a chemical defence (reviewed by Brower 1984; see Bowers and Wiernasz 1979).

The adaptive significance of cryptic coloration has been demonstrated by numerous experimental studies using a variety of organisms (Endler 1978, 1980, 1986). Such work has usually shown that colour forms which more closely resemble their resting backgrounds have a higher survival rate than more conspicuous ones. Although two studies on European species have involved colour dimorphism in butterfly pupae (Baker 1970; Wiklund 1975), experiments have not been performed with adults. These would be most valuable not least with regard to the extensive studies on wing pattern variation in *Maniola jurtina*, *Coenonympha tullia*, and the speckled wood *Pararge aegeria* (see chapter 9). Indeed, these studies frequently invoke a selective influence on relative crypsis of the different phenotypes by visually-hunting predators as the major process producing geographical variation in wing pattern. At present there is only circumstantial evidence for this rather than direct experimental demonstration. The most striking example of the association of habitat with changes in predation and selection for crypsis is given by industrial melanism in many moths (reviewed by Brakefield 1987*a*). Black forms of many species, including the peppered moth *Biston betularia*, spread through most cities in Britain during and after the industrial revolution in response to a blackening of their resting backgrounds by air pollution. They are more difficult for birds to find than the paler, wild-type forms. Many predation experiments, beginning with those of Kettlewell (1973) involving direct observation of predation, have confirmed the qualitative switch in survival of the forms from rural to industrial environments. Some present-day declines in the frequencies of the dark forms are associated with reduced air pollution and changes in resting backgrounds.

Crypsis involves a variety of components of colour and pattern, which have only recently been fully recognized. It involves being inconspicuous to a predator which effectively does not recognize the received image as being potential prey. Endler (1978) provides a useful definition: 'a pattern is cryptic if it resembles a random sample of the background perceived by the predator at the time and age, and in the microhabitat where the prey is most vulnerable to visually-hunting predators'.

Crypsis is maximized in relation to colour and pattern when the image of the insect represents a random sample of the components of its resting background. This randomness must extend to components of colour, size, shape, and contrast. If, for example, the elements of the pattern are on average larger than those of the background, the insect will stand out even if these same elements match it in terms of colour, shape, and contrast. Examples of these effects are illustrated in Fig. 5.2. The matching of insect and background should be maximized with respect to the resting background encountered when the risk of predation is highest. Predation on many populations of butterflies is likely to vary diurnally. Thus, Rawlins and Lederhouse (1978) and Lederhouse *et al*. (1987) found that predation on the black swallowtail *Papilio glaucus* occurred during evening or weather-prolonged roosting. Kingsolver (in Bowers, *et al*. 1985) found that predation on pierids took place in the early morning as butterflies were warming up and unable to fly.

The effectiveness of crypsis also depends on the behaviour of the insect, for example on its choice of resting background, resting posture, and activity pattern, and on the behaviour and hunting tactics of its predators. If a butterfly always rests inactively with its wings closed then the ventral wing surface is likely to be more cryptic than the dorsal surface. Thus, crypsis is not merely an attribute of pattern but is an interaction between the insect's colour pattern, resting background, behaviour, and activity, and the behaviour of predators (see Dennis and Shreeve 1989; Shreeve and Dennis 1991).

A frequent component of cryptic colour patterns is the presence of disruptive markings or bands crossing the wings and bisecting their edges. These are thought to function in breaking up the outline of the insect when at rest (see Cott 1940). They are particularly well developed in the admirals, including *Ladoga camilla* in Britain. This species has white bands crossing both dorsal and ventral wing surfaces. They contrast most highly with the background colour on the dorsal surface which is exposed in dorsal basking behaviour (see chapter 2). Silberglied *et al*. (1980) examined the proposed function of disruptive bands in a tropical nymphalid, *Anartia fatima*, a study made on the assumption that the white band was conspicuous rather than a normal part of the background. Silberglied and co-workers, captured individuals in a natural population releasing them again after either painting out the white band or adding a black 'control' stripe over

(a) (b) (c) (d) (e) (f)

Fig. 5.2 Some components of crypsis illustrated by seasonal forms of the evening brown *Melanitis leda* taken from the Shimba Hills of Kenya. The three on the left are representative of the excellent background matching of the brown-coloured dry season form of the butterfly when in its usual resting place among dead brown leaves on the forest floor. (b) Shows an individual of the dry season form resting on the floor of a conifer plantation. There is an obvious loss of crypsis due to a mismatching between the shape and size of the butterfly and the background of dead pine needles. The matching, in terms of the generally brown colour of the wings and the needles, remains close. The four individuals shown of the dry season form illustrate the characteristic variability in the details of the colour pattern in this form which may reduce the chance of a predator developing a search image for it. (d) and (f) are of the wet season form with a finely striated pattern and submarginal eyespots. The latter may be prominent to a searching predator when this form is at rest on dead brown leaves because of the absence of this type of feature in the background (d). When this form is at rest among the carpeting green herbage typical of the wet season (f), crypsis is poor because of a mismatch in colour and the eyespots probably serve a deflective function.

the dark ground colour. Unfortunately there was no difference in survival or in the incidence of wing damage between the experimental and control butterflies. There is therefore, no direct evidence supporting a disruptive function. The lack of positive results may have been due to the conversion of the pattern to that of a locally aposematic species (Waldbauer and Sternburg 1983) or from one cryptic pattern to another (Endler 1984). Endler suggests that it would be interesting to repeat such an experiment using an additional type of control with the band painted out as before but with an additional one of similar dimensions and colour pattern painted along the wing edge. This would help to determine whether such a band is conspicuous and disruptive or represents a normal part of the background. Such an experiment could be done with *L. camilla*.

The above discussion of crypsis has referred to cases of a general resemblance to the background. Cott (1940) and others make a distinction between such general resemblance and examples of specific resemblance to a particular component of the background such as leaves, twigs, and so on. Specific resemblance means mimicry to some people but not to others (Endler 1981; Pasteur 1982). All overwintering adults in Britain (e.g. comma, *Polygonia c-album*, peacock, *Inachis io*, brimstone, *Gonepteryx rhamni*) are leaf-mimics, in pattern and colour, if not shape, and resemble the background of their overwintering sites. To some extent they have other, secondary, defence mechanisms which operate when the primary mechanism of leaf-mimicry fails. Vane-Wright (1986) documents the hissing sound produced by wing-rubbing in overwintering *I. io*. This may startle a predator or even mimic the sound of a predator, such as a weasel or owl. Dennis (1984*a*) describes the way in which disturbed adult *G. rhamni* remain inactive when picked up—'playing possum' (thanatosis)—with a rigid body and tucked in legs. This behaviour may make prey recognition less likely. In this, and other overwintering species, the chance of predators failing to recognize the adult as prey may be enhanced by wing-toughness. This behaviour may also cause handling problems for the predator, thus facilitating escape.

The resting posture of a butterfly may be a compromise between crypsis and other behaviours. The grayling *Hipparchia semele* may provide an example. Findlay *et al.* (1983) demonstrated the importance of resting posture, involving body-orientation and wing-tilting, in relation to exposure to incident solar radiation and thermoregulation. Longstaff (1906) believed that the behaviour of tilting the closed wings closer to the ground aided concealment by minimizing the shadow. Observations by Findlay *et al.* suggest that thermoregulation is the primary function but they did not exclude the possibility of interactions with crypsis. Shreeve (1990) demonstrated that in this species crypsis may not be compromised when the butterfly is cool and most vulnerable to predation. In cool conditions the butterfly basks on the warmest substrates available and it is these against which the hindwing phenotype is most cryptic. When the butterfly is fully active and warm, it tends to settle on other substrates against which it is less cryptic and exploits other defence mechanisms, such as secondary flash coloration involving the forewing eyespots together with evasive flights. In most cases movement by a resting butterfly will tend to attract a searching predator and so diminish the effectiveness of crypsis. Since daytime activity is necessary for reproductive success in butterflies there must, for species which are generally camouflaged, be some balance between effective crypsis at rest and a breakdown of crypsis during spells of activity.

Endler (1984) measured and compared the cryptic wing patterns of a community of North American woodland moths with that of their resting backgrounds from analysis of photographs of each species taken at rest in nature. He showed that moths are on average more cryptic at the period of their normal flight time on their normal resting backgrounds than on other backgrounds available in their habitats. Specialist species which rest on a single background are more cryptic than generalist species and species which rest under leaves and are not visible from above are not very cryptic. It would be fascinating to apply this technique to butterfly communities such as those of grasslands or woodlands in Britain or to the satyrine species, discussed in chapter 9, which have variable wing patterns.

The vulnerability of a resting individual may also be related to the microclimate and position of its roosting site. At night, ground temperatures are generally lower than at sites higher in the vegetation. Sites near the ground are also more humid than those higher up. Thus any individual which roosts in high vegetation remains potentially active for longer in the evening and is likely to become active earlier in the day than one which roosts low down. Therefore, the use of particular roosting sites can influence how long an individual is exposed to predators when it is inactive. Dennis (1986*a*) suggested that the wall brown *Lasiommata megera* roosted in high positions to enable evening and early morning activity and also to facilitate escape, by dropping from the perching site, if detected. Similar arguments may apply to the use of tall vegetation by the orange-tip *Anthocharis cardamines* and prominent grass stems by species such as the Glanville fritillary *Melitaea cinxia* (Willmott 1985).

If crypsis is not successful in overcoming the prey-recognition of a searching predator then some butterflies may have a second line of defence which can be effective even if they are unaware of the predator and cannot or do not escape by flight. For example, small, often not fully differentiated, eyespots on the margins of the wings of many species may be components of crypsis except when the predator is very close. If the prey is located then such eyespots are thought to function by deflecting the attack of the predator away from the vulnerable body towards the wings (see Fig. 5.2). The butterfly may then escape, albeit having lost a portion of wing-tissue. The relatively thin wing-margins of such species may facilitate escape (Dennis *et al.* 1984). Erratic flight, exhibited by many species, will also reduce the probability of any further attack. Many British butterflies exhibit submarginal spots on the ventral wing surface which is exposed when at rest. More fully differentiated and larger eyespots are often hidden at rest. They may be suddenly or quickly exposed on disturbance, startling or confusing a predator and causing it to withdraw (Blest 1957; Coppinger 1969, 1970). Some Lepidoptera show this type of 'flash' display associated with patches of bright colour rather than eyespots, though large eyespots are frequently associated with bright patches of colour. The striking eyespot pattern of *Inachis io* is of this type and may be exposed by suddenly opening and closing the wings. This display is also associated with an audible scraping or hissing sound made by the inner margins of the forewings rubbing against the costal margin of the hindwings.

Predation by birds has been implicated as an important selective influence on intraspecific variation in *Maniola jurtina* and *Coenonympha tullia* (chapter 9). Figure 5.3 illustrates a model developed for *M. jurtina* but it is more widely applicable (Brakefield 1984). The balance of selection on the development of wing spots is considered to depend on interactions between several factors. In habitats dominated by linear shapes (i.e. uniform grassland) spots may be relatively conspicuous and so increase the risk to resting insects by acting as a cue to searching predators. In more heterogeneous habitats this constraint may be lessened. More active individuals which frequently change their position may tend to attract predators through movement. In this type of situation the deflective function of eyespots may be paramount. The more strongly expressed ring of submarginal spots exhibited by male *M. jurtina* is consistent with their higher activity levels than females (Brakefield 1982*a*). The larger and more contrasting forewing eyespot of females is hidden at rest. It may sometimes be exposed on disturbance in a similar way to that of *Hipparchia semele* (Tinbergen 1958). It may then be associated with both deflective and startling functions. The optimum spot pattern thus depends not only on the nature of resting backgrounds but also on insect activity rhythms and their exposure to predators. The model for species showing variable spot patterns suggests that shifts occur between populations in the balance between 'passive' crypsis and reliance on eyespots in deflection and other 'active' anti-predator functions. Work by Young (1979, 1980) on Neotropical *Morpho* butterflies suggests that those species with well-developed, ventral eyespot rings feed on rotting fruit on the ground and are thus exposed to a particular type of high intensity predation. Species that seldom rest on the ground have much smaller eyespots, presumably because larger ones would disrupt their crypsis when at rest on foliage.

Another probable type of deflective device occurs in several British hairstreaks. These have wing tail filaments at the anal edge of the hindwing. A strong clue to their deflective function comes from research on more fully developed systems of this type occurring in many tropical lycaenids (Fig. 5.4). In the most striking examples, longer pairs of tail filaments and well-marked radiating lines occur together with small eyespots close to the anal corner. This is the

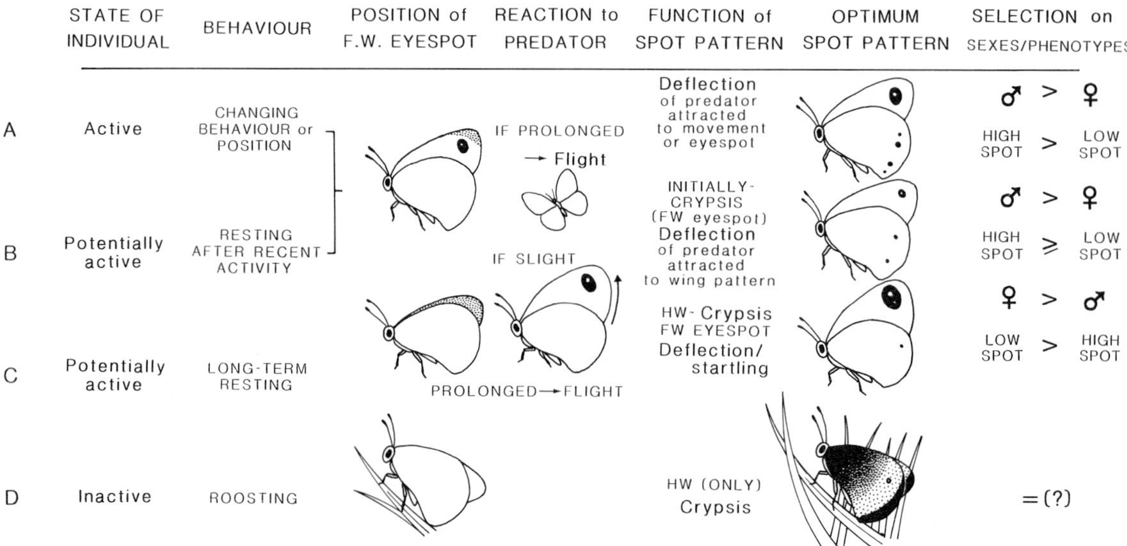

Fig. 5.3 A model to account for the phenotypic variation in the pattern of wing spotting in the meadow brown *Maniola jurtina* and other satyrinae. Four activity states are labelled A to D on the left. An outline of butterfly behaviour and of the proposed mechanism of visual selection involving predators follows for each state, from left to right. The final column indicates predictions about the relative fitness of the wing pattern of the sexes and of spotting phenotypes of *M. jurtina* in relation to the model. (After Brakefield 1984; courtesy of Academic Press.)

so-called 'false-head' pattern considered to involve false antennae and eyes with the radiating lines attracting attention towards this image. The field studies of Robbins (1980, 1981) provide strong support for the effectiveness of the false-head in combination with associated patterns of adult behaviour on alighting from flight in deflecting predator attacks away from the body and true head. Riley and Loxdale (1988) have demonstrated that the twisted structure of the tail, as in the tropical butterfly *Arawacus aetolus*, better simulates antennal movement in air currents than an untwisted form, thus accentuating false-head deception. A recent experimental study by Wourms and Wasserman (1985) involving analysis of video recordings of the responses of captive jays to pierid butterflies painted with different combinations of the various elements of the false-head system provides especially compelling evidence for the proposed functional components of the pattern. Figure 5.4 includes a subset of their data which demonstrates that a small eyespot in the outer-anal part of the hindwing produces a significant increase in attacks to this wing area (i.e. deflections). This element and other components of the pattern also changes the way in which the butterfly is handled by the bird. Butterflies with the complete pattern were more likely to escape during handling by the predator.

The false-head device of the British woodland hairstreaks is far less developed than in other species with wing-tails. It presumably has some deflective function at very short distances while complementing crypsis at longer distances. It may also be poorly developed in British species because of constraints imposed on escape by cool weather and the tensile strength of a tail filament (R. Dennis, personal communication). Butterflies with longer tails may be more vulnerable to capture and less able to escape when inactive in cool conditions.

5.1.3 Chemical defences and warning coloration

The adults of many species of butterfly are protected by a chemical defence often involving more than one type of compound. Two important groups of noxious chemicals are cardenolides and pyrrolizidine

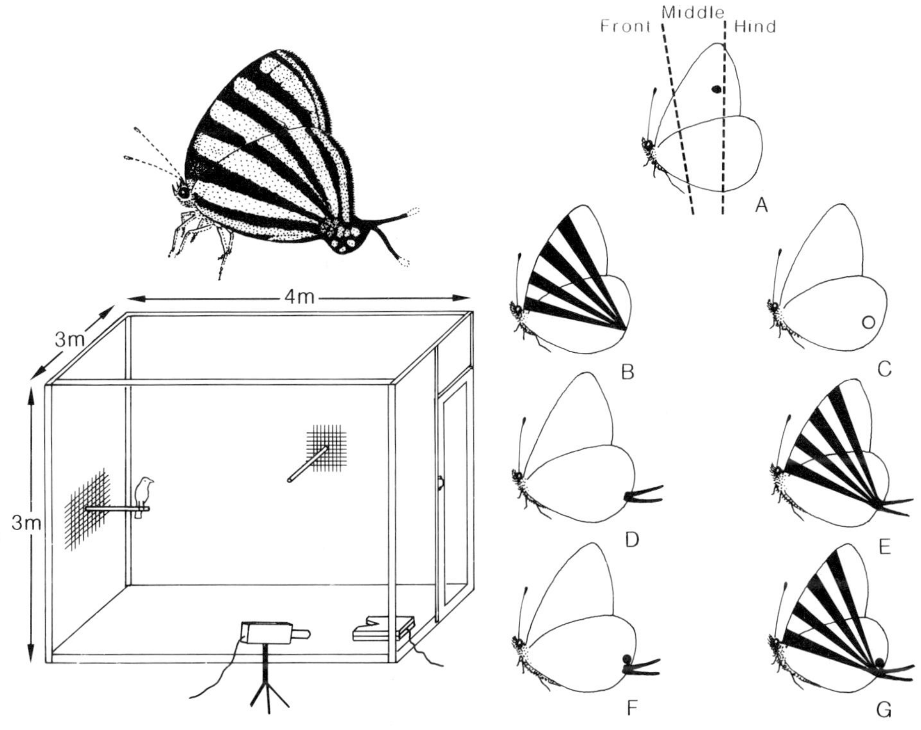

Type	Wing marking	Areas of attacks (%)			Areas of handling (%)		
		Front	Middle	Hind	Front	Middle	Hind
A	Controls	49	35	16	51	35	14
B	Lines	41	34	25	25	32	43
C	Eyespots	59	4	37	28	33	39
D	Tails	33	26	41	33	26	41
G	Complete	36	32	32	16	19	65

Fig. 5.4 An experimental analysis of some anti-predator functions of butterfly wing markings during feeding by birds. The components of the 'false-head' pattern were analysed by Wourms and Wasserman (1985). This wing pattern is commonly found in lycaenids including the tropical species *Arawacus aetolus* which is illustrated. Eyespot, convergent lines, tail and anal-spot components were painted on white *Pieris rapae* in the combinations A to G as indicated. The dead butterflies were fed singly to five blue jays *Cyanicitta cristata* using special presentation devices in aviaries of the design shown. The positioning on the wings of feeding attacks by the birds and the birds' subsequent handling of the butterflies in their bills was recorded on videotape for later analysis. A subset of the results of this analysis is shown in the table below. The difference in the areas of the wings attacked between the painted butterflies and the unpainted controls (type A) was only statistically significant for butterflies having an added eyespot. All such differences were significant when the location of strike points during handling was analysed. (After Wourms and Wasserman 1985, courtesy of the Society for the Study of Evolution; Robbins 1981, courtesy of the *American Naturalist* and Chicago University Press.)

alkaloids. The defence may act either before (as in many unpleasant volatile compounds) or after ingestion of the butterfly. Some species are able to synthesize defensive chemicals while many others rely, at least in part, on sequestering them or their precursors from larval foodplants or sometimes adult food sources. Most of the knowledge about such defences comes from an analysis of species not resident in Britain (see Brower 1984; Boppré 1986).

Butterflies that have chemical defences are frequently warningly coloured, or aposematic. They have bright and contrasting colour patches, usually in simple patterns, which are prominently displayed both at rest and in flight, which advertise their distasteful nature to potential, visually-hunting predators. This enables the predators to learn rapidly to avoid them once experienced. Experiments by Lane and Rothschild (Lane 1957) using a tame flycatcher suggested that among British pierids, *Gonepteryx rhamni* is palatable, the small white *Pieris rapae* is distasteful but edible, and *Anthocharis cardamines*, the large white *Pieris brassicae*, and green-veined white *P. napi* are the most unpalatable. Thus, there seems to be a spectrum of palatability among the pierids. Aplin *et al.* (1975) identified probable noxious chemicals—glucosinolates—in different life stages of *P. brassicae* and, at lower concentrations, in *P. rapae*. They also occur in their brassica foodplants. Work is lacking on other groups in Britain such as the fritillaries and nettle-feeding nymphalids which may have some chemical protection and associated warning coloration.

Many brightly coloured insects show striking similarities in colour pattern to other species. Such similarities are characteristic of the phenomenon of mimicry (reviewed by Turner 1984). Two basic forms of mimetic resemblance have long been recognized. Müllerian mimicry occurs between two or more species (called a mimicry ring), each of which is warningly coloured and protected by some noxious compounds. Müllerian mimics exhibiting the same warning coloration derive mutual benefit in terms of lower mortality. Their resemblance to each other increases the efficiency of predators in learning to avoid the unpalatable prey because their rate of encounter with that particular colour pattern will be higher than if each of the different prey species had different colour patterns. In contrast, Batesian mimicry involves one (or more) species that exhibit warning coloration and another that has the same colour pattern, although it is perfectly palatable to predators. The latter species derives some protection because predators sometimes mistake it for the unpalatable prey. The former species acts as a model for the latter, which is the mimic. Some Batesian mimics, such as the African mocker swallowtail *Papilio dardanus* are polymorphic, with forms mimicking different unpalatable models (see section 8.2). In practice the distinction between Müllerian and Batesian mimicry is probably not always clear cut, being based on an underlying continuous spectrum of palatability (Turner 1984; 1987). Furthermore, predators differ in their response to the chemical defence of particular prey species; some are able to tolerate high levels of particularly noxious chemicals. The chemical defence may also vary in quality between individuals of a particular species (Brower 1984).

The British butterflies have received little attention with respect to mimicry. The white pierids may represent a rather loose form of Müllerian mimicry ring, although the apparent variability in chemical protection suggests that some species may be deriving protection more akin to Batesian mimicry (see above). The satyrine marbled white *Melanargia galathea* may also be a Batesian mimic of pierids, as it has a similar colour pattern and behaviour to the chemically defended pierids (Dennis and Shreeve 1988). The classic studies of mimicry in butterflies have concerned tropical species, especially *Papilio* spp. in Africa and Asia, and *Heliconius* spp. in South America (Turner 1984, 1987). Müllerian mimicry in Britain is probably best illustrated by groups of day-flying burnet moths (Rothschild 1984), bumble-bees (Plowright and Owen 1980), wasps, and ladybird beetles (Brakefield 1985). Hoverflies are also good examples of Batesian mimics of bees (Heal 1982).

Predators such as spiders (Moore 1987) and dragonflies, and parasites such as ceratapogonid midges (Lane 1984) are likely to cause substantial mortality in many populations of adult butterflies. In many cases they are able to eat with impunity species that are highly unpalatable or toxic to birds. Some form of visual recognition is likely to be employed by stalking or hunting predators such as crab spiders and hornets. However, it seems unlikely to involve details of wing patterns. There is little knowledge of what appears warningly coloured or cryptic to such species, but this may provide a fertile area for future research. Despite extensive knowledge of chemical defences against vertebrate predators the role of chemicals against invertebrates is little understood. Most understanding is based largely on identification of chemicals and hypothetical inferences about their effects (see Brower 1984). A partial exception is chemomimetic manipulation of ants which probably includes adults of those species (such as the large blue *Maculinea arion*) that must emerge from pupae within the ant colony (see sections 5.4 and 7.3.3). *Ogyris* females in Australia are also able to oviposit in the midst of ants, apparently by producing an ant dispersal pheromone (Atsatt 1981*a*,*b*).

5.2 The defences of eggs

5.2.1 Egg-placement as a defence mechanism

To a large predator the small size of eggs offers little reward unless they are numerous and easily located. Consequently the majority of their enemies tend to be other invertebrates (Dempster 1984) most of which rely on scent, particularly that of the hostplant and, in limited cases, touch, to locate food.

The immobility of eggs poses several problems, most obviously the egg has no behavioural mechanisms to facilitate escape. Egg placement, which results from the egg-laying behaviour of females (see chapter 3), is critical to survival because it determines their availability to particular types of predators and parasites. Among the grass-feeding butterflies (Satyrinae and some Hesperiidae) there is considerable variation in the placement of eggs (Wiklund 1984; Table 5.1). Most of those species (e.g. *Maniola jurtina*, *Melanargia galathea*) which use open habitats that are maintained by grazing herbivores, tend to attach their eggs either to the basal parts of hostplants or they drop them on the ground, on litter, or in vegetation adjacent to suitable grasses. By contrast those species (e.g. Lulworth skipper *Thymelicus acteon*, *Pararge aegeria*) associated with grasses growing in areas where grazing is infrequent tend to attach their eggs higher on hostplants. While egg placement by such species may have evolved partly in relation to grazing, the pattern of placement exposes eggs of different species to different enemies. Eggs on the ground or in litter are vulnerable to free-roaming generalists, such as some beetles, bugs, and ants, but those attached to leaves may be more vulnerable to more specialist enemies. For instance, the eggs of the small skipper *Thymelicus sylvestris* and *T. lineola* which are placed in the rolled leaf sheaths of flowering stems of several grasses may be inaccessible to ant and beetle predators but perhaps not to specialized or small enemies such as *Trichogramma* wasps which can locate them in the leaf sheath.

Dennis (1983*c*) suggests that eggs of the wall brown *Lasiommata megera* which are deposited on exposed leaves may be more likely to be washed off the plant by rain than eggs that are hidden in the recesses at the margin of grass clumps. Once washed off the eggs would probably have a low probability of survival when faced with ground-foraging predators. In other species that deposit their eggs on leaf-litter such a position for the egg may be advantageous. Both the pearl-bordered fritillary *Boloria euphrosyne* and small pearl-bordered fritillary *B. selene* lay eggs on dry leaves adjacent to hostplants as well as on the plants themselves. By placing eggs in a variety of microclimates they may ensure a variable period of egg development. Such a strategy could be a form of risk spreading, exposing eggs (and emergent larvae) to enemies for a variable time and also to different suites of enemies. Eggs on plants may be less exposed to free-roaming and ground-dwelling predators while those on the ground or on the hostplant may be more likely to escape from parasitoids and predators which locate their prey by searching the larval hostplant.

Eggs that are attached to a substrate have a form of defence which is unavailable to those more generally scattered by females. Ants can be important predators of many eggs, but attached eggs may pose problems of removal. For instance, eggs of *Pararge aegeria* on the underside of the leaves of cocksfoot grass *Dactylis glomerata* are more easily levered off by individual foraging workers of the ant *Lasius niger* than are those attached to the underside

Table 5.1 Egg placement of British grass-feeding butterfly species in relation to possible grazing effects

Species	Egg-placement	Height (cm)	Hostplant sites	Vulnerability to grazing
Carterocephalus palaemon	Attached singly to leaf underside	5–20	Ungrazed edges	Low
Thymelicus sylvestris	Small batches within rolled leaf sheaths	10–30	Ungrazed woodland and scrub edges	High
Thymelicus lineola	Small batches within rolled leaf sheaths	10–30	Ungrazed woodland and coastal marshes	High
Thymelicus acteon	Small batches within rolled leaf sheaths	10–30	Ungrazed grassland	High
Hesperia comma	Attached singly to leaf blades, stems and shoots	0–2	Grazed and broken areas of chalk grassland	Medium
Ochlodes venata	Attached singly to leaf blades	5–20	Ungrazed edges, open grassland	High
Pararge aegeria	Attached singly to leaf underside	1–25	Ungrazed woodland in sun and shade	Low
Lasiommata megera	Attached singly, mostly to leaf underside and basal shoots	1–25 (most low)	Recess areas in sun	Medium
Erebia epiphron	Attached singly to leaf blades	1–5	Grazed moorland	High
Erebia aethiops	Attached singly to blades within tussocks	5–20	Ungrazed areas of damp grassland	Low
Melanargia galathea	Unattached, dropped by settled female	0	Field and woodland edge sites, some sites grazed	Low
Hipparchia semele	Attached singly in low-medium density patches, on shoots, roots and litter	0–2	Dry dune-slacks, scree, downland	Medium
Pyronia tithonus	Attached singly to leaf blades	5–20	Sheltered tall, shaded, edges	Low
Maniola jurtina	Attached singly or dropped in basal plant parts	0–5	Various, includes tall and short grassland	High
Aphantopus hyperantus	Unattached in basal plant parts	0–5	Open, part shaded, tall grassland	Low
Coenonympha pamphilus	Attached singly to leaf blades	0–5	Short, grazed grassland	High
Coenonympha tullia	Attached singly to leaf blades	5–20	Ungrazed or part grazed moorland and mosses	Low

See Appendix 1 for hostplants.

of wood false-brome *Brachypodium sylvaticum* (Shreeve, personal observation). This is probably because the wider spaced and less prominent veins on the leaves of *B. sylvaticum* permit firmer attachment than do the closer, more pronounced veins of the former. However, strength of attachment may not always be of significance in deterring ants; they are reported eating through the top of eggs of *Maniola jurtina* and leaving the base (Brakefield, personal observation). Strength of attachment and the nature of the substrate to which the eggs are attached may also be critical to the survival of eggs of species with a long, overwintering egg stage, such as the white-letter hairstreak *Satyrium w-album*, purple hairstreak *Quercusia quercus*, and high brown fritillary *Argynnis adippe*.

5.2.2 Physical and chemical defences of eggs

Egg placement is important for escaping detection but egg colour or contrast is only relevant to visually-hunting enemies. This relationship between egg colour, placement, and the method of detection by natural enemies may partly explain why long-lived eggs are placed singly and tend to be visually cryptic (as in black hairstreak *Satyrium pruni* and in *Argynnis adippe*). Such eggs may well be apparent to invertebrate predators but these have minimal winter activity. Eggs of many species which are laid singly and develop directly, tend to resemble the general background colour and pattern of the area in which they are deposited. Similarly, clustered eggs with no secondary defences, which offer a rich reward to all types of enemies, are often located on leaf undersides, where they are difficult to detect by eye because of their position and, when newly laid, their green colour. This minimizes predation from those predators that detect their prey visually.

The surface structures and shapes of eggs are highly complex (see Table 3.1). Those of some, such as the Pieridae are finely ribbed and elongate, while those of the Lycaenidae are reticulated and doughnut-shaped. Although this complexity is probably significant structurally and also important in the control of water balance (Hinton 1981; see chapter 3), it is possible that it may facilitate escape from parasitoids and perhaps predators. The commonest egg parasitoids are the minute Chalcid wasps, usually species of *Trichogramma*, which may be as small as 0.25 mm long. Before laying their eggs these wasps first assess the size of the host egg, usually by walking from the apex to the base (Schmidt and Smith 1987). Then they use the surface of the host egg for a foothold as they insert their ovipositor. It is possible that some surface structures and egg shapes may prevent certain parasites assessing egg size correctly or gaining a foothold before inserting their ovipositor. Surface patterns and structures may also provide a limited form of defence against the piercing mouthparts of bugs and the biting jaws of beetles, and in combination with egg chemistry provide some form of protection from fungal attack.

Egg clustering is described as an adaptation which increases the probability of individual eggs escaping detection by predators and parasitoids (i.e. Fisher 1930; Stamp 1980; Chew and Robbins 1984); firstly, because it may increase the time predators have to search before encountering any of the prey items, thereby increasing the chance that individual enemies move elsewhere, and secondly, because the clustering of eggs has sometimes been associated with the evolution of distastefulness and warning-coloration in this and in later stages (see below). Birds are not generally believed to be major predators of eggs but it is known that those of *Pieris rapae* and *P. brassicae* are eaten by the house sparrow *Passer domesticus* and garden warbler *Sylvia borin*. Baker (1980) reports that 20 per cent of eggs of *P. rapae* can be taken by these birds but fewer *P. brassicae* are consumed. This difference is related to the advantages and disadvantages of clustering in the presence or absence of egg-toxins. Eggs of *P. brassicae*, but not *P. rapae*, contain allylisothiocyanate, an irritant compound, making them unpalatable (Aplin *et al.* 1985). Clustering by *P. brassicae* enhances the effect of this compound because fewer eggs within batches will be eaten once a predator experiences the protective chemical. The eggs of *P. rapae* which are not protected in this way probably rely on a combination of dispersal, scarcity, and greater crypsis for protection.

Egg-clusters may protect eggs from parasitoids and predators by their spatial arrangement. Innermost eggs of large masses are less accessible to parasitoids and invertebrate predators and those in the centre of single layer clusters escape attack because of the searching and host-locating patterns of their enemies (analogous to the 'edge-effect' of butterfly egg-laying in chapter 3). No British species lays its eggs in strings as does the map butterfly *Araschnia levana* on mainland Europe. Such an arrangement may minimize the chance of detection by invertebrate predators because an egg string occupies the same area as a single egg when viewed from above. It is therefore harder to find than a batch which covers a larger area of leaf. It may also make the handling of the eggs more difficult once detected.

5.3 The defences of larvae

5.3.1 The enemies of larvae

Larvae are potentially mobile, and like all stages their basic problem is to avoid being eaten or parasitized. They are the growth stage and have the additional problem of needing to consume food without being detected, a behaviour which can render them conspicuous by their association with hostplants, by their feeding damage, and frass. As larvae consume food they increase in size, and their value to particular enemies changes throughout their growth. Thus large larvae tend to have different enemies to newly emerged larvae. Additionally, the larval stage is the overwintering form of many species, a quiescent period with its own unique problems.

Most larvae have multiple defence mechanisms to cope with a variety of enemies, but however effective any particular defence mechanism is against one class of enemy there may be others with alternative strategies that cannot be circumvented. For example, ichneumonid and braconid wasps are common larval parasitoids, which usually lay their eggs in the host body by inserting the ovipositor through the larval cuticle. Others lay their eggs on the host body and the emergent larvae then burrow through the cuticle. Larvae may have defences, either physical (i.e. spines and hairs) or behavioural (e.g. wriggling) which lessen the chance of successful parasitoid egg-laying, but Tachinid flies, that have larvae which are also parasitic may invade the host by an alternative mechanism. Some species lay their eggs on the hostplant and larvae consume these eggs when they ingest the leaf material. Eggs then hatch in the larval gut and penetrate the gut wall to develop within the host body cavity. Once a parasitoid has penetrated the larval body it may fail to develop further because of the additional defence of encapsulation. In this process the developing eggs or larvae of the parasitoid become melanized by the haemocyte cells that surround them, restricting their oxygen and nutrient supplies.

Different enemies search for prey items at different times of day, for example birds are daytime searchers and some mammals are crepuscular. Therefore, the activity of a larva can influence its exposure to different kinds of enemies, as can its position on the hostplant, since different enemies may search different parts of a plant.

5.3.2 Larval concealment

Among British butterfly species some form of crypsis in the larval stages is extremely common. This may extend from appearance to behaviour. For example, the larvae of the purple emperor, *Apatura iris*, resemble the leaf on which they feed, but the effectiveness of this crypsis is dependent on the orientation of the larva. All of the Satyrinae and grass-feeding Hesperiidae are cryptically coloured and counter-shaded, being green or brown (Table 5.2). Those species which are green tend to be located in the wetter environments where the background grass-colour throughout the larval period is green, but those of drier habitats, such as *Hipparchia semele*, are brown in colour. Inconspicuousness in these species is also maintained by behaviour—most rest during the day in the basal parts of grasses, adopting night-time or crepuscular feeding. This behaviour may prevent their detection by visually-hunting predators, but may also minimize their risk to diurnal-feeding herbivores and day-time active invertebrates which locate their prey by searching the hostplant or detecting feeding damage. Inconspicuousness may also be enhanced by manipulating the hostplant, for example, large skipper *Ochlodes venata* larvae construct a tube of leaves from which they emerge only to feed. All of the Hesperiidae enhance their inconspicuousness by

Table 5.2 The colour of larvae of grass-feeding British butterflies in relation to their resting sites, feeding times, and the general nature of their habitats and hostplants

Species	Larval colour	Resting site	Feeding	Significance of colour
Carterocephalus palaemon	Grass-green striped dark green, white	In tube of rolled leaf blade	Day and night	Cryptic when feeding, concealed at rest
Thymelicus sylvestris	Pale green striped dark green	In tube of rolled leaf blade. Old larvae out of tube	Day and night	Cryptic when feeding, concealed at rest. Old larvae cryptic at rest
Thymelicus lineola	Pale green striped dark green and yellow	In tube of rolled leaf blade	Day	Cryptic when feeding, concealed at rest
Thymelicus acteon	Pale green striped cream, dark green	In tube of folded leaf	Day and dusk	Cryptic when feeding, concealed at rest
Hesperia comma	Olive-green	In spun tent of leaves	Day	Cryptic when feeding, especially on *Festuca ovina*. Concealed at rest
Ochlodes venata	Blue-green striped dark green and yellow	In tube of folded leaf blade	Day	Cryptic when feeding, concealed at rest
*Pararge aegeria**	Pale green, striped dark and light green	Leaf underside near feeding site	Day and night	Cryptic at all times especially on pale leaves of *Brachypodium sylvaticum*
Lasiommata megera	Bluish-green, striped white	Vertical, near feeding site	Night	Cryptic at all times especially on dark green plants
Erebia epiphron	Olive-green, striped white buff tails	Within tussocks	Night	Cryptic when at rest, less so when moving to feeding sites
Erebia aethiops	Grey-green, striped light green	Basal plant parts	Night, some day feeding	Cryptic when at rest, especially against *Molinia caerulea*
Melanargia galathea	Variable, brown to lime-green striped dark green, pink tails	Basal plant parts	Night	Different forms may be cryptic when at rest, dependent on composition of resting site
Hipparchia semele	Dark, striped brown, white, yellow	Basal plant parts and dry litter	Night, some day feeding	Cryptic when at rest on brown and bleached dead leaves and roots
Pyronia tithonus	Green-grey or green-brown, striped dark green	Low parts of hostplant clumps	Day, then night	Cryptic when resting on dark coloured grasses
Maniola jurtina	Bright green striped pale and darker green	Basal plant parts	Day, then night	Cryptic when at rest

Table 5.2 (*cont.*)

Species	Larval colour	Resting site	Feeding	Significance of colour
Aphantopus hyperantus	Pale brown striped pink and dark brown	Low leaf blades	Night	Cryptic at rest, against dry leaf sheaths and dead leaf blades
Coenonympha pamphilus	Grey-green, striped white, pink tails	Low parts of hostplant	Day	Cryptic at all times, especially on reflective grasses
Coenonympha tullia	Dark green striped dark and light green	All parts of upright leaf blades	Day	Cryptic at all times, stripes resembling leaf veins

* Excludes first instar larvae.

Note: many larvae may feed at different times of day in the final instar. In such circumstances crypsis may break down but individuals may gain some protection by their relative scarcity, reducing the likelihood of particular predators forming an effective search image.

ejecting their frass some distance away from their feeding site with the aid of an anal comb. This may render them less conspicuous to those enemies that rely on vision to detect their prey and also to those which associate the scent of frass with prey items.

Most of these palatable larvae have a secondary defence mechanism that can be employed when they are detected. If disturbed when at rest on a leaf, these larvae will release their grip on the plant, drop to the ground, and roll into a ball. Such behaviour is probably effective and involves little risk in losing the hostplants that are low growing and frequently abundant. This secondary defence mechanism may also be effective against some parasitoid attacks. All violet-feeding fritillaries rest away from their hostplants, a behaviour which may be related to the searching patterns and recognition methods of some generalist enemies that search damaged hostplants. However, this behaviour may also be related to thermal constraints on activity (see p. 92).

Those larvae (e.g. *Anthocharis cardamines* and swallowtail *Papilio machaon*) which feed on high growing parts of their hostplants cannot wander from the hostplant when not feeding, and must rely on crypsis or some other mechanism to avoid detection. Second and later instars of *A. cardamines* are greenish-white in colour and counter-shaded. They resemble the seed-pods of the cruciferous hostplants on which they feed and their counter-shading enhances their crypsis when seen from different angles. First instar larvae are orange and conspicuous; thus the probability that they are recognized by egg-laying females is increased and larval competition reduced (see chapter 3). When larvae of *A. cardamines* are small, visually-hunting predators are not important as most deaths at this stage are related to larval competition (see chapter 7) and the mechanical problems of chewing through a developing seed-head (Courtney and Duggan 1983). Larger larvae are often consumed by birds and therefore cryptic coloration reduces the chances of detection.

5.3.3 Chemical and physical defences of larvae

Small larvae of *Papilio machaon* closely resemble small bird droppings, but after the third instar they are remarkably different in coloration, with green and black stripes. This pattern is cryptic at a distance but conspicuous and warningly-coloured when seen at close quarters. Presumably predation by visual-hunters (like birds) is important at all stages but for various reasons mimicking a non-food item may be ineffective when larvae are large, the advantage of warning-coloration relative to crypsis being greater for large larvae than small. Size itself may not be the most important factor in determining the relative advantages of the different defence mechanisms. After the first moult all the larval stages of *Polygonia c-album* increasingly resemble bird-droppings. All larval stages of *Papilio machaon* have a secondary defence mechanism involving a large, orange organ (the osmaterium) which is extruded from behind the head when the larva is disturbed. This organ gives off a smell when extruded which has been described as being like pineapple (Dempster *et al*. 1976) and unpleasant or acrid (Frohawk 1934; Thomas 1986). The secretions of this organ (isobutyric and 2-methylbutyric acids) are probably an irritant, and are effective against ants (Eisner and Meinwald 1965). In the related North American *Eurytides marcellus* these secretions are also effective against small salticid spiders (Damman 1986), but not against larger spiders or wasps, which also attack *P. machaon* (Dempster *et al*. 1976). These secretions may not be effective against some birds. Although reed bunting *Emberiza schoeniclus*, sedge warbler *Acrocephalus schoenobaenus* and bearded tit *Panurus biarmicus* take larvae (Dempster *et al*. 1976), aviary experiments with great tits *Parus major*, done by Jarvi *et al*. (1981), and with *P. major* and blue tits *P. caeruleus* by Wiklund and Jarvi (1982) revealed that members of these species were unwilling to eat larvae and rejected those they pecked. Both intact larvae and those without osmateria were avoided, and Jarvi *et al*. and Wiklund and Jarvi concluded that warning-coloration alone was sufficient for protection. However, these experiments were not done in the field where predatory birds may show different behaviours.

Larvae which feed on noxious plants are able to accumulate or sequester plant chemicals which can be used in defence. Such a mechanism is most effective if larvae are warningly-coloured. Larvae of *Pieris brassicae*, for example, are conspicuous and feed on the upper leaves of their cruciferous hostplants. Their gregarious behaviour may enhance defence because it provides a more effective visual warning. Living in aggregations may also enhance escape from parasitoids and invertebrate predators which are not deterred by chemical defences (see below and p. 105). However, Stamp and Bowers (1988) document the effect of predatory wasps on gregarious larvae of the moth *Hemileuca lucina*. Those larvae harassed by wasps move to less favourable sites in cool microclimates, where they grow more slowly. In such circumstances escape from wasps may expose larvae to different enemies and also affect larval fitness.

Some aggregations of larvae (e.g. marsh fritillary *Eurodryas aurinia*, *Aglais urticae*) may defend themselves by spinning webs; such structures may entangle or prevent access of some small enemies. Clustering of dark-coloured larvae may also speed up development by elevating body temperature within the cluster (see p. 10), minimizing the time that larvae are exposed to attack. All nymphalid larvae (and those of some other species) show a jerking response if lightly touched. This may be effective against a wide range of predators and parasitoids, and in clustering species disturbing one individual sets off a chain reaction in the whole cluster. This could be triggered by mechanical disturbance of the substrate on which larvae are settled or perhaps by the release of an alarm-pheremone.

In the nettle-feeding larvae and some satyrines, disturbance also causes the larvae to regurgitate their stomach contents. Nettles themselves are defended from herbivores by hairs which secrete acids when touched, but the leaves apparently have no chemicals which can be used by larvae for defence. However, the gut contents of the nymphalids which feed on this plant are distasteful (Shreeve, personal observation) and perhaps provide some defence from avian predators. Spines on the larvae of these and other species may also make handling by predators difficult, acting as a physical irritant and also increasing the effective size of the larvae. They may also prevent penetration by some parasitoids by their length, particularly when they form an effective mesh over the whole body (e.g. late larvae of *Inachis io*).

5.4 Adaptations to living near ants

Ants are so dominant in most terrestrial biotopes that numerous organisms have evolved ways of living alongside them. Adaptations range from simple protective devices, such as thick or toughened skins, to special organs such as nectaries (Hölldobler and Wilson 1990). The latter secrete sugars or amino acids, and in return for this food, ants not only exempt the organism from attack but protect or even cultivate it (Donisthorpe 1927; Way 1962; Heads and Lawton 1984). Butterflies are among the many insects that possess such devices; ants incessantly tend certain lycaenids through the whole pupal period and during all larval instars, except usually the first. It is impossible to determine how widespread this relationship is among the Lycaenidae, for only a small proportion of species has been observed undisturbed in the wild when young. Nevertheless, regional studies suggest that the vast majority will prove to have a relationship at least somewhere within their geographical ranges (e.g. Geiger 1987; Thomas and Lewington 1991). Since over one-third of the world's 18 000 butterfly species belong to this family, an association with ants must therefore be regarded as a major feature of butterfly ecology.

This strange phenomenon was first noticed 200 years ago in a continental lycaenid, Reverdin's blue *Lycaeides argyrognomon*, and by 1793 it had also been reported for the green hairstreak *Callophrys rubi* and silver-studded blue *Plebejus argus* (Hinton 1951). On present knowledge, the larvae or pupae of 16 out of 17 British Lycaenidae are known to attract ants in captivity, and an association in the wild has been observed in 14 species (Table 5.3); there is little doubt that the rest live with ants in some parts of Europe, if not in Britain. Most of the important adaptations that enable larvae and pupae to live with ants had been described by the end of the nineteenth century. These, and more recent discoveries, are reviewed by Hinton (1951), Malicky (1969), and Cottrell (1984). The commonest adaptations are illustrated in Figs 5.5 and 5.6 and are described below; their ecological significance, and the additional adaptations of aphytophagous species, are described in chapter 7.

The dorsal nectary organ (DNO) or honey gland has been described by many authors; Kitching and Luke (1985) give photographs from the chalk hill blue *Lysandra coridon* (see Fig. 5.6), Adonis blue *L. bellargus* and common blue *Polyommatus icarus*. It consists of a large slit-like depression across the top of the larva's seventh abdominal segment, surrounded by slightly raised lips that are clearly visible through a hand lens (Fig. 5.6a). The slit leads down to a sac into which collect liquid secretions from four glands situated deep in the body of the larva. The whole organ is surrounded by curious knob-like mechanoreceptors on the outside (Fig. 5.6b). Ants drum these and the lips of the slit, beating them very rapidly for long periods with their antennae. In response, the slit opens, exposing a droplet of liquid honeydew, which is squeezed out as the sac is raised. In many species, there are short hairs pointing inwards round the edge of the slit, which holds the droplet in place. This, however, is often unnecessary, for the honeydew is highly attractive to ants and generally drunk the moment it appears. Droplets from the Provence chalk hill blue *Lysandra hispana* have been analysed, and found to consist of a 13–19 per cent sugar solution, containing glucose and sucrose, with traces of trehalose, protein, and the amino acid methionine (Maschwitz *et al*. 1975). This represents a 6–10 fold concentration by this gland of the normal sugars found in the larval haemolymph. The secretions of *P. icarus* are very similar, as are those of *Jalmenus evagoras*, an Australian hairstreak studied by Pierce (1984). However, Pierce also found high concentrations of the amino acid serine in the honeydew, and at certain times of day the concentration of sugars reached 55 per cent. This also varies with diet: *J. evagorus* females preferentially oviposit on specimens of their foodplant that have high concentrations of nitrogen in their leaves, and the resulting larvae secrete particularly attractive secretions for ants (Baylis and Pierce, 1991); *P. icarus* secretions are richest when larvae are fed on their natural foodplants (Fiedler 1990*a*). The DNO usually appears first in the second or third larval instar, and is particularly active in full grown larvae just before pupation. It is absent or rudimentary on some British Lycaenidae (Table 5.3) and is absent from all pupae.

Tentacle organs are curious, finger-like organs that occur as a pair either side of the larval honey gland, one segment further back (the eighth). They are normally kept retracted within the body, but are periodically everted, being blown up in unison by haemolymph, each unrolling from the centre (Figs 5.5a,b; 5.6c,d). As the tip finally unfolds, a group of spiny hair-like threads springs out to form a rosette around the tip; in some species these rotate. The structure of these 'tentacles' differs from species to species, but many look rather like a chimney sweep's brush (Clark and Dickson 1956). After a second or

Table 5.3 A summary of the relationships between British Lycaenidae (including Riodininae) and ants

Species	Larva						Pupa				
	Honey gland	Tentacles	Cupola	Sound produced in captivity	Attraction to ants in captivity	Relationship observed in the wild	Cupola	Sound organ observed	Sound heard in captivity	Attraction to ants in captivity	Observed tended by ants in wild
Hamearis lucina	×	×	✓								
Thecla betulae	×	×	✓		✓	✓	✓			✓	✓
Quercusia quercus	×	×	✓		✓		✓	✓	✓	✓	✓
Satyrium pruni	×	×	✓		✓		✓	✓			
Satyrium w-album	✓	×	✓		✓	✓	✓	✓	✓		
Callophrys rubi	†	×	✓		✓		✓	✓	✓	✓	✓
Lycaena phlaeas	×	×	✓		✓		✓	✓	✓		
Lycaena dispar	×	×	✓		✓	✓	✓				
Maculinea arion	✓	×	✓	✓	✓	✓	✓			✓	✓
Cupido minimus	✓	×	✓	✓	✓	✓	✓			✓	
Lysandra coridon	✓	✓	✓		✓	✓	✓	✓	✓	✓	✓
Lysandra bellargus	✓	✓	✓	✓	✓	✓	✓			✓	✓
Celastrina argiolus	✓	✓	✓		✓	✓			✓	✓	
Aricia artaxerxes	✓	✓	✓		✓	✓	✓		✓		
Aricia agestis	✓	✓	✓		✓	✓	✓		✓	✓	✓
Polyommatus icarus	✓	✓	✓		✓	✓	✓	✓	✓	✓	✓
Plebejus argus	✓	✓	✓		✓	✓	✓		✓	✓	✓

×, no organ present, but the absence of a tick does not necessarily mean that no organ or relationship exists.

†, rudimentary organ.

Based on Hinton (1951); Downey (1966); Malicky (1969); Kitching and Luke (1985); Baylis and Kitching (1988); P. J. DeVries, personal communication; N. W. Elferrich, personal communication; J. A. Thomas, unpublished data.

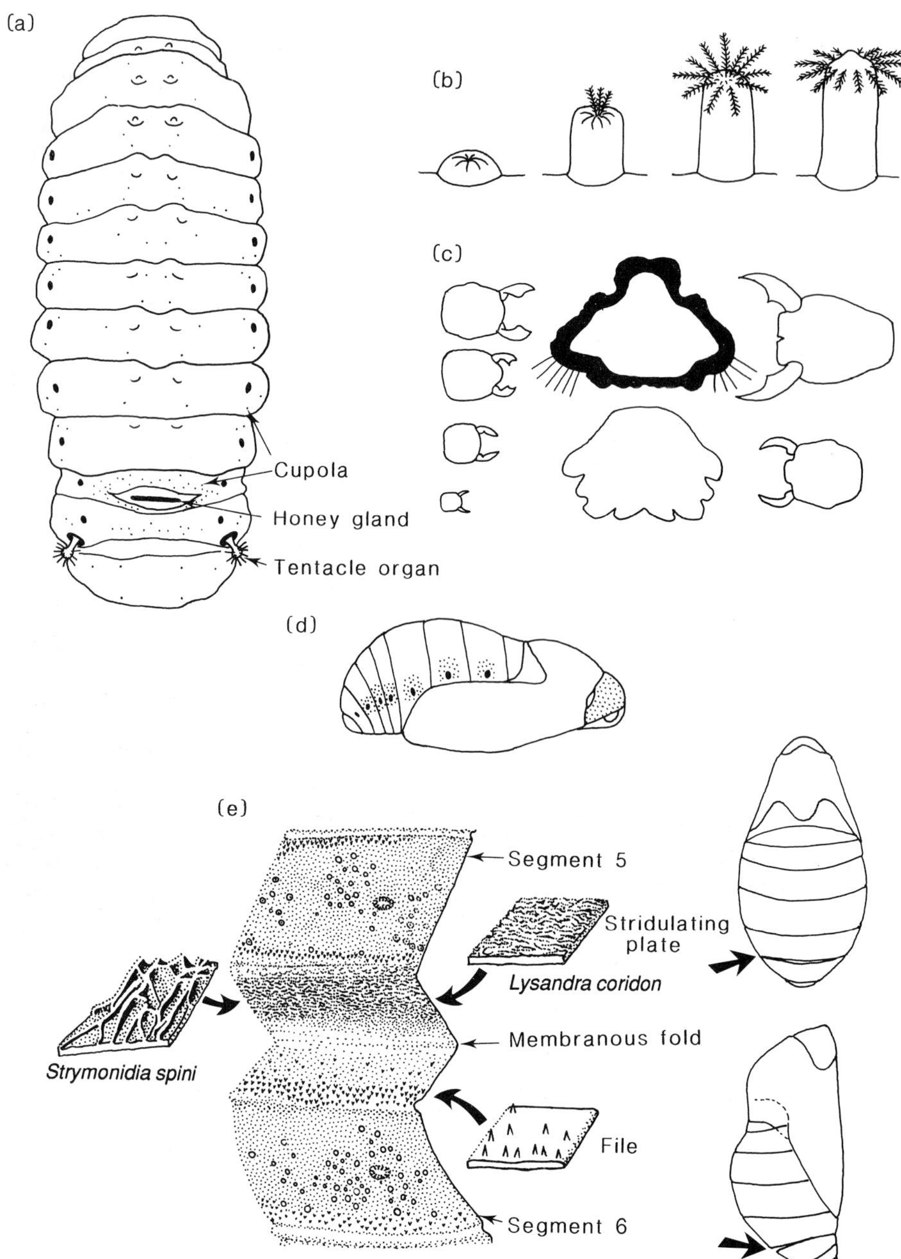

Fig. 5.5 Organs on the larvae and pupae of Lycaenidae and Riodininae that enable them to live in association with ants. (a) Dorsal view of the final larval instar of the chalk hill blue *Lysandra coridon*, showing the honey gland, everted tentacle organs, and cupola. (b) Stages in the eversion of a tentacle organ. (After Malicky 1969; courtesy of *Tijdschrift voor Entomologie*.) (c) Cross-section of a typical lycaenid larva showing thick integument and fringe of hairs (top), and cross-section of other families of butterfly larvae (bottom). The heads of various European ants are shown in outline on both sides. (After Malicky 1969; courtesy of *Tijdschrift voor Entomologie*.) (d) Distribution of cupola on a typical lycaenid pupa. (After Malicky 1969; courtesy of *Tijdschrift voor Entomologie*.) (e) The stridulating organ of a lycaenid pupa. Enlargements are shown of the stridulating plate of the two species, *Lysandra coridon* and the blue-spot hairstreak, *Strymonidia spini*. (After Downey 1966; courtesy of the Lepidopterists' Society, USA.)

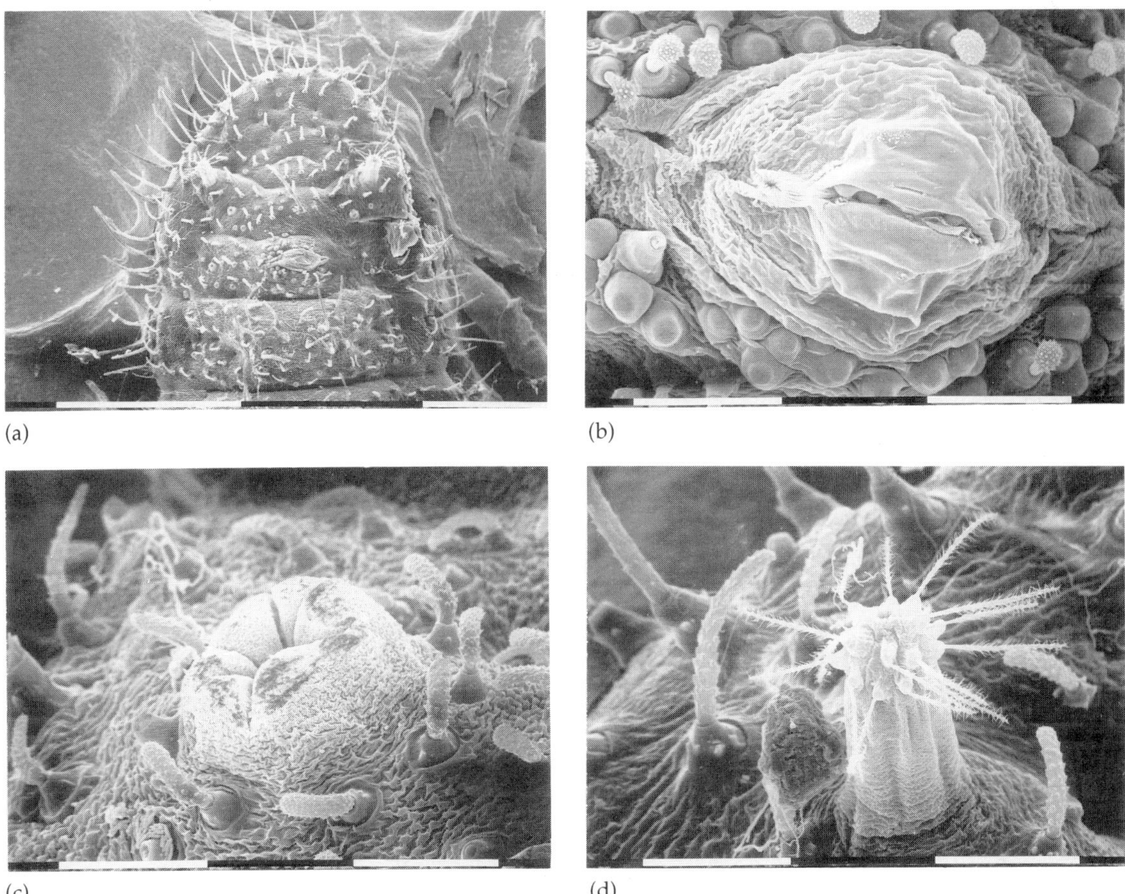

Fig. 5.6 Scanning electronmicrographs of organs on the larvae of the chalk hill blue *Lysandra coridon* that enable them to live in association with ants. (Courtesy of Barbara M. Luke.) (a) The last four segments (6–9) of a full-grown larva of *L. coridon*, showing a central honey gland on section 7 and two everted tentacles on section 8. Scale bar: 1 mm. (b) Close up of the honey gland of *L. coridon*. Note the knob-like mechanoreceptors surrounding this organ, which probably detect the drumming of the ant's antennae. Scale bar: 0.1 mm. (c) The tentacle organ of *L. coridon*, just beginning to unroll outwards from the centre. Scale bar: 0.1 mm. (d) The tentacle organ of *L. coridon*, fully everted. Scale bar: 1 mm.

two, the tentacles are withdrawn into the body by a special muscle attached to the cuticle on the lower side of the larva.

There has been much argument about the function of this organ, which may well differ from species to species. Ants seldom touch everted tentacles, and often seem oblivious to them. However, there is no doubt that the tentacles are everted much more frequently—every two or three seconds—when a larva is moving from one part of its foodplant to another, and also when the ants have wandered a little way off (Thomas, personal observation). The consensus is that they release a cocktail of volatile pheromones which alarm and activate the ants (Fiedler 1988; Fiedler and Maschwitz 1988*a*; De Vries 1988). Ants that are not already milking the caterpillar are whipped up into a frenzy when the tentacles are extruded, and are then more prone to investigate nearby objects, which brings them into contact with the larva. It is possible that this organ also releases pacifying pheromones to supplement those secreted by the pore cupola (see below) and thus further reduce the danger that the ants might attack the larva itself; there is little doubt that agitated ants are more prone to attack other potentially dangerous organisms in the vicinity that do not have pacifying pheromones. Tentacle organs are not possessed by all lycaenid larvae (Table 5.3) but, if present, usually appear in the second, third, or fourth instar, depending on the species.

Pore cupola are the commonest organ on lycaenid larvae and the only secretory organ found on pupae or first instar larvae; 52 out of 53 European Lycaenidae and Riodininae tested by Malicky (1969) possessed them, the exception being the Duke of Burgundy *Hamearis lucina*; however, Kitching and Luke (1985) located cupola on the last three abdominal segments of *H. lucina*, and have published a convincing photograph. These organs consist of specialized hairs that have become modified and reduced to form microscopic pores (Fiedler 1988). They are found in clusters mainly around the DNO, spiracles, and head of the larva and pupa (Fig. 5.5a,d). It is now certain that they too produce secretions, as larvae that possess neither tentacles nor eversible glands are coated in a mixture of amino acids, of which serine was the main component in one Australian lycaenid that has been analysed (Pierce 1984). In addition to producing a film of food over the body, there is strong evidence that these secretions also contain pheromones that appease the ants or mimic ant pheromones (Fiedler 1990*a*). The latter pay great attention to those parts of the body where cupola are concentrated, and in some cases spend as much time licking the cupola as they do in milking the DNO (Thomas, J. A., personal communication).

A few other ant-attracting organs have been described in butterfly larvae, including 'dish organs', which are saucer-like depressions found on the thorax of some riodinines and lycaenids, and along the dorsal midlines of species of *Aphnaeus* (Cottrell 1984). These secrete a fluid that attracts ants and which presumably contains sugars or amino acids for, if not removed, it becomes mouldy.

Since pupae do not feed, they are unable to secrete large quantities of food to attract ants, and merely possess pore cupola. But they compensate by having the additional ability to produce sounds (Høegh-Guldberg 1971*b*; Downey and Allyn 1973). The stridulating organ usually lies on the dorsal side of the abdomen, in the groove between the fifth and sixth segments back from the head. Here there is a large V-shaped fold of exoskeleton that penetrates deep into the body of the pupa (Fig. 5.5e). Closer examination reveals that the posterior wall of this fold contains a series of hard projecting teeth while the opposite side, nearest the head, has a toughened surface that is criss-crossed by an intricate pattern of tubercules, ridges, and reticulations, together known as the stridulating plate. Thick muscles are attached to the membrane connecting these two plates, which draw the teeth across the stridulating plate in a rapid series of contractions.

The first report of this phenomenon was made in 1774 by Kleeman, who noticed that the pupa of *Callophrys rubi* made squeaking noises when disturbed (Downey 1966). This species, and *Quercusia quercus* (Thomas and Lewington 1991), are among the few lycaenids that have stridulations that are clearly audible to the human ear. Since then, stridulating organs have been found on the pupae of a large number of blues, coppers, hairstreaks, and riodinines, including several species that live in Britain (Table 5.3). Moreover, recent recordings of live lycaenid pupae suggest that all the European species can emit noises which, when amplified, sound rather like the drumming of a woodcock or the drilling of a woodpecker (N. W. Elfferich, personal communication). Each species has its own characteristic song, which no doubt reflects the fact that there are considerable differences between the patterns on the stridulating plates of different species, as well as in the size and spacing of the teeth (Fig. 5.5e from Downey 1966). However, this organ has not been found on all pupae that sing, and some other device must be present, for example, on alcon blue *Maculinea alcon* pupae, which produce powerful and frequent noises.

Elfferich has recorded the songs of many European Lycaenidae, and has made tapes of pupae from within ant nests. Ants themselves communicate to one another by stridulating, and it is extraordinary to hear pupae emitting bursts of song and seemingly being answered by the ants. This becomes particularly frenzied just before the adult butterfly emerges. It seems likely that these songs, like pheromones, simultaneously alarm and pacify ants, so that they are on hand in an agitated state near the pupa, ready to attack enemies but not the pupa itself (Wilson 1971). Predators may even associate pupal sounds with unpleasant encounters with ants (Brower 1984). It has recently been discovered that some—possibly many—larvae also stridulate to attract ants (De Vries 1988, 1991*a*,*b*). For example, some riodinines have vibratory papillae: small hard blades that project forward from the front edge of the prothorax and which vibrate very rapidly against a hard patch or the epicranium on each downstroke and against two hardened horns on the upbeat. Similar low amplitude sounds are produced by some lycaenid larvae, including *Lysandra bellargus*, *Cupido minimus* and all the European *Maculinea* (De Vries 1991*a*; De Vries, Cocroft and Thomas, unpublished data). In all cases, the song is transmitted as vibrations through the substrate rather than as audible noise through the air. Singing is particularly pronounced when the larvae are on the move, and probably reinforces the tentacular organs, where present, in agitating and hence attracting ants.

All the above organs serve to agitate, attract, pacify, and feed ants at various times during the larval and pupal periods. However, ants are dangerous and unpredictable; even workers from the same colony will often attack each other if they become severely agitated. The bodies of lycaenid larvae have therefore evolved other adaptations with which to enhance their chances of survival. Nearly all larvae, including all British species, have an exceptionally tough and rubbery outer skin (integument) that is 20–60 times thicker than the integument in other butterfly families (Fig. 5.5c). Although this can be penetrated by prolonged attacks, it is not pierced if the ant nips the larva or tries to pick it up; at worst a small nick is made in the side. A few tropical Lycaenidae that live in close association with aggressive species of ant have an armour of stout hairs modified to form studs over the dorsal surface, whilst *Liphyra* species have hard leathery cuticles that are also impervious to attack (Cottrell 1984).

The characteristic 'woodlouse-like' shape of the lycaenid larva also helps to protect it (Fig. 5.5a). The mouth parts, legs, and other organs that cannot be cushioned, if they are to function properly, are kept hidden beneath a flattened body, the edges of which are fringed with stout downward-pointing hairs that seal off the vulnerable underside. All that the ant can see is the humped dorsal surface, with its array of attractants, agitants, and protective devices.

5.5 Pupal defences

5.5.1 Pupation sites

Pupae are relatively large and immobile. Their position results from site selection by the prepupal larvae and this is influenced by predictable patterns of hostplant seasonality, weather, and the duration of the pupal stage. However, unpredictable events and enemies are the major mortality factors. Of the 11 resident species that overwinter as pupae (Table 5.4), most pupate low in the vegetation. This includes those larvae that feed on the high parts of their foodplant, as for example *Anthocharis cardamines*, *Papilio machaon*, green hairstreak *Callophrys rubi*, holly blue *Celastrina argiolus*. High and exposed pupation sites could increase mortality through several effects, especially winter loss of aerial parts, and long-term conspicuousness to predators. Some individuals of *A. cardamines* and *P. machaon* may sometimes pupate high up but in such cases the pupation site is sheltered and the pupae cryptic.

Detection of, and access to, pupae of some species (e.g. *Ochlodes venata*, red admiral *Vanessa atalanta*) by predators is minimized as a result of their being concealed, for instance contained within rolled or folded leaves or within a loose tent of silk spun by the final larval stage before pupation (e.g. *Thymelicus sylvestris*, silver-spotted skipper *Hesperis comma*). However, silk may also prevent the access to pupae by some parasitic insects. When a predator encounters a pupa that is attached to a substrate, it must be able to remove it in order to consume the whole. Baker (1970) found that predation by birds on *Pieris brassicae* and *P. rapae* differed. The more firmly attached pupae of *P. brassicae* are less frequently eaten. However, *P. brassicae* pupae are protected chemically (containing sinigrin), more so than are *P. rapae* (Aplin *et al.* 1975) and this is also likely to influence predation by birds (see p. 105).

Many species pupate on or near the ground, influencing the ease with which they can be detected. *Hipparchia semele* buries itself in sites with

Table 5.4 The usual pupation sites and major defences of British butterfly species, including common breeding migrants

Species	Pupal position and height				Form of visual protection		
	attached	suspended	litter	buried	background matching	resemblance to plant parts	
						dead	live
Species with overwintering pupae							
Pyrgus malvae	Low in cocoon of foodplant					+	
Papilio machaon[1]	Low				+	+	+
Leptidea sinapis	Low				+	+	+
Pieris brassicae[1]	Low				+	+	+
Pieris rapae[1]	Low				+	+	+
Pieris napi[1]	Low				+	+	+
Anthocharis cardamines	Low				+	+	
Callophrys rubi[2]	Low		+	+	+	+	
Celastrina argiolus[1,2]	Low		+		+	+	
Pararge aegeria[1]		Low			+		+
Species which form pupae at other times of the year							
Carterocephalus palaemon		Low in loose grass tent				+	
Thymelicus sylvestris		Low in cocoon in grass tent					+
Thymelicus lineola		Low in cocoon in grass tent					+
Thymelicus acteon		Low in cocoon in grass tent					+
Hesperia comma		Low in cocoon in grass tent					+
Ochlodes venata		Low in cocoon in grass tent					+
Erynnis tages			In silk web				+
Colias croceus	Low				+		+
Gonepteryx rhamni		Low			+		+
Thecla betulae[2]			+	+	+		
Quercusia quercus[2]			+	+	+		
Satyrium w-album[2]	High				+		+
Satyrium pruni[2]	High				Resembles bird dropping		
Lycaena phlaeas	Low				+		+
Lycaena dispar		Low			+	+	
Cupido minimus[2]		Low	+	+	+	+	
Plebejus argus[2]				+			
Aricia agestis[2]			+	+	+		
Aricia artaxerxes[2]			+	+	+		
Polyommatus icarus[2]			+	+	+		
Lysandra coridon[2]			+	+	+		
Lysandra bellargus[2]			+	+	+		
Maculinea arion[2]				+			
Ladoga camilla		High			+		+
Apatura iris		High			+		+
Vanessa atalanta		Low			+		
Cynthia cardui		Low			+		
Aglais urticae		Low			+		

Species	Pupal position and height				Form of visual protection		
	attached	suspended	litter	buried	background matching	resemblance to plant parts	
						dead	live
Nymphalis polychloros		High			+		
Inachis io		Low			+		
Polygonia c-album		Low			+		
Boloria selene		Low	+		+		
Boloria euphrosyne		Low	+		+		
Argynnis adippe	Low in loose tent of vegetation						+
Argynnis aglaja	Low in loose tent of vegetation						+
Argynnis paphia		Low			+	+	
Eurodryas aurinia		Low			+		
Melitaea cinxia	Low in loose tent of vegetation					+	
Mellicta athalia		Low			+		
Lasiommata megera		Low			+		+
Erebia epiphron	Low in loose tent of foodplant						+
Erebia aethiops			+	+	+		
Melanargia galathea			+		+		
Hipparchia semele				+	+		
Pyronia tithonus		Low			+		+
Maniola jurtina		Low			+		+
Aphantopus hyperantus			In loose cocoon		+		
Coenonympha pamphilus		Low			+		+
Coenonympha tullia		Low			+		+

[1] Has one or more generations per year and pupates at other times.
[2] Ant attended, pupae may be buried by attendant ants.
Low < one metre; high > one metre.
Background matching refers to resemblance to pupation site background; those species which also have a resemblance to plant parts have a general resemblance to the coloration of specific sites where they pupate.

loose soil before pupating underground, or moves under stones where detection is impossible. The larvae of *Celastrina argiolus* and *Quercusia quercus* also move to underground sites to pupate, burying themselves in ground cracks or under leaves and moss (Thomas 1986) and in the latter species being taken into ant's nests (Heath *et al*. 1984). With the exception of the lycaenids, other ground-pupating species are located on the soil surface or among litter, relying on background matching for protection (Table 5.4). Most lycaenid larvae pupate in sites where ants frequently occur (see section 5.4).

5.5.2 Defences from visually-hunting predators

Pupae of British species are nearly always extremely cryptic. They are located in hidden sites and/or rely on background matching or resemblance to a non-food item. Like larvae, the colour of pupae is related to the types of enemies normally associated with pupation sites and the types of protection that the pupae can use after detection. Some, such as that of *Satyrium pruni*, resemble bird droppings and occur in sites where such droppings would be found, for example on the upper surfaces of twigs or leaves. Others which pupate within green vegetation, such as *Lasiommata megera* and *Pararge aegeria* are green in colour and are suspended from the underside of green grass blades which they generally resemble. Equally common is some form of brown or black colour often enhanced with elaborate shading and highlights or spangling with gold spots (e.g. *Aglais urticae*) which resembles dry vegetation. Even in those species which appear to have strikingly obvious pupae when seen close to, as in *Eurodryas aurinia* and the heath fritillary *Mellicta athalia*, individuals closely resemble the resting background when seen from a distance. In the case of *E. aurinia* the pupae are white with black and orange markings, resembling the underside of bleached leaves of dead vegetation. Both these species may appear warningly-coloured when seen close to.

In the three British *Pieris* species, and in *Anthocharis cardamines*, *Papilio machaon*, and *Leptidea sinapis*, pupal colour is variable. This variation, or polyphenism, is under environmental control (e.g. Smith 1978, 1980; and see pp. 188–91), individuals of these and related species tending to be of the colour form which best matches their resting background (Wiklund 1975; West and Hazel 1982). This phenomenon is probably of importance to pupal survival at different times of the year. Individuals least likely to survive detection will be those which contrast with their background and the ability to vary colour enhances survival when backgrounds vary in colour, especially with season (see section 8.5).

5.5.3 Secondary mechanical defences of pupae

Independent of primary defences, pupae have a range of structural and behavioural characteristics which can deter attacks, particularly by pupal parasites. Many satyrine and nymphalid pupae are capable of restricted movement, and will jerk when disturbed. Although this jerking can prevent attack by vertebrate predators it is probably of more importance to escape parasitoid attacks. It is most rapid when pupae are touched by parasitoids and slowest when they are gripped in a manner similar to a bird or rodent attack (Cole 1959). Newly formed pupae of *Aglais urticae* were found to be easily stabbed by Ichneumonid species (*Apechthis* spp.) but older pupae, because of their rapid vibration and hard burnished cuticle make it impossible for the parasitoids to grip the pupa or press the ovipositor through the cuticle. Similarly, newly formed pupae of *Pararge aegeria* were stabbed by *Apechthis resinator* but the smooth waxy cuticle and slight abdominal movements of pupa 24 hours old or more prevented attack. Not all pupae are so protected. While a smooth cuticle may prevent or hinder attack by making the parasitoid itself vulnerable to disturbance or attack if handling time per pupa is long, rough surfaced pupae can be more vulnerable. Pupae of *Pieris brassicae* have a cuticle which facilitates grip by parasites, being easily handled and penetrated by species of *Apechthis* (Cole 1959) and by *Pimpla instigator* (Picard 1922). However, a rough cuticle, especially if associated with spines, may deter vertebrate predators.

5.6 Butterflies and multiple defence mechanisms

All stages have a variety of natural enemies, which may be specialists or generalists. Most defence mechanisms, such as crypsis, are effective against a variety of enemies. Because every stage is liable to be attacked by a variety of enemies, which differ in their prey-locating and handling mechanisms, butterflies have multiple primary and secondary defence mechanisms (Fig. 5.7). Adaptations which reduce attacks by generalist enemies are usually less specific and less highly evolved than those which operate against specialist enemies. A small adjustment may prevent attacks by some of the former because these enemies have prey-locating and handling mechanisms which cope with a variety of prey items. More complex adaptations are required to circumvent specialist attacks since the biology of these enemies is more closely tied to the biology of a limited number of prey types.

The defence mechanisms of the stages of particular species are related to the predictability of enemies. The most effective defences will, therefore, be those that reduce the probability of attacks from enemies which have the potential to cause the most mortality. These enemies may not be of significance in all years (see Chapter 4) and their activity is in part related to the abundance and activity of other members of the community in which the butterfly species lives.

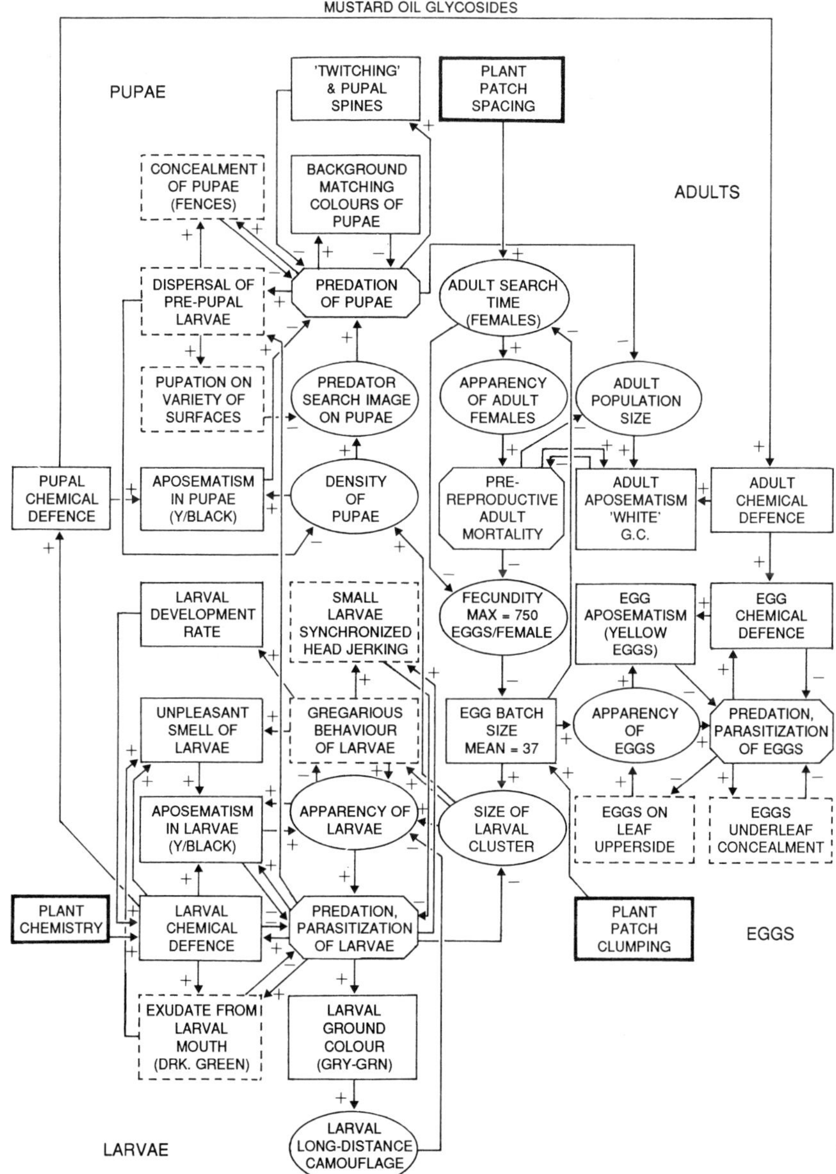

Fig. 5.7 A hypothetical model for some of the morphological, physiological, behavioural, and ecological consequences (i.e. various defences) of cluster egg-laying in the large white, *Pieris brassicae*. Influences and responses associated with climate (e.g. reduction of desiccation by underleaf egg placement), hostplant quality, and size (e.g. starvation and egg-load assessment by females) are excluded. Boxes with bold outlines, hostplant–habitat precursors of cluster egg-laying; boxes, morphological and physiological adaptations; boxes with pecked outlines, behavioural adaptations; octagons, predation and/or parasitization; ellipses, other variables. For the interpretation of signs, see **process-response models** in the glossary. (Data from J. Feltwell 1981*a,b*, and personal communication; B. O. C. Gardiner, personal communication; Aplin *et al*. 1975; Rothschild *et al*. 1977.) Chemical defence is transmitted from larvae, through pupae and adults, into the next generation of eggs. Pupae typically conceal themselves by background colour matching, but are also distasteful to predators. Most overwintering British pupae are brown and the summer pupae are green. However, Rothschild *et al*. 1977, believe that the bright green diapause (overwintering) pupae of the reared Cambridge strain are aposematic. (After R. L. H. Dennis)

6

Monitoring butterfly movements

Tim G. Shreeve

6.1 The components of movement

Whereas the success of a mated adult female depends on its ability to locate and use appropriate egg-laying sites, that of a male depends on its ability to find and mate with fertile females. To fulfil these functions adults must use resources in the habitat, such as food and roosting sites, and in the process of doing so maintain an adequate heat budget to be active as well as to avoid predators. Thus, the area occupied by a population of adults must consist of at least four functional categories of habitat: mating areas, egg-laying areas, foraging areas, and refuges or roosting sites.

For some populations these resources occur together within habitats but for the majority they are to some extent spatially segregated, necessitating essential within-habitat movements for reproductive success. Many species of butterfly occupy temporary, or seral, habitats in which case the survival of an individual's offspring requires adults to find new habitats when the existing habitat becomes unsuitable. Additional factors such as competition for limited resources or interference between individuals, resource-lifespan and distribution, may also influence individual movement.

All butterflies are potentially mobile, but those which occur in long-lived habitats where all their requirements occur together and that use hostplants which remain suitable for many generations tend to be sedentary (e.g. black hairstreak *Satyrium pruni*, small blue *Cupido minimus*). These individuals rarely move from one colony to another. Because of their low mobility such species have closed population structures. Individuals of such species which remain within a single habitat may nevertheless spend a substantial amount of their time in flight and the sum total of all their daily movements or 'trivial' flights may be considerable. However, their net displacement, or dispersal, is nil if they remain in a single area (Fig. 6.1).

Species such as the large white *Pieris brassicae* and the peacock *Inachis io* which use temporary habitats and short-lived hostplants have more open population structures. The boundaries of any population are difficult to define because of the movement of individuals between habitats. In such species this movement results in the displacement of individuals from their emergence sites (Fig. 6.1). For any individual the sum total of all flight distances, or movements is usually greater than the displacement, or range travelled over the whole lifespan, but the ratio of the total flight distance to displacement is much less for butterflies with an open population structure than for sedentary individuals of colonial species.

Populations of even the most sedentary species produce some individuals that move, or disperse, from the habitat in which they developed. These individuals usually show no predictable direction in their movement, and as far as is known the majority do not attempt to return to their original habitat. Butterfly migration can be distinguished from dispersal because the former involves individuals moving in a particular direction at one time or season, whilst dispersal is unpredictable in both direction and timing. A reverse movement, usually by individuals of a subsequent generation, may be associated with butterfly migration. Some of these definitions of migration and dispersal differ from those of Baker (1978) but may be more familiar to the entomologist. Perhaps the best known examples of butterflies which migrate to and from Britain are the red admiral *Vanessa atalanta* and painted lady *Cynthia cardui*, which both arrive in spring and may

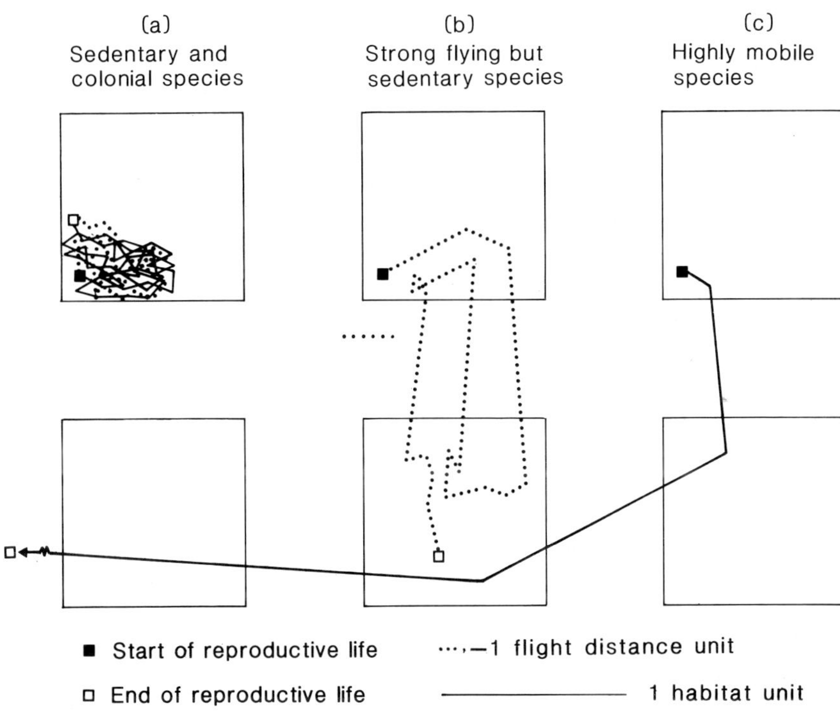

Flight parameters	Species type		
	Sedentary and colonial	Strong flying but sedentary	Highly mobile
Total flight distance during lifespan	50 distance units	50 distance units	50 distance units
Number of turns during lifespan	50	17	3
Area over which offspring are distributed	<1 square habitat unit	2 square habitat units	>4 square habitat units
Maximum range*	5 distance units	15 distance units	*c*.50 distance units
Distance between turns (flight steps)	Small	Medium	Large

A flight distance unit is the maximum distance moved in one time unit.
* Distance between furthermost points of capture.

Fig. 6.1 A hypothetical comparison of the components of movement using a mark–release–recapture technique on three species with different population structures and mobilities.

Table 6.1 Frequent and infrequent migrant butterfly species to the British Isles, during the years 1980–90

Regular migrants incapable of overwintering, but which breed in the summer	Resident species[1] supplemented by migrants	Occasional migrants,[2] some of which may form temporary populations
Colias croceus	*Pieris brassicae*	*Nymphalis polychloros*
Colias hyale	*Pieris rapae*	*Nymphalis antiopa*
Colias alfacariensis	*Pieris napi*	*Araschnia levana*
Vanessa atalanta[3]	*Inachis io*	*Issoria lathonia*
Cynthia cardui[3]	*Aglais urticae*	*Iphiclides podalirius*
		Papilio machaon[4]
		Lampides boeticus
		Danaus plexippus
		Melitaea didyma

[1] May include other common species, but infrequent migrants may be indistinguishable from resident species.
[2] Infrequent migrants may be unrecorded and the variety of migrant species may therefore be greater than that recorded. However, some records may also represent introductions (see Cribb 1986).
[3] May overwinter in some years (see Gardiner 1986; Pollard *et al.* 1986).
[4] Subspecies *gorganus*, widespread in Continental Europe.
From data collected by Bretherton and Chalmers-Hunt 1981–90.

have a small reverse movement in the autumn. These species are not unique; other common species (e.g. *Pieris brassicae*) located in the British Isles are often supplemented by immigration of individuals from continental areas (Table 6.1).

The movement of individuals from one population to another can have significant effects on the gene pool of the 'invaded' population. Large scale movements of fertile individuals can result in the breakdown of genetic isolation between individual colonies and demes (see chapter 9); in this situation local aggregations form part of a much larger integrated unit (metapopulation). By contrast the absence of gene flow, most frequently as a result of habitat fragmentation and destruction, can have important consequences for local adaptations including those which affect flight patterns. Dempster *et al.* (1976), studied morphometric characters of adult swallowtails, *Papilio machaon* from Wicken Fen and the Norfolk Broads from 1820 to 1940 and 1960 respectively, recording changes in wing and body characters which affect flight capacity. Changes in the Wicken population occurred after it became isolated and its habitat was reduced in size (see section 11.2). This, it is suggested, was the result of both isolation and selection against mobility. Similar, but later, changes in the Norfolk Broads population may also be the result of selection against mobility because of habitat loss and fragmentation.

6.2 Variability in butterfly movement

Adults of some species, such as *Cynthia cardui* and *Vanessa atalanta*, regularly reach Britain from as far away as southern Europe and North Africa, yet others, such as *Satyrium pruni* and the silver-studded blue *Plebejus argus* show no such spectacular movement. Instead, adults of these species tend to remain in restricted areas throughout their lifespan, moving as little as 50 metres from where they emerged from the pupa (Thomas 1984*a*; Thomas, C. 1985*a*). Between these two extremes different species exhibit a range of movements. For example, the nettle-feeding small tortoiseshell *Aglais urticae* and *Inachis io* rarely remain in particular areas for long periods of time. On the other hand the white admiral *Ladoga camilla* and speckled wood *Pararge aegeria*, are apparently sedentary, most adults remaining in specific woodland areas throughout their lives. However, their recorded range expansions in the 1950s belie the sedentary nature of these last two species, a phenomenon that must have involved individuals moving from existing populations to form new colonies.

No two individuals of the same species behave identically. Different individuals have unique requirements dependent on their previous activities, their age, sex, and genotype. Thus, in addition to differences between species, there are differences within and between populations of the same species. Studies of such variability may reveal those environmental and genetic factors which determine the nature of individual movements and population structure.

Butterflies that overwinter in the adult stage have different requirements in autumn than in spring, consequently they tend to have different behaviour patterns in different seasons. All overwintering adults are in reproductive diapause, becoming reproductive in spring when conditions are appropriate for mating and egg-laying. The brimstone *Gonepteryx rhamni* is the longest lived adult butterfly resident in Britain, and in late summer and autumn adults spend almost their entire time feeding, in order to build up reserves of lipids for successful overwintering. In some parts of Britain breeding sites and overwintering sites are spatially segregated. Feeding individuals move from breeding sites to woodland or woodland edge habitats where they feed on floral nectar, especially from thistles *Cirsium* spp. and black knapweed *Centaurea nigra*. Once feeding commences they are relatively immobile and will then move short distances to suitable overwintering sites, such as bramble and ivy thickets, often within woodland (Pollard and Hall 1980). In spring, both sexes become more mobile. Females search for suitable egg-laying sites, bushes of purging buckthorn *Rhamnus catharticus* and alder buckthorn *Frangula alnus*, but do not remain long in such areas. Instead, they constantly move in search of new sites, infrequently returning to localities where they have previously laid their eggs, thereby distributing their offspring over a large area. Males continuously search for unmated females which, because of their mobility, tend to be widely distributed.

Within populations there may be differences in the mobility of various individuals of the same sex. In the case of the highly sedentary *Cupido minimus*, male mobility is related to the number of females they encounter (Morton 1985). Males perch in wait of females in warm areas where nectar sources and the hostplant, kidney vetch *Anthyllis vulneraria*, occurs. Those individuals that perch in areas where they encounter few females are more mobile than those that perch in areas where more females occur. Age and reproductive status were found to have a significant influence on female mobility. Unmated females forage in areas where there is a high probability of mating, but mated females leave these areas by making longer flights, thereby reducing the probability of male harassment and maximizing the time available for egg-laying and feeding.

Individuals of the meadow brown *Maniola jurtina* with the most spots on the underside hindwings have a tendency to move further than those individuals with fewer spots (Brakefield 1982*a*). The function of this difference is not clear, but it does indicate that there is a genetic component to variation in mobility within this species, because spot frequency is in large part genetically determined. In another satyrine butterfly, *Pararge aegeria*, differences between the mating behaviours of males with 3- and 4-spot upper hindwing phenotypes have been identified (Shreeve 1987; chapter 2); 3-spot males tend to patrol in search of females and 4-spot males tend to perch. Not only do 3-spot males engage in more flight activity but they also remain in particular localities for shorter periods than do 4-spot males; consequently they are more mobile and have a greater potential to move out of existing habitats.

Because the two sexes have different activities and behaviour patterns there are often marked differences between their respective movements. In species that adopt perching mate-locating behaviour, males are less mobile than females because males, having located suitable perching sites, tend to remain there, whereas females move in search of egg-laying sites. No such clear patterns emerge in species that adopt patrolling behaviour. In the orange-tip *Anthocharis cardamines*, there is some evidence to suggest that males tend to remain and patrol in particular areas but that females are more vagrant, rarely returning to areas previously visited (Wiklund and Åhrberg 1978; Dennis 1982*b*, 1986*b*). In contrast, male *Pieris brassicae* tend to move more than females (Baker 1978; see p. 129), as do male wood whites *Leptidea sinapis* in the first few days of their life (Wiklund 1977*a*; Warren 1984).

Movement patterns are intricately linked with the distribution and predictability of resources. Since different populations of the same species are located in dissimilar sites, it is not surprising that their movement patterns are frequently found to differ. However, for most species in which movements differ between populations it is not known whether these differences have evolved in response to different environments or whether flight behaviour is plastic and can be modified by the individual in response to the environment (Ehrlich 1984). Small scale transplant experiments done with the chequer-spot *Euphydryas editha* in North America (Gilbert and Singer 1973) in which larvae from a mobile population were moved to the site occupied by a sedentary population, suggested that the emergent adults were more mobile than the resident individuals which fly earlier. However, this could be related to diminishing nectar supplies, necessitating longer flight distances, rather than to a genetic basis for mobility (Ehrlich 1984). Morton (1985) transferred individuals from mobile and sedentary populations of *Cupido minimus* to a new, but previously unoccupied, habitat and concluded that any genetic differences between populations were insufficient to override the influence of the abundance and distribution of nectar and hostplants in determining the mobility of females and also possibly of males.

In southern Britain habitats which are suitable for *Anthocharis cardamines* are more abundant than in northern Britain. There is some evidence that individuals from southern areas may be more mobile than those from northern populations (Palmer and Young 1977; Baker 1978; Courtney 1980; Dennis 1982*b*, 1986*b*; Dempster 1989). This may be related to immediate reproductive gains and potential risks associated with moving from existing areas to new areas, and the model (Fig. 6.2) proposed by Dennis (1982*a*) to account for this variation is based on genetic variation and selection for dispersal genes. In northern areas mobility may be less than in southern areas because the risks of leaving one habitat and not finding an alternative may be greater than any potential advantages to be gained from finding new habitats. In southern areas the abundance of suitable habitats promotes mobility because the risks of moving are less than the reproductive losses from remaining in particular areas, especially as larvae are cannibalistic. At high population density a significant proportion of an individual's offspring may die as a result of intraspecific competition for limited larval resources (Courtney and Duggan 1983). Immediate gains and risks may also be related to the time available for flight activity: northern areas have lower temperatures and fewer sunshine hours, which reduces the time available for flight (see section 2.2) and increases the risks associated with locating new habitats.

Most species are behaviourally flexible (see chapters 2 and 10; Dennis and Shreeve 1988) and this may enable populations of many species to cope with a variety of habitats of different qualities. Behavioural flexibility may well mask underlying genetic differences in flight behaviours of those species which have been studied, but the significance of a genetic component should not be discounted in butterflies, especially as it is involved in flight duration and variation in the African armyworm moth *Spodoptera exempta* (Parker and Gatehouse 1985).

6.3 Local movements of butterfly adults

6.3.1 Studying butterfly movements

Whilst it is possible to describe broad differences between the mobility of species and also between individuals, one of the major difficulties of studying mobility is in determining just how far butterflies travel. The most commonly applied methods tend to

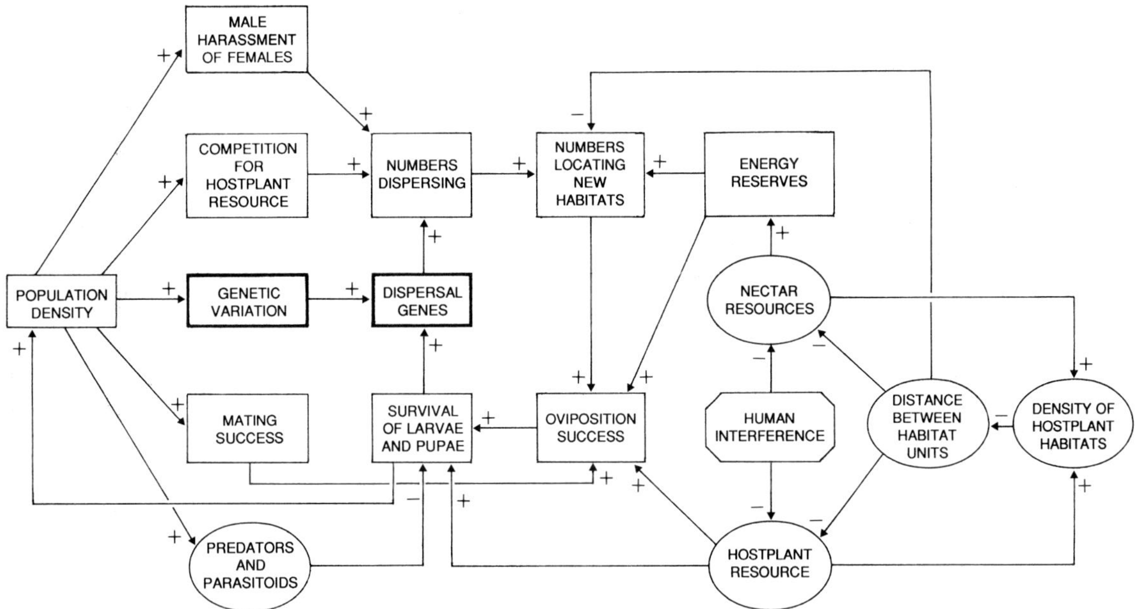

Fig. 6.2 A model of the basic factors which affect the mobility of the orange-tip *Anthocharis cardamines*. Central to the model is a genetic component to dispersal. +, an increase (decrease) in one variable results in an increase (decrease) in another variable; −, an increase (decrease) results in a decrease (increase) in another variable. Rectangles, butterfly variable; rectangles with bold margins, genetic variables; octagons, ecosystem regulators; ellipses, habitat variables. (After Dennis 1982*a*; courtesy of *The Entomologist's Gazette*.)

underestimate both the frequency of long distance movements and the distances moved during longer flights because such movements are often the hardest to detect. If movements between populations are undetected then the role of gene flow will be underestimated. The significance of movements to population structure and their role in gene flow can also be overestimated if movements involve post-reproductive adults of either sex.

Two main methods of study are open to the investigator: (i) marking individuals and recapturing them at subsequent times in known locations (mark–release–recapture MRR), and (ii) following individuals. Both methods can produce interesting and valuable data but both are subject to assumptions and errors. The major assumption of marking is that the method of marking and subsequent re-identification has no effect on the behaviour of the marked individuals. Such assumptions may be easily disputed (Begon 1979; Morton 1982), marked individuals being more or less likely to be recaptured than unmarked species. The most common problems associated with marking are those which affect behaviour, notably when individuals engage in escape flights after handling, and those which increase predation on the marked sample. If the marked and recaptured sample is not representative of the whole population then hypotheses based on what becomes unrepresentative population sampling may not be valid. MRR techniques are usually carried out in a predetermined area, and only those movements within the area are recorded. Marked individuals that move out of the area are effectively lost. Without the appropriate use of such alternative methods (i.e. tracking individuals) such a technique will possibly underestimate mobility of the whole population. Furthermore this type of study, whilst appropriate for determining movements and periods of residence within the study area, will show only the distances moved between recapture times. Such methods are not suitable for revealing how individuals move at times other than those chosen for sampling (Fig. 6.3).

An alternative method to MRR is to follow

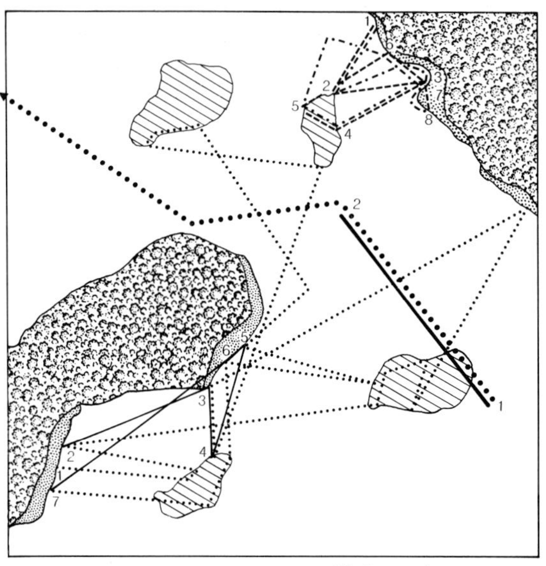

Fig. 6.3 A comparison of movement data obtained from mark–release–recapture with that obtained from tracking using species with three types of population structure within one habitat area.

Mobility parameters	Probable errors during period of survey	
	Periodic mark—recapture	Individual tracking
Total distance moved	Underestimated for all species types	Quantified for all but most mobile species
Habitat use	Poorly quantified	Well known for followed individuals
Area over which offspring are distributed	Underestimated	Well known for followed individuals
Range	Underestimated for all but most sedentary species	Quantified for followed individuals only
Individual behavioural differences	Partially detected by range differences	Detected in followed sample
Sampling intensity	High	Low
Sampling difficulty	Little to great	Moderate to very great

individuals for as long as possible, in an attempt to discover the 'life-time track' or daily movements of individuals. Such a method avoids marking and handling errors and should be seen as complementary to MRR. The method is extremely time consuming but can reveal much about how individuals behave and move within or between habitats (Fig. 6.3), facilitating extremely detailed mapping of the movement of individuals, particularly when every landing point and behaviour is recorded (see Morton 1985). By using this method the number of individuals on which data are obtained is often small, and MRR and individual observation should be used together, the first to obtain data on a large sample and the latter to relate movements to other types of behaviour. However, followed individuals may be easily 'lost' either when they cross obstacles or even when they interact with other individuals, and it is rarely possible to follow any one individual for long periods.

Other methods have been used with considerable success in studying local movements. Jones *et al.* (1980) successfully tracked the movements of egg-laying females of the small white *Pieris rapae* by a variant on MRR. Reared larvae were fed on a diet incorporating a dye (Sudan Black or Sudan IV), which enabled the movements of released adult females to be tracked by locating their abnormally grey and pink coloured eggs. Dennis (1982*a*) mapped the distribution of eggs of *Anthocharis cardamines* in part of the Bollin Valley in Cheshire. By using such a method, which indicates where females have been, it was possible to plot the distribution of satellite populations and examine the degree to which the hostplant resources were being exploited.

6.3.2 Factors affecting local movements

For a species of butterfly to persist in a particular location adult butterflies must be able to locate both larval and adult resources within that area. Thus species with the least specialized requirements and the most widespread resources may be expected to be more mobile than the most specialized species. For example, the chalk hill blue *Lysandra coridon* and the Adonis blue *Lysandra bellargus* rarely move from existing colonies where the sole larval foodplant, horseshoe vetch (*Hippocrepis commosa*), and the ant species *Myrmica sabuleti* and/or *Lasius alienus*, which attend the larvae, are located in short and warm calcicole vegetation. By contrast, species with less specialized requirements, such as *Anthocharis cardamines*, that deposit eggs on the developing seed-pods of a range of cruciferous plants in a variety of open habitats, tend to be more mobile.

One of the least mobile butterflies resident in Britain is the lycaenid *Plebejus argus*, with numerous local races or forms (de Worms 1949; Dennis 1977). Using a mark–recapture technique C. Thomas (1985*a*) monitored the movements of adults from six different populations, two located in heathland and the remainder in limestone areas. Most individuals moved less than 20 metres per day and movements over several days were no greater than those made in a single day. There were also no significant differences between the mobilities of individuals from the various sites, despite differences of habitat size and resource types. These results raise several interesting questions about the nature of local butterfly movements. In heathland habitats, populations of this butterfly persist until successional changes of the vegetation eliminate sites suitable for egg-laying and larval development, a process which occurs between 5 and 10 years after an area becomes suitable. By contrast populations on limestone are known to persist for 80 years or longer (C. Thomas 1983). Here, the vegetation is maintained in a suitable condition by the instability of the soil which is on very steep slopes.

Southwood (1962) postulated that insect mobility is closely related to the lifespan of their habitats. Species which occupy temporary habitats must be more mobile than those in long-lived habitats because habitat impermanence will select against the least mobile organism. The apparent lack of mobility of the heathland population of *Plebejus argus* is argued by C. Thomas (1985*a*) as contradicting this hypothesis. Several explanations can be offered to account for this, which need not be mutually exclusive. First, the method employed did not recover all the marked individuals. Those that were never recaptured may have moved out of the area used in the experiment. In this study it is reasonable to assume that a greater proportion would do this in the temporary heathland than in the limestone habitats, but evidence for this was lacking (Table 6.2). Even if only a small number of individuals leave

Table 6.2 The distances moved by adult *Plebejus argus* in long-lived limestone habitats and one temporary heathland habitat during 1983

Site		Number of individuals and distances (metres) moved between captures					
		male			female		
		<20	20–50	>50	<20	20–50	>50
Limestone grassland							
Great Orme		7	2	0	3	0	0
Dulas Valley,	A	33	4	0		No data	
North Wales	B	18	6	0		No data	
	C	12	3	0	1	0	0
	D	7	2	0		No data	
Heathland							
Holy Island,	A	37	1	0	19	0	0
off Anglesey	B	8	0	0	3	1	0

Data for limestone grassland are for same-day captures, those for heathland over a period of two days.
After Thomas, C. D. (1985*a*).

the habitat, the rate of emigration may be sufficient to establish new colonies. Other observations of the butterfly in North Wales (see Dennis 1977, pp. 130 and 173) suggest that it is capable of forming colonies at a distance of at least 1 km from the founder population. At the time of sampling the heathland habitats may have been ideal; C. Thomas (1983) cites some evidence to suggest that mobility increases as habitat quality deteriorates. An alternative explanation may be related to the evolutionary time scale. Suitable heathland habitats are created by disturbance—grazing or furze burning—and were probably more common than they are at present. Consequently short flight distances may have evolved in a habitat mosaic where suitable areas were separated by short expanses of unsuitable habitat over which *P. argus* could easily fly. Scarcity of habitat is a recent phenomenon to which the butterfly may have made no adjustment. As *P. argus* is single brooded there have been between 50 and 100 generations since habitat isolation became a significant factor in the ecology of the species.

The heath fritillary *Mellicta athalia*, is another species which occupies temporary habitats. In woodland areas these are short-lived clearings (5–10 years). Where suitable sites (see section 11.4.4) are available and are in close proximity individuals are able to fly between them, the maximum recorded range of males being 1120 m and of females 1000 (Warren 1987*b*). Critical to the relationship between habitat lifespan and mobility is the fact that some individuals leave existing populations (Fig. 6.4). Such movements are scarce, most individuals having a range of less than 150 m. The factors influencing the rate of movement from any existing habitat patches were not precisely determined, but were thought to be dependent in part on both the size of the site and its suitability as a breeding area. There is some indication that the greatest rates of dispersal occurred from the smallest sites and those which were the least suitable (Table 6.3).

When populations are located in habitats where resources are spatially segregated, individuals must be more mobile than in sites where resources are co-located. While this may mean that individuals move greater distances between resources within a habitat, it does not necessarily imply that the net displacement, or range, of individuals is greater when resources are segregated. Warren *et al*. (1986) using MRR, and Wiklund (1977*a*) using observational data found that the mean range of *Leptidea sinapis* in pure woodland sites where nectar sources and egg-laying

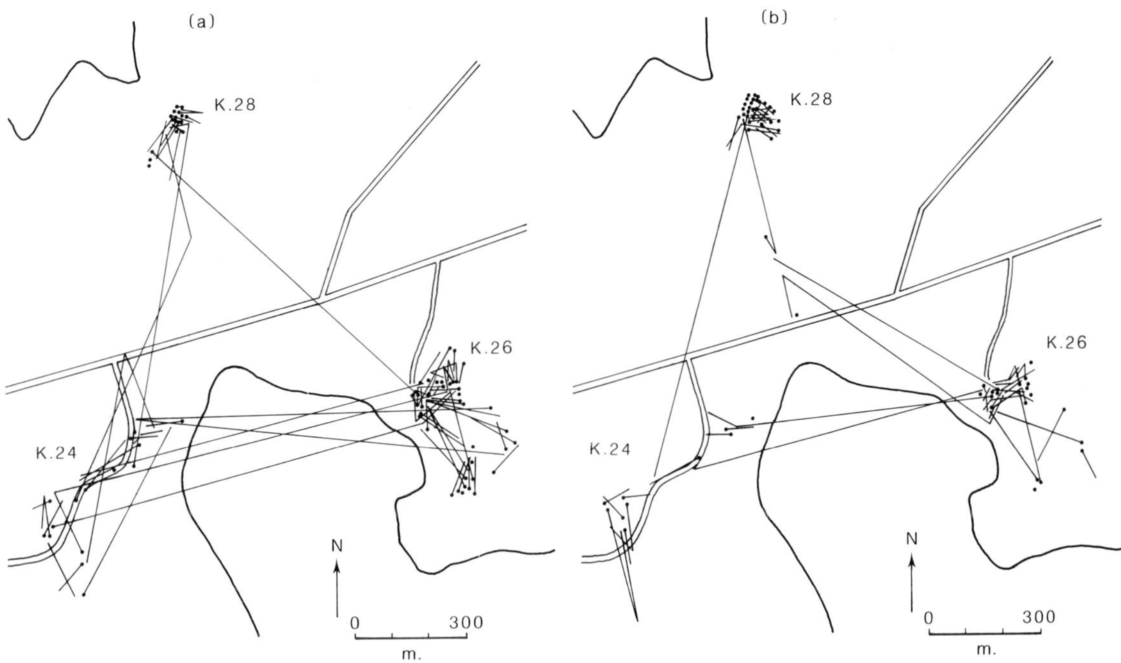

Fig. 6.4 Recorded movements of heath fritillary *Mellicta athalia* within and between three coppice areas of Blean Woods, Kent, in 1982. (a) males; (b) females. Most movements are within coppice areas, but a small proportion of movement in this sedentary species was between three coppice areas. Flights to other areas may be undetected but are unlikely to result in the formation of new colonies because of a scarcity of suitable habitats. (After Warren 1987*b*; courtesy of *Journal of Applied Ecology* and The British Ecological Society.)

sites occur together was broadly similar to that in a meadow and woodland site where nectar and egg-laying sites are segregated.

As an individual butterfly ages its physiological state changes, and if it has engaged in previous reproductive activity (mating or egg-laying) the relationship between the values of the costs and benefits of moving to new areas changes. For example, the potential loss of total reproductive output associated with moving to a new habitat is less for a female that has already laid some eggs than for a female that has yet to lay any eggs, providing offspring success is assured in the original habitat. This may account for increasing mobility of female *Mellicta athalia* with age (Warren 1987*b*), though physiological processes may also be involved. On emergence, females have large abdomens which are heavy with eggs. This egg weight may restrict their capacity for flight and it is not until the first large egg-batch is laid some two or three days after emergence that their abdomen becomes lighter and flight activity increases.

Baker (1978) has found that female *Pieris rapae* and *P. brassicae* are more mobile during the first two or three days of adult life than later. This, it is argued, is closely related to the ephemeral nature of their egg-laying sites, which are located in early successional habitats and therefore remain suitable for larval development for only short periods. New areas of habitat are therefore likely to be better sites for laying eggs than those where individuals developed, especially when hostplant localities are suitable for only one or two generations. On emergence, females are not capable of immediate reproduction. Movement to new hostplant areas during this period increases the time available for egg-laying because females will have already located suitable new areas before or when they can commence egg-laying. Whilst this may explain changes of female mobility with age it does not offer a suitable explanation for

Table 6.3 Movements of *Mellicta athalia* in relation to population size, habitat area and suitability in three woodland coppice areas in Blean Woods, Kent, recorded between 3 and 13 July 1982

	Site					
	A		B		C	
Habitat size	Large		Intermediate		Small	
Successional stage and suitability	Slowly deteriorating, suitable		Becoming shaded, low suitability		Becoming shaded, low suitability	
	Male	Female	Male	Female	Male	Female
Estimated population size during marking period	323	273	411	72	198	342
Estimated population size for whole generation	615	508	659	134	590	639
Number of individuals marked	301	87	73	53	111	95
Number of marked individuals moving						
to site: A	—	—	3	0	0	1
B	2	2	—	—	0	1
C	0	0	1	0	—	—
Total	2	2	4	0	0	2
Estimated minimum number of adults moving						
to site: A	—	—	18	0	0	11
B	12	8	—	—	0	5
C	0	0	10	0	—	—
Total	12	8	28			16
(%)	(4)	(3)	(7)			(5)

Note: A is K.26, B is K.24 and C is K.28 in Fig. 6.4
After Warren (1987*b*.).

male mobility patterns in these species. Dennis (1982*b*) found that males of the green-veined white *Pieris napi* move most when receptive females are scarce. This species has a similar reproductive pattern to the other two pierid species and also uses short-lived hostplants. Male mobility may therefore be related to the probability of finding receptive females. Because females are not immediately receptive on emergence and move from their emergence sites before becoming so, males must be equally mobile to locate mates. Individuals of these species can be mobile throughout their entire lifespan and, for female *P. brassicae* at least, movement after egg-laying has commenced is correlated with the number of eggs to be laid on the following day (Baker 1984). Such continuous movement may be associated with risk spreading; the use of a large number of host-plant patches increases the probability that some offspring will survive when offspring mortality is unpredictable.

Resource distribution has also been identified as being of significance in the mobility of the wall brown *Lasiommata megera* (Dennis and Bramley 1985). At Brereton Heath, in Cheshire, males moved from the sites where they emerged to nearby perching and patrolling sites at the boundaries of different vegetation types and patches of bare earth. Females dispersed further than males; once mated they moved to sites where males were scarce in order to avoid male harassment while egg-laying. The sites that females used also tended to be near the most abundant nectar sources. Some evidence is presented to suggest that interactions between males are also responsible for male flight patterns. Removal of all resident males from one particular area resulted in their being replaced by other males within 30 minutes. Thus, it is likely that when male density at favoured mate-locating sites is high, some males are forced to move elsewhere. At low density in these sites, males will move in to occupy the vacant sites.

Flight is facilitated when the thoracic muscles are at an optimal temperature. Although this temperature differs between species (see section 2.2) there is a tendency for butterflies to engage in more frequent and more prolonged flights with increased air temperature and sunnier conditions, factors which can also influence local movements. In cool weather, flights of *Lasiommata megera*, *Pararge aegeria* and *Maniola jurtina* are made less frequently than in warm weather (Dennis 1982*d*; Shreeve 1984; Dennis, unpublished data). Consequently males, particularly those that engage in patrolling behaviour, fly shorter distances. Similar changes in behaviour have been recorded for male small heath *Coenonympha pamphilus* (Wickman 1985*a*). When the weather is cool, males compete for perching territories and are sedentary once they have located a territory. As temperatures rise, males adopt patrolling behaviour and fly in search of receptive females. Because flight activity increases with rising air temperature, male displacement also increases with rising temperature. Flights of female *P. aegeria*, are similarly reduced in cool, dull weather (Shreeve 1985) and these conditions result in females distributing their eggs in smaller areas than in hot sunny weather.

In order to remain within a particular area an individual must be able to recognize the boundaries of the population or the habitat, or be restricted by certain features of the habitat. There is a wealth of anecdotal evidence to suggest that certain colonial species such as the gatekeeper *Pyronia tithonus*, *Maniola jurtina* and many of the Lycaenidae will change their flight direction when they meet obstacles such as hedges or other tall vegetation. However, the sensory mechanisms whereby individuals recognize these barriers are not known, and could involve any of the senses of vision, scent, orientation, and possibly temperature recognition. Alternatively, individuals may learn the location of barriers to their habitat, and flight distances and directions could be related to aspects of the population, in particular density.

Learning is of importance in the behaviour of male *Pararge aegeria*, which hold territories in woodland. Wickman and Wiklund (1983) found that particular individuals displayed a strong fidelity to particular areas of sunlit understorey vegetation whilst others avoided these areas. Their explanation was that particular individuals learnt the location of the best mate-locating sites, presumably by some form of spatial map of the habitat. Learning may also be associated with habitat use and mobility of the large skipper *Ochlodes venata* (Dennis and Williams 1987) and *Coenonympha pamphilus* (Wickman 1985*b*) but its importance in other species is a matter of conjecture.

In a study of the mobility of *Anthocharis cardamines*, Dennis (1986*b*) observed that males differed in the amount of time that they remained in certain parts of the Bollin valley, Cheshire; where hostplants were abundant, individuals remained for longer periods making more frequent turns in their flight direction, but where hostplants were scarce flights tended to be more direct with fewer turns. Some individuals returned to and remained in particular sections of the valley, a finding which suggests that these individuals retained a spatial map of particular areas of habitat. This study also demonstrated the effect of a large barrier, the M56 motorway, on movement patterns (Fig. 6.5). Individuals were found to avoid crossing the motorway, making changes of flight direction when encountering shade and lower temperatures in the vicinity of its embankment. Dennis (1986*b*) estimated that only about two per cent of individuals crossed this obstruction, but noted from MRR studies that those that did were the most mobile individuals which are also the least

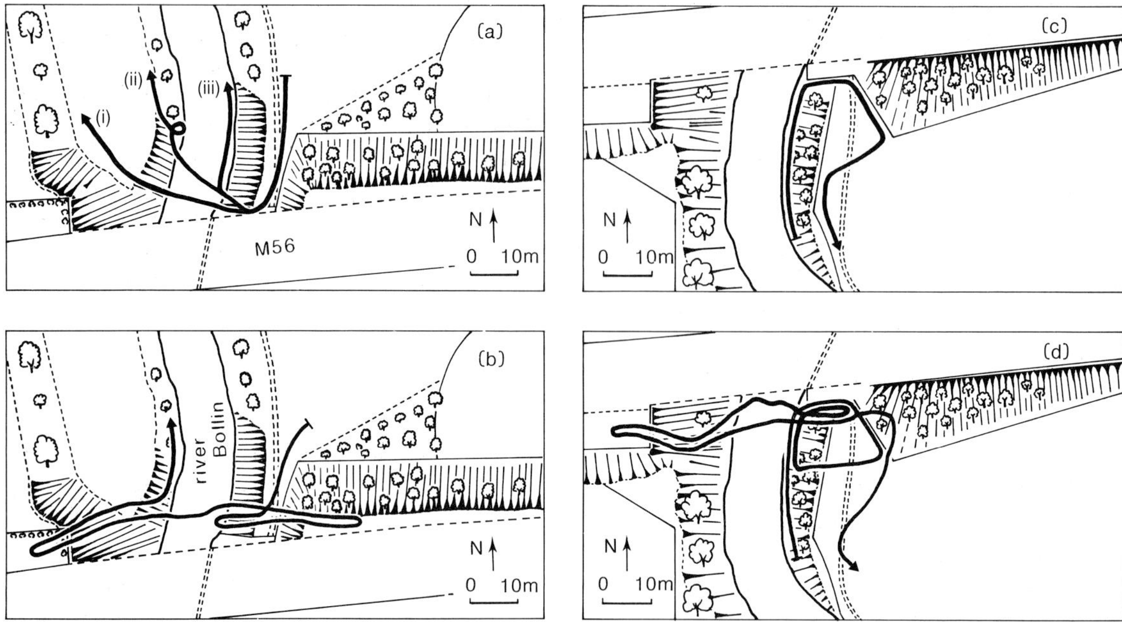

Fig. 6.5 The influence of the M56 motorway and its associated earthworks on the movement of male orange-tip *Anthocharis cardamines* in the Bollin Valley, Cheshire. Maps (a) and (b), north side of M56; maps (c) and (d), south side of M56. (a) Tracks taken at different times of day (BST): (i)<12.00, (ii) 12.00 to 15.00, (iii) >15.00. (After Dennis 1986*b*; courtesy of The Amateur Entomologist's Society.)

likely to be recorded with this method. However, the movement of even this small proportion of the total number of individuals may be sufficient to prevent the genetic isolation of populations on either side of this barrier. Such a barrier may be more likely to be of significance to those species which are less mobile than this butterfly. On the Isle of Tean, in the Isles of Scilly, windswept areas of short vegetation were found to be sufficient barriers to isolate populations of the common blue *Polyommatus icarus* and *Maniola jurtina* (Dowdeswell *et al*. 1940, 1949), even though these areas separated populations by only 150 to 300 m (see Fig. 9.2).

Population density has been demonstrated as being of major importance in determining the movements of two North American species Harris' checkerspot *Chlosyne harrisii* (Dethier and MacArthur, 1964) and the checkered white *Pieris protodice* (Shapiro, 1970), mobility increasing with population density. In other species no simple relationship between density and mobility has been observed. If density is used as a cue to recognize the boundary of a population, and if it is disadvantageous for a butterfly to leave an existing population, then this should be evident in flight behaviour. The number of turns made during a flight should increase and the distance travelled between turns should decrease with increasing population density. To date there has been little detailed analysis of flight behaviour in relation to density in species of butterflies.

Morton (1985) proposed that aggregations of male *Cupido minimus* are related to the distribution of perching sites and female distribution patterns are related to the distribution of the larval hostplant *Anthyllis vulneraria*. Aggregations were found to be maintained by a reduction in the distance travelled between turns but population density apparently had no influence on the mobility of either sex of this highly colonial species. The only butterfly in which such a mechanism has been reported to operate is *Euphydryas editha* of North America. For successful larval development two hostplants are required, but the secondary hostplant is not evident to the adult female (Singer 1972). It is argued that the best way for an individual to locate areas where this hidden resource occurs is to remain in areas where other individuals are found (Gilbert and Singer 1973). Since the butterfly can occur in parts of a uniform habitat, occupied and unoccupied areas being visually similar, the cue used to determine the boundaries of a population are not features of the habitat but the presence of other individuals. Thus, without physical boundaries, the association of individuals with particular areas must be maintained by increases in turning rates or shorter flight steps when population density becomes low at the population boundaries. To what extent population density acts as a population boundary in any species resident in Britain is a matter of conjecture, but it is worthy of investigation. The recognition of other individuals in determining population boundaries may operate in colonial species such as the marbled white *Melanargia galathea*, the small pearl-bordered fritillary *Boloria selene*, the marsh fritillary *Eurodryas aurinia*, and perhaps some of the lycaenid species.

6.4 Migration and dispersal

6.4.1 Measuring migration and dispersal

Migration and dispersal are similar in that both involve individuals leaving the habitat in which they developed. Unlike migration, individuals which disperse leave their habitat in an unpredictable manner with no seasonal pattern in the timing and direction of dispersal. The phenomenon of migration involves predictable seasonal movements in the direction that individuals move (Williams 1958; Baker 1969). However, since both involve movements away from habitats they may both be governed by the same underlying principles.

As with the problems of studying local movements, the major difficulty in investigating long-distance dispersal and migration is discovering just how far individuals travel. Although MRR techniques have been most frequently used to study 'within-habitat' movement they have also been successfully employed to study large-scale movement, often with the help of an alerted public. Urquhart (1960) was able to track the migratory movements of the monarch *Danaus plexippus* between Mexico and North America using marked individuals and citizen help. Similarly, Roer (1961, 1962) was able to track long distance movements of *Pieris brassicae* and *Aglais urticae* in Europe.

Alternative methods are also available, in particular following individuals in the study of dispersal, though the result of such studies often seem to be dependent on the method applied. In studying the movement of *Pieris rapae* in Britain, Baker (1978) followed individuals for as long as possible. When an individual was lost the study was resumed with the next individual that was observed flying in the same direction as the one which was lost. Using this method an average speed of the butterfly in sunshine was calculated as 0.8 km per hour and since the general direction of movement was uniform it was estimated that during an individual's lifetime it would move between 100 and 200 km. However, Jones *et al.* (1980), by using a dye technique, analysed the movement of this butterfly in Australia by recording where individuals laid eggs in cabbage fields. They concluded that for an average lifespan of 16 days an individual would move 2 km. Such a disparity is almost certainly related to Baker concentrating on between-habitat movement and Jones *et al.* on within-habitat movement. In Japan, Ohsaki (1980) has used an MRR technique, the findings of which suggest that once a female of this species finds an area containing suitable hostplants it remains there, a result similar to that of Jones *et al.* However, casual observations of *P. rapae* in its habitats in Britain, especially in gardens and allotments, tend to support the observation by Baker, since individuals rarely remain for longer than one to two hours in any one location.

MRR techniques and lifetime tracking are of use in determining where individuals go but are of less value in discovering from where they have originated if the potential sources are numerous. Techniques relating the movements of individual moths to large-scale wind systems have been applied with considerable success (Davey 1985). Such a technique may be applied with some success to butterfly migration and dispersal (Fig. 6.6), though wind-assisted movements

Fig. 6.6 Backtracks of four records of the Camberwell beauty *Nymphalis antiopa* from Britain in 1976. (Redrawn from Chalmers-Hunt 1977; courtesy of *The Entomologist's Record*.) These are constructed by calculating the surface windspeed and direction at the time and place of recording and tracking in the reverse direction for a specified time period. A new windspeed and direction vector is then calculated for the terminus of the first backtrack. The process is repeated until a probable source of origin is located. The major assumption of backtracking is that the first record is as near as possible in both time and space to the first landfall. Undetected local movement over a period of days may produce spurious results, particularly if there is a major change in wind direction between the time of landfall and first observation. These backtracks strongly suggest that the probable source of the massive invasion of *N. antiopa* in the summer of 1976 was the Scandinavian region. The four symbols illustrate different individuals.

may be less easily determined and predictable for the regular butterfly immigrants to Britain (see below), documented by Bretherton and Chalmers-Hunt (1981–1991).

During larval feeding individuals accumulate trace elements such as titanium, zinc, and copper in their body tissues. These, and other elements accumulate in different proportions and quantities, depending on the elemental composition of the plants on which they feed (McLean and Bennett, 1978; Bowden *et al.* 1984; Dempster *et al.* 1986). Providing the variation in the elemental composition of hostplants within a site is less than the variation between sites it is possible to determine where an individual has come from by analysing its elemental composition. Dempster *et al.* (1986) applied this technique to *Gonepteryx rhamni*, at Wicken and at Monks Wood in East Anglia, but could not partition either population into resident and immigrant individuals because elemental composition rapidly changed with adult feeding. Although this technique was of only limited success when applied to *G. rhamni*, it may well prove to be a valuable tool in the study of movement of other species.

6.4.2 Models of dispersal and migration

Movement between habitats is a regular feature of species with an open population structure (e.g. *Pieris brassicae*, *Aglais urticae*). It may be less frequent in more colonial species, but even individuals of species such as the pearl-bordered fritillary *Boloria euphrosyne* and purple hairstreak *Quercusia quercus* must leave habitats because colonization events by these and other species are recorded. The precise factors which govern the dispersal of individuals from existing habitats are for the main unknown. Various explanations have been offered, all of which may have some relevance.

When an individual cannot locate an essential resource (i.e. mates and hostplants) within a habitat it must move elsewhere for any chance of success. Therefore, the dispersal of an individual from an existing habitat can be described as a response to deteriorating conditions in a particular environment (Williams 1958). Such an explanation may well account for the increased mobility of *Cupido minimus* in response to a decline in the abundance of suitable egg-laying sites (Morton 1985), and also the higher rates of movements of *Mellicta athalia* from old coppice areas than from new coppice workings (Warren 1987*b*). However, not all species appear to become increasingly mobile when habitats deteriorate. For example, *Satyrium pruni* is extremely slow to colonize apparently suitable areas of blackthorn within a single woodland area, even when existing sites within the same area deteriorate (Thomas 1984*a*). Introductions of this butterfly to suitable areas are frequently successful, indicating that it is a lack of mobility rather than habitat suitability which has been the cause of its extinction from some woodland sites (see section 11.4.3). Although habitat suitability is of some importance to dispersal other factors must also intervene. Numerous observations have been made of, for example, *Pieris brassicae*, *P. rapae* and *Anthocharis cardamines* flying away from apparently suitable habitats in a more or less predictable manner (Baker 1968*b*).

Different species use habitats which remain suitable for varying periods of time, dependent in part on the successional stage in which their resources are located (see sections 7.5.3 and 10.4). Species which use long-lived resources such as trees and shrubs in late successional stages (e.g. *Satyrium pruni*) and those which use widespread and abundant resources such as certain grasses (e.g. small mountain ringlet *Erebia epiphron*) tend to be immobile compared to others that depend on short-lived resources in early successional habitats, such as many of the Pieridae which use cruciferous host-plants. Southwood (1962, 1977) suggested that the probability of insects leaving existing habitats is related to the longevity of these habitats. If this is so, species in similar habitats should have similar ranges of mobility. While habitat lifespan is of importance it does not explain observed directionality in movements of, for example, the clouded yellow *Colias croceus* and *Pieris brassicae* (Fig. 6.7), nor seasonal changes of direction displayed by *Vanessa atalanta* (Baker 1969). Neither does it explain differences in the mobility of individuals within populations,

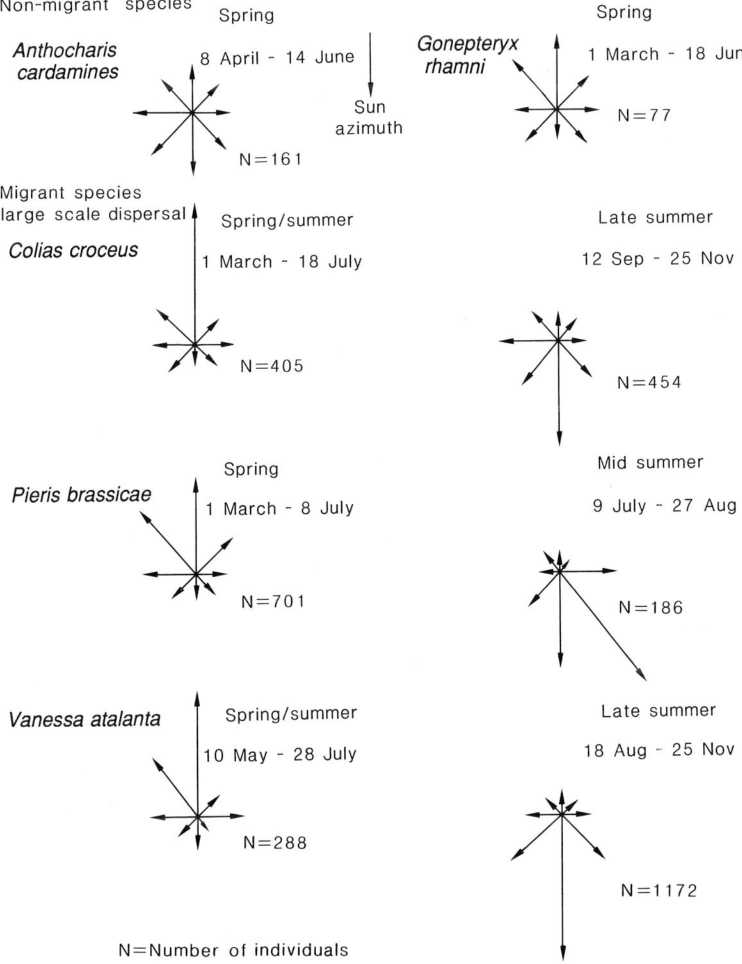

Fig. 6.7 The flight directions of selected resident and migrant butterfly species, recorded between 1964 and 1968. The two single brooded non-migrant species have no significant directionality in their flight, but the migrant species do. In spring their movements are predominantly in a northerly direction but the movement of later emerging generations is southwards. (From data in Baker 1969.)

described for many species, including *Maniola jurtina* (Brakefield 1982*a*) and *Pararge aegeria* (Shreeve 1985).

One theoretical framework which may explain much about variation in dispersal and migration is that developed by Baker (1969, 1984). It incorporates habitat predictability, intraspecific effects including competition for larval and adult resources, and the probability of locating new resources. It also accommodates differences between the mobility of individuals and the mechanism of movement. It is argued that for any individual butterfly a particular habitat has a certain suitability as a breeding site. This depends on its predictability in time (incorporating Southwood's model), the activity of the remainder of the population, and the previous activity of the individual butterfly. For every individual there is a threshold value, measured as habitat suitability, which determines when it will move from the habitat. Different individuals may well have different threshold values, dependent on their remaining reproductive potential or genotype. Habitat suitability is influenced by the activity of the individuals which exploit the resources within the habitat. For females any hostplant is a finite resource that can only support a given number of larvae and once an individual lays eggs on that hostplant its value for future exploitation declines. If the value of the resource falls below the migratory threshold of the individual which is exploiting it, then that individual should move on. The arrival of any butterfly at a resource patch decreases its value to all others and those with the lowest migratory thresholds should move first. Once these individuals have gone, the value of the resource increases, until other individuals arrive and lower its value. Individual threshold values change with reproductive activity and the risks associated with movement. Baker (1984) argues that young females have lower thresholds than old females because their future activity determines the success of a greater proportion of their offspring than does the activity of old females. They may also have a greater colonizing ability because of their potentially longer lifespan.

Movements away from any habitat has risks and costs, associated with the scarcity and distribution of resources and the energetic cost and risks of predation during flight (see Fig. 6.2). When habitat suitability declines the ratio of habitat suitability to movement–cost decreases and movement becomes more likely, even in commonly sedentary species. Baker (1984) cites the movement of the ringlet *Aphantopus hyperantus* and *Melanargia galathea* during the 1976 summer drought as occurring because their normal habitats declined drastically in suitability, though increased flight activity may have been weather related, hot sunny weather allowing more time for movement.

6.4.3 Dispersal and migratory behaviour

When a butterfly is in a suitable habitat it may adopt a flight behaviour which keeps it within the habitat. Such flights have been described for *Pieris rapae* (Jones *et al*. 1980), the American sulphur *Colias philodice eryphile* (Stanton 1982) and *Maniola jurtina* (Baker 1978), in which individuals make frequent turns and have short flight steps. When a butterfly leaves a habitat it will maximize its future reproductive effort by reaching a new habitat as quickly as possible, without repeatedly revisiting areas that it has encountered before. The most obvious way of doing this is to travel in a straight line out of a habitat. Baker's (1978) observations of a variety of species traversing unsuitable habitats in straight lines would appear to confirm that this is the behaviour adopted. However, other species may behave differently and other observers have recorded individuals following landmarks such as hedgerows, stream edges, and road verges, tending to remain close to these edges rather than moving away from them.

Migration and dispersal can be distinguished by an apparent lack of predictable direction in dispersal movements but predictable directions and return flights in migration. Baker (1978, 1984) argues that this can be related to the number of generations a butterfly has during a year and the capacity of a species for flight. Weak to moderate fliers with a single generation a year will not evolve a predictable flight direction for two reasons. A weak flying species with a single generation will only be able to move within one climatic zone within a single generation. Any genotypes which adopt a single flight direction will be selectively eliminated because they will eventually move out of the range of the species into unsuitable geographic locations. This hypothesis is supported by observations of the flight directions of the single-brooded *Anthocharis cardamines* by Baker (1978) and Dennis (1986*b*), no peak flight direction being recorded.

In seasonal environments habitat and hostplant suitability varies with the time of year. In some species unfavourable periods are avoided by a diapausal stage (see chapters 4 and 10) but in others by long-distance movement or migration. Thus, butterflies with more than one generation a year which travel long distances (e.g. *Cynthia cardui*, *Vanessa*

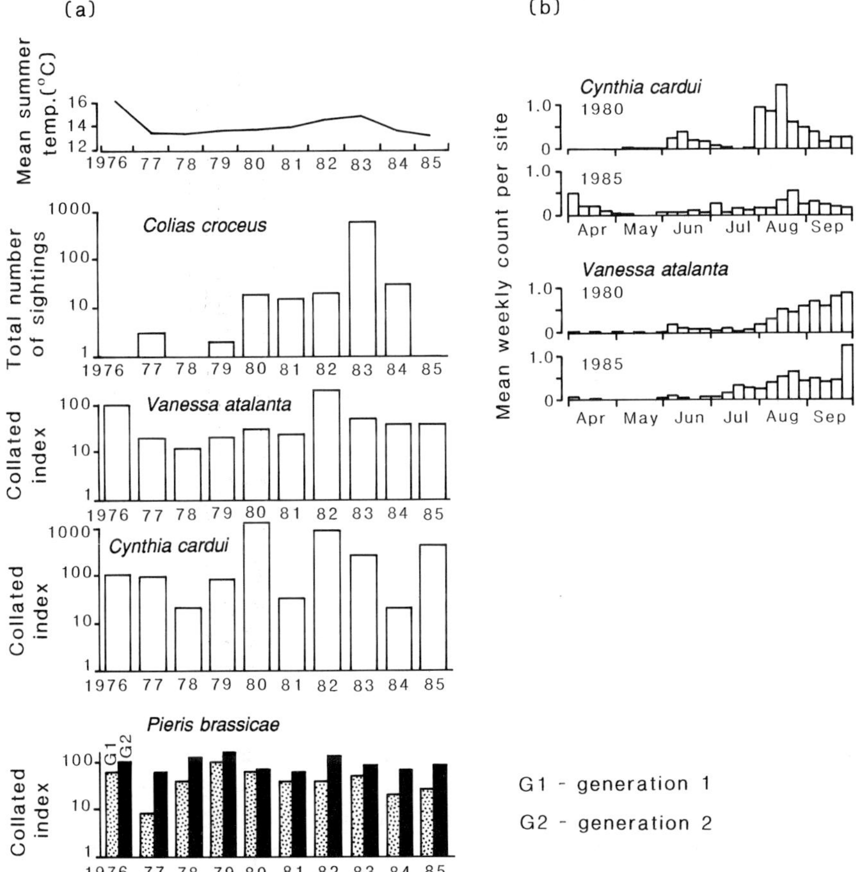

Fig. 6.8 Recent migrations of butterflies into Britain. (From Pollard *et al*. 1986; courtesy of the Nature Conservancy Council.) (a) The numbers of commoner migrant butterflies recorded between 1976 and 1985 at all sites in the Butterfly Monitoring Scheme, together with temperature data for the same period: (i) clouded yellow *Colias croceus*; (ii) red admiral *Vanessa atalanta*; (iii) painted lady *Cynthia cardui*; and (iv) large white *Pieris brassicae*. A consistent pattern is not apparent, the peak abundance of different species occurring in different years. This may be due to meteorological conditions at the source of origin (temperature, moisture, wind directions) affecting the number of arrivals and also to the variable breeding success of the different species. *P. brassicae* is included since numbers may be supplemented by migration (Baker 1978). However, large-scale changes of numeric abundance are absent, a characteristic that may be expected if migration is a predominant factor. Data are based on the collated index of abundance from all sites on the butterfly monitoring scheme. Each graph starts from an arbitrary value of 100 in 1976. (b) The numbers of *Cynthia cardui* and *Vanessa atalanta* recorded at weekly intervals in 1980 and 1985 at Butterfly Monitoring Scheme sites. The pattern of changes in abundance of the two species differ, indicating that different factors regulate the numbers of each which reach Britain, the latter being more regular and predictable than the former.

atalanta), encounter habitats which may well vary in quality with season and location. For a multi-brooded butterfly, then, optimal habitats will be in different compass directions at different times of year (i.e. for different broods). In temperate regions there is a tendency for the best habitats to be located at increasingly northerly latitudes as spring progresses, but in autumn northern habitats decline in suitability. Baker (1969) has demonstrated that such circumstances can generate peak flight directions in multi-brooded species which are seasonally different in direction, and proposed that directional ratios are maintained by selection. These theoretical arguments are supported by direct observations of the movement of *Pieris rapae* throughout Europe. In spring and summer the movement is in a north-northwest direction but in late summer it is southward. Breeding experiments suggest that the directional ratio was inherited, and further experiments demonstrated that the direction of movement was controlled by the number of hours of darkness and temperature experienced by the adult (Baker 1968*b*). When temperatures experienced by adults in autumn fall below a critical level the flight-direction becomes southerly.

In moving between habitats or making a migratory track, a butterfly must be able to maintain a constant direction of movement to decrease the time spent crossing unsuitable areas. Observations of pierid butterflies, nymphalid butterflies, and *Danaus plexippus* in America (Baker 1968*b*, 1969; Kanz 1977), suggests that these butterflies orientate using the sun as a compass. However, some time compensation may take place, as occurs in bees, so that a constant direction is maintained despite a changing sun angle.

Not all long distance movements and migration need involve peak flight directions. Pollard *et al.* (1986) cite evidence for the movement of *Cynthia cardui* in Europe which indicates that its movements are less predictable than would be expected from Baker's model (1972*b*). Weather can influence the number of individuals which engage in long distance movements, either by increasing the number of potential migrants or by facilitating movement itself, either by wind or by warm sunny weather. Variation in temperature and sunshine have been identified as of importance in determining the number and abundance of some migrant butterflies which reach Britain (Fig. 6.8), though not *C. cardui*, whose arrival in Britain is apparently independent of these factors (Pollard *et al.* 1986). Strong winds can prevent flight but they can also assist in movement. Observations of *Danaus plexippus* in America (Urquhart 1960; Gibo and Pallet 1979) demonstrate that this species at least changes its flight height in response to wind direction. With adverse headwinds it flies close to the ground where wind speed is minimized, but with a tail wind it flies at a greater height, taking advantage of the wind.

6.5 Unresolved issues

Clearly, the factors involved in local movements, dispersal, and migration in butterflies remain understudied. Although Baker's work on dispersal and migration has done much to further understanding and raise important issues, fundamental questions, such as the roles of genetic diversity, weather, and learning in flight behaviour, remain unanswered. A major concern is the lack of knowledge about basic patterns of mobility of even the most sedentary species, knowledge on colonization and habitat use, which would be of importance to conservation (see chapter 11). Reports of stray individuals of typically sedentary species frequently occur in literature but what is not known is whether or not these are chance occurrences or regular phenomena (Shreeve and Dennis 1991). If the latter, the techniques used in studying mobility will fail to record them, and the capacity of species to reach new areas may be seriously underestimated.

7

Butterflies and communities

Keith Porter with Caroline A. Steel and Jeremy A. Thomas

7.1 Biological communities

7.1.1 What is a community?

This chapter examines interactions between butterflies and other organisms such as other butterflies, animals, and plants, in communities. 'Communities are groups of species that interact, or have the potential for interacting, with one another' (Strong *et al*. 1984). The concept infers that individuals of one species influence the fitness of other individuals of the same or different species, either directly or indirectly. This could involve 'competition' for food or physical space but does not necessarily require that two individuals come into physical contact. Consequently communities are often difficult to verify in the field, especially as they can be of any size.

7.1.2 Patterns within communities

Well-defined associations and relationships between organisms determine communities, not random assemblages of species. The most obvious links involve the food chain—hostplants, phytophagous insects, and their parasitoids and predators. Although different butterflies are frequently associated at different sites, for instance the dingy skipper *Eryynis tages*, common blue *Polyommatus icarus*, and small copper, *Lycaena phlaeas*, there is little evidence, as yet, that they strictly form a community (Fig. 7.9).

In other organisms, species-associations have been shown to be relatively constant. This has led to the phytosociological classification of plant communities which reflects on the complex interactions among plant species and the influence of climate, soil, state of management, stage reached in natural succession, and other factors at a site. Affinities among plant communities are generally determined from site samples which record the abundance and constancy of individual species between samples. Although these data do not strictly define communities they provide the basis for analysing associations between species (Rodwell 1982–89; Huntley 1988). It may be possible in future years to construct an insect community classification based upon similar methods and one presumes that butterflies would form a part of such a classification (see Balletto and Kudrna 1985). Such a task is made difficult because 'communities' are constantly changing, either as a result of vegetation succession, or because of changes in land use.

7.1.3 Studying communities

Even the most simple community is an extremely complex system and it is rarely possible to study one in its entirety. Much knowledge has been gained on butterfly community biology from autoecological studies (see Chapter 4; MacArthur 1972), but this approach does not usually reveal how groups of closely related species, such as the meadow brown *Maniola jurtina* and the gatekeeper *Pyronia tithonus* share the same habitats. On a practical level, there are two other basic approaches to the study of a butterfly community: (i) investigating species which share a common resource, and (ii) studying interactions between selected species.

The study of guilds—groups of species that utilize a common resource—can highlight interactions among related organisms (Gilbert 1984). Examples within the British butterfly fauna include the nettle-feeding nymphalids and the violet-feeding fritillaries. In a true guild each species usually interacts strongly with its co-members but less strongly with other species in the community. Few studies of guilds in butterflies have been made, but do occur for other animal groups such as birds (Cody 1974). Any 'rules' derived from these non-insect studies must be treated with caution but may give direction when planning butterfly guild studies.

An alternative, but complementary approach is to concentrate on the interactions which occur between a variety of organisms within a community (Fig. 7.1). These interactions may involve members of the same or different species, and may benefit or adversely affect one or both organisms. Such interactions lie at the centre of community ecology and are given priority in the following sections.

		Effect of species A on the fitness of species B		
		Increase	Neutral	Decrease
Effect of species B on the fitness of species A	Increase	+/+	+/0	+/−
	Neutral	0/+	0/0	0/−
	Decrease	−/+	−/0	−/−

Fig. 7.1 Categories of interaction between two species. +/+, interactions are to the mutual benefit of both species (e.g. Müllerian mimicry, symbiotic ant–lycaenid relationships); +/0, commensalism in which one species benefits without reducing the fitness of the other (e.g. various forms of mimicry, see Turner 1987); −/0, amensalism, one species loses out in an interaction but the fitness of the other is unaffected; −/+, situations typical of predator–prey, parasite–host relationships and Batesian mimicry; −/−, forms of interspecific competition. (After Begon and Mortimer 1981; courtesy of Blackwell Scientific Publications.)

7.2 Resource partitioning and the niche

7.2.1 Living together within a community

In order to coexist in a community, plants and animals must each have a specific role and 'living space', often referred to as its niche. Hutchinson (1957) described the niche as a 'multidimensional hypervolume', defined by the sum of all the interactions of an organism and its environment. The term can be better understood by dividing the multidimensional space into its separate dimensions, for example into niches for larval feeding, adult nectaring, mate-location, roosting, and so on.

When individuals have certain common resource requirements, and if the available resources are limited, competition may occur (Fig. 7.1). Competition between members of the same population (intraspecific competition) is frequent, since their resource requirements are most similar. Although less common, interspecific competition may occur between individuals of species sharing the same resource; this is often referred to as niche overlap. Competition is potentially harmful to reproductive output and species tend to evolve out of competition (Begon and Mortimer 1981).

Intraspecific interactions

Butterflies belonging to one species may compete for a range of resources, such as larval hostplants, nectar sources, and sites for mate-location, roosting, and hibernation. The activities of one individual may reduce the fitness of others. The most obvious instances are provided by males contesting territories (e.g. Duke of Burgundy *Hamearis lucina*;

Porter, unpublished data). In the speckled wood *Pararge aegeria* and large skipper *Ochlodes venata*, these territorial interactions between males have even been observed to result in physical contact (see p. 38).

Competition for larval hostplants may be observed from the ovipositional behaviour of females, particularly with respect to avoidance of plants already holding eggs (see section 3.2.4 and 3.3.1), and between individual larvae, as exemplified by the cannibalistic behaviour of orange-tip *Anthocharis cardamines* larvae. This latter example has been described as the evolutionary consequence of intraspecific competition for larval food (Courtney 1980). Egg-load assessment is of interest as an indicator of the reality of competition and as a factor in dispersal. Shapiro (1981) notes that in crucifer-feeding pierid butterflies, members of the inflorescence-feeding guild lay red or orange eggs and engage in egg-load assessment, the so called 'red-egg syndrome', while members of the leaf-feeding guild lay yellow eggs and do not. This behaviour has been observed in *Anthocharis cardamines* (Wiklund and Åhrberg 1978; Courtney 1980; Fig. 3.3). Intraspecific competition for nectar sources is discussed in section 7.4.3.

Interspecific interactions

Competition between species is more difficult to detect than competition within species. The problem is that many interactions that may be attributed to interspecific competition can also be explained by alternative hypotheses not involving competition (Strong *et al*. 1984). Competition clearly occurs when one individual prevents a number of another species from exploiting a resource. However, most larvae competing for food are affected more by the reduction in food quality and the difficulty in finding alternative intact resources; see section 7.4. Competition may be indicated by patterns of resource utilization in habitats where potentially competing species are found. The use of control populations in which one or other of the species is absent is essential for these comparisons. In some cases the presence of species 'A' may reduce the fitness of species 'B' to such a degree that species 'B' may become locally extinct; it suffers competitive exclusion. This may be observed when a species is introduced onto an island or into a new geographical area.

Although no clear cut case of competitive exclusion is apparent among British butterflies, the accidental introduction of small white *Pieris rapae* into North America in 1860 provided an opportunity for it elsewhere. The North American subspecies of the green-veined white *Pieris napi oleracea*, was observed to decline in the subsequent decades. These species share common hostplants and it has been suggested that before the introduction of *P. rapae*, *P. napi* was able to exploit a wider range of habitats, especially those influenced by humans (Shapiro 1975*b*). *Pieris rapae* is better adapted to these artificial or man-made habitats and may have displaced *P. napi* from such habitats in North America. However, Chew (1981) could find no evidence of reproductive interference between *P. rapae* and *P. napi* and explained the local extinction of *P. napi* in terms of changes in hostplant species distribution resulting from land-use changes, rather than through competitive exclusion. This example illustrates the difficulty of demonstrating interspecific competition in a field situation.

Nevertheless, competitive exclusion has been observed in other animal groups by manipulating one of the species involved (Begon and Mortimer 1981). Inouye (1978), working with two species of bumble-bee, *Bombus appositus* and *B. flavifrons*, has shown that if one of the species is temporarily removed from the habitat, then the other species will expand its niche to utilize a wider range of flower species. *Bombus appositus* has the longer proboscis of the two species, and is able to extract nectar from larkspur *Delphinium barbeyi* with such efficiency that *B. flavifrons* cannot compete. The converse is also true in that *B. flavifrons* excludes *B. appositus* from flowers of monkshood *Aconitum columbianum*.

Mimicry and the palatability spectrum

Niche overlap for the same predators can result in diverse forms of defence guilds involving mimicry between unrelated species. Mimicry of aposematic distasteful models provides an effective escape from predators, but the protection gained by the mimic and model depends on their relationship to one another. Mimics are classed as Batesian when the model species is distasteful to predators and Müllerian when the mimic is also distasteful (see section 5.1.3). Turner (1987) has demonstrated the reality of these classes despite the existence of a

Table 7.1 Summary of distinctions between Batesian and Müllerian mimicry, with respect to butterfly predation and evolution

Batesian characteristics	Müllerian characteristics
1. Palatable	1. Unpalatable
2. Model suffers	2. Model benefits
3. Polyphenic	3. Normally monophenic
4. Males non-mimetic	4. Males mimetic.
5. Mimic converges, model diverges	5. Mutual convergence
6. Mimicry improved by selection for tightly linked blocks of modifier genes (supergenes)	6. Mimicry improved by selection for weakly linked modifier genes

Adapted from Turner (1987).

palatability spectrum for mimics and models, and other factors influencing their phenology and population dynamics (Table 7.1).

Batesian and Müllerian mimics have different effects on their model species. In the case of Müllerian mimicry, both species benefit from their joint aposematic pattern and distastefulness, because predation on the species falls as the form becomes commoner. Selection results in the convergence of pattern in the Müllerian mimics and each can exist at high density. In contrast, Batesian models are perpetually at a disadvantage, since the deterrence value of their pattern is undermined as palatable mimics increase in density. Thus, while the Batesian model tries to escape by evolving away from the mimic, the mimic converges on the model. However, the model, which is closer, by definition, to the optimum pattern, evolves at a slower rate; thus the mimic wins this evolutionary 'arms race' in a process called advergence (Brower and Brower 1972). As Batesian mimics require the protection of their model, they cannot exist at high density (Rothschild 1971).

The evolution of mimicry is believed to occur in two stages. An approximate resemblance is achieved by a single mutation of large effect. Convergence of pattern is then achieved by selection of modifier genes. As mimicry can only evolve when patterns of the two species are sufficiently close to be bridged by a single mutation, this may explain the coexistence of several distinct Müllerian mimicry rings (e.g. as in butterflies of the genus *Heliconius*). The fine tuning of mimetic forms often involves tightly linked blocks of genes, or 'supergenes', as in the great mormon *Papilio memnon* (Turner 1987; section 8.2). Mimicry is unusual among British butterflies, but the Pieridae are believed to form a Müllerian ring (section 5.1.3).

7.2.2 Resource partitioning

If two or more species with similar ecological needs are to coexist in a community, they must generally do so by occupying different niches. This process can occur through divergence in two main ways: (i) by competition in sympatry; and (ii) by allopatry and a fine-tuning of ecological distinctions in sympatry. In Britain it is true to say that any given community will contain species which originated in habitats different from those of today. Consequently most of the niche divergence occurred before 'modern' communities developed (see chapter 10). Owen (1959) noted the following niche differences among British butterflies:

(1) larval food or hostplant;
(2) part of hostplant used;
(3) time of appearance (phenology) and number of generations (voltinism); and
(4) habitat and flowers visited by adults.

A later review (Gilbert and Singer 1975) added parasite- and predator-escape.

Hostplant partitioning

Not only has hostplant–insect evolution led to taxonomic distinctions in hostplant use (see section 7.4.4; Table 7.6), but closely related butterflies also tend to have preferences for different plant species. For example, although the pearl-bordered fritillary *Boloria euphrosyne* and small pearl-bordered fritillary *B. selene* both use the common dog-violet *Viola riviniana* and the marsh violet *V. palustris* at sites where both species of butterfly and violet occur, it appears that *B. selene* prefers to oviposit on *V. palustris* and *B. euphrosyne* on *V. riviniana* (Warren and Thomas, personal communication). However, it is difficult to separate the partitioning due to hostplant from the effect of microclimate and habitat partitioning, for *V. palustris* is found on the wetter parts of a site and *V. riviniana* on the drier areas. The Essex skipper *Thymelicus lineola* and small skipper *T. sylvestris* also use different hostplants (see Appendix 1); this is believed to reflect the relative tightness of the leaf sheaths used as oviposition sites.

Our knowledge concerning hostplant preferences of grass-feeding species is still relatively unsatisfactory. Satyrine larvae are apparently able to feed on a variety of grass species, especially in captivity (Table 7.2; Bink 1985). Although a large proportion of the grass-feeding larvae need the spring flush of succulent growth, it is questionable whether grass as a resource is a limiting factor. It is possible that factors other than competition for food, such as microclimate and predation, produce the observed partitioning. Shreeve (1986*a*) was able to show preferences in oviposition of *P. aegeria* which relate to microclimate and this behaviour may also relate to larval feeding.

Partitioning of parts of plants

A plant provides several resources, for example, roots, stems, old leaves, young leaves, buds, flowers, and fruits. In certain instances, butterfly species which exploit the same hostplant species coexist by using different parts of the plant. Female *Pieris napi* lay their eggs on the leaf undersurface of several species of crucifer (Appendix 1) and it is here that the larvae feed. Female *Anthocharis cardamines* lay their eggs on the carpels or perianth of the flowers of the same species and their larvae feed on developing fruits. Lees and Archer (1974) show that, in captivity,

Table 7.2 Hostplant preferences of selected grass-feeding butterfly species

	Species of very poor grassland		Species of rather poor grassland				Species of richer grassland			
	Hesperia comma	*Hipparchia semele*	*Aphantopus hyperantus*	*Coenonympha pamphilus*	*Coenonympha tullia*	*Melanargia galathea*	*Lasiommata megera*	*Maniola jurtina*	*Pararge aegeria*	*Ochlodes venata*
Grass species of all kinds of soils										
Agrostis stolonifera		2	2							
Agrostis capillaris		2		2						
Festuca rubra		2				2		2		
Poa pratensis								2		2
Grass species of poor soils										
Corynephorus canescens	2			2			2			
Deschampsia flexuosa			—	1				1		—
Festuca ovina	2	2		2				2		
Molinia caerulea					2					2

Table 7.2 (*cont.*)

	Species of very poor grassland		Species of rather poor grassland				Species of richer grassland			
	Hesperia comma	*Hipparchia semele*	*Aphantopus hyperantus*	*Coenonympha pamphilus*	*Coenonympha tullia*	*Melanargia galathea*	*Lasiommata megera*	*Maniola jurtina*	*Pararge aegeria*	*Ochlodes venata*
Nardus stricta					2			2		—
Sieglingia decumbens					2		2	3		3
Grass species of rather poor soils										
Anthoxanthum odoratum			1			2		2		2
Brachypodium pinnatum								2		2
Briza media								—		—
Bromus erectus						2	2	2	2	
Deschampsia caespitosa								3	2	2
Festuca gigantea									2	
Glyceria fluitans								3		3
Milium effusum								3	2	3
Phleum p. bertolonii								2	2	
Poa palustris										2
Grass species of rich soils										
Alopecurus pratensis								2		2
Arrhenatherum elatius								—	—	1
Cynosurus cristatus		2				2	2		2	
Dactylus glomeratus								3	2	2
Elytrigia repens									3	
Festuca arundinacea								3		2
Holcus lanatus								2		2
Lolium multiflorum									2	
Lolium perenne								2		2
Phleum p. pratense			2					2		2
Non-grass species										
Carex nigra			2							
Carex ovalis					2	3				
Carex pilulifera								2		
Carex rostrata					2					
Eriophorum angustifolium					2					
Luzula multiflora								1		

— = refused; 1 = disliked; 2 = accepted; 3 = preferred.
After Bink (1985).

some female *P. napi* appear to oviposit more frequently on crucifer flowers than on the leaves, and that larvae can be reared successfully on the fruits of some of these plants. It would be inadvisable to deduce that competition has been responsible for this behaviour but it may lead to its maintenance.

Detailed studies often reveal a lack of any competition. This is true of the two European lycaenids, the scarce large blue *Maculinea teleius* and the dusky large blue *M. nausithous*, which both fly together at the same time, in the same habitat, and use the same foodplant, great burnet *Sanguisorba officinalis*. Superficially they appear to have identical needs; however they differ in two important respects (Thomas 1984*c*). *Maculinea nausithous* only lays its eggs on the large terminal flower buds, whilst *M. teleius* invariably lays on the very young buds which lag two to three weeks behind in growth. The two species were also found to differ in the ant species necessary for the completion of their life cycle. Similarly, Warren (1985*b*) found that the heath fritillary *Mellicta athalia* and five related species of plantain-feeding fritillary in south-east France tend to use different sizes of hostplant. The average size of ribwort plantain *Plantago lanceolata* eaten by larvae of Glanville fritillary *Melitaea cinxia* was far smaller than those eaten by *M. athalia* and the spotted fritillary *M. didyma*. Although the latter two species are ecologically very similar, competition for the same hostplants may not be critical as *P. lanceolata* is often abundant in their habitat.

Partitioning by time

Many species of butterfly may avoid competition for resources by virtue of their phenology. This could be true of the chalk hill blue *Lysandra coridon* and Adonis blue *L. bellargus* which both use horseshoe vetch *Hippocrepis comosa* as their hostplant and occur together in large populations at several sites in Britain (Thomas 1983*a*). These two species differ in voltinism and phenology (Thomas and Lewington 1991). Experiments in the laboratory which have synchronized the emergence of the two species, have demonstrated a selective advantage for *L. coridon*. When equal numbers of the larvae of the two species were placed in the same feeding containers, *L. coridon* larvae were able to feed for longer periods of time and consume more food than *L. bellargus*; the latter grew slowly and several starved (Morton, personal communication). This may be an example of time partitioning in response to interspecific competition. However, *L. bellargus* also requires higher temperatures for the development of its two annual broods and selects warmer microclimates for oviposition (Thomas 1983*a*).

Phenological divergence could also evolve to reduce the chance of cross-mating between similar species. This may apply to *Boloria euphrosyne* and *B. selene*, that have flight periods separated by two or three weeks (Thomas and Lewington 1991). Males could waste much time and energy courting the wrong species, and this could significantly influence mating success. Such 'distracting' models can reduce the time spent on profitable behaviour. For example, small mountain ringlet *Erebia epiphron* males have been observed to approach and investigate tufts of dark brown sheep wool, presumably in mistake for females (Porter, personal observation).

Differences in phenology and times of peak emergence may reduce competition for nectar. This idea was expounded by Clench (1967) who grouped 11 species of skipper in western Pennsylvania, USA, into five sets which flew at different times and used different nectar plants. He argued that these patterns have emerged from competition for limited resources. However, his conclusions are open to other interpretations (Shuey 1986). One requisite for competition-driven temporal partitioning is that the community is stable enough for adaptations to develop from interactions and long-term associations between species. It may be argued that this has not been the case for Clench's skippers' as these communities inhabit moist to dry fields that are artificially maintained in early successional states and are often dominated by exotic species.

As the phenologies of most butterflies and plants are influenced by temperature, this could have consequences for competition between species. Whereas generally weeks of peak numbers for butterflies differ, in cooler years they may coincide (Brakefield 1987*d*). Species may then have to use the same nectar sources. Thus, there is the potential for weather, acting via emergence time, to alter the fecundity of members of these species if they use a common resource and if this resource is limited. It would be interesting to make detailed comparisons at different sites to determine whether in certain years there is any competition for resources such as nectar.

Partitioning by habitat

Apparently, small differences in habitat can play an important part in ecological segregation, as in the case of *Lysandra coridon* and *L. bellargus* discussed earlier. Chalk grassland and scrub in southern England support up to six satyrine species and factors such as height of turf and degree of shelter determine their local distribution (Fig. 7.2; see Heath, Pollard and Thomas, 1984). Examples of local habitat differences between species occur among the pierids. In addition to using cultivated brassicas, larvae of *Pieris rapae* also feed on wild crucifers such as hedge garlic *Alliaria petiolata*, and hedge mustard *Sisymbrium officinale*, which are also hostplants of *P. napi*. However, whereas *P. rapae* prefers sunny, fairly dry habitats, *P. napi* usually lays its eggs on quite small plants growing in damp habitats (Petersen 1954; Dennis 1985*b*). The larvae of *P. napi* develop more rapidly at a lower temperature than do *P. rapae* and they also prefer to feed on thinner leaves that have a higher water content; such leaves are usually found on plants growing in shaded or damp habitats (Courtney 1988). Petersen (1954), Ohsaki

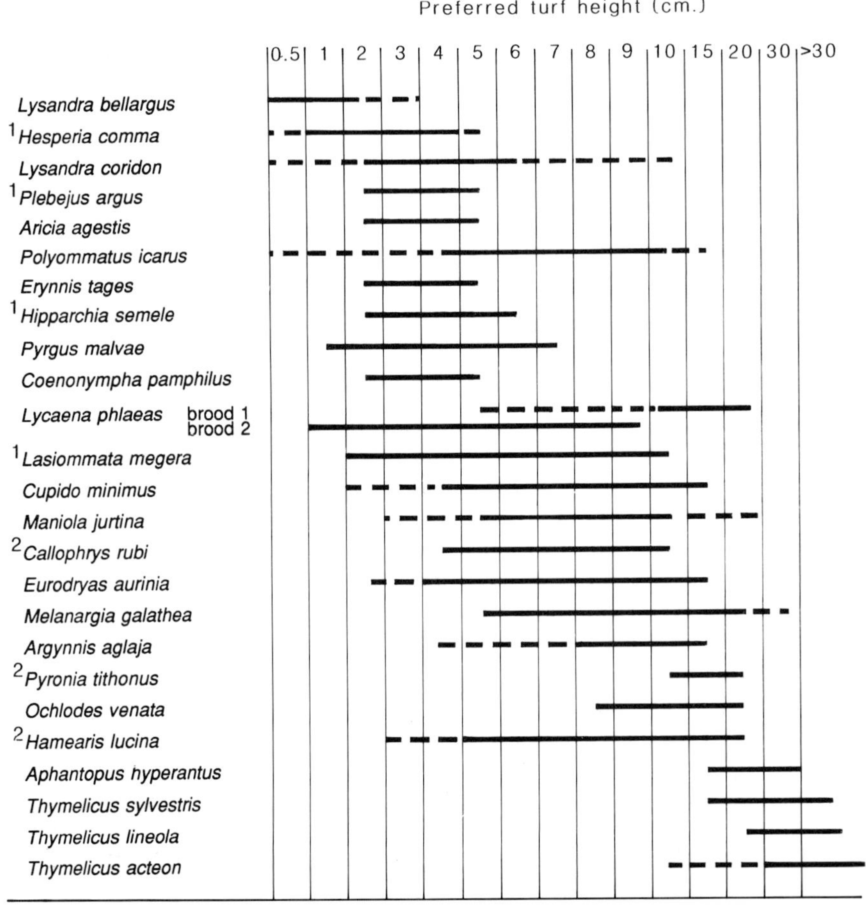

Fig. 7.2 The preferred turf heights of butterflies breeding on typical chalk downland sites. Note, if the turf is sparse, much taller heights are tolerated. (After BUTT 1986; courtesy of Nature Conservancy Council.)

(1979), and Courtney and Chew (1987) have also shown that pierid butterflies appear to be separated ecologically more by habitat than by hostplant choice.

Hostplant specialization is often related to the temporal stability of the hostplant habitat and hostplant density. Hostplant species that have numerically stable populations are more likely to be used than those with unstable populations; this stability is correlated with habitat type, a relationship demonstrated for pierids by Ohsaki (1979) and Courtney and Chew (1987).

Habitat partitioning may also occur in the nettle-feeding nymphalids, peacock *Inachis io*, small tortoiseshell *Aglais urticae*, comma *Polygonia c-album*, and red admiral *Vanessa atalanta*. Common nettle *Urtica dioica* is now a common perennial. However, prior to human changes in land use, *U. dioica* would have been far more localized and there could have been competition for egg-laying sites (Dennis 1984*c*; section 10.4). Both *I. io* and *A. urticae* prefer to lay on young, tender shoots; but, whereas in Britain *I. io* and *P. c-album* are closely associated with woodland, *A. urticae* is more common in open habitats, especially on disturbed ground. *V. atalanta* overlaps considerably with both *A. urticae* and *I. io*, but is a migratory species which usually does not survive the British winter.

In a large woodland it is possible for all the species of violet-feeding fritillaries to coexist and even share the violet species. At a Dorset site, *Boloria euphrosyne* is found in the warmest, driest conditions and breeds in a one- or two-year-old coppice. *Boloria selene* does not need such high temperatures and will breed in coppice plots with two- to four-year-old regrowth (Thomas, J. A. personal communication). Female silver-washed fritillary *Argynnis paphia* select hostplants at the edge of old coppice, young coppice regrowth, or in light woodland where a third to half of the ground flora is shaded. The dark green fritillary *A. aglaja* occurs in the most overgrown areas of rank grassland and scrub but may still select warm, south-facing slopes. *Boloria euphrosyne* and high brown fritillary *A. adippe* seem to have similar ecological requirements in that they both need exposure to sunshine in the immature stages but are segregated in their phenology. *A. aglaja* and *A. adippe* were thought to be very similar in habitat requirements; both Ford (1945) and Owen (1953) noted that where one species was absent the other tended to occupy its habitat but they provided no definitive evidence of competitive exclusion.

Partitioning adult nectar resources

Adult butterflies have far less specific requirements for food than do larvae, and may compete among themselves and other insect species for the same resources. There are a number of means by which butterflies can avoid competing for nectar. For instance, they may differ in phenology and weeks of peak emergence (see section 7.4.3). Feeding at different times of day, using different sugar concentrations or different flower species are alternative strategies, as suggested by observations made in Bernwood Forest, Buckinghamshire (Peachey 1978, 1980). Although nectar can clearly be in limited supply and greatly influence the population ecology of butterflies (Murphy 1983), competition between species may be avoided when resources are abundant.

7.3 Interactions between butterflies and other animals

7.3.1 Butterflies as part of a community

Ecological differences between closely related butterflies often provide insights into the ways species avoid competing with each other. Habitat appears to be the primary dimension partitioned between two or more species. Hostplant partitioning occurs mainly as an indirect result of habitat choice. Nevertheless, communities may comprise a variety of invertebrate species using the same hostplants. For example, oak (*Quercus* species) is a hostplant for over 400 British insects (Kennedy and Southwood 1984). However, despite the short period when food of appropriate quality is available on this and other plants (see section 7.4.2), there are few convincing examples of interspecific competition between these phytophagous insects, even between species with considerable niche overlap (see Rathke 1976; Strong 1982). Populations are generally kept below the carrying capacity of the habitat by abiotic factors such as weather, and biotic ones including predators, parasites, and symbionts (see Chapter 4; Hassell 1978; Dempster 1983).

7.3.2 Associations with parasitoids

Adult and larval stages of butterflies are, in theory, competing for the same kinds of enemy-escape niches (Gilbert and Singer 1975; see chapter 5 and section 7.2.1). Hence, there may be selection for different means of avoiding a predator or parasitoid which is common to both species. Microhabitat selection by ovipositing females may play an important part in parasitoid avoidance (Table 5.1), as exemplified by Shaw's (1977) study of the distribution of satyrine larvae at a coastal sand-dune site in relation to their ichneumonid parasite *Ichneumon caloscellis* (Fig. 7.3). This parasitoid can use grayling *Hipparchia semele*, *Pyronia tithonus*, and *Maniola jurtina* larvae as hosts, but was clearly associated with one vegetation zone. Shared parasitoids could also represent a kind of selective pressure organizing species in time. The black-veined white *Apo crataegi* and *Pieris brassicae* share the braconid wa *Apanteles glomeratus*, but their differences in phen logy and voltinism could be one means of escapi heavy mortality from the parasite.

Any given species of butterfly may act as host several species of parasitoid, which may also atta other hosts. The larvae of the marsh fritillary *Eu dryas aurinia* are parasitized by at least two species braconid wasp, *Apanteles bignellii* and *A. melit arum*, and two species of Tachinid fly, *Erycia cine* and *Tachina larvarum* (Van Emden 1954; Por 1981). The two braconid wasps are believed to host specific, although *A. melitaearum* has be reported from *Mellicta athalia* (M. S. Warren, pe sonal communication); the tachinids also atta other hosts. The braconids avoid interspeci competition for the same host by differences in th distribution; *A. bignellii* is only known from popu tions of *E. aurinia* south of Lancashire and *melitaearum* is only known from populations Cumbria and Scotland. The interaction betwe parasitoids and host does lead to a high degree

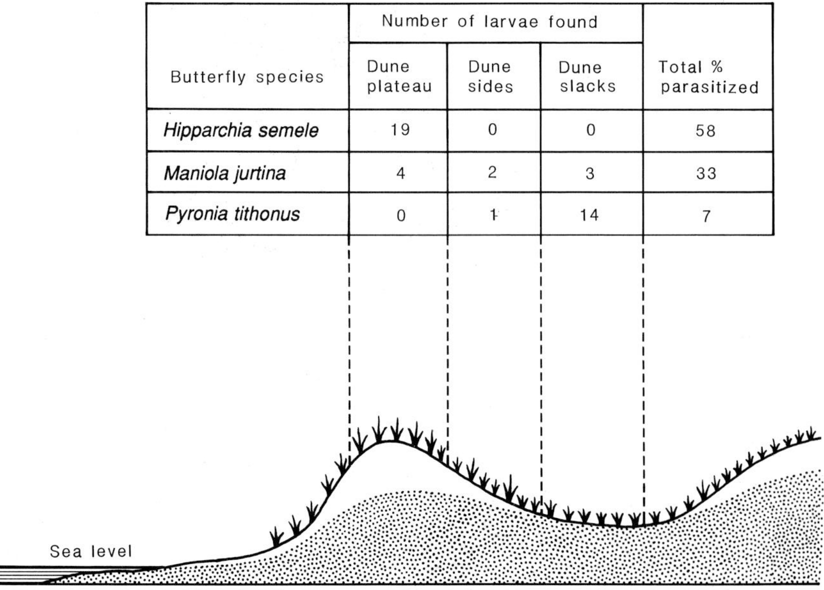

Butterfly species	Number of larvae found			Total % parasitized
	Dune plateau	Dune sides	Dune slacks	
Hipparchia semele	19	0	0	58
Maniola jurtina	4	2	3	33
Pyronia tithonus	0	1	14	7

Fig. 7.3 The distribution of larvae of three satyrines in three zones on sand-dunes at Ainsdale National Nat Reserve and their relative parasitism by *Ichneumon caloscellis* Wesmael. The dune summits are dominated by marr grass *Ammophila arenaria*, the slack zone by shorter grasses. Water table (stippled zone) influences the vegetat and local climate of the dune system. (Drawn from data in Shaw 1977.)

control of host population size, in conjunction with weather (Porter 1983; Fig. 7.4). The primary parasitoids are themselves part of a community, with their own parasitoids or hyperparasitoids. *A. bignelli* is attacked by at least one species of cryptine ichneumonid *Gelis* spp. and in turn the latter has been found to be parasitized by the chalcid wasp *Spaniopus dissimilis*; a hyper-hyperparasitoid. The potential interactions between butterflies and their parasitoids can be very complex and little is known about how such interactions affect butterfly numbers in the field.

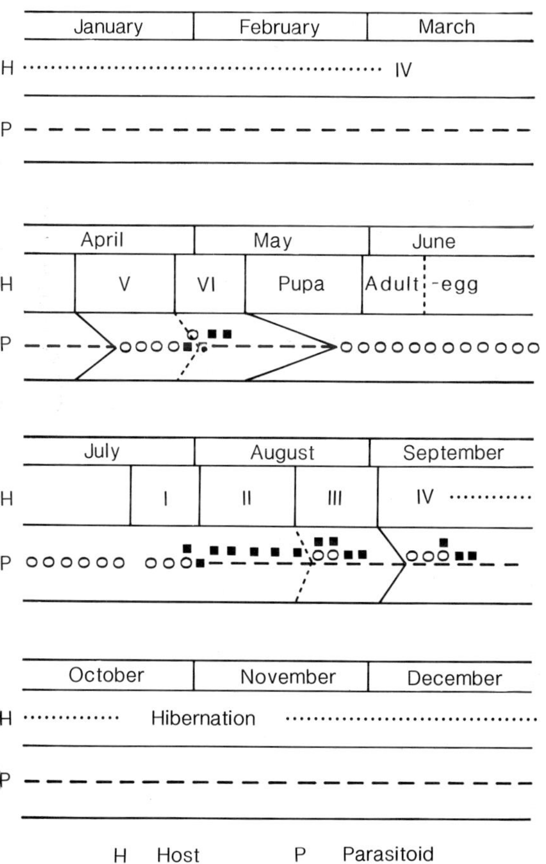

Fig. 7.4 Host–parasitoid synchronization between the marsh fritillary *Eurodryas aurinia* and *Apanteles bignellii* Marshall. This braconid parasitoid has up to four generations within one host generation and maintains synchrony by having overwintering diapause larvae within the host, and a protracted pupal stage whilst *E. aurinia* is in the host phase. Host stages are given as egg, pupa, adult, and larval instars (I–VI). Parasitoid stages are shown as larva (–), pupa (o) and adult (■). (After Porter 1983; courtesy of Kluwer Academic Publishers.)

7.3.3 Relationships between butterflies and ants

Several adaptations are described in chapter 5 that enable lycaenid larvae and pupae to live in close association with ants. There are disappointingly few descriptions of the relationship in the wild, yet Hinton (1951) proposed various classifications based on different types of life-style. Since these can vary in different instars of the same species, it is perhaps sensible, on present knowledge, to divide species merely into those that are wholly phytophagous and those that are aphytophagous for some stage of their lives.

Phytophagous species

Sixteen of the 17 British Lycaenidae belong to this group, as do 95 per cent of European species. Their larvae feed solely on plants, but they possess some or all of the adaptations described in section 5.4, and are accompanied by ants to a greater or lesser degree. At one extreme are species like the Adonis blue *Lysandra bellargus*, which is tended incessantly from the start of the second larval instar until it ecloses as an adult (Thomas 1983*a*): the caterpillar emerges from beneath low-growing *Hippocrepis comosa* in the morning to feed on leaves, and is then usually smothered by up to 30 ants for the first hour or two, after which the attendants fall off leaving usually four to six workers that milk the larva all day long; Fiedler and Maschwitz (1988*a*) have recently shown that it is the dorsal nectary organ (DNO) that is particularly effective in ensuring that fresh ants are recruited to attend its close relative, the chalk hill blue *L. coridon*. The Adonis blue larva appears oblivious to the ants, apart from producing honey dew and repeatedly everting its tentacles. Whenever it stays still for a long period, for example at night and during the two to four days taken over each moult, the larva is buried in a thin layer of soil which is sometimes worked by the ants into a crusty earthen cell. Several larvae may occur together inside these, and are again constantly attended by perhaps 12 ants, day in day out. Some pupae are formed in crevices, and are again earthed up and attended by ants until they hatch four weeks later; others have been found among ant brood, deep inside the ant nest.

Over a thousand *Lysandra bellargus* larvae have been watched on a wide range of sites in the wild, but only once were they found unattended by ants (J. A. Thomas, unpublished data). This was during an exceptionally cold wet spring, when the air temperature was below the threshold for foraging workers. It is perhaps significant that this is the only occasion that larvae collected in the wild were parasitized, by *Apanteles arcticus*, which is probably prevented from stinging larvae when ants are present.

The chalk hill blue *Lysandra coridon*, silver-studded blue *Plebejus argus*, brown argus *Aricia agestis*, and presumably the northern brown argus *A. artaxerxes* have a similar, or even greater association with ants. *A. agestis* produces such copious secretions that it sometimes dies of mould in the absence of ants, while rows of *P. argus* pupae have been found lining the passages of *Lasius* nests (J. Pontin, personal communication). Even more remarkable is the report of an emerging adult crawling up the vegetation to expand its wings, carrying several black ants on it, all drinking droplets of (presumably) honeydew that soaked the butterfly's body (Thomas, C. 1983). Others have confirmed this strange phenomenon in *P. argus*, but it is not known if any other Lycaenidae are similarly tended and protected during this dangerous time of life. It had also long been rumoured that ants pick up and carry the young larvae of certain phytophagous Lycaenids, such as *L. coridon* (Donisthorpe 1927). This is now confirmed in the case of *P. argus*. Both *Lasius niger* and *L. alienus* often carry the hatchling larvae into their nests, although it is not known if they feed there; later instars are usually found above ground feeding on foliage. *Lasius* workers are pugnacious in defence of all *P. argus* larvae and, if provoked, pick up and retreat with any that are small enough to carry (C. Thomas, in press). The relationship seems similar to that between Weaver ants and the Malayan lycaenid *Anthene emolus*: first instar larvae are carried into ant nests where they eat the leaves forming the nest sides, while older larvae are taken outside the nest to feed, and then carried back to the safety of the ant colony (Fiedler and Maschwitz 1989).

Other lycaenids are less frequently attended, either because they exert a weaker attraction or because they live in places where ants are uncommon. For example, although ants are abundant on dry *Polyommatus icarus* sites, only three-quarters of final instar larvae found in Britain were attended and no younger instars had ants present (Heath *et al.* 1984). It is even rarer to find small blue *Cupido minimus* larvae with ants in Britain, but this is because few British ants climb the 10–20 cm tall stems of *Anthyllis vulneraria* to reach the flower-heads on which this species feeds: further south in Europe several species of ant climb flowers, and in France, Spain, and Yugoslavia the final instar of *C. minimus* has never been found without ants present (J. A. Thomas, personal observation).

For the same reason, in Britain it is unusual for Lycaenidae that live on bushes or trees to be found with ants present. Ants may nevertheless be important to their survival. The larva of the purple hairstreak *Quercusia quercus* feeds on oak leaves and lives at all heights on the canopy, but is rarely if ever attended by ants. However, the full grown larva descends to pupate on the ground, and pupae have only been found deep in the brood chambers of *Myrmica* ant nests (Thomas and Lewington 1991). Since roughly 80 per cent of a *Q. quercus* population dies between leaving the tree and emerging as an adult, it is likely that this is the key period that regulates the abundance of this butterfly on different sites (Thomas 1975*b*). If so, it is possible that the protection afforded by ants is crucial to this species, and that the comparative sizes of *Q. quercus* colonies may depend more on the structure and microclimate of the woodland floor, allowing high densities of ants to develop in some areas but very few in others than on any properties of the oak trees (Thomas and Lewington 1991).

It seems certain that the relationship between pupae and ants has been grossly overlooked in the wild. Buried pupae are inconspicuous, but it is likely that all European Lycaenidae that have brown or speckled pupae live on the ground, and since ants invariably bury sweet immobile objects that are too large to carry to their nests, probably all these are found and hidden by ants. It is not yet known, however, whether the wide range of pupae that have been found inside ant nests (Thomas 1986) were dragged underground as prepupal larvae, or ha followed ant trails, or had settled by chance in the associated loose earth. It is noteworthy, however, that many larvae are especially attractive to ants just before pupation (Mallicky 1969; J. A. Thomas, personal observation), and that their fleshy lobed bodies, which can easily be gripped by ants, are just light enough for determined workers to move if no resistance is offered by the larva.

In the tropics, various species of lycaenid and riodinine species secrete very different cocktails of carbohydrates and amino acids, and it is likely that particular mixtures have evolved to suit the tastes of different species of ant (De Vries and Baker 1989). To date, little specificity has been found in Britain among the ants that attend the Lycaenidae, beyond a general predominance of species of *Myrmica* and *Lasius*, which are especially dependent on animal and plant secretions in their diets. In general, the commonest species in the habitat tends the butterfly; in the case of *Polyommatus icarus*, the same larva may be attended by different species of ant at different times in the same day (Thomas and Lewington 1991). However, *Plebejus argus* is an exception (C. Thomas, in press): it shows a clear-cut specificity for the black ants *Lasius alienus* and *L. niger*, and will oviposit only on young foliage that contains the scent of these ants. Similarly, the Australian hairstreak *Jalmenus evagoras* lays preferentially on *Acacia* bushes attended by *Iridomyrex* species of ant (Pierce and Elgar 1985).

Scientists have argued for many years about whether the relationship between phytophagous butterflies and ants had any greater ecological significance other than that the former had developed mechanisms to reduce their chances of being killed by ants. In support of this theory were the facts that ant attendance was very rarely seen in the wild and that captive larvae developed perfectly well without ants present, apart from the few cases where their secretions grew mould. In fact, ant attendance is far commoner than was generally supposed, but most entomologists who search for larvae fail to observe this because the act of searching usually frightens away the ants long before a larva is found.

Recent work by Pierce, De Vries, Maschwitz, Fiedler, and their co-workers proves conclusively that the relationship can be truly symbiotic: in some habitats, ants not only provide vital protection to young Lycaenidae from parasites and invertebrate predators, but do themselves obtain substantial and significant quantities of food from the butterfly.

The best example of enhanced survival when ants are present is Pierce and Easteal's (1986) study of *Glaucopsyche lygdamus*. By making a series of exclusion experiments in the wild, they found that larvae were apt to fall off their bushes if ants were prevented from attending them and that a larger number were killed by predators and parasites; in all, 9.9 per cent of individuals survived from egg to adulthood when ants were present, but only 0.7 per cent survived when ants were excluded. Population models based on life tables of other lycaenids indicate that the first figure is a comparatively high rate of survival, indicative of a population that is viable in the wild, whereas any population experiencing a 99.3 per cent mortality, as occurred when ants were absent, would rapidly become extinct (calculation based on Duffey 1968; Thomas 1974, 1975*b*, 1977, and unpublished data). Although there are no comparable figures for the effect of ant exclusion on British butterfly species, anyone who has watched ants attending larvae and pupae, or who has seen them fiercely repelling invertebrates and fighting to the death with other ants over them (Thomas and Lewington 1991), is left in no doubt that ants play a major role in the survival of wild populations. This may explain why some sites with an abundance of foodplants support large colonies whereas others do not.

That lycaenid secretions can be an important source of food to ants was demonstrated on limestone grassland in Germany. Fiedler and Maschwitz (1988*b*) recorded the density of *Lysandra coridon* larvae present and the extent to which they were attended by ants. They found that mature larvae produced an average of 31 droplets an hour from their honey glands. By combining these figures with the known size and sugar content of droplets from the closely related *L. hispana*, they estimated that carbohydrates providing a total of 1.1–2.2 kJ of energy were consumed by ants in each square metre of grassland from mid-May to late June. This, moreover, was a conservative estimate of honeydew secretions, nor did it take into account the contributions of other lycaenids, such as *Polyommatus icarus*, *Lysandra bellargus*, and *P. daphnis*, which were secreting at 'the same order of magnitude' on the same slopes. Nevertheless, the lowest estimates based on *L. coridon* alone account for up to 25 per cent of the energy needs of the indigenous ants at that time of year, and are clearly a major source of food. If, as seems certain, the closely related species *L. bellargus* produces similar quantities of honeydew, then the period when lycaenids make a substantial contribution to ant resources is extended from March to late-August, which accounts for all but one of the months in the year when temperate species of ant forage above ground.

Further evidence of the importance and complexity of butterfly secretions in an ant's diet comes from De Vries: in at least one Panamanian riodinine species the larva adds insult to herbivory by also drinking at its foodplant's nectaries, thereby depriving the ant of a rival food-source and at the same time obtaining substantial nourishment for itself (De Vries and Baker 1989). It seems certain that ever more complicated relationships between phytophagous butterflies, ants, and plants will be uncovered as this research progresses.

Aphytophagous species

Many of the symbiotic relationships described above enable a lycaenid to live in safety inside or near an ants' nest. Another characteristic of phytophagous lycaenidae is an aptitude, in their early instars, to attack and eat other larvae that are encountered on the same foodplant, including those of their own species (e.g. Cottrell 1984). It is a short evolutionary step for the first relationship to become one-sided, with the butterfly larva plundering the ants' resources whilst still being shielded by its various protective devices. This has evolved independently in several subfamilies of Lycaenidae, although the total number of species involved is small, accounting for perhaps three per cent of this family and one per cent of all butterfly species (calculations based on Cottrell 1984). These specialists live as larvae either inside ant nests, feeding on ant brood, worker ant regurgitation or detritus, or more often outside the nest where they prey on the ants' herds of domestic insects, such as coccids and aphids (Fig. 7.5). Many also have a short initial period of feeding on a plant.

Only the large blue *Maculinea arion* among British species belongs to this group, although there are another four species of *Maculinea* in Western Europe, respectively accounting for more than one

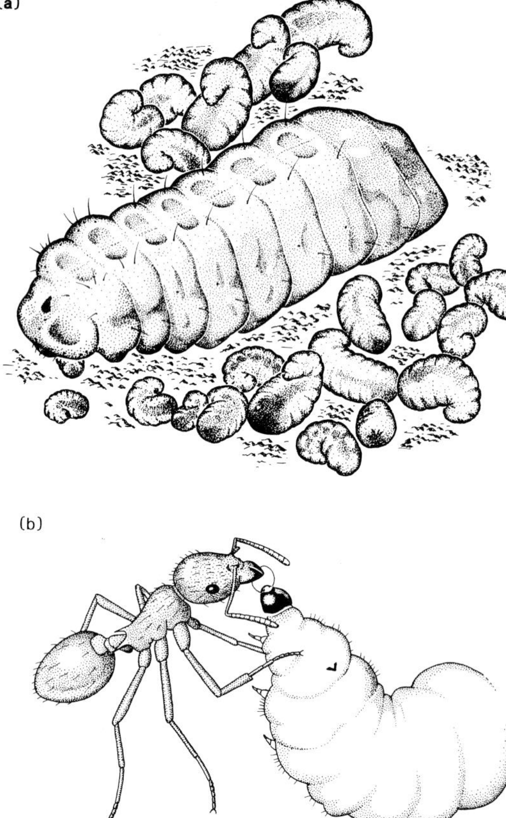

Fig. 7.5 Contrasting methods of feeding by *Maculin* caterpillars inside *Myrmica* ant nests. (a) *Maculir arion*, a primitive species, usually rests apart from t ants, but periodically glides over the brood to fe stealthily on ant larvae. (b) *M. rebelli*, an advanc species, closely mimics ant larvae, and it is fed wi regurgitation by worker ants whenever it begs for foc This is a much more efficient method of feeding. Frc original drawings by R. G. Snazell, from (a) Thom and Wardlaw (1990) and (b) Elmes *et al*. (1991).

per cent of British and European butterfly speci The adaptations of *M. arion* are among the mc thoroughly studied, and are a good example of t kind of specializations that have evolved in th group. *M. arion* lives in closed populations and univoltine, the adults emerging in late June and Ju Eggs are laid among the flowerbuds of *Thymus* a *Origanum*. They soon hatch and the young lar feeds for about three weeks on the flowers a developing seeds, killing other larvae on the same flowerhead. After three weeks the larva has developed through three instars on the plant, but has acquired only about one per cent of its final body weight (Thomas and Wardlaw, in press). It then enters the fourth and final instar, which lives for nine months underground in a red (*Myrmica*) ant's nest (Thomas 1977).

The freshly-moulted fourth instar larva has both cupola and a well-developed DNO. Moulting occurs throughout the day but the small larva remains hidden on its foodplant until early evening, when it flicks off onto the ground. It then hides close to the soil to await discovery by ants. By this procedure, the larva greatly increases its chances of being encountered by *Myrmica* rather than by ants of other genera, for, unlike most ants, *Myrmica* workers forage mainly at ground level and avoid the heat of the sun (Thomas 1977). Frohawk (1924) claimed that the larvae actively sought ants' nests and perhaps followed their trails, but his observations were made in unnaturally exposed conditions, and the larvae were probably just crawling away to hide; several hundred observations have recently been made of undisturbed wild larvae at this stage, and all waited to be discovered by ants, even when this took over 12 hours (Thomas 1977, unpublished data). Schroth and Maschwitz (1984) also claimed that the scarce large blue *Maculinea teleius* followed artificial ant trails, at least in the laboratory. This, again, has never been observed in the wild, where *M. teleius* behave similarly to *M. arion* (Thomas 1984*c*; Fielder 1990*b*). This much-quoted observation of trail following has now been shown to be an experimental anomaly, in which starving *M. teleius* larvae followed a trail of pheromones that were unfortunately contaminated by their food, ant haemolymph (Fiedler 1990*b*).

Wild *Maculinea arion* larvae do not attract ants by pheromones, but are soon discovered by foraging *Myrmica* workers. The ant initially behaves as if it had encountered a phytophagous Lycaenid, licking the DNO and pore cupola, then usually rushing back to its nest to recruit about 10 more workers (Frohawk 1924; Thomas and Lewington 1991). These disperse after a few minutes, leaving just the original ant in attendance. 'Milking' continues for up to four hours more, until the larva rears up on its prolegs, contorting its body into a figure S and squeezing the body so that the thoracic segments swell up like a balloon (Fig. 7.6). At this, the frenzied ant picks up the larva and runs to her nest; she then carries the larva underground and places it among the ant brood.

This adoption process was first observed and described by Purefoy in Frohawk (1924). The most likely explanation is that, by contorting its body, the larva mimics an escaped ant grub and tricks the ant into retrieving it. *Myrmica* workers recognize their own brood by a combination of four stimuli: size, hairiness, scent, and texture (Brian 1975). The *Maculinea arion* larva already fulfils the first two criteria and, after the prolonged milking and drumming, must also smell of that particular ant's colony, for every nest has its own individual mix of pheromones which is transmitted through the colony, dissolved in surface waxes, by workers tapping one another with their antennae. The fourth criterion is fulfilled when the larva distorts its body, for the wrinkled, flexible skin—which must eventually expand to contain a body mass one hundred times greater—is blown up, in the thoracic region at least, to a firmness similar to that of a plump ant grub (J. A. Thomas, personal observation). Moreover, adoption does not occur if the ant fails to touch this swollen area while the larva is rearing up, and must wait for the whole process to be repeated, perhaps 10–30 minutes later.

Fig. 7.6 A *Maculinea arion* larva signals its readiness to be adopted by a *Myrmica sabuleti* ant by rearing up and expanding its thoracic segments. The ant is still milking the dorsal nectary organ. Photographed in the wild, Devonshire, 1976.

Cottrell (1984, and personal communication) has suggested that, in rearing up, *Maculinea arion* larvae may be exposing a touch-pheromone secreted from between the thoracic segments. This seems unlikely when one has observed the behaviour first hand, and it does not explain why it takes so long for the larva to release its putative pheromone, nor why it often fails; the long milking period is more convincingly explained by the need for the larva to acquire a thorough coating of the ant's scented waxes before it allows itself to be taken underground. *M. teleius* induces adoption in the same way as *M. arion* (Chapman 1919; Thomas 1984*c*; Fiedler 1990*b*), but the other, more advanced, European *Maculinea* may indeed have evolved pheromones to aid their adoption. *M. nausithous* contorts its body into a dumbell shape within about 30 seconds of being found (Elmes and Thomas 1987), whereas the alcon blues *M. alcon* and *M. rebeli* are picked up and carried to a nest the moment they are encountered, with none of the rigmarole indulged in by *M. arion* (Elmes *et al.* 1991).

Once inside the nest, *Maculinea arion* feeds on ant brood and rapidly grows. It hibernates when half grown, and resumes feeding in late April or May. Pupation occurs near the top of the nest, and in possessing cupola and (probably) the ability to stridulate, the pupa has no more than the normal adaptations of a phytophagous lycaenid. *M. arion* larvae are generally ignored in the ant nest, but are prone to attack and usually experience heavy mortalities when underground (Thomas 1977). Three factors account for this: starvation, queen effect, and host specificity. The first often occurs in smaller nests or when more than one larva has been adopted, for the larva is so voracious a predator that it often eats out the ant brood and dies of starvation (Thomas and Wardlaw, in press). Queen effect occurs when a larva is adopted by a nest that contains queen ants. These often induce workers to attack quite large grubs that would otherwise develop into rival queens, and many *M. arion* larvae are mistakenly killed when they reach that size (Thomas and Wardlaw 1990). Host specificity is the most serious mortality: up to five species of *Myrmica* forage on *M. arion* sites, and all adopt larvae with equal readiness. However, survival is much higher with one species—*Myrmica sabuleti*—than with others, so much so that population models indicate that a colony can survive only if 60–70 per cent of larvae are adopted by that particular species of *Myrmica* (Thomas 1977, 1991).

Recent work on the other European *Maculinea* has revealed that each has a similar dependence on a single, but in each case different, species of *Myrmica* (Thomas 1983*e*; Thomas *et al.* 1989). Thus it appears that this genus of butterflies has speciated to parasitize each of the four common species of *Myrmica* in the Palearctic, plus, in the case of *Maculinea rebeli*, a fifth *Myrmica* that is common in the xerophitic habitats where its initial foodplant grows. Each butterfly has almost certainly evolved to mimic its specific host, perhaps through behaviour, pheromones, and sound, which, in turn, would diminish its chances of survival in nests of other *Myrmica* species. However, some *Maculinea* are clearly more advanced than others: *M. rebeli* and *M. alcon* larvae have the additional ability to beg and be fed directly by worker ants on regurgitated food (Fig. 7.5*b*). This is a much more efficient method of feeding, and the same sized *Myrmica* nest can support about six times as many larvae of *M. alcon* or *M. rebeli* than predacious species of *Maculinea* (Elmes *et al.* 1991; Thomas and Wardlaw, in press).

Similar aphytophagous relationships have evolved elsewhere in the world, and include other adaptations. The large African genus of *Lepidochrysops* has evolved similar adaptations, and has a similar life-style to the *Maculinea*. But larvae of the Australian genus *Liphyra* inhabit nests of a much fiercer genus of ants, *Oecophylla*, where they probably feed on a combination of ant brood, domestic coccids, and perhaps leaves (Cottrell 1984). Their adaptations include an extraordinary flattened 'pork pie shaped' armoured carapace, that enables the larva to withstand frequent onslaughts, whilst the emerging adult escapes by being smothered with modified scales that stick to the attacking ants' jaws, allowing the butterfly to emerge unmolested. Most other aphytophagous Lycaenidae feed on aphids or coccids in or near ant nests, although little is known of the behaviour and ecology of many of these; an excellent review of current literature is provided by Cottrell (1984).

7.4 Interactions between butterflies and plants

7.4.1 Herbivores and communities

Plants provide a number of diverse resources for butterflies, the most obvious being food for larvae and adults. Plants also make up the physical structure of a habitat which butterflies use for mate-location, courtship, roosting, and hibernation (see chapter 2), and modify the microclimate, influencing where eggs are laid or larvae feed (see chapters 3 and 5). Thus the butterfly–plant relationship can be obvious or subtle, yet both are crucial to an understanding of butterfly community biology (see section 10.1).

Despite the obvious influences that butterflies and plants have on each other, the evolutionary consequences of such interactions, remains a controversial topic. All but one species of British butterfly, *Maculinea arion*, have entirely phytophagous larvae and therefore the implications for herbivores and plants is central to community biology.

7.4.2 Plant defences and butterflies

Plants such as the *Quercus* spp. undergo dramatic seasonal changes in their nutrient value, physical features, and chemical composition (Feeny 1970). The young leaves of *Quercus*, used by *Quercusia quercus*, are high in nitrogen, low in toxins, and have thin cuticular layers. As the season progresses the leaves become tougher in texture and the level of toxic chemicals increases. Under natural conditions the larvae of *Q. quercus* are fully grown by early June. If the larvae are fed on a diet containing just one per cent tannin, the main toxin which accumulates in the oak leaves, they grow slowly and produce undersized pupae, if at all. As the larvae bite into the oak leaf the tannin and protein combine and the digestibility of the leaf protein is greatly reduced. *Q. quercus* largely avoids tannins by feeding during the 'undefended' season, but this exacts a price from the butterfly; it can only produce one generation per year. When reared on artificial diets with 0.5 per cent tannin, about the natural level in April, *Q. quercus* can produce three generations a year under the same conditions of temperature and daylength (Morton, personal communication). A similar situation occurs with respect to the nitrogen level and hence nutritional value in *Urtica dioica* for the nymphalids that use this as a foodplant. *U. dioica* regrowth, following cutting, is of higher nutritional value to both *Aglais urticae* and *Inachis io* (see section 8.6). If sufficient nettle regrowth is available, then these species may have more generations per year than populations on uncut and, ultimately, senescent nettle (Pullin 1986*a*,*b*).

Chemical defences

All plants contain chemicals which do not appear to be essential for their normal metabolic growth and may be by-products or waste products (but see Chew and Rodman 1979). Such chemicals (secondary plant compounds) may act as toxins, damaging, killing, or otherwise thwarting potential herbivores and therefore protect the plant from attacks by herbivores. The secondary plant compounds can be either a benefit or a deterrent to butterflies. They dictate foodplant specificity and prevent every plant being equally palatable to every butterfly species. Butterflies have overcome these chemical defences designed to deter generalist herbivores, by feeding behaviour and by detoxifying the chemicals. Thus, some larvae of tropical Lepidoptera cut a circular trench in the leaf before feeding, so preventing the release of toxins into the area of leaf on which they feed. Butterflies react to plant chemical defences in one of two basic ways: (i) they avoid plants with such compounds; (ii) they use these plants by neutralizing the toxins, or by diverting them into alternative biochemical pathways.

The brimstone *Gonepteryx rhamni* avoids host-plants with toxins (McKay 1988). The females preferentially lay their eggs on the leaves and buds of young alder buckthorn trees *Frangula alnus* in exposed positions rather than on older trees. This behaviour is correlated with differences in secondary plant substances, believed to be phenolic compounds; it appears that females avoid trees with high levels of some of these compounds. Each tree also has wound-induced changes in leaf chemistry and *G. rhamni* larvae avoid damaged leaves and the adjacent foliage.

Many species can feed on potentially toxic plants by immobilizing the chemicals using a group of enzymes called mixed function oxidases. These enzymes convert specific toxins into water-soluble molecules which can be excreted safely. Most populations of bird's-foot trefoil *Lotus corniculatus* include plants which produce cyanide when their leaves are damaged and this is believed to be a non-specific deterrent aimed at snails and slugs (Briggs and Walters 1969). These plants can reduce the growth rate of species not adapted to coping with the toxin (e.g. *Erynnis tages*), but *Polyommatus icarus* has no difficulty in detoxifying the cyanide as the larvae produce an enzyme called rhodanese which converts it into the harmless thiocyanate (Parsons and Rothschild 1964). A similar reaction is used in the clinical treatment of cyanide poisoning in humans.

Of the 30 000 or so, known secondary compounds perhaps the most familiar are the alkaloids, including poisons such as curare, strychnine, and colchicine, and stimulants and soporifics, such as morphine, quinine, and caffeine. There are no confirmed examples of British species of butterfly being deterred or attracted by plant alkaloids. They are often used as chemical cues during egg-laying as in the North American blue butterfly *Glaucopsyche lygdamus* which selects between plants of *Lupinus* (Dolinger *et al*. 1973).

Glycosides form another important group of secondary plant substances. These include the cardiac glycosides and the iridoid glycosides which are the active deterrents used respectively by the monarch butterfly, *Danaus plexippus* and the North American *Euphydryas* butterflies against larval and adult predators (Bowers 1979; Brower 1983; see chapter 5). These substances are acquired from the foodplant or occasionally by the adult from nectar or damaged leaves (Boppré 1984). However, the glycosides used by tropical Ithomiinae may be produced by the butterfly and not acquired intact from their hostplant (see Ackery 1988). Insects usually advertise their possession of these toxic chemicals by displaying bright colour patterns and 'overt' behaviour (see chapter 5). The chemicals function as emetics in the usual target predators, birds, and act as powerful modifiers of behaviour; forming the logical basis for mimicry (see chapter 5 and section 7.2.1).

Evidence for the use of glycosides by British butterflies is circumstantial, but nevertheless strong in *Eurodryas aurinia*. The usual foodplant of this species, devil's bit scabious *Succisa pratensis*, a member of the Dipsacaceae, contains the aucubic glycoside known as cephalarioside (Gibbs 1974). In the wild, larvae of *E. aurinia* occasionally use honeysuckle *Lonicera periclymenum* (Goss 1874; Ford and Ford 1930), and in captivity will readily accept snowberry *Symphoricarpos rivularis*. These plants are in the Caprifoliaceae which is closely related to the Dipsacaceae. *Lonicera* contains several toxic glycosides including loganin, kingiside, and morroniside (Plouvier 1964), whilst *Symphoricarpos* contains secologanin (Jensen *et al*. 1975); glycosides belonging to the same iridoid glycoside group as those which occur in the North American *Euphydryas* species. The behaviour of *E. aurinia* also implicates the use of chemicals in defence. *E. aurinia* larvae are gregarious and form conspicuous black clusters in four out of their six instars (Porter 1982). The adults are brightly coloured and highly apparent as they feed or bask in their habitat. Even so, the inspection of some 2500 individuals in a three-year study of a colony of *E. aurinia* produced little evidence of bird predation (Porter 1981). In feeding trials using chicks, M. S. Warren (personal communication) reports that the chicks rejected larvae of *E. aurinia*. Clearly, more work is required on both field studies of bird predation on adults and larvae, and laboratory studies of the secondary compounds present in butterfly foodplants.

Physical defences

Physical defences against herbivores vary from those which are obvious, such as the spines of gorse *Ulex* spp., to others which are hidden, for example the retention of silica spicules in grasses. These defences are often directed towards animals other than insects but nevertheless can have an important influence on foodplant choice by butterflies.

The surfaces of plants are rarely smooth, they are covered in a variety of hairs, spines, or hooks the function of which may relate to a plant's metabolic needs such as aiding water retention or preventing freezing. However, some may have a direct anti-predator role. *Urtica dioica* has a very sophisticated defence system of hardened 'hypodermic needles' containing formic acid, a powerful irritant. These spines deter grazing animals (Pollard and Briggs 1984), yet are ineffective against larvae of *Aglais urticae*, *Inachis io*, *Polygonia c-album*, and *Vanessa atalanta*. These butterflies, along with other insects in the nettle-patch community, can walk over or eat the spines. Similarly, thistles (*Cirsium* spp.), are protected from most grazing herbivores by stout spines, but the larvae of the painted lady *Cynthia cardui* eat all the soft portions of the leaf and avoid the spines.

A critical stage in the life cycle of any butterfly occurs when the newly hatched larva begins feeding. The waxy layer on the surface of seed-pods of Dame's violet *Hesperis matrionalis* presents an obstacle to newly emerged larvae of *Anthocharis cardamines* and causes a lower rate of survival than on other alternative foodplants such as cuckooflower *Cardamines pratensis* (Courtney 1980, 1981, 1982*a*). The older seed-pods of crucifers such as hedge garlic *Alliaria petiolata* are too tough for the earliest instars of *Anthocharis cardamines* to penetrate. These factors influence the recruitment of individuals and host-plant choice in subsequent generations. Damaged plants can also deter herbivores by producing excessive quantities of sticky 'gum' (Edwards and Wratten 1980). This can immobilize small insects and the mouthparts of larger insects, but little is known of its impact on British butterflies.

Finally, plants may integrate physical and chemical defences. This has been demonstrated by rearing larvae of *Erynnis tages* on two groups of *Lotus corniculatus* plants—one undamaged and one nibbled by slugs (A. Morton, personal communication). Larvae reared on the latter plants suffered higher mortality rates than those reared on the former. This common response to many plant defence systems (Rhoades 1983; Fowler and Lawton 1985; Edwards and Wratten, 1980; Leather *et al*. 1987) is not unlike antibody sensitization in mammals, and may arise from feeding by the larvae themselves or through feeding by other herbivores in the community.

7.4.3 Butterflies and flowers

The interaction between butterflies and flowers depends primarily on the reward offered by the flowers and the matching of flower and butterfly structure. The outcome also depends on a variety of subsidiary factors, particularly coincidence in flight period and flowering, habitat, competitors, and the array of opportunities offered to butterflies.

Butterflies as pollinators

When visiting flowers to obtain the nutrients in floral nectar butterflies may pick up pollen on their probosces and become potential pollinators. Unlike the tropical Heliconiinae, British butterflies are not known to use pollen as food, despite its containing rich supplies of protein (Boggs *et al*. 1981). Through evolution, plants have generally adopted one type of pollen vector which can be wind, water, or mobile animals such as flies, bees, butterflies, humming-birds, bats, or other mammals (Proctor and Yeo 1979). Close relationships between plants and specialist pollinators are frequently cited as examples of 'coevolution'. There are advantages and disadvantages for plants with specialist pollinators. The plant avoids having to compete with other plant species for pollinators and there is a greater probability that all its flowers will be found pollinated. However, the plant becomes dependent on the population dynamics of its specialist pollinator. In return, the specialist pollinator has access to the food source without competition. The advantages and disadvantages for the pollinator are similar to those affecting specialists and generalists of hostplants (see sections 3.4.1 and 10.4.2). Most plants fall into a wide spectrum of access and insect use (see Table 7.3). For example, the deep corolla of bugle *Ajuga reptans* prevents access to the nectar from all but long-tongued insects such as bees and butterflies. In contrast, the open nectaries of bramble *Rubus fruticosus agg.* are exposed to many butterflies and other insects.

Do butterflies act as pollinators? Pollen has certainly been found on the probosces of butterflies but the amount depends on the flower shape. Butterflies which feed at the flowers of violets *Viola* spp., bluebell *Hyacinthoides non-scriptus*, and stitchworts *Stellaria* spp., pick up very few pollen grains (Wiklund *et al*. 1979; Courtney *et al*. 1982). It seems that the open nature of these flowers enables the butterfly to probe past the anthers and 'steal' the nectar without collecting pollen (Wiklund *et al*. 1979). *Leptidea sinapis* obtain nectar from the vetch, *Lathyrus montanus*, by probing between the wing and standard petals of the flower, thereby avoiding contact with anthers and stigma in the flower keel (Wiklund *et al*. 1979). This could be construed as theft rather than pollination. Pollen may

Table 7.3 Nectar resource utilization in British butterflies showing some commonly used plant species

	HESPERIIDAE								PAPILIONIDAE	PIERIDAE							LYCAENIDAE														
	Pyrgus malvae	*Carterocephalus palaemon*	*Erynnis tages*	*Ochlodes venata*	*Thymelicus sylvestris*	*Thymelicus lineola*	*Thymelicus acteon*	*Hesperia comma*	*Papilio machaon*	*Gonepteryx rhamni*	*Pieris rapae*	*Pieris napi*	*Anthocharis cardamines*	*Leptidea sinapis*	*Pieris brassicae*	*Colias croceus*	*Celastrina argiolus*	*Callophrys rubi*	*Lycaena phlaeas*	*Aricia agestis*	*Lysandra bellargus*	*Cupido minimus*	*Polyommatus icarus*	*Satyrium pruni*	*Aricia artaxerxes*	*Plebejus argus*	*Lysandra coridon*	*Maculinea arion*	*Quercusia quercus*	*Satyrium w-album*	*Thecla betulae*
Buttercups, *Ranunculus* spp.	●		●									●		●			●		●					●							
Daisy, *Bellis perennis*											●								●												
Cuckooflower, *Cardamines pratensis*												●	●																		
Greater stitchwort, *Stellaria holostea*											●	●	●	●	●																
Thyme, *Thymus praecox*																				●			●		●		●	●			
Bramble, *Rubus fruticosus* agg.		●		●	●								●	●			●							●					●	●	●
Tormentil, *Potentilla erecta*																															
Privet, *Ligustrium vulgare*				●													●	●						●						●	
Hogweed/Angelica, Umbelliferae																								●					●	●	●
Ivy, *Hedera helix*																	●														
Yarrow, *Achillea millefolium*																			●												
Ragwort, *Senecio jacobaea*											●	●			●	●			●	●	●		●							●	●
Knapweeds, *Centaurea* spp.				●	●					●					●	●							●								

Field scabious, *Knautia arvensis*				•											•												•				
Marjoram, *Origanum vulgare*					•					•						•				•	•		•							•	
Carline thistle, *Carlina vulgaris*								•		•					•								•				•	•			
Hemp agrimony, *Eupatorium cannabinum*																													•		•
Heather, *Calluna vulgaris/Erica* spp.						•													•							•					
Water mint, *Mentha aquatica*																	•														
Teasel, *Dipsacus fullonum*									•						•																
Fleabane, *Pulicaria dysenterica*					•	•				•	•	•			•	•			●				•								
Dandelion, *Taraxacum* agg.				•						•	•		•		•	•			•												
Primrose, *Primula vulgaris*										•																					
Bluebell, *Hyacinthoides non-scriptus*									•	•	•	•	•		•																
Bugle, *Ajuga reptans*	•	•	•	•							•	•	•	•	•		•						•					•			
Ragged robin, *Lychnis flos-cuculi*	•		•	•					•	•	•	•	•	•	•																
Selfheal, *Prunella vulgaris*										•																	•				
Betony, *Stachys officinalis*					•					•		•																			
Hawkweeds, *Hieracium/Hypochoeris*											•		•						•												
White clover, *Trifolium repens*						•														•			•								
Red clover, *Trifolium pratense*					•	•			•		•			•					•												
Kidney vetch, *Anthyllis vulneraria*				•																		•	•								
Vetches, *Vicia* spp.			•	•	•					•		•	•	•		•							•								
Red campion, *Silene dioica*										•	•	•	•																		
Bird's-foot trefoil, *Lotus corniculatus*	•		•	•	•						•			●				•				•	•				•				
Sanfoin, *Onobrychis viciifolia*				•							•				•																
Thistles, *Cirsium* spp.				•	•	●		•		●	●	•			●	•	•		•				•				•			•	•
Devil's-bit scabious, *Succisa pratensis*				•	•				•	•					•																•
Honeydew/Sap																		•						●				●			●

Table 7.3 (*Cont.*)

	RIODININAE	NYMPHALIDAE																SATYRINAE										
	Hamearis lucina	*Aglais urticae*	*Inachis io*	*Polygonia c-album*	*Boloria euphrosyne*	*Eurodryas aurinia*	*Melitaea cinxia*	*Boloria selene*	*Cynthia cardui*	*Mellicta athalia*	*Ladoga camilla*	*Argynnis adippe*	*Argynnis paphia*	*Apatura iris*	*Argynnis aglaja*	*Vanessa atalanta*	*Nymphalis polychloros*	*Pararge aegeria*	*Lasiommata megera*	*Coenonympha pamphilus*	*Aphantopus hyperantus*	*Maniola jurtina*	*Coenonympha tullia*	*Melanargia galathea*	*Erebia epiphron*	*Hipparchia semele*	*Pyronia tithonus*	*Erebia aethiops*
Buttercups, *Ranunculus* spp.	•				•	•				•										•		•						
Daisy, *Bellis perennis*																			•									
Cuckooflower, *Cardamines pratensis*			•			•												•										
Greater stitchwort, *Stellaria holostea*		•																		•								
Thyme, *Thymus praecox*		•						•																•	•		•	
Bramble, *Rubus fruticosus* agg.		•		•						●	•	•	●					•		•	•	•				•	•	
Tormentil, *Potentilla erecta*	●				•	•				•										•			•		•			
Privet, *Ligustrium vulgare*		•	•	•					•		•		•			•		•			•	•					•	
Hogweed/Angelica, Umbelliferae											•																	
Ivy, *Hedera helix*		•		•					•							•												
Yarrow, *Achillea millefolium*			•																•	•		•		•				
Ragwort, *Senecio jacobaea*		•	•						•				•					•	•	•	•	•					●	•
Knapweeds, *Centaurea* spp.		•		•		•			•	•		•	•		•				•			•		•				•
Field scabious, *Knautia arvensis*		•																										•
Marjoram, *Origanum vulgare*		•	•																•		•	•		•		•		•
Carline thistle, *Carlina vulgaris*		•							•						•							•		•		•	•	

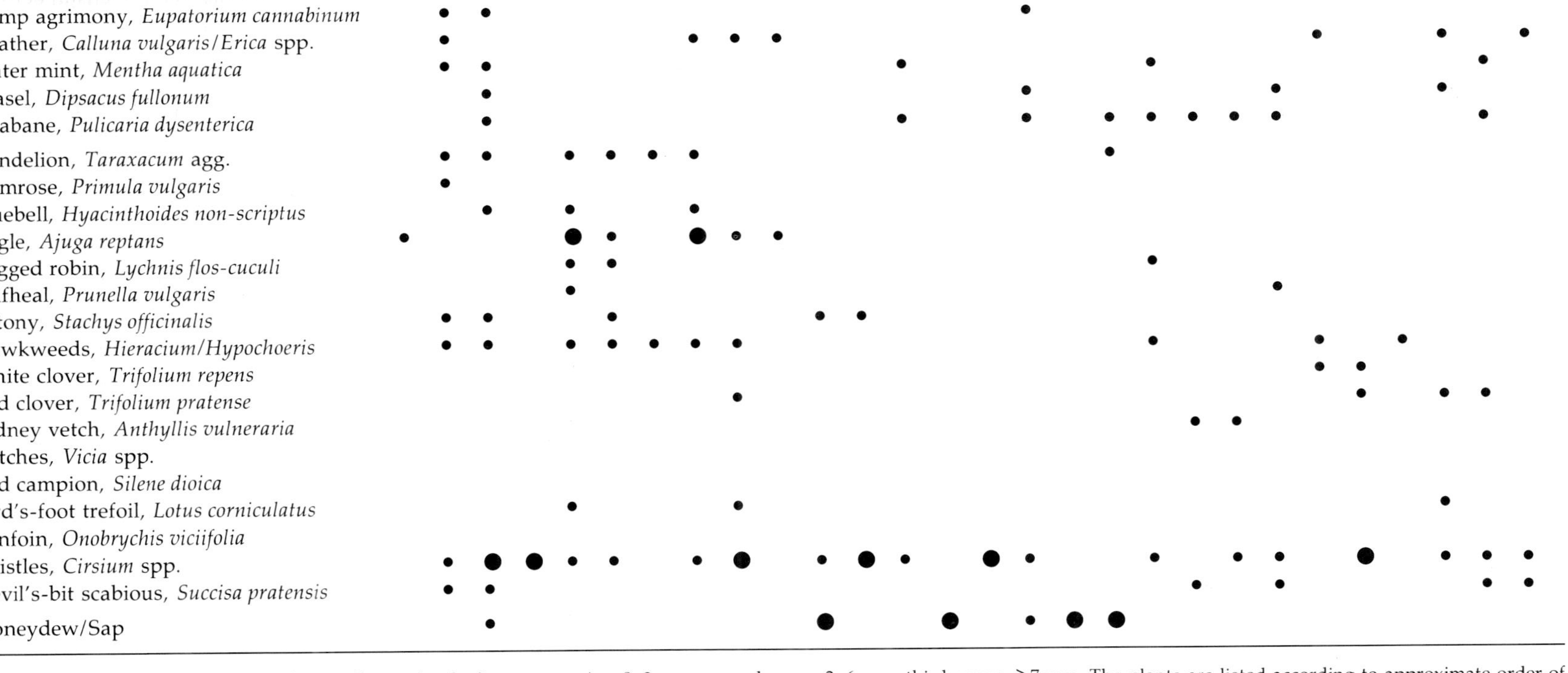

The nectar sources are ranked according to flower depth: first group, *circa* 0–2 mm; second group 3–6 mm; third group, >7 mm. The plants are listed according to approximate order of flowering. The butterfly species are divided into families and ordered according to time of appearance. The records are by no means exhaustive, but give some idea of the range of plant species visited for nectar. ● Nectar sources which are considered major sources for a given species.

Based upon Stallwood (1972–73); Peachey (1980); Waring (1984); K. Porter, unpublished observation.

be transferred by other body organs (Murphy 1984). Courtney *et al.* (1982) found significant amounts of dandelion *Taraxacum officinale agg*, and *Hesperis matronalis* pollen on *Anthocharis cardamines*, *Pieris napi*, *P. rapae*, *P. brassicae*, and *Aglais urticae*, mainly located within the facial cavity, with small amounts on the eyes, thorax, and legs. But it is still questionable whether this pollen contributes much towards pollination. Wiklund (1981) maintains that butterflies are adapted to 'steal' nectar from flowers aimed at bee visitors, and that butterflies are poor pollinators. This view is reinforced by the findings of Courtney *et al.* (1982) who suggest that butterflies play a minor part in pollination. However, for some plant species, butterflies may play an important role in pollination. One group with close connections with insect pollinators are the orchids. The flower structure of the fragrant orchid *Gymnadenia conopsea*, pyramidal orchid *Anacamptis pyramidalis*, and the two butterfly orchids, *Platanthera* spp., is such that the proboscis of a butterfly or moth may act as a stimulus which will release the orchid's pollinia (Proctor and Yeo 1979). Careful observation of butterflies feeding at pyramidal or fragrant orchids reveal that some of these butterflies will have a pair of tiny club-shaped pollinia attached to their probosces. However, the role of butterflies in pollinating these orchids is open to question as they are commonly pollinated by large bees and some butterflies become 'gummed-up' by the pollinia, rather than pass them on to other orchid blossoms. The wild gladiolus *Gladiolus illyricus* may depend upon butterflies for pollination. During 1987, in the New Forest, Hampshire, 27 of 33 pollination events were due to *Ochlodes venata* and *Pyronia tithonus* (Stokes 1988). Other examples may exist and further studies are needed.

Interactions on flowers

One of the factors that determines which flowers are visited is the length of the butterfly's proboscis relative to the depth of the flower corolla (Table 7.4). Proboscis length in British butterflies ranges from 5 mm in the hairstreaks and *Cupido minimus*, to 19 mm in the swallowtail *Papilio machaon*. Wingspan is no real guide to proboscis length; for example, *Ochlodes venata* has a longer proboscis than the much larger *Argynnis paphia*. Foraging butterflies are faced with a wide range of flowers differing in structure, colour, and suitability as nectar sources.

Although few studies on flower feeding in British butterflies have considered the full complement of species available (Proctor and Yeo 1979; Stallwood 1972–73); different butterflies seem to have a bias for particular flower colours, shapes, or types. There is evidence that butterflies can learn the location of suitable nectar sources (Gilbert 1975). In a study of the butterfly community of Bernwood Forest, Buckinghamshire, Peachey (1980) found that 51 species of plant were visited by up to 24 species of butterfly. However, 85 per cent of visits were to only 13 species of plant (25 per cent of total plants available). Of these 'preferred' plants 23 per cent of visits were to bramble; 14 per cent to marsh thistle *Cirsium palustre*; 10 per cent to creeping thistle *C. arvense*; and 10 per cent to ragwort *Senecio jacobaea*. These figures suggest that certain species of plant do attract more butterflies than others, or they would but for the question of relative abundance of each flower and nectar (Fig. 7.7a).

Differences in resource use can occasionally be determined from differences in flower structure between plant species (Fig. 7.7a). *Rubus fruticosus agg.* is rarely used by *Thymelicus sylvestris*, but it is a major nectar resource for *Maniola jurtina* and *Pyronia tithonus*. The converse is true of red clover *Trifolium pratense* which is a major source of nectar for *T. sylvestris*. This difference between the species may be partly related to proboscis length, as that of *T. sylvestris* is 50 per cent longer than that of *M. jurtina* and *P. tithonus*. However, other aspects of butterfly behaviour could result in bias for subsets of the same habitat. Whatever the reason, differences were found to characterize each species. In selecting different flowers, butterflies are known to use colour and ultraviolet reflectance patterns as cues (Ilse 1928; Mazokhin-Porshnyakov 1969; Eisner *et al.* 1969; Wiklund 1977*a*; Silberglied 1984).

The selection of nectar sources by butterflies depends much on opportunity. Although some 73 per cent of all feeding observations of *Boloria euphrosyne* on one site were on *Ajuga reptans* this masks the fact that there were relatively few flowers available to it at that time of year. Evidence for this notion is found in the use by butterflies of different flowers through the season (Fig. 7.7*b*). The results of this study pose questions as to how

Table 7.4 Butterfly proboscis length relative to wingspan. The relationship between proboscis length and butterfly family highlights the long probosces of skippers and the short probosces of hairstreaks

Species	Average wingspan (mm)	Average proboscis length (mm)	$\frac{\text{Wingspan}}{\text{proboscis length}} \times 100$	Average tongue length for group
HESPERIIDAE				
Thymelicus sylvestris	30	14	47.6	
Ochlodes venata	33	16	47.6	
Thymelicus lineola	27	13	47.6	
Hesperia comma	33	15	45.5	
Thymelicus acteon	25	11	43.5	
Erynnis tages	29	10	34.5	
Carterocephalus palaemon	29	10	34.5	
Pyrgus malvae	27	8	29.4	
PAPILIONIDAE				
Papilio machaon	80	19	23.8	
PIERIDAE				
Pieris rapae	48	15	31.3	
Gonepteryx rhamni	58	16	27.8	
Anthocharis cardamines	46	12	26.3	
Colias croceus	58	15	25.6	
Pieris brassicae	63	16	25.6	
Leptidea sinapis	42	10	23.8	
Pieris napi	50	11	26.3	
LYCAENIDAE				
Lysandra coridon	38	10	26.3	
Lysandra bellargus	38	10	26.3	
Plebejus argus	30	8	26.3	
Aricia agestis	29	7	24.2	
Aricia artaxerxes	29	7	24.2	
Polyommatus icarus	35	8	23.3	
Lycaena phlaeas	33	7	21.3	
Celastrina argiolus	35	7	20.0	
Cupido minimus	25	5	20.0	
Maculinea arion	43	8	18.5	
Callophrys rubi	33	6	18.2	

Table 7.4 (*Cont.*)

Species	Average wingspan (mm)	Average proboscis length (mm)	$\frac{\text{Wingspan}}{\text{proboscis length}} \times 100$	Average tongue length for group
Satyrium w-album	35	5	14.3	
Satyrium pruni	37	5	13.5	
Quercusia quercus	38	5	13.2	
Thecla betulae	39	5	12.8	
RIODININAE				
Hamearis lucina	30	5	16.7	
NYMPHALIDAE				
Inachis io	63	17	27.0	
Aglais urticae	53	14	26.3	
Argynnis aglaja	66	16	24.4	
Apatura iris	75	17	22.7	
Polygonia c-album	55	12	21.7	
Cynthia cardui	65	14	21.7	
Mellicta athalia	42	9	21.3	
Boloria selene	43	9	20.8	
Argynnis adippe	64	13	20.4	
Melitaea cinxia	44	9	20.4	
Boloria euphrosyne	46	9	19.6	
Vanessa atalanta	69	13	18.9	
Ladoga camilla	64	12	18.8	
Nymphalis polychloros	67	12	17.9	
Argynnis paphia	74	13	17.5	
Eurodryas aurinia	48	8	16.7	

SATYRINAE			
Lasiommata megera	48	13	27.1
Erebia aethiops	48	10	20.8
Melanargia galathea	58	12	20.8
Coenonympha tullia	41	8	19.6
Maniola jurtina	52	10	19.2
Hipparchia semele	58	11	18.9
Pyronia tithonus	43	8	18.5
Coenonympha pamphilus	38	7	18.5
Erebia epiphron	38	7	18.5
Pararge aegeria	48	8	16.7
Aphantopus hyperantus	48	8	16.7

Data from Knuth (1906); K. Porter, unpublished data.

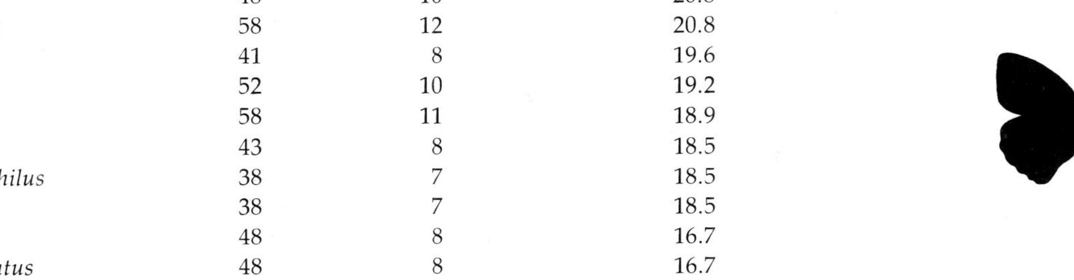

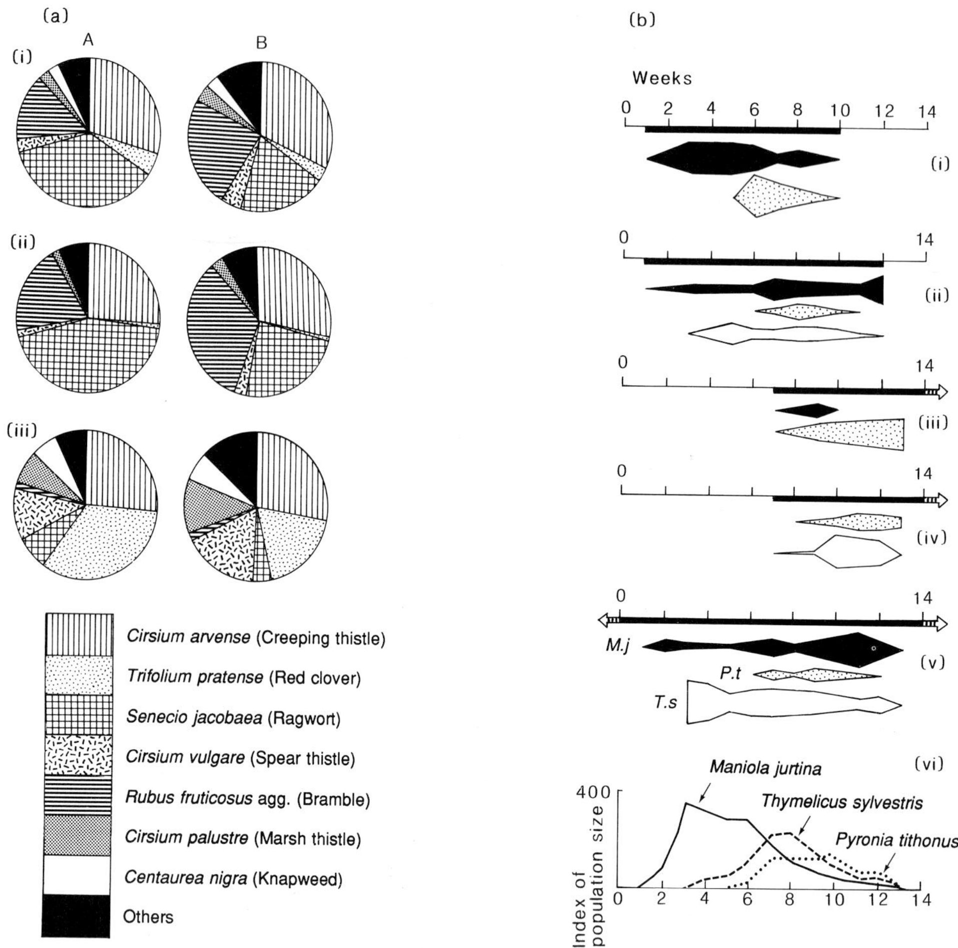

Fig. 7.7 The effect of interactions between flowers, butterflies, and other insects on the utilization of nectar sources. (a) Observations on three butterfly species—(i) meadow brown *Maniola jurtina*, (ii) gatekeeper *Pyronia tithonus* and (iii) small skipper *Thymelicus sylvestris*—at a single site on the same day. A, direct observations; B, observations corrected for plant abundance. The butterflies visited a total of 19 plant species but the majority of visits were made to seven species. Most visits by *M. jurtina* and *P. tithonus* were made to ragwort *Senecio jacobaeae* and by *T. sylvestris* to red clover *Trifolium pratense* (A). These observations disregard the relative abundance of each species of flower plant and if the data are weighted to account for this the conclusions are different: *M. jurtina* and *T. sylvestris* appear to prefer creeping thistle *Cirsium arvense* and *P. tithonus* prefers bramble *Rubus fruticosus agg*. (B). (b) Nectar sources change in both availability and quality through the season. Observations on the same three butterfly species over their flight periods show the degree of interchange between the species of butterfly and plant. The 'kites' (total area equal to 100% visits) represent the proportion of observed visits made to (i) bramble, *R. fruticosus agg*., (ii) creeping

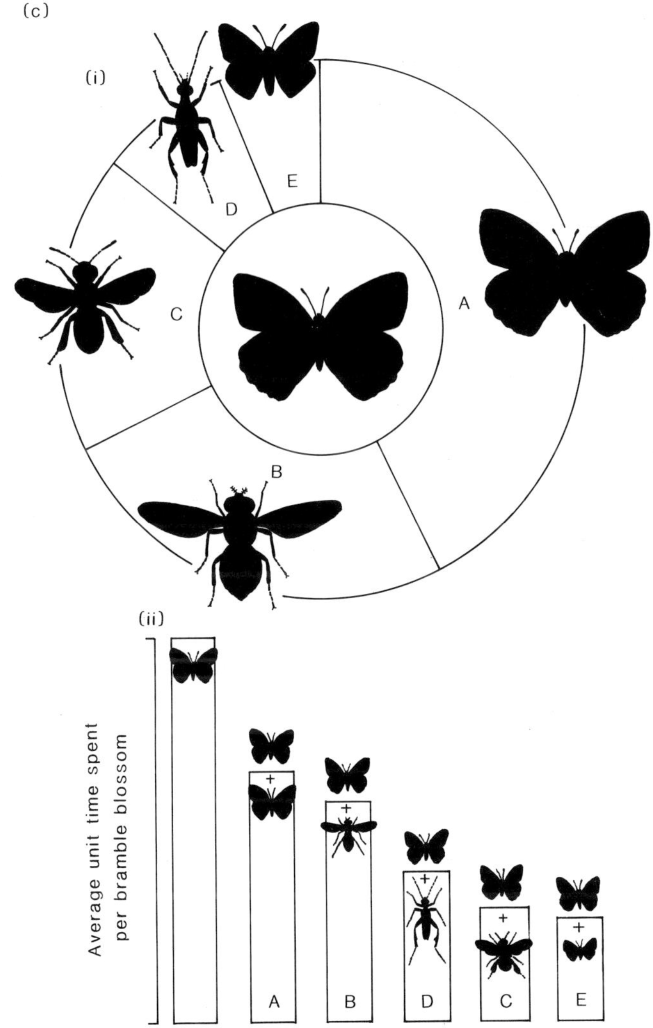

thistle *C. arvense*, (iii) ragwort *S. jacobaeae*, (iv) devil's-bit scabious *Succisa pratensis*, and (v) 'other' species. The flowering period of each species is indicated by a black bar. (vi) These data are related to adult butterfly population size, as measured by standard transect counts per week. (Modified from Peachey 1980.) (c) The time that a butterfly spends feeding at a flower is influenced by other insects at that flower. (i) In a series of observations of *M. jurtina* on *R. fruticosus agg*. blossoms, all made at the same site between 10.00 and 13.00 BST, the most frequent disturbing influence is that of other *M. jurtina* (A), followed by flies, especially hoverflies (B), bees (C), beetles (D), and other butterfly species (E). (ii) The influence of each class of visitor is assessed by comparing the average time spent by *M. jurtina* on a blossom (36.9 ± 3.2s) with the average time spent on the blossom if another insect is present (A–E). The greatest reduction of time spent by *M. jurtina* per blossom is in the presence of another species of butterfly, and least when sharing the blossoms with another *M. jurtina*. (Porter, unpublished data.)

butterfly communities interact over nectar seeking. The fall in the frequency of visits made to *Rubus fruticosus agg.* flowers by *Maniola jurtina*, following the emergence of *Pyronia tithonus* may suggest some degree of interaction. As *R. fruticosus agg.* blossoms decline in frequency, *P. tithonus* shifts away from them towards *Senecio jacobaea*, while *M. jurtina* moves to 'other' flowers. Despite the fact that interactions within a butterfly community constantly change as the season progresses, useful findings can be made from such studies.

Some flowers may provide pollen but no nectar (e.g. wood anemone *Anemone nemorosa*) and it is logical that butterflies do not visit them, but other reasons exist as to why certain flowers are not visited by butterflies. Extensive carpets of *Hyacinthoides non-scriptus* in shaded woods are rarely visited by butterflies, but the same species on woodland edges are visited by a wide variety of butterflies (Table 7.3). Butterflies make up a small part of the total number of flower-visiting insects and may have to compete with other insects which are far more efficient foragers (Table 7.5). Competition for nectar occurs at two levels: butterfly versus butterfly, and butterfly versus other insects. Nectar feeding can have major implications for reproductive success in butterflies (see chapters 2 and 3). Other insects influence this by reducing the time butterflies are able to feed, as a series of observations on *Maniola jurtina* demonstrate (Fig. 7.7c), underlining the importance of considering the whole community of nectar-feeders when feeding strategies in adult butterflies are investigated.

Three types of conspecific interaction occur at flowers:

(1) male–male
(2) male–female and
(3) female–female.

Attempted courtship by males of unreceptive females nectaring on flowers (harassment) may reduce the ability of females to feed and develop eggs, despite their adoption of mate-refusal postures. In some butterflies, such as *Anthocharis cardamines* (Wiklund and Åhrberg 1978), females escape harassment by foraging in areas remote from the males (see sections 2.4 and 6.3). Males can also be interrupted when feeding; this response in *Maniola jurtina* is presumably related to their opportunistic mate-location behaviour (see chapter 2). The

Table 7.5 The range of insect orders using flowers as a food resource and their potential as competitors for butterflies

Group	Insect visitors to flowers: Reason for being on a flower			Competitor for butterflies?
	Nectar	Pollen	Herbivore	
Collembola		○		No
Orthoptera				
Crickets			○	(Yes)
Cockroaches	○			(Yes)
Hemiptera			●	No
Thysanoptera			●	No
Dermaptera			●	No
Diptera	●	●		Yes
Coleoptera	●	●	●	Yes
Lepidoptera (moths)	●	○		(Yes)
Hymenoptera	●	●		Yes

Modified from Proctor and Yeo (1979); Faegri and Van der Pijl (1979). Open circles and brackets indicate infrequent occurrence.

effect of male–male or female–female encounters are more subtle; although no apparent interference is seen, the average time spent on flowers by individuals of *M. jurtina* is shorter than if either was feeding alone (K. Porter, personal observation).

Interspecific interactions between two butterflies may also be subtle, but can involve 'agonistic' behaviour. The most efficient way of reducing the 'cost' is for the butterfly approaching a flower already occupied to avoid it and move on to an alternative blossom. More 'aggressive' behaviour has been demonstrated by numbers of *Maniola jurtina* which have been observed to mob *Arygnnis aglaja* attempting to feed from the same flowerhead of marsh thistle *Cirsium palustre*. A similar observation has been made on *Cynthia cardui* which was attempting to feed at a *Buddleia* bush (C. Steel, personal observation). Robust insects, notably large bumble-bees such as *Bombus terrestris*, the smaller but more persistent honey-bee *Apis mellifera*, flower-visiting cerambycid beetles such as *Stranglia maculata* and large hoverflies like *Volucella pellucens* often displace rather than merely disturb butterflies (see Morse 1977, 1979).

Clearly, much more work needs to be done to determine how the butterfly community in a given habitat interacts with the general flower-visiting insect community. It appears that butterflies are often the 'losers' in butterfly–other insect interactions and that the 'winners' of butterfly–butterfly encounters depends upon the participant species and the sexes involved. Intraspecific contests and sexual harassment make up the majority of such interactions.

7.4.4 Butterfly–host plant 'coevolution'

The origins of the flowering plants and of insect groups which are associated with flower-visiting lie together on the geological scale of evolutionary time (see chapter 10). The proliferation of flowering plants is believed, by some, to be the result of an intimate coevolution of these two groups of organisms (Ehrlich and Raven 1965; Feeny 1977). More recently there has been reaction to the concept of coevolution, particularly as there is little real evidence that insects have been responsible for speciation in angiosperms (Jermy 1976, 1984; Bernays and Chapman 1978).

Much of the current interest in butterfly–plant 'coevolution' can be traced to Ehrlich and Raven's (1965) model of reciprocal interactions between butterflies and their hostplants. They envisaged a sequence of events analogous to the development of weapons and defences by opposing military forces; in fact, a coevolutionary 'arms race' (Dawkins and Krebs 1979). This model can also be expanded to consider the evolutionary interactions of butterflies and their predators and parasitoids (Fig. 7.8). In this analogy of an arms race the weapons are the biochemical arsenal and physical deterrents of the plant, butterfly, parasitoid, or predator. The defence mechanisms of the plant are believed to be effective on all but a few species of specialist herbivore which have evolved mechanisms to overcome them. A similar situation exists for predators on parasitoids which feed on adult or larval butterflies.

The central feature of butterfly–plant 'coevolution' is the presence of so-called secondary compounds—chemicals characteristic of plant taxa (Brues 1924; Feeny 1975; Futuyuma 1983). Once latched on to a particular plant compound, or class of compounds, insects have tended to remain with the same group of foodplants. This is the basis for foodplant specificity in egg-laying females (chapter 3) and larval preferences (Appendix 1). The chemical deterrents eventually become kairomones or stimulants to adapted species of herbivore. This is believed to be true for the mustard oil glycosides present in crucifers, the hostplants of *Pieris brassicae* (Thorsteinson 1960) and *Pieris rapae* (Renwick and Radke 1983); the larvae will even eat filter paper or flour impregnated with these chemicals. As a species of herbivorous insect specializes to overcome a particular set of chemical defences in a group of plants, then it foregoes the ability to overcome the defences of other plants. Foodplant choice is thus a balance between the costs and benefits of generalist and specialist strategies. Hostplant specificity permits more efficient hostplant exploitation, with potential for habitat specialization, but limits the range of food available for use (see section 10.1).

Particular groups of butterflies tend to be associated with specific plant groups (Ehrlich and Raven 1965; Ahmad 1983). Among British butterflies, for example, there exists a general correlation between pierids and crucifers, satyrines and grasses, and nymphalid subfamilies which feed on violets and

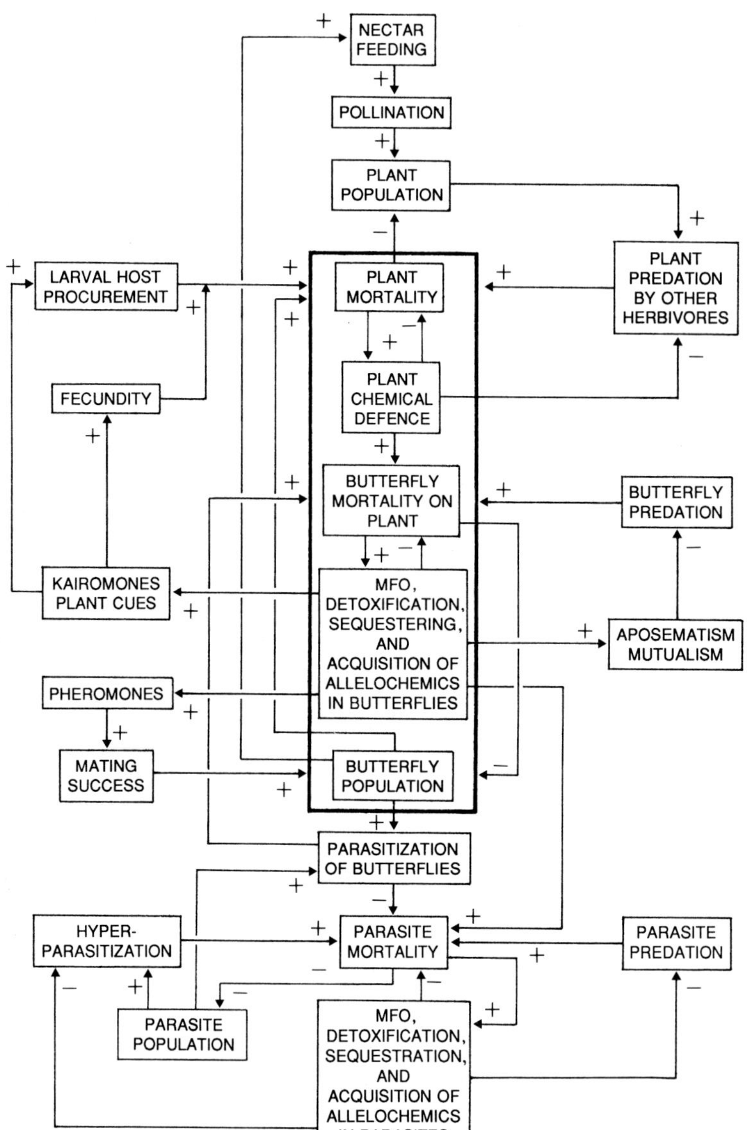

Fig. 7.8 Potential process-response and feedback in hostplant, butterfly, and parasitoid evolution. The hypothetical 'arms race' arena between plants and butterflies (Ehrlich and Raven 1965) is outlined as a black box. MFO, mixed-function oxidase system, a major detoxification system in insects. (After R. L. H. Dennis.)

nettles. The implication is that all butterflies can be grouped at different taxonomic levels according to their foodplants (Ackery 1988; Table 7.6) which in turn reflects on the similarity of secondary compounds within each plant group. The evolution of these associations has depended on the acceptance by butterflies of novel hostplants. Two obstacles occur: (i) the plant may not provide the necessary stimuli to attract females and (ii) it has chemical or physical barriers which reduce the fitness of those individuals attracted to it. The main requirement is that individuals arise in a population that are genetically capable of surmounting both obstacles. The initial step depends on changes in oviposition behaviour (Dethier 1970; Feeney *et al*. 1983). In an isolated population with an abundance of a potential inferior host and rarity of its normal host, selection will favour attraction to the new host, even if its defensive chemistry reduces survival. Once the behavioural barrier is overcome, selection will favour physiological adaptation to the toxic properties of the plant (Futuyuma 1983, 1986). It is not difficult to see how new species of both plant and butterfly would evolve through geographical isolation and local chemical differences between plant populations. This concept of plant species creating a 'breathing space' by avoiding herbivores, through chemical defence, has led to the phrase 'adaptive zone' (Ehrlich and Raven 1965). Each new chemical change provides an opportunity for evolutionary radiation of both plant and butterfly.

Although accepting that 'coevolution' with plants has been important in the evolution of butterflies, Vane-Wright (1978) suggests that Ehrlich and Raven's model is oversimplistic. In particular, non-exclusive hostplant families point away from extreme coevolutionary dependence. Furthermore, butterfly diversity can be accounted for by factors other than strict foodplant species diversity. Jermy (1984) cites evidence that denies reciprocal evolution altogether, specifically the existence of closely related species in most groups of phytophagous insects which live on distantly related plants, and the lack of any correlation between the evolutionary age of plant groups and that of insects living on them. He also demonstrated the speed with which insects can adapt to new hostplant niches. Instead, Jermy envisages a process of sequential evolution, where ongoing radiation in plant defences is traced by adaptations in phytophagous insects which have little influence on their hostplants. Yet, exceptions do exist as in the relationship of figwasps (Agarnidae) and figs (*Ficus* spp.).

Insect–hostplant specialization makes for ecological, if not for biochemical, dependency of insects on their hostplants and habitats (section 10.1). This is reflected in the restricted use of hostplants, as in the case of *Eurodryas aurinia* discussed earlier, and is a feature of local butterfly races (section 10.4.2). Although the larvae of *E. aurinia* will readily accept *Lonicera periclymenum*, *Symphoricarpos rivularis* and teasel *Dipsacus fullonum* in captivity, they are rarely found on any species other than *Succisa pratensis* in the wild. *Lonicera periclymenum* is a climber of shady woodland habitats and does not receive sufficient sunlight to meet the larval needs (Porter 1981); *Symphoricarpos rivularis* is an alien species to Britain, again often found growing in woodlands; *Dipsacus fullonum* is an early successional species which has features of growth-form (spines and water-traps) and habitat which make it unsuitable as a foodplant for *E. aurinia* larvae despite appropriate chemistry (Kerner 1878). An understanding of butterfly–plant 'coevolution' is a small step towards learning more about the evolution of communities but to achieve realistic models of insect–plant 'coevolution' requires extensive research on the whole spectrum of herbivores.

7.5 Diversity, ecological succession, and butterfly communities

7.5.1 Measures of community richness

The number of species present in a community is regarded as useful indicator of environmental potential and of ecosystem structural complexity. However, as explained earlier (see section 1.3.3) the number of species alone is insufficient to measure species diversity and several indices have been developed to express species–abundance relationships; these determine different aspects of the partition of abundances between species (Shannon and Weaver 1949; Simpson 1949; Loya 1972; Hill 1973; Owen 1975). The use of indices has its limitations; Shapiro (1975*a*) believed that because of the great differences in population structures and vagility of butterflies, gathering suitable data on the equability component of diversity was virtually impossible. No single plot size was found to be satisfactory for all the species in any area. Indices also fail to take account of 'desirable' species from a habitat point of view; a woodland may have a high diversity without containing any true woodland species. But to determine the status of species in diversity studies, some sort of population estimate is essential (see section 4.2).

7.5.2 Case studies of butterfly diversity

Bernwood Forest, Buckinghamshire, is one of the few sites in Britain where diversity studies have been

Table 7.6 The association of British butterflies with hostplant families in Britain and Europe. Hostplants used by species in Britain are shown by ● and in Europe by ○

	Hesperiidae		Pa[1]	Pieridae			Lycaenidae			Nymphalidae	
Plant family	Hesperiinae	Pyrginae	Papilioninae	Dismorphiinae	Coliadinae	Pierinae	Theclinae	Lycaeninae	Riodininae	Five subfamilies[2]	Satyrinae
Gramineae	● ○										● ○
Cyperaceae											● ○
Leguminosae		● ○		● ○	● ○		● ○	● ○			
Rosaceae		● ○	○			● ○	● ○	● ○		● ○	
Compositae		○								● ○	
Labiatae		○						● ○			
Malvaceae		○								● ○	
Umbelliferae			● ○								
Rutaceae			○								
Rhamnaceae					● ○		● ○	● ○			
Ericaceae					○		● ○	● ○			
Cruciferae						● ○					
Berberidaceae						○					

Cornaceae	● ○	● ○		
Fagaceae	● ○			
Ulmaceae	● ○			● ○
Betulaceae	○			● ○
Polygonaceae	○	● ○		○
Caprifoliaceae		● ○		● ○
Araliaceae		● ○		
Aquifoliaceae		● ○		
Celastaceae		● ○		
Geraniaceae		● ○		
Gentianaceae		○		
Crassulaceae		○		
Primulaceae			● ○	
Dipsacaceae				● ○
Cannabaceae				● ○
Plantaginaceae				● ○
Salicaceae				● ○
Scrophulariaceae				● ○
Urticaceae				● ○
Violaceae				● ○

[1] Papilionidae
[2] Limenitinae, Apaturinae, Nymphalinae, Argynninae, and Melitaeinae
Data derived from Higgins and Riley (1983); Heath *et al*. (1984); reduced taxonomic divisions from Emmet and Heath (1989).

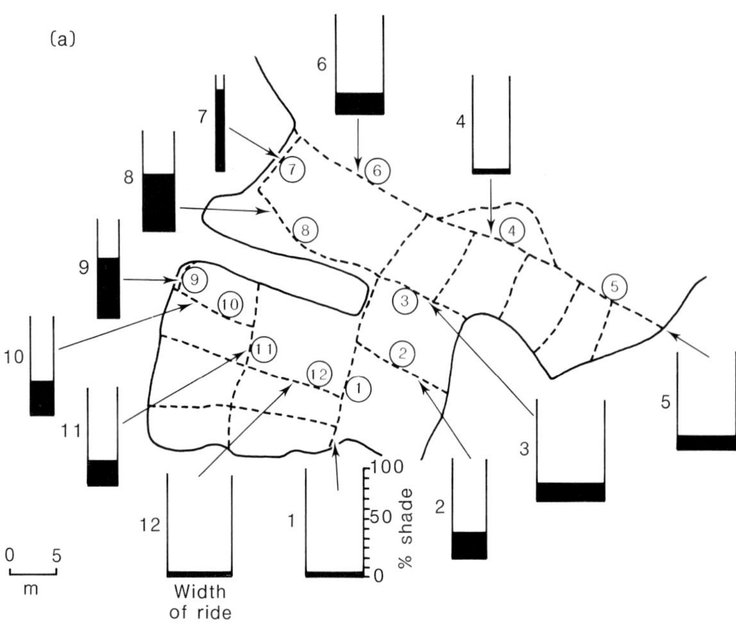
(a)
% shade
Width of ride
m

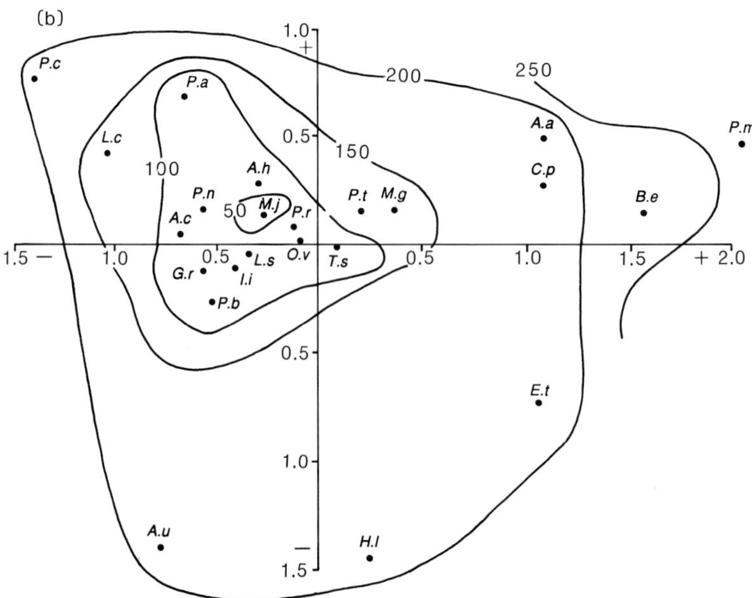
(b)
P.c
P.a
L.c
A.h
P.n
M.j
P.r
A.c
O.v
L.s
G.r
I.i
P.b
T.s
P.t
M.g
A.a
C.p
B.e
P.m
E.t
A.u
H.l

Ride no.	No. of species	No. of individuals per 100 m	Diversity index E(H′)
1	30	365.7	2.361
2	21	299.6	2.222
3	21	427.4	2.220
4	20	466.6	2.299
5	25	415.0	2.279
6	27	427.4	2.047
7	3	99.3	0.591
8	14	182.0	1.371
9	13	164.3	1.713
10	12	104.4	1.286
11	13	136.6	1.985
12	24	229.7	2.173

Based on 33 transects over 20 weeks from May 8 to September 24, 1978

Fig. 7.9 Factors influencing butterfly species diversity in forest rides in Bernwood Forest, Buckinghamshire. (Adapted from Peachey 1980.) (a) The physical characteristics of rides and butterfly diversity statistics as measured by the Shannon–Weaver index H' (see also Fig. 11.8). $E(H') \approx -\Sigma_{i=1}^{s} p_i \mathrm{Log}\, p_i$, where p_i is the percentage of individuals belonging to species i and s is the number of species in the sample. E is the expected value (see Peet 1974). (b) Computer plot (non-metric scaling) illustrating the affinities between butterfly species based on their relative frequencies in rides (axes have arbitrary values). Isolines give coefficients of variation; low scores indicate species that are spread over numerous rides and high values those that are concentrated in few rides. Species in the right upper part of the plot inhabit wide rides with short vegetation, whilst those in the left upper part of the plot frequent increasingly shaded rides with taller vegetation. Note, several species are not residents in the rides (e.g. small tortoiseshell *Aglais urticae*; peacock *Inachis io*; large white *Pieris brassicae*; brimstone *Gonepteryx rhamni*). *T.s.*, *Thymelicus sylvestris*, *O.v.*, *Ochlodes venata*, *E.t.*, *Erynnis tages*, *P.m.*, *Pyrgus malvae*, *L.s.*, *Leptidea sinapis*, *G.r.*, *Gonepteryx rhamni*, *P.b.*, *Pieris brassicae*, *P.r.*, *Pieris rapae*, *P.n.*, *Pieris napi*, *A.c.*, *Anthocharis cardamines*, *H.l.*, *Hamearis lucina*, *L.c.*, *Ladoga camilla*, *A.u.*, *Aglais urticae*, *I.i.*, *Inachis io*, *P.c.*, *Polygonia c-album*, *B.e.*, *Boloria euphrosyne*, *A.a.*, *Argynnis aglaja*, *P.a.*, *Pararge aegeria*, *M.g.*, *Melanargia galathea*, *P.t.*, *Pyronia tithonus*, *M.j.*, *Maniola jurtina*, *A.h.*, *Aphantopus hyperantus*, *C.p.*, *Coenonympha pamphilus*. Transect counts for some species (*P.c.*, *A.c.*, *A.u.*, *H.l.*, and *E.t.*) are low and this may influence their placement in the plot. (Analysis from R. L. H. Dennis and W. R. Williams, unpublished material.)

attempted on butterflies. Despite a drastic habitat change in the mid-1950s involving clear felling, spraying, and replanting with conifers, it retained most of its species into the late 1970s (Peachey 1980). Butterfly diversity at Bernwood has been found to correspond to the physical structure of rides and vegetational features, such as ride width, percent shade, and plant diversity (Fig. 7.9a). Wider rides support more species and individuals than narrow ones; they have greater structural diversity and resources. The number of butterfly species correlates highly with the increased variety of nectar sources ($r = 0.81$), which in turn corresponds with higher light levels ($r = 0.79$) and ride width. Analysis of the Bernwood data using the multivariate technique of non-metric scaling (Fig. 7.9b) also reveals groups of species with different physical tolerances; thus, for example, one group of species comprising the small heath *Coenonympha pamphilus*, *Argynnis aglaja*, *Boloria euphrosyne*, and the grizzled skipper *Pyrgus malvae*, required the most open rides, whilst another group including *Pararge aegeria* and the white admiral *Ladoga camilla*, prefer shady conditions.

Warren (1985*a*) made similar findings in his survey of two woodlands in Yardley Chase, Northamptonshire, and Monks Wood, Cambridgeshire, and was able to quantify the levels of light and shade tolerated by different species. For example, *Leptidea sinapis* was found to be most abundant in rides with moderate direct shade of between 20–50 per cent (see Fig. 4.5). Warren's study identified the potential influences of light and shade on adult and larval resources. In the case of *L. sinapis* at Yardley Chase, abundance of nectar and hostplants of sufficient size for larvae, corresponded with light conditions for which most adults were recorded.

The existence of an extensive ride system within Bernwood Forest (Fig. 7.9a) contributes greatly towards its richness and the survival of the butterfly fauna, in spite of the drastic habitat alteration. However, structural diversity is only one of several factors which determine the number of species a habitat contains (Price 1975). Other factors include the size of the habitat, the time it has been available for colonization, the number of potential colonizers, and diversity of nearby vegetation types, all of which evoke island biogeography concepts discussed in section 1.3.3. Bernwood Forest is a remnant of the extensive forests that once covered the basin of low-lying clays in the East Midlands belt between Peterborough and Oxford. Thus the antiquity of Bernwood Forest and the high number of potential colonizers within several kilometres, are two important reasons accounting for the current high species diversity.

7.5.3 Vegetation succession and butterfly communities

Succession results in both qualitative and quantitative changes in the flora and fauna. From exposed rock, vegetation communities are believed to proceed towards a climatic-climax, a state of dynamic homeostasis between vegetation and abiotic conditions. The climax vegetation may support a richer fauna, as in tropical rain forests, or one which is remarkably species-poor by comparison with the early successional stages, as in the case of the taiga, simulated in British coniferous plantations. Moreover, since each stage in the succession supports characteristic species, particular plants and animals will be lost from a site and others gained in the course of ecological succession. In Britain, over three-quarters of our butterflies use low-growing herbaceous hostplants which are associated with early or mid-successional seres (Appendices 1 and 2). Only 10 species use woody shrubs or trees. Thus, even in woodland habitats, most butterflies depend upon open clearings and rides which support their low-growing hostplants. The climax vegetation in Britain is deciduous forest (section 10.2); and, therefore, many British 'woodland' species which require clearings, depended on both natural and man-made disturbance agents (Dennis 1977). As humans started to clear larger areas of forest to obtain new agricultural land, a mosaic of successional patches was created, providing ideal breeding habitats for many of the 'woodland' butterflies (see chapter 10).

Grassland is not uniform but is often a mosaic of successional patches. The richest grassland, in an ecological sense, are those that have a patchwork of grass heights, with short turf on thin-soiled slopes, medium and tall grass areas, and patches of scrub. In southern England a large site containing all these components, such as Old Winchester Hill National Nature Reserve in Hampshire, can support over 33 species.

Erhardt (1985) has investigated the patterns of

diversity associated with ecological succession on abandoned sub-alpine meadows in Switzerland. As expected, he found that the early stages of abandoned grassland, dominated by grasses and herbs, are the richest habitats for butterflies—butterfly diversity reflects plant diversity. The later successional stages, which were dominated by dwarf shrubs and with a reduced plant diversity, supported fewer butterfly species. Erhardt also compared species richness in these abandoned meadows with cultivated meadows. Again, butterfly diversity was correlated with plant diversity; significantly, both plant and butterfly diversity decreased with increased fertilizer application.

In Britain today, only a small proportion of semi-natural habitats exist, due largely to changes in agricultural and forestry practices. Thus, apart from nature reserves and other sites where suitable management can be implemented to maintain the early successional stages, many biotopes in Britain are becoming increasingly species-poor (see chapter 11). Butterfly community interactions are complex and it is clear that much more work has to be done to determine the biotic and abiotic factors which affect butterfly diversity and community composition. In carrying out such studies it is essential that butterflies are viewed as part of the spectrum of herbivores, myrmecophages, and flower-visitors.

8

Diversity within populations

Paul M. Brakefield and Tim G. Shreeve

Within any butterfly population no two individuals are identical. This applies both to their genes or genotype and to their phenotype or set of individual characteristics resulting from interactions between the genes and the environment during development. The reproductive success or fitness of an individual depends on how well its phenotype is adapted or fitted to the environment which it experiences. Some species of butterfly exhibit discrete genetic forms which differ in wing pattern and which occur together within populations. Research on such polymorphisms has contributed much to our understanding of the processes of adaptation to the environment. Having first introduced some basic concepts, this chapter describes work on polymorphisms in butterflies together with recent insights into how wing patterns are developed. Rare aberrations much beloved by butterfly collectors (see Russwurm 1978) are not considered in detail here since they are unlikely to be significant in evolutionary terms even when they have a genetic basis (see Fig. 8.1).

Early studies of the occurrence of discrete forms of particular species of butterflies began to draw attention to the ubiquitous nature of genetic variation and its evolutionary significance (see Ford 1975). However, most variation in the genetic material does not influence the external phenotype in any obvious way. One example which will be described is that of variation between individuals in the structure of particular enzymes. Somewhat more obvious effects are those that give rise to continuous phenotypic variation which can be measured quantitatively (chapter 9). An example of such variation in the life histories of butterflies is examined later in this chapter.

8.1 Genetic variation, natural selection, and evolution

The processes which can lead to differences in the reproductive success of individuals within a population involve random effects, such as the incidence of weather resulting in density-independent mortality (see chapter 4), and natural selection, in which some component of the environment produces a predictable difference in fitness between genotypes.

Evolution occurs when there is some net directional change or cumulative change in the characteristics of a population and in gene frequencies over many generations. This definition excludes the type of change associated with an unpredictable event such as catastrophic weather which by chance influences a group of populations in a single generation. Such an event may nevertheless have a strong influence on the genetic make-up of the populations and their future evolutionary responses to a changing environment. A similar effect may occur if a population is reduced to a very small size and by chance suffers a loss of some variants of the genes or alleles within it. Such a loss of genetic variation may then restrict the scope of subsequent evolution. Similar random changes in genetic variation can be associated with populations which are consistently small, or with the founding of a new colony by a few individuals. Such events involving chance rather than selection are described as random genetic drift. The effects will tend to be erratic over time although sometimes they will appear more progressive.

Natural selection is often thought of as resulting from differential mortality. This is so in many cases but it may also involve other differences such as those in fecundity, mating success or the ability of individuals to obtain the fittest mating partners (the last two are referred to as sexual selection). The *process* of natural selection involves:

(a) variation between individuals in some trait: *variation*;
(b) a consistent relationship between the trait and mating ability, fertilizing ability, fertility, fecundity and/or survivorship: *fitness differences*;
(c) transmission of genetic influences on the trait from parents to offspring: *inheritance*.

Given these conditions then:

(1) the frequency distribution for that trait will differ among age classes or life history stages, beyond that expected from development;
(2) if the population is not at equilibrium and hence is evolving, the distribution of that trait in all offspring in the population will be predictably different from that in all parents (after Endler 1986).

As a result of this process, the distribution of a trait may change in a predictable way over many generations. However, to be able to make such a prediction for a natural population one needs to have a complete knowledge of the inheritance of the trait and obtain rigorous estimates of relative fitnesses of the various genotypes determining the trait for the environments involved. In practice this is extremely difficult because of the complexity of populations, environments, and interactions between genes. The differences in fitness may also be very small and hence very difficult to detect or measure. The conditions for natural selection show that differential mortality will lead to evolutionary change only if it results in differences in reproductive success among genotypes. In other words, if the mortality involves only phenotypes which differ because of some effect of the environment and not because of some contribution from heritable variation, there can be no change in gene frequencies between generations.

Evolution can therefore occur only when genetic variation exists within populations. Such variation ultimately has its source in the processes of mutation which give rise, at very low rates, to new alleles of the genes. Some of the rare aberrations found in butterflies are due to such mutations occurring in single genes (Fig. 8.1; see Tubbs 1978). If the new allele or mutation is advantageous it may increase in frequency within a population due to the influence of natural selection. This type of change, in association with a novel environment, was the basis of the evolution of industrial melanism in the peppered moth *Biston betularia* (Brakefield 1987*a*). The great majority of mutations are, however, deleterious and will therefore be selected against. Occasionally a mutation which is not too disadvantageous may increase in frequency in a particular population due to random genetic drift. New combinations of genes also occur as a result of recombination processes during the replication and assortment of chromosomes carrying the genes in the production of sperm or eggs. Such new combinations may produce novel phenotypes because of the interactions which frequently occur between different genes. Thus they may also be important in enabling an evolutionary response to a changed environment. However, if such a mutation or genetic combination does not arise, or is not available at the time of such a change, then the population may become extinct. This illustrates the importance of not only generating genetic variation but also maintaining it within populations. The central role of natural selection in these processes has been investigated using polymorphisms in butterflies.

8.2 Polymorphism and wing pattern forms

Ford (1940) defined genetic polymorphism as the occurrence together in the same locality (population) of two or more discontinuous forms of a species in such proportions that the rarest of them cannot be maintained merely by recurrent mutation. This excludes geographical races and continuously varying characters that are influenced by many genes, both of which are discussed in the next chapter. Genetic polymorphisms that produce discrete phenotypes are controlled by major genes with large effects or in some cases by so-called supergenes consisting of a number of genes which are closely linked on a chromosome so that they are effectively inherited as a single unit.

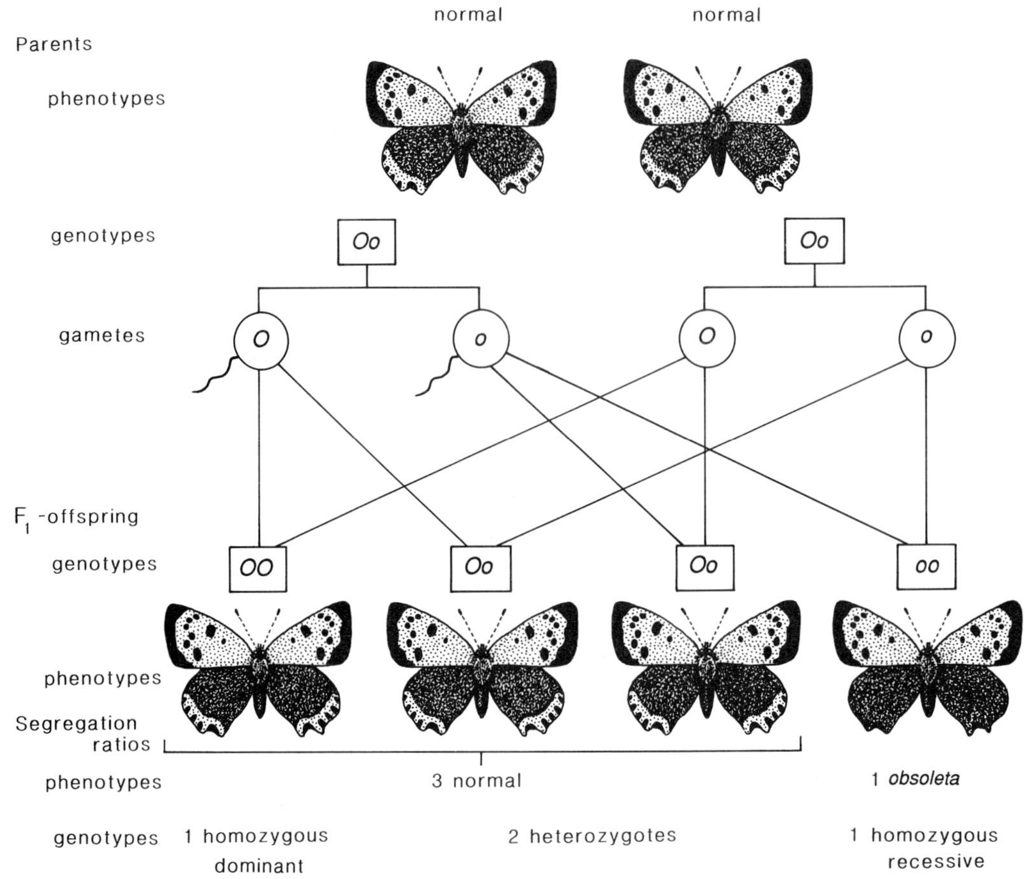

Fig. 8.1 The genetic control of the form *obsoleta* of the small copper *Lycaena phlaeas* by a recessive allele. The expected segregation of phenotypes and genotypes resulting from a cross between two heterozygotes is shown. The normal or wild-type is fully dominant over *obsoleta*.

An individual butterfly has two copies of each of its genes. This complement is referred to as the genotype. Figure 8.1 illustrates how in the simplest case of a polymorphism involving a major gene with two different alleles and exhibiting complete dominance there will be three genotypes and two phenotypes. Where dominance is incomplete the heterozygote carrying one copy of each allele will have a third phenotype which is to some degree intermediate between the phenotypes of each of the homozygotes. Some more complex polymorphisms are known to be controlled by allelic variation at two genes, sometimes with interactions between them. An example of the inheritance of such a polymorphism involving epistasis (interaction) is shown in Fig. 8.2. The figure includes a possible explanation of the epistasis in terms of a metabolic chain of pigment synthesis. Variable characters or traits which are determined by larger numbers of genes are usually more continuous in nature and thus do not exhibit a small number of discrete phenotypes. Those discussed in chapter 9 are controlled by comparatively large numbers of genes each having a small effect on the phenotype.

A number of British butterflies exhibit colour polymorphisms (Ford 1945). The two most striking examples, each involving two alleles and complete dominance, are the clouded yellow *Colias croceus* and the silver-washed fritillary *Argynnis paphia*. The circumstances associated with them are remarkably

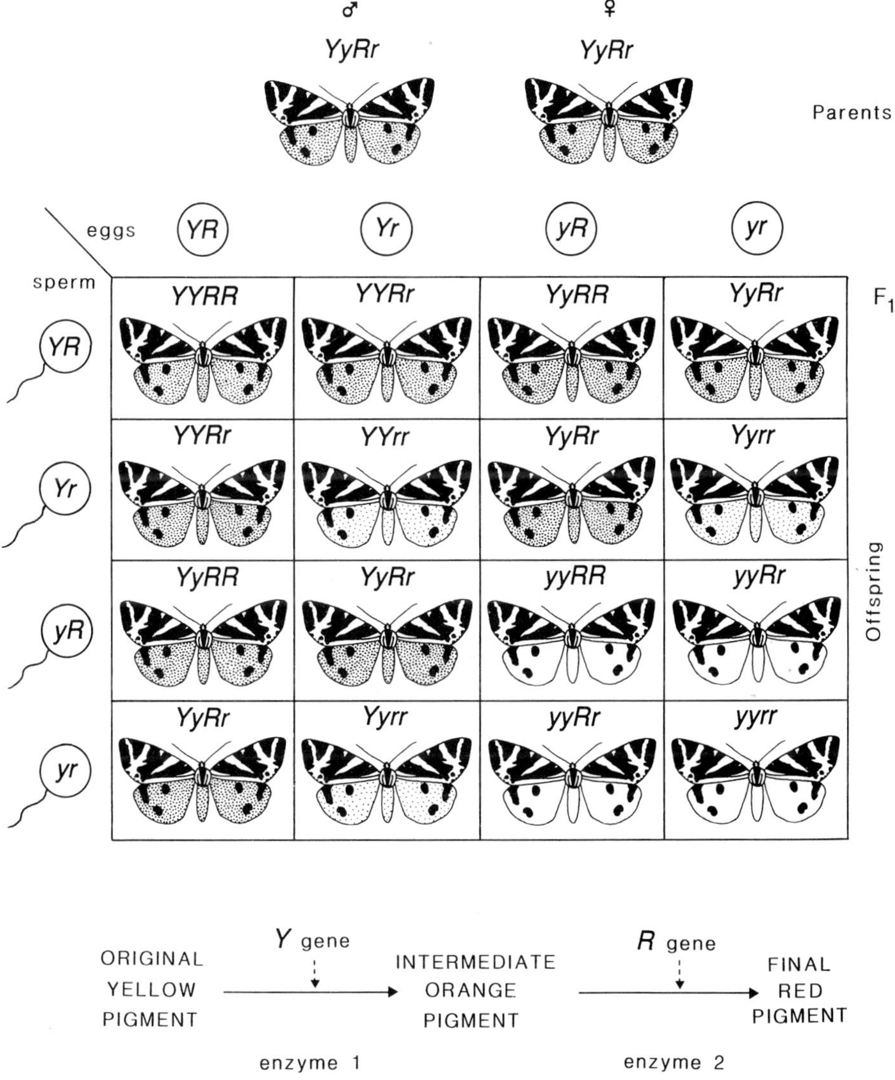

Fig. 8.2 An example of variation in wing pattern of the jersey tiger moth *Callimorpha quadripunctaria*, controlled by alleles at two gene loci on different chromosomes but with an interaction or epistatic relationship between them. The variation is that of the hindwing colour found in populations of *C. quadripunctaria* in south-west England, the Channel Isles and north-west France (Brakefield and Liebert 1985; Liebert and Brakefield 1990). The chequerboard diagram shows the genotypes and phenotypes expected from a cross between double heterozygotes. The expected segregation ratio of the three possible phenotypes in such a cross is thus 9 red : 3 orange : 4 yellow. Two families reared by Liebert from this type of cross yielded ratios of 28 : 16 : 12 and 25 : 8 : 14 which approximate to the expected ratios. At one gene locus the dominant allele *Y* controls the expression of a colour other than yellow. If *Y* is present then the other gene locus *R* determines whether the colour is orange or red. The *Y* and *R* genes probably control consecutive steps in a metabolic pathway of pigment synthesis as shown in the lower diagram. When the first step is inhibited in the *yy* genotype, because of a particular enzyme, the second process controlled by the gene *R* cannot occur.

similar. The males of each species are all alike irrespective of genotype. The rarer phenotypes, the pale *helice* and dark *valezina* forms of *C. croceus* and *A. paphia* respectively, are limited to females. These forms are both apparently inherited as sex-controlled dominants (Ford 1945): the allele controlling each form is expressed only in the wing phenotype of females (Fig. 8.3).

The species differ in their habits, especially in that *Colias croceus* is migratory, colonizing Britain in large numbers only in exceptional summers (see chapter 6) while *Argynnis paphia* is a non-migratory, mainly woodland species. Ford (1945) gives details of three large samples of *C. croceus* from different areas of England in years when the species was abundant. These suggest that the frequency of *helice* females is of the order of 5 to 10 per cent. It is interesting that there are numerous other species of *Colias* with orange males in which the females are dimorphic with male-like and rarer, paler female forms. The British distribution of *A. paphia* includes southern and western England, Wales, and most of Ireland. However, *valezina* reaches an appreciable frequency indicative of polymorphism only in certain areas, including parts of southern England and also, of continental Europe (Ford 1975). Here it constitutes about 10 per cent of females. However, a population study over a number of generations would be most valu-

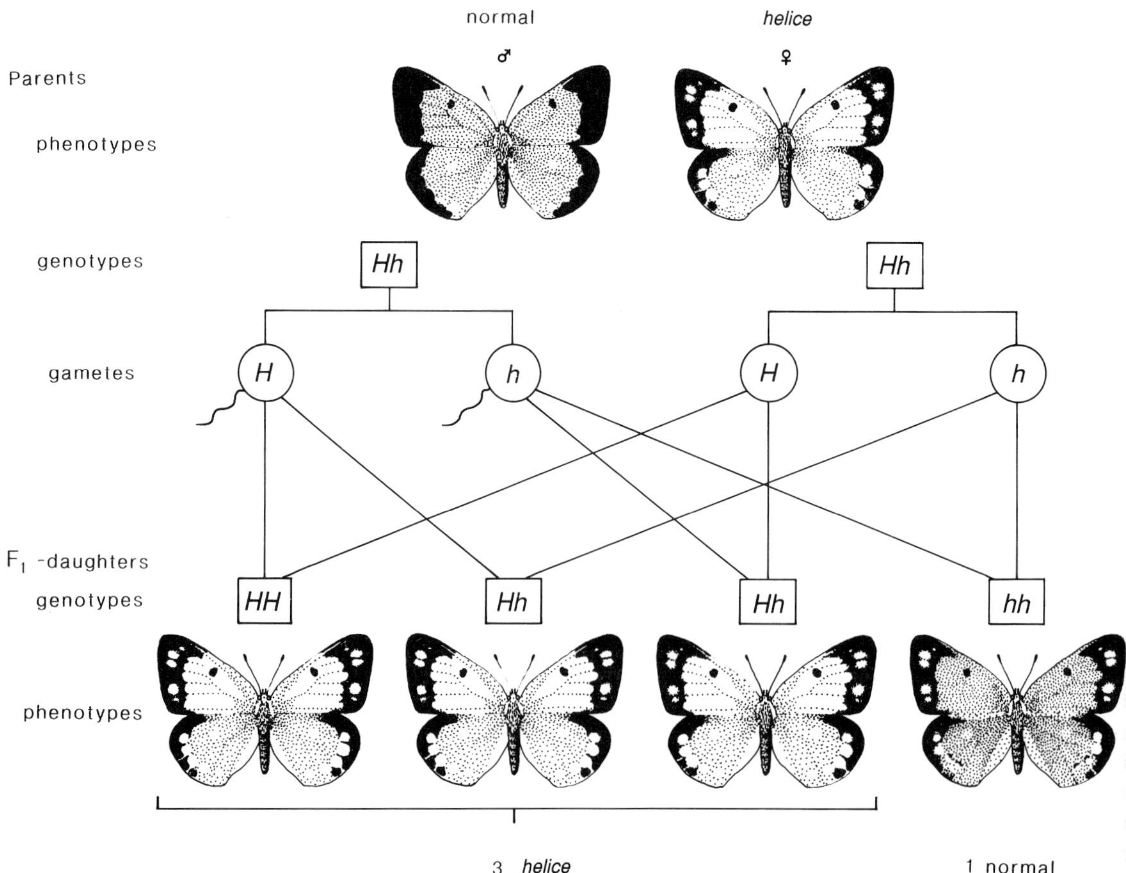

Fig. 8.3 The inheritance of a sex-controlled dominant as illustrated by a cross involving the pale *helice* form of the clouded yellow *Colias croceus*. Male offspring, which are all a pale orange colour irrespective of their genotype, are not shown. The dominant *H* allele, although carried by both sexes, is only expressed in females. The normal or wild-type females with the same colour as the males are all recessive homozygotes.

able to substantiate this, to determine whether, as has been suggested, frequencies are higher in warmer summers and to examine the survival and movement of the forms.

Ford (1975) points out that such genetic polymorphisms may be of two types. They may be transient, in the case where one of the two alleles is in the process of replacing the other, or balanced, when there is some form of equilibrium resulting from a balance of advantage and disadvantage. The present examples are probably of the latter type as there is no evidence that they are recent in appearance or that the forms are changing in abundance. Ford (1945, 1975) argues vigorously that they are likely to be examples of heterozygous advantage in which the heterozygote carrying one of each allele is of higher fitness and at a selective advantage over each of the homozygotes. This pattern of relative fitnesses, as occurs in the case of sickle cell anaemia in humans in regions where malaria is endemic, will give rise to an equilibrium maintaining both alleles in the population. Such an advantage is likely to be based on some non-visual effect of the interaction between alleles in the heterozygote which produces a higher viability during development. However, there is no direct evidence for this mechanism being involved in maintenance of the polymorphisms in the butterflies. Additional genetic crosses beween heterozygotes which demonstrated a deficiency of homozygotes of *helice* or *valezina* would provide strong support for a low viability of these genotypes and suggest that Ford is correct.

Interestingly, a study of the North American *Colias eurytheme* has demonstrated that the pale *alba* phenotype (*Aa* genotype), which lacks nitrogeneous wing pigment, develops more rapidly in the pupae than the orange-coloured wild type (*aa*), especially under cold conditions (Graham *et al.* 1980). Plant tissue is low in available nitrogen and this is usually considered to limit the growth of phytophagous insects. Since the yellow pteridine wing pigment of *Colias* butterflies represents several per cent of the nitrogen budget of larvae, *alba* genotypes may be able to develop more rapidly because nitrogen is not diverted from growth into pigment synthesis. This is not the only consequence of the diversion of nitrogen resources to wing pigment synthesis which is not directly related to the colour difference itself. Thus *alba* females tend to retain in their fat-bodies more of the resources derived by the larvae, to mature their eggs more rapidly and to be more fecund on a daily basis than orange females (Graham *et al.* 1980; Gilchrist and Rutowski 1986). This demonstrates the type of non-visual effect which can be associated with colour polymorphisms.

The visual differences associated with these polymorphisms in female butterflies appear to influence the chance of obtaining a mate, a form of natural selection known as sexual selection. Thus Graham *et al.* (1980) and Gilchrist and Rutoswki (1986) report that the *alba* phenotype of *Colias* butterflies, including *C. eurytheme*, is less attractive to males than the normal orange coloration. Some observations made by Magnus (1958) also provide some evidence for similar differences in *Argynnis paphia*. He demonstrated experimentally that the normal brown coloration of the female in combination with rapid wing movement in flight must be provided to ensure visual recognition by a distant male. The greenish hue of *valezina* females is less effective. There is, therefore, probably a longer average time lag between emergence and copulation for *valezina* females. These may tend to mate during chance encounters while feeding at flowers when sexual scents (see chapter 2) may aid species recognition over short distances. This delay could adversely affect their lifetime fecundity by reducing the time available for egg-laying. It could also account for the apparent association of substantial frequencies of *valezina* with populations at high density, in which delays in mating are less likely because the probability of chance encounters with potential males is higher. Such selection against *valezina* can account for the inability of the *valezina* allele to spread through populations to fixation replacing the wild type allele. It can, however, play no part in maintaining the polymorphism. Indeed, the reverse is the case since such a disadvantage to the phenotype will tend to lead to a progressive reduction in the frequency of the *valezina* allele. Thus some other mechanism such as heterozygous advantage is still required to account for the persistence of polymorphism.

Both this polymorphism in *Argynnis paphia* and those in *Colias* butterflies have also been associated with behavioural differences between the phenotypes which may in combination with variability or heterogeneity in the environment also play an important role in the selection regimes influencing them. Thus the pale forms of *Colias* species may show different levels of activity as a result of different efficiencies of absorption of solar radiation (Hovanitz 1948; Watt 1968, 1969, 1973; Kingsolver and Watt 1983*a*,*b*; see section 2.2) and *valezina* females are said to prefer shadier conditions (Ford 1975). As explained below, the rarer colour form in populations of *A. paphia* may tend to be relatively less often taken by insectivorous birds than the more abundant form. Furthermore, the colour forms may differ in their conspicuousness to birds. The direction of this difference may depend on whether the woodland they are flying in is mostly shaded or more open and sunlit. Such differences in exposure to predation may be especially critical when females are laying their eggs on tree trunks (Dennis, personal communication). Figure 8.4 indicates how these various factors may interact with one another to influence the frequency of the two alleles in populations of the butterfly. Mark–release–recapture experiments carried out by Gilchrist and Rutowski (1986) suggest that orange females of *C. eurytheme* may be more likely than *alba* females to move away from agricultural fields of their foodplant, alfalfa. This may occur partly because of greater harassment by males rather than as a by-product of different thermal properties mentioned above. The ephemeral nature of the fields of alfalfa in combination with differential dispersal behaviour may tend to facilitate persistence of this polymorphism in such habitats.

Another important means by which genetic variation can be maintained within populations is illustrated by extensive colour polymorphism in females of the African mocker swallowtail *Papilio dardanus* (Ford 1975). The various female forms that are controlled by major genes are palatable Batesian mimics of different unpalatable model species (see sections 5.1.3 and 7.2.1). Many populations of *P. dardanus* exhibit polymorphic females. These polymorphisms are probably maintained by natural selection which acts in the following, frequency-dependent, way. When a particular colour form is rare and its model is common, individuals with the mimetic wing pattern will be well protected from predators which will usually tend to mistake it for the unpalatable model. As the palatable form increases in frequency relative to its model species, its protection will decline because predators will have relatively fewer encounters with the unpalatable model and will, therefore, be less likely to avoid the palatable mimetic form. There will come a point when selection will begin to operate against this particular colour form, driving it down in frequency again. This illustrates how a polymorphic equilibrium, in which the frequency of an allele oscillates around a value between 0 and 1, can be maintained by frequency-dependent selection. Experimental work has recently failed to support the suggestion that this mode of selection might influence the polymorphism in *Colias* butterflies because of the similarity in the appearance of the *alba* form to many white pierids, some of which are unpalatable to birds (Ley and Watt 1989; Watt *et al.* 1989). Thus *Colias* and pierids from Colorado, USA, were not distasteful to birds and they do not co-occur sufficiently to support a mimetic relationship. In the case of the colour forms of non-mimetic butterflies, which probably include *Argynnis paphia* (see chapter 5, p. 102), the rarer a colour form is, the more protected it will be because birds will be less likely to form a 'search image', and thus to concentrate their hunting activities on it rather than the alternative form (see Fig. 8.4). This type of effect also introduces a frequency-dependent component of selection which will tend to maintain variation.

For the mimetic *Papilio dardanus* it is more complex when one considers the different female forms. As one particular colour form declines in frequency another form, controlled by a different allele and which mimics a different unpalatable species that has become more abundant, may independently increase in frequency. There may, however, be some interaction between the frequency of the different forms, for example because of competition for larval food or patterns of non-random mating (see e.g. Gibson 1984; Smith 1984). These polymorphic colour forms are probably restricted to females because the general features of the non-mimetic male wing pattern are important releasers of mating behaviour in females during courtship (see Silberglied 1984). Thus males which depart from the ancestral, non-mimetic pattern may be at a disadvantage because they are less likely to elicit mating in females. This hypothesis is discussed, together with new evidence, in detail by Clarke *et al.* (1985). A similar explanation may also account for the fact that the non male-like

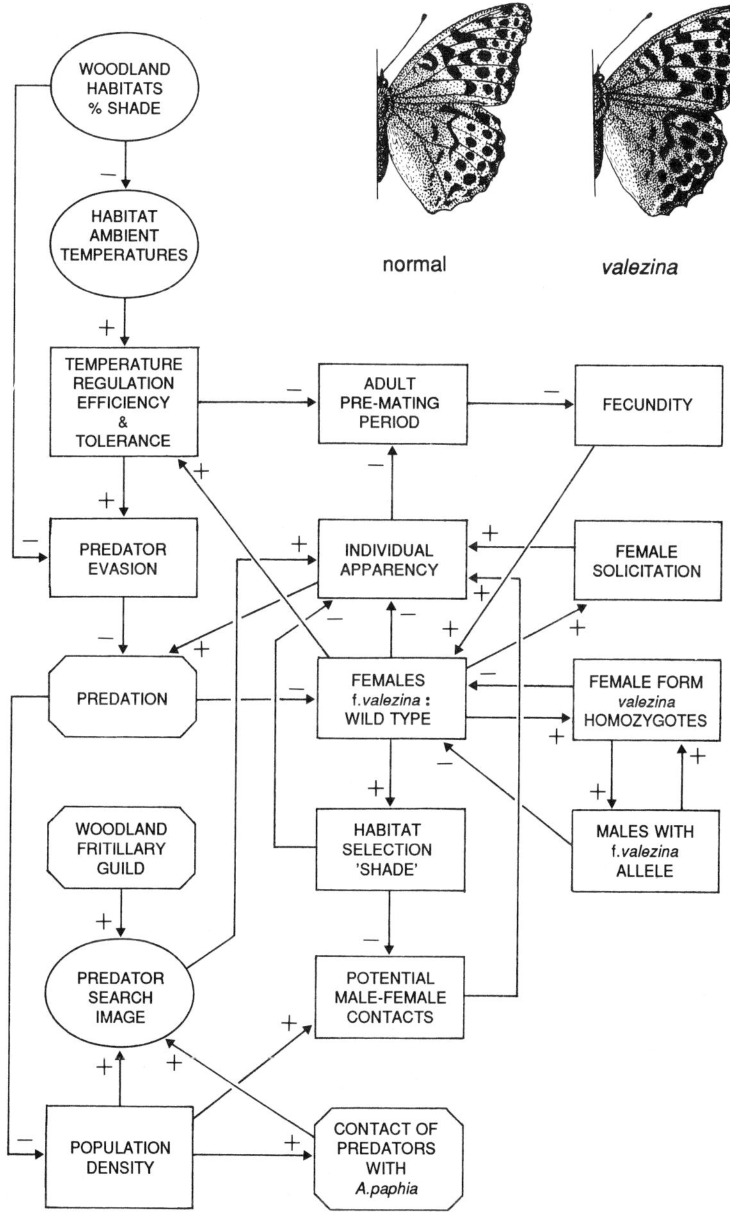

Fig. 8.4 The interaction of factors suspected to be involved in maintaining the female-limited polymorphism, form *valezina*, in the silver-washed fritillary *Argynnis paphia*. Rectangles, *A. paphia* variables; ellipses, factors influencing *A. paphia*; octagons, extrinsic system regulators. (After R. L. H. Dennis, from data in Magnus 1958; Ford 1975, and Brunton *et al*. 1990.) Form *valezina* may be at a disadvantage in populations which exist at low density because of their failure to emit appropriate mate-location signals and possibly because of a lower viability of individuals homozygous for the *valezina* gene. This may be counterbalanced by the reduced apparency of the *valezina* morph to predators in woodland, especially where search images of predators are cued into a wider, brightly coloured, fritillary guild. Other advantages may be the wider thermal tolerance and the mate-solicitation behaviour of form *valezina* females.

helice and *valezina* forms are only expressed in females.

It is possible that this type of sexual selection is also involved in the variation exhibited by females of many of the blues (Lundgren 1977; P. Brakefield, unpublished observation). Females of species such as the chalk hill blue *Lysandra coridon* and the common blue *Polyommatus icarus* show more or less continuous variation from phenotypes which are blue to others which are brown due to varying densities of blue-reflective wing scales. The bluest phenotypes are male-like and thus may tend to be favoured by sexual selection eliciting more rapid courtship and copulation. In contrast, the brownest females tend to be more inconspicuous and may, therefore, be favoured by increased crypsis leading to higher survival rates. Such circumstances could be involved in a balance of sexual and visual components of natural selection maintaining the underlying genetic variation (see Dennis and Shreeve 1989). Climate and thermoregulation may also be involved (see chapter 2 and Fig. 10.7b). Although some genetic crosses have been performed, usually with more extreme phenotypes (see Ford 1945; Robinson 1971), more are necessary to establish the modes of inheritance of this variation.

8.3 The development of wing patterns

Variation in wing pattern within species is central to many studies of processes of natural selection and evolution in butterflies. Features of wing pattern are often key characters in distinguishing between related taxa. Some recent research has greatly increased our understanding of how the elements of wing pattern are determined during the development of the wings beginning in the late larval period or very early in the pupae.

Colour pattern determination can be considered to be a three-step developmental process (Nijhout 1985, 1991). First is the determination of the position and properties of signalling sources on the wing-epidermis. This is followed by an activation of these sources and the establishment of 'information gradients' in the chemical substances (morphogens) or signals they produce. The final step is the 'interpretation' of these gradients by surrounding wing-scale building-cells and their determination to synthesize a specific colour pigment. In butterflies there are no experimental data describing the first step, but modelling studies (Bard and French 1984) and comparisons of homologies of pattern elements in species of *Charaxes* and other groups (Nijhout and Wray 1986) provide some support for the involvement in comparative morphology of changes in the properties of sources and in their number and distribution on the wings. Evidence for the presence of signalling sources which establish the information gradients is discussed below. The main element of the third step is assumed to be a threshold-sensing function to such a signal by cells surrounding sources.

The basic unitary model for wing pattern development described fully by Nijhout (1985) incorporates three sources or foci in each wing cell controlling the patterns of the central banding system, the submarginal eyespots and the chevrons of the wing margins. Nijhout (1980; 1986) has provided evidence for the presence of foci controlling eyespot development in the North American buckeye butterfly *Precis coenia*. This species has a large, well-developed eyespot of concentric rings of pigment on the dorsal surface of the forewing (which immediately underlies the pupal cuticle). The position of the focus corresponds roughly to the central white pupil of the developed eyespot and can be mapped precisely on the developing wing-epidermis. If the focus is removed or killed by inserting a fine needle and applying heat (microcautery) soon after pupation no eyespot develops. If it is removed at progressively later times increasingly larger eyespots develop, until at about 48 h (at 29 °C) there is no difference in average size between the experimental eyespot and that on the untreated wing of the emerged control adult. Thus in Nijhout's system, the longer the presumptive focus is active, the larger the subsequent eyespot will be. He also showed that the small focus could be grafted to another area of the wing to produce an eyespot in a novel position. Closely similar effects, to those discovered by Nijhout occur for eyespots on the forewing of the African satyrine *Bicyclus safitza* (French and Brakefield, unpublished data).

The data from these experiments with *Precis* and *Bicyclus* butterflies are consistent with propagation of a signal by simple diffusion of a morphogen with constant rates of diffusion and decay. This type of model is illustrated in Fig. 8.5a. The concentric rings of pigment are produced in response to certain thresholds of the morphogen. Development of the model illustrates how the shape and characteristics of the eyespot can vary with the form of the threshold responses of surrounding cells from the simplest pattern of concentric rings to ellipses or other shapes (Fig. 8.5b). If no cells are sensitive to the amount of morphogen produced then the eyespot will not develop. A V-shaped chevron can occur when the plane of response is tilted giving sensitiv-

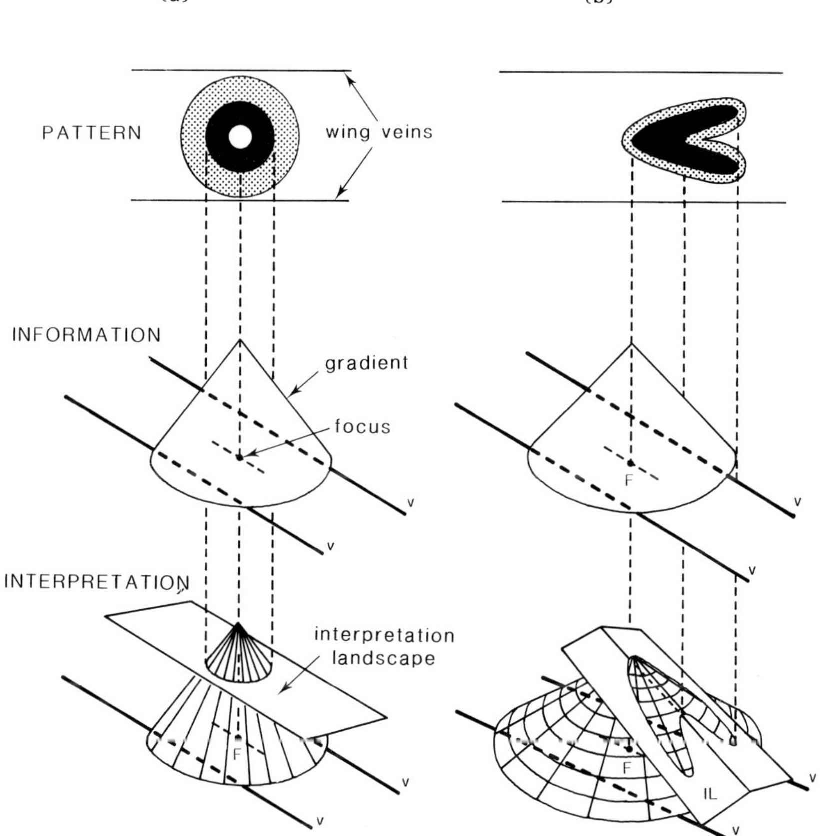

Fig. 8.5 Nijhout's model for the determination of two common wing pattern elements in butterflies. (a) The development of a simple eyespot comprising three concentric rings of different colour pigments. (b) The possible transformation of such a pattern with the loss of the white centre to give a simple chevron. Cones represent the concentration-profile of a substance or morphogen diffusing away from a point source or focus (F) within the area of the wing bordered by the two veins (V). A linear gradient is shown for ease of illustration. The lower diagrams illustrate interpretation landscapes (IL). These involve variation among cells on the wing surface between the veins in their response to the concentration of the gradient. Planes indicate the critical concentration at which cells register the position of the outer edge of the black area of the pattern. This response to the gradient then specifies which colour pigment the scale cells will develop later to give the patterns in the top diagrams. (After Nijhout 1985, 1986; courtesy of H. F. Nijhout and Academic Press and *Bioscience*.)

ity only proximally or distally. Further complexity of pattern is introduced by involving sinks or different patterns of distribution and shape of sources and sinks (Nijhout 1985). However, much research remains necessary to establish the reality of the components of this type of model. Such studies in developmental biology will enable a clearer insight into the complex interactions between genes and phenotype and into constraints on the evolution of wing patterns.

8.4 Enzyme polymorphism

Proteins called enzymes catalyse the synthesis of the various pigments in the wing scales which underlie the great diversity of wing patterns found in butterflies. A large number of different enzymes catalyse the many other biochemical reactions of metabolism and development. Different alleles of a gene produce changes in a particular enzyme. The development of electrophoretic techniques in the 1960s for separating the different forms of an enzyme (called allozymes) according to their mobility in an electrical field on a gel, demonstrated the existence of substantially more heterogeneity in the genetic material of individuals than had previously been thought. This led to controversy about whether this genetic variation was largely adaptive and influenced by natural selection or whether it was mostly neutral and influenced by random genetic drift (Lewontin 1974). Studies of variation in particular enzymes have sought evidence for its adaptive nature but the overall question is still unresolved.

Some surveys of enzyme variation have been made in butterflies, although rarely on British or European species (but see examples concerning *Pieris* by Geiger 1980, 1982; Courtney 1982*b*; Lorkovic 1986). The most thorough study of a single enzyme system is that by Watt and his co-workers on phosphoglucose isomerase (PGI) in North American *Colias* butterflies. This enzyme converts fructose-6-phosphate to glucose-6-phosphate and vice versa. It plays a role in the supply of these metabolites from nectar and more generally in the mobilization of carbohydrate energy reserves.

Watt (1977) showed that at least four to six PGI alleles were present in natural populations. Functional differences were identified between purified allozymes which related to their heat stability and their kinetic function or efficiency in catalysis. These differences were consistent with the frequency data for the alleles. For example, certain heat-sensitive allozymes are at high frequency in tundra populations of *Colias meadii* where overheating of adults is quite rare while relatively heat-resistant ones are more common in *C. alexandra* flying in the warmest period of the Colorado summer. Further studies have substantially strengthened the case for direct natural selection on the PGI gene alleles (Watt 1983, 1985; Watt *et al.* 1983, 1985). Survivorship and flight activity differs among the genotypes but the direction of such differences can change between temperatures. For example, certain heterozygotes which are kinetically most effective at low temperatures begin to fly earlier in the day and are active over a wider environmental range than other genotypes. These genotypes in males also tend to obtain higher numbers of matings. At high temperatures such genotypes may, however, survive less well than thermally stable ones.

The overall results suggest that heterozygous advantage in combination with changes in selection associated with temporal fluctuations in the environment are responsible for maintaining this enzyme polymorphism. The particular role of PGI in the metabolic pathways may account for its apparent position as the focus of intense natural selection (Watt 1983, 1985). Much of the variation observed in other enzymes is unlikely to be as strongly influenced by selection (see section 9.1.4).

8.5 Seasonal polyphenism

The phenotypic expression of the various alleles controlling a genetic polymorphism is dependent solely or largely on genetic factors. There is little if any influence of the environment during development. There is, however, another way in which discrete phenotypes or discontinuous forms can arise in a population; that is by some environmental factor acting in a switch-like manner to control which of two or more developmental pathways are followed. Such instances are called polyphenisms and involve the phenomenon of phenotypic plasticity (for more information see Schlichting 1986). Their basic difference from genetic polymorphisms is that each individual in a population has the genetic capability to produce each of the possible phenotypes. It must be emphasized that the ability to exhibit phenotypic plasticity is under genetic control which is itself influenced by natural selection. Thus the form of the response to the particular range in environment experienced by a population is under genetic control.

In many polyphenic butterflies the different phenotypes are associated with different seasonal environments (Shapiro 1976). Although this phenomenon has been investigated in butterflies for over a century its adaptive value is poorly understood. It is only in those species of Pieridae that show increased melanin deposition (a quantitative character) in adults during the cool season that this has been achieved (Watt 1969; Shapiro 1976). In some of the more rigorous experiments Douglas and Grula (1978), working with the North American *Nathalis iole*, demonstrated that the higher density of melanic scales increases the efficiency of absorption of solar energy. This appears to facilitate a higher intensity and duration of activity (see Roland 1982; Guppy 1986; section 2.2).

Phenotypic plasticity provides a means of adaptation to environmental variation, whether in time or space, when some external factor experienced during development can act as a predictor of later environmental conditions. For a sedentary or short-lived species in which any one individual experiences only a subset of the variable conditions during a particular developmental stage, genetic polymorphism would be a wasteful means of adapting to such heterogeneity since in any one environment a proportion of individuals would be mismatched.

In tropical environments seasonal polyphenisms represent an important ecological alternative to migration or complete cessation of development (diapause) for coping with the extreme environment of a dry season. Thus many species of satyrines in the Old World tropics are characterized by different forms occurring in the generations of the wet and dry seasons (Brakefield and Larsen 1984). The most striking feature of these forms is that large, well-differentiated eyespots are found in wet season butterflies, but not in the dry season when they are dramatically reduced or missing (see Fig. 5.2). All individuals can respond to some environmental cue to develop the appropriate form. Temperature during the final larval instar has a major role in at least two African species of *Bicyclus*. Larvae that develop at constant high temperature produce the wet season form while those that develop at a lower temperature produce the dry season form. In the seasonal environment inhabited by this species, higher temperatures occur in the rainy season. Well-developed eyespots and banding patterns are favoured in the wet season when butterflies are active and are 'forced' by habitat change to rest on green herbage thus disrupting their crypsis. In contrast, butterflies are quiescent in the dry season resting on brown leaf-litter after the dieback of herbage. A highly cryptic pattern without distinctive eyespots is favoured to avoid being picked out at rest by browsing predators, especially lizards (Brakefield and Larsen 1984; Brakefield 1987*b*,*c*; see further discussion of spot patterns in chapters 5 and 9).

In general, polyphenisms in adult butterflies are less striking and less frequent in species of the temperate regions. A conspicuous example is, however, the occurrence of seasonal forms in the European map butterfly *Araschnia levana* which is common in many regions of continental Europe (Fig. 8.6). As in most examples of this phenomenon in butterflies from temperate regions, the polyphenism is controlled by day length which changes in a predictable way with season (see Shapiro 1976). Day-length is less likely to be an effective predictor in the tropics. *A. levana* has a fritillary-like phenotype in the late spring generation but has one similar to the white admiral in mid-summer (Fig. 8.6). Studies of differences in behaviour and ecology between butterflies of these generations within colonies would be fascinating. Does the spring form behave like a fritillary with patrolling flight patterns? Observations (P. Brakefield, unpublished data) suggest that the summer form is rather similar to the white admiral *Ladoga camilla* with regard to the predominance of perching behaviour (see chapter 2). It has been suggested that it mimics this group of *Limenitis* butterflies (Shapiro 1976). Of the British butterflies, the comma *Polygonia c-album* and the speckled wood *Pararge aegeria* (see chapter 9) exhibit seasonal changes in wing pattern, although to a less marked degree than occurs in the tropical satyrines or the map butterfly.

Polyphenisms which are at least partly unrelated to season occur in many butterflies at the pupal stage. This phenomenon enables pupae of these species to match the colour of their resting background which varies in space rather than in time. It has been especially well studied in the whites, *Pieris rapae* and *P. napi*, and swallowtails including *Papilio*

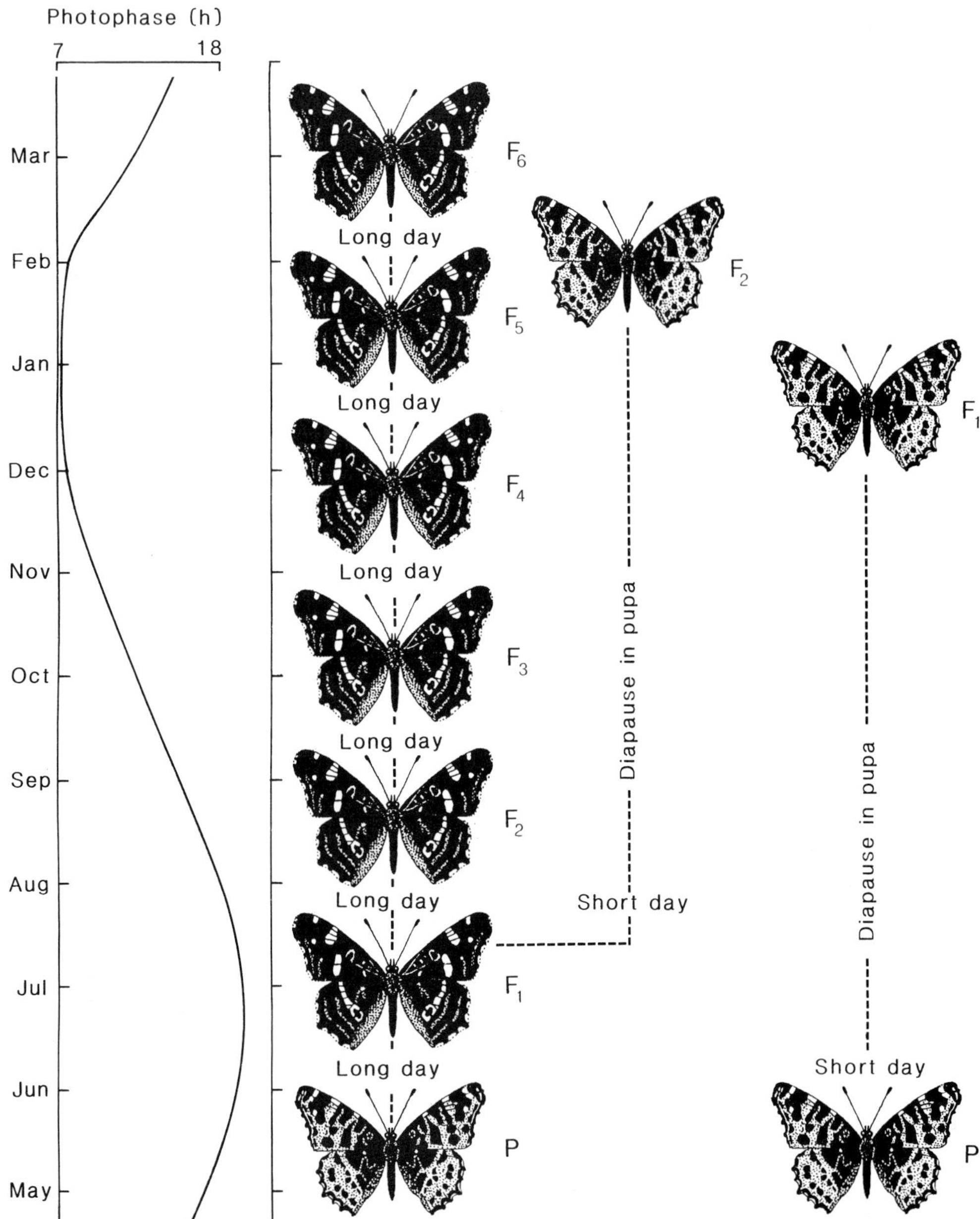

Fig. 8.6 The environmental control of the seasonal forms of the map butterfly *Araschnia levana*. The primary mechanism involves sensitivity to photoperiod; the longest and shortest days in the northern part of the butterfly's distribution are about 16 hours 30 minutes and 7 hours 15 minutes, respectively. The dark form (*prorsa*) which flies in the summer or autumn is produced when caterpillars are exposed to long day-lengths in early summer and do not diapause. The pale form (*levana*) flies in late spring and develops after caterpillars have been exposed to shortening day-lengths in the autumn and have experienced a prolonged winter diapause in the pupae. The figure illustrates how each form can be produced in the laboratory under appropriate conditions of day-length from a stock of either parental form. P, parental generation, F, filial generation in the sequence as indicated by the number. The original breeding work was carried out by Müller (1955). (Redrawn from Wigglesworth 1964; courtesy of Weidenfeld & Nicolson Ltd.)

machaon. Two major sorts of environmental cues induce this phenotypic plasticity in pupal colour: optical and textural (Smith 1978, 1980; Hazel and West 1979, 1983; West and Hazel 1985). Optical cues include the colour of the site of pupation or of its surroundings. In some species photoperiod, temperature, and relative humidity may also influence pupal colour. Insect hormones and other similar factors are involved in mediating the biochemical differences in pigmentation between the colour forms as for example in the peacock *Inachis io* (Maisch and Buckmann 1987). *P. machaon*, as in many other swallowtails has dimorphic, green or brown pupae. West and Hazel (1979) released larvae of three North American species of *Papilio* into their natural habitats just before pupation. The larvae were marked with UV-fluorescent paint so that their pupation sites could be identified. Pupae were usually found to match their resting background—with green pupae on green stems or foliage and brown pupae on bark and dead stems. One forest species, *P. glaucus*, pupated very close to the ground in the leaf-litter and had monomorphic brown pupae. The pupae of all the butterflies which exhibit polyphenisms are cryptic and rely on resemblance to their resting background for survival during the period of pupation which may occur overwinter (section 5.5; see also West and Hazel 1982; Hazel and West 1983). Smith (1978) found that green pupae in *P. polytes* result from prepupae which select a smooth substrate to pupate on, while brown ones occur on rough surfaces. Green and brown substrates in nature are dominated by relatively smooth and rough surfaces, respectively. Breeding studies have demonstrated that heritable variation controls the nature of the response to environmental factors in swallowtails (Clarke and Sheppard 1972; Hazel 1977; Hazel and West 1982; Hazel *et al.* 1987). Selection on a laboratory stock can change the frequency distribution of larvae which pupate on different substrates.

8.6 Life history variation

Polyphenism is a trait with a genetic basis which facilitates success in seasonal environments. Seasonality in resource availability, for example in host-plants for larvae and suitable weather for adults, imposes constraints on the number of generations per year (voltinism) and their time of appearance (phenology). A certain amount of variation in phenology is environmentally determined. In general, warm weather speeds development and promotes early completion of life history stages (Brakefield 1987*d*) but timing also involves a genetic component. In most butterfly species of seasonal environments which have discrete generations and in which females usually mate only once, it is usual for males to emerge before females (protandry). Early emerging males are at an advantage over those emerging later since they have the greater chance of encountering fresh, unmated females (Wiklund and Fagerström 1977). In territorial species this may occur because first-emerging males are more likely to establish territories at locations where the chance of encountering receptive females is maximized (Wickman and Wiklund 1983).

8.6.1 Variation of the number of generations

In Britain some species (e.g. brimstone *Gonepteryx rhamni* and wall brown *Lasiommata megera*; see Table 4.3) have a constant number of generations per year throughout their range. Others (e.g. small copper *Lycaena phlaeas* and *Polyommatus icarus*) are more variable, tending to have more generations in warmer regions and in warmer years. This flexibility gives them the opportunity to exploit suitable sites and seasons, though there may be risks associated with such a strategy. The mechanisms whereby variation in generation number is maintained are discussed below.

The small tortoiseshell *Aglais urticae*, which feeds on nettle (*Urtica dioica*), is normally bivoltine in southern Britain, with emergences of adults in midsummer and in the autumn. The first generation reproduces immediately whereas the second overwinters as diapausing adults. Another nettle-feeding butterfly, the peacock *Inachis io*, is normally univoltine in southern Britain. Although spring adults breed at the same time as *A. urticae*, offspring emerge later and form the overwintering generation. In northern Scotland both species are univoltine but with increasing distance south in Europe the number of generations increases. Thus, in central France *A. urticae* is trivoltine and *I. io* is bivoltine. In both, there is a critical day-length or photoperiod above which development is direct but below which individuals enter diapause. In southern and central Britain a small proportion of second generation individuals of *A. urticae* do not enter diapause, but produce a small third generation (Pullin 1986*a*). This third generation is rarely successful since the larvae are usually in the second or third instar at the onset of winter, which they cannot survive. In years when June–August is exceptionally warm it can be successful, producing diapause adults which emerge early enough to feed before entering hibernation. Two possible mechanisms may produce this third generation. Larvae and pupae of the second generation could develop beyond the stage sensitive to a diapause-inducing photoperiod before that critical day-length occurs and, therefore, develop into adults which are capable of immediate reproduction. Alternatively, the response of individuals within particular populations to specific day-length may vary; at any given photoperiod a certain proportion of individuals will enter diapause and others will develop directly.

Pullin (1986*a*) has shown that the duration of development in *Aglais urticae* is comparatively uniform for populations from different geographic areas (Table 8.1) and that the stage sensitive to photoperiod is the late fourth and fifth instar. This stage is close to that of the adult, enabling individuals to predict adult conditions accurately from a prior environmental cue. The average response of populations from different geographic areas were dissimilar (Fig. 8.7), being the most flexible in southern areas and enabling these populations to produce another generation in favourable years. These results are similar to those of Lees and Archer (1980), in which offspring of the green-veined white *Pieris napi* from sites at different latitudes in Britain produced contrasting proportions of diapause and non-diapause pupae, indicating genetic differentiation. However, in this species there appears to be an interaction with temperature. Constant photoperiods but different temperatures produce different proportions of diapause and non-diapause pupae from the same populations.

Insect photoperiodic responses are usually controlled by one of three mechanisms. Diapause induction may be controlled by single dominant genes, segregating in simple Mendelian ratios, by a number of closely associated genes which also segregate as a single unit, or by a system of polygenic inheritance (see Tauber *et al.* 1986 for full discussion). In the first two cases matings between individuals with different diapause responses will not usually produce offspring with a response intermediate between those of the parents. A cross between *Aglais urticae* from central Britain and

Table 8.1 The mean development time from egg to adult for *Aglais urticae* and *Inachis io* from different localities

Temperature (°C)	Mean development time (days)					
	Aglais urticae		*Inachis io*			
	Burren 53° 8′	Oxford 51° 46′	Burren 53° 8′	Oxford 51° 46′	Southampton 50° 55′	Obernai (France) 48° 28′
15.0	62.5	59.0	89.5	85.0	88.0	86.5
20.0	30.5	29.5	40.5	43.0	37.5	43.0
25.0	21.0	22.0	30.5	32.5	29.5	31.5
27.5	19.0	18.5	25.5	26.0	27.0	26.5

The latitude is given for each locality.
Experiment conducted under 12:12h photophase
From Pullin (1986*a*).

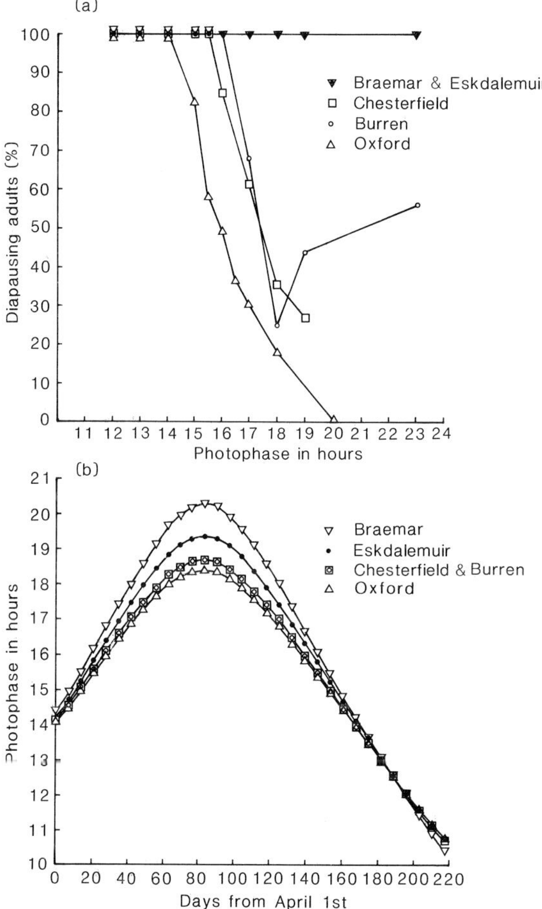

Fig. 8.7 Diapause induction curves and photoperiods for the small tortoiseshell *Aglais urticae* from different localities in Britain (Pullin 1986*a*; courtesy of A. S. Pullin). (a) Diapause induction curves; (b) photoperiods. For all sites the greatest proportion of individuals enter diapause at the shortest day-lengths, but those from northern latitudes require longer photoperiods than those from southern latitudes to develop directly. Such a requirement reduces the number of generations in northern latitudes. Latitudes: Oxford 51° 46′; Burren, 53° 8′; Chesterfield, 53° 14′; Eskdalemuir, 55° 12′; Braemar, 57° 2′.

central France produced an intermediate diapause response, suggesting that diapause induction in this species is under polygenic control (Pullin 1986*a*). Such a system can produce a rapid and smooth response to changing conditions (see chapter 9).

Variation of diapause response within any population affords flexibility. The range of responses within particular populations is probably maintained by variation in temperature and sunshine between years (Dennis 1985*d*; Pullin 1986*a*, 1986*c*). For example, in an extended warm summer, larvae will develop quickly, permitting time for a third generation. In cool summers, however, there will be insufficient time for an extra generation and environmental conditions will select against third generation individuals. Genetic variation may be maintained by variation in weather conditions between years.

Diapause responses within populations are complicated by the relationship between the species and the quality of its foodplant. As with many other species, individual *Aglais urticae* which feed on foodplants low in nitrogen develop into small individuals that produce few eggs, have smaller energy reserves and are poor competitors for mates (Pullin 1986*b*). These individuals will be at a disadvantage relative to larger ones. In the absence of grazing and cutting, the nitrogen content of *Urtica dioica* is higher in spring than in late summer and first generation adults are larger than those of the second (Dennis 1985*d*; Pullin 1987). In response to cutting and trampling the plant produces regrowth leaves which have a high nitrogen content similar to that of spring leaves, making them more suitable for larval feeding. Females will deposit most of their eggs on such regrowth (Dennis 1984*c*, 1985*d*,*e*). Pullin (1986*a*) suggests that the voltinism of this butterfly may well have changed in response to grazing and cutting because second generation adults which have fed on regrowth leaves are larger and probably overwinter more successfully than individuals which have fed on older leaves. Furthermore, conditions for a third generation are more favourable now because suitable hostplants are more abundant. In Britain, *Inachis io* shows responses to photoperiod similar to *A. urticae* but because development takes longer in this larger butterfly there is insufficient time for a second generation (Table 8.1); consequently the critical photoperiod for the onset of diapause is longer than in *A. urticae*. In both species there are costs and benefits associated with the period of development. Individuals which develop in the warmest and sunniest conditions grow most rapidly but are the smallest (Baker 1972*b*; Dennis 1985*d*; Pullin 1986*a*). Consequently there is a latitudinal gradient of size, the largest and most fecund individuals being located in the most northerly regions.

The small heath *Coenonympha pamphilus* also has a complex life cycle with a variable number of generations. It has a single brood in the north but as many as three in southern Britain, though different individuals can have different development rates within single populations (Lees 1962). At 12 °C about two per cent of a predominantly univoltine Yorkshire population developed directly in captivity to produce a second annual generation. The remainder developed more slowly and went into diapause in the fourth larval instar. By contrast, approximately 14 per cent of individuals from a normally biovoltine south Devon population developed directly in similar conditions. At 16 °C the proportion of such individuals from both populations increased. Further experiments (Lees 1965) in which directly developing individuals were selected as parents, thereby substantially increasing the proportion of such individuals over three generations, indicated a polygenic control (see chapter 9). The control of the number of generations in this species is related to variation in development rates in response to temperature rather than to a variable photoperiodic response.

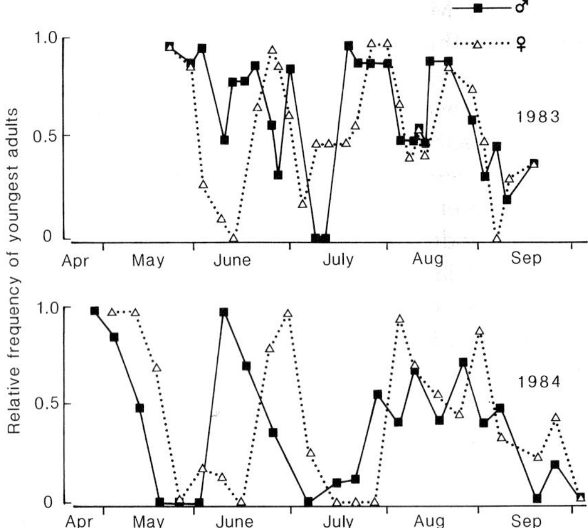

Fig. 8.8 The pattern of adult emergence of the speckled wood *Pararge aegeria* in 1983 and 1984 in Bernwood Forest, Buckinghamshire. Throughout each year there is a succession of adult emergences. This pattern does not correspond to a synchronous emergence of successive generations but represents a more or less continuous emergence throughout the year following two spring emergences. (From Shreeve 1986*b*; courtesy of The Royal Entomological Society.)

8.6.2 Variation in larval and pupal development rates

The complexity of life history patterns involving overlapping rather than discrete generations is well illustrated by the speckled wood *Pararge aegeria* (Fig. 8.8). In central Britain the duration of the larval stage varies considerably, some individuals developing quickly with four instars and others more slowly, with five (Shreeve 1986*b*). Rapid development produces smaller adults and there is considerable variation in development rates within single families. In the first two larval stages development time is relatively uniform but this becomes more variable in the later stages, particularly for females. In central Britain and southern Sweden the butterfly can overwinter either as pupae or larvae (Shreeve 1986*b*; Nylin *et al.* 1989) but in northern Sweden it overwinters only as pupae (Wiklund *et al.* 1983).

This variability results in an unusual and diverse life cycle since the pattern of adult emergence is controlled by temperature, which influences the development rate of all stages, and by photoperiodic induction of diapause. A day-length of 11 hours produces diapausing pupae and also slows larval growth (Lees and Tilley 1980) but a photoperiod of 16 hours does not (Shreeve 1986*b*). Nylin *et al.* (1989) detail the complexity of response to day-length in this butterfly; both day-length and temperature are involved in the induction of diapause and larval aestivation. Although the stage which is sensitive to photoperiod in Britain is not precisely known there is a time between mid-August and early October when individuals which are sensitive will produce diapausing, overwintering pupae. Cool weather during this period will reduce the number of adults emerging in late summer because a greater proportion of individuals will enter diapause than develop directly. In warm summers more individuals will have developed beyond the stage sensitive to diapause induction. They will then develop directly and emerge as adults in late summer, extending the flight period.

This complex life cycle of *Pararge aegeria* raises many questions about its causes and adaptive significance. Most stages may be present throughout the year. Of the larval stages, probably only the third instar can survive cold winters (Shreeve 1986*b*). Cool autumns slow development and second instar larvae can be present in mid-November. These larvae die at the onset of severe winter cold, but they may survive a mild winter. Warm autumns speed up development, individuals reaching the third instar before winter. Variation in development time within populations can produce high weather-related mortality but it is probably maintained by variation in weather between years, in a similar way to that in the nymphalids (p. 193). It may also be influenced at the individual level by females adopting a 'risk-spreading' strategy. For example, pupae are attacked in the autumn by the ichneumonid *Apachthis resinator* (Cole 1967), but this parasite is active only when air temperature is higher than 11 °C. Therefore, early pupating individuals are more likely to be attacked than later-formed pupae, though the latter individuals run a higher risk of weather-related mortality.

8.6.3 The genetic control of size

Adult size of the nettle-feeding nymphalids and of *Pararge aegeria* is partly under environmental control. The greatest influences are temperature and food quality, though there must also be a genetic element related to development rate. Variation in adult size within populations is universal, but in some species (e.g. Pierinae) it is more marked than others (see Dennis and Shreeve 1989; Dennis, in press). Extremely small or dwarfed individuals from many populations of most species are recorded in the literature: these are most frequently the result of undernourishment and are usually environmentally determined.

In seasonal environments the period when the hostplant is suitable for feeding is limited, though the time available for male and female larval development is the same. Differences between male and female sizes are therefore related to development duration—those remaining in the larval stages the longest being the largest. In the majority of species resident in Britain females are larger and heavier than males. Those species in which male size is greater than or equal to that of females (e.g. some Lycaenidae and Pieridae) tend to be those which engage in patrolling mate-locating behaviour (see section 2.4) or those which have a long pupal stage.

Singer (1982) has associated the relative difference between male and female size to the advantages and costs of reduced development time in males. In species with distinct generations and in which females mate once after emergence it is usually advantageous for males to emerge early (see p. 34), but this early emergence necessitates curtailment of body size (e.g. purple emperor *Apatura iris*, *Lasiommata megera*). When there is a long pupal period, as over the winter (e.g. *Anthocharis cardamines*), or when adults do not mate after eclosion (e.g. *Gonepteryx rhamni*, *Inachis io*) there is no advantage for males to sacrifice larval development time to gain early emergence and the difference in size of the two sexes is slight. Similarly, in species with overlapping generations (e.g. *Pararge aegeria*), or when populations may be periodically supplemented by individuals from other areas (perhaps the small white *Pieris rapae*) there is no advantage in early eclosion of males. In those few species (chiefly Lycaenidae) in which males are larger than females other factors must influence size, such as the need for male mobility to locate females. Whilst the factors which may influence size are readily identifiable, the genetic basis for the control of body size in the majority of species is not known.

Gilbert (1984*a*,*b*,*c*, 1986, 1988) has demonstrated that size in *Pieris rapae* is under polygenic control. In this species large females can lay approximately 20 per cent more eggs than small females. Genetically large individuals have a lower threshold temperature for development and grow more quickly at low temperatures than do genetically small individuals. At intermediate temperatures large and small individuals grow at the same rate although at high temperatures small individuals grow most rapidly, but such conditions rarely occur. From development rates alone large size should be favoured at almost all times but size is variable within populations and the distribution of size remains uniform with time. Therefore, some other mechanisms must operate to prevent the increase of adult body size of successive generations. Intermittent catastrophically high temperatures select against large size because of reduced hostplant availability, but this alone cannot account for the maintenance of uniform size distributions because there is little change of body size between such events. The mortality of larvae produced by large females is greater than that found for small females, particularly when larval hostplants are limited. Large adults have higher mortality rates than do small adults. Therefore, the advantage of large size in terms of fecundity is at least partly balanced by a shorter period in which to lay eggs and lower larval survival. Smith *et al.* (1987) argue that these trade-offs are sufficient to maintain any particular distribution of genetically determined sizes. Gilbert (1984*c*, 1988), however, offers an alternative explanation, suggesting that genetic recombination can effectively balance advantages for large size. This, it is argued, is the result of gamete production and 'disassortative' mating, in which individuals are as likely to mate with an individual of a dissimilar size as a similar size. In this manner, genetic combinations which produce large individuals are disrupted.

The extent to which variation in body size within populations of other species is under genetic control is largely unknown. Where such variation among offspring of individuals is an advantage, as in areas where weather is upredictable and where different sized individuals have different activity levels (see chapter 2), there may well be a genetic component to size.

The interpretation of the significance of genetic variation is fraught with difficulty, primarily because interactions with the environment are, by their nature, complex, but also because the degree to which variation in certain characters is genetically or environmentally determined is usually not known. However, genetic variation is of major importance in facilitating success in diverse environmental conditions as, for example, the capacity to produce seasonal forms and polymorphisms. Genetic variation is also essential to produce a response to changes in climate and habitats. It is the key to speciation and the production and maintenance of local races and forms. These topics are the subject of chapters 9 and 10.

9

Case studies in evolution

Paul M. Brakefield and Tim G. Shreeve

Genetic variation is all important to evolutionary change. It forms the basis of adaptation to specific or novel environments and also of divergence between populations and speciation. This chapter uses work on three European satyrine butterflies to illustrate the processes involved in such divergence. Firstly, studies of local differences, and of progressive clinal changes between populations are described. These may reflect changes in adaptation to the environment. Secondly, examples of more extensive phenotypic and genetic divergence characteristic of disjunct races or subspecies are discussed. Such divergence may be the prelude to complete reproductive isolation. Once this stage has been reached and genes cannot be exchanged by hybridization between two groups of populations, they have become distinct species. The same processes involved in adaptation to novel environments or in local divergence between populations are also the ultimate basis of speciation events.

Reproductive isolation may involve differences at many genes resulting in the inviability of hybrids. It may also be associated with more specific changes in the genes underlying the mating systems which lead to successful sexual contacts within a population. Broadly speaking such effects are thought of as involving postmating and premating isolating mechanisms, respectively. In practice most speciation events are likely to involve some combination of both. Unfortunately research on British butterflies has yet to contribute significantly to the understanding of premating isolating mechanisms. Some recent interpretations of the fossil record have described a pattern of comparatively short periods of rapid speciation and adaptive radiation of new lineages interspersed with periods of apparent stasis or relative stability (Eldridge 1987). This pattern of evolutionary change is known as 'punctuated equilibrium'. It is not, however, necessary to invoke any new processes to account for such a pattern. The work on the three species of satyrines has largely been concerned with phenotypic variation in wing pattern.

9.1 The meadow brown: continuous variation and adaptation

The meadow brown *Maniola jurtina* has long been recognized as a particularly variable butterfly comprising a wide range of different forms, races, and subspecies (Thomson 1973). It is a common species, forming more or less discrete populations in a wide variety of grassland habitats throughout Britain and most of continental Europe (Pollard 1981; Brakefield 1982*a*,*b*; Heath *et al.* 1984). Adult emergence in Britain ranges from about 100 to 2000 per hectare with an expectation of life of some five to twelve days. The butterfly is rather sedentary in areas of favourable habitat with individuals, on average, ranging over about one hectare during their lifetime. A wide range of nectar sources are used by adults and larvae feed on a variety of grasses.

In early work on the Isles of Scilly, Dowdeswell *et al.* (1949) noted that the number of small black spots near the margins of the ventral surface of each hindwing of *Maniola jurtina* varies from 0 to 5, rarely six (Fig. 9.1). They chose to use this variation as an index of the fine adjustment and adaptation of populations. Although care is needed in scoring the presence of these small spots (Brakefield and Dowdeswell 1985), hindwing spot-number provides

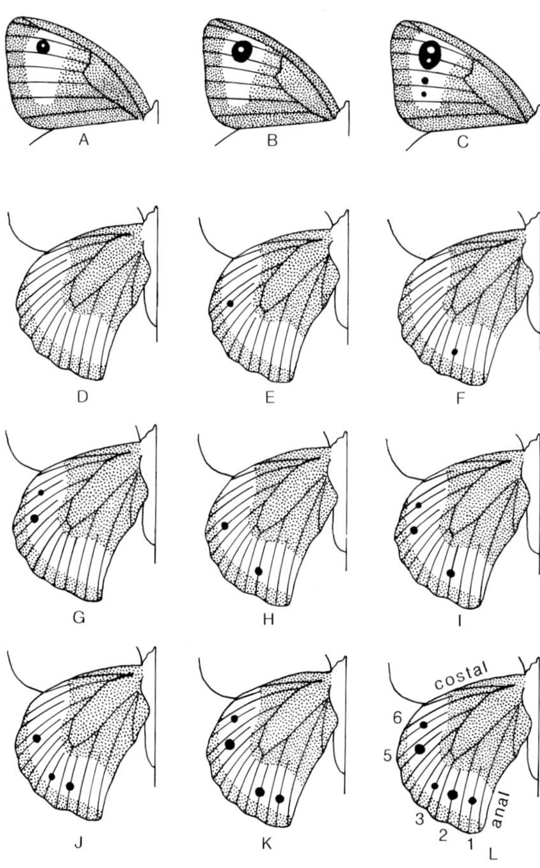

Fig. 9.1 Diagram of variation in the spot pattern on the wing ventral surface of the meadow brown *Maniola jurtina*. Top row shows variation in the forewing eyespot (unshaded area is of brighter fulvous coloration): A, small black eyespot with single white pupil (characteristic of males); B, large spot with single pupil (characteristic of females); C, a more extreme female phenotype showing a very large eyespot with two pupils (f. *bioculata*) and with two additional spots (f. *addenda*). D–L illustrate nine of the thirteen commonly occurring hindwing spot phenotypes: D, nought spot, (0); E, costal 1 (C1); F, anal 1 (A1); G, costal 2 (C2); H, splay 2 (S2); I, costal 3 (C3); J, median 3 (M3); K, splay 4 (S4); L, all 5. Not shown: A2, A3, C4 and A4. The reference numbers of the spots are indicated. Different sized hindwing spots present an idea of changes in relative (not absolute) spot size. (After Brakefield 1990*a*.)

a simple means of scoring their total area. The size of each indivdiual spot is actually a threshold character in that when it is present it varies in a truly continuous manner; butterflies with no spots represent all individuals below the threshold at which spots are expressed phenotypically. Such quantitative characters are usually jointly determined by the interacting effects of a number of genes or polygenes, each of small effect. The same phenotype can be determined by different combinations of these polygenes. Differences between populations, and the results of processes of natural selection influencing quantitative characters, are usually recorded in terms of the population mean (or spot average) and variance. They cannot be represented in terms of gene frequency changes as is possible for Mendelian characters such as polymorphisms controlled by one or a small number of major genes (see chapter 8). Systems of polygenes which determine continuous variation in phenotypic characters are critical in evolution since they provide the basis for smooth adaptive change.

9.1.1 Field surveys

A large number of populations of *Maniola jurtina* in Europe have now been surveyed to compare the frequency distributions for hindwing spot-number which are referred to as spot frequencies (reviews by Ford 1975; Dowdeswell 1981; Brakefield 1984). Some examples of spot frequencies are shown in Fig. 9.2. Males tend to be more highly spotted than females so that the sexes must be analysed separately (Fig. 9.2a). The following discussion is based on some patterns of variation in the most intensively sampled regions of the species range. The significance of these patterns is discussed in relation to processes which are involved in generating and maintaining them.

The earliest survey work by Ford and his co-workers (Ford 1975) involved populations on the Isles of Scilly archipelago situated off the south-west coast of England (Fig. 9.2). Their study rose to prominence in the 1960s because of the controversy among evolutionary biologists over the relative contribution to geographical and population differentiation of natural selection and adaptation to particular environments on the one hand, and on random genetic drift on the other (discussed further

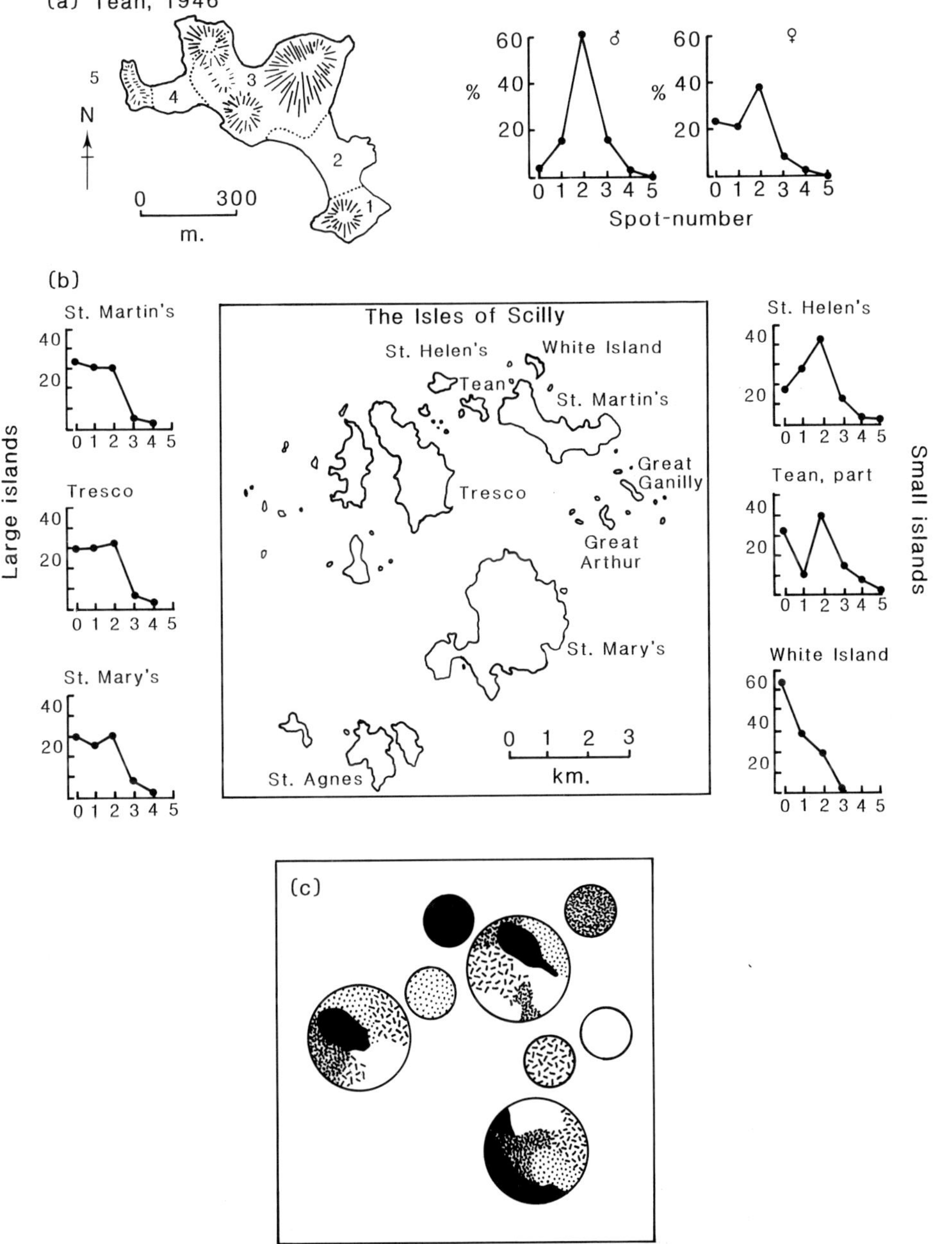

Fig. 9.2 Variation in the number of spots on the ventral surface of the hindwings of the meadow brown *Maniola jurtina* in the Isles of Scilly. (a) A summary of the frequency of males and females with different spot-numbers collected by Ford and his co-workers in their original study on Tean. Note the difference between the sexes. Their mark–release–recapture experiments showed that the population inhabiting areas 1, 3, and 5 of the island were largely isolated from one another by unfavourable habitat in areas 2 and 4. (Adapted from Ford 1975.) (b) Map of the Isles of Scilly showing the spot-frequency distributions for females found on three large islands and on three small islands. (Adapted from Ford 1975.) (c) Diagram of the hypothesis put forward by Ford to account for the dissimilarity among small island populations and the uniformity on large islands. The proposed occurrence of specialized environments on small islands and of mixed habitats on large ones is illustrated by different shadings. (After material in Ford 1975; Berry 1977; Dowdeswell 1981 and Shorrocks 1978.)

below). Islands are ideal study sites because for sedentary species these processes can be examined under conditions of very low individual movements and gene flow between islands.

Most female populations on the three large islands (> 275 ha) of the Isles of Scilly showed a 'flat-topped' spot frequency distribution with similar numbers of 0-, 1-, and 2-spot butterflies (Fig. 9.2b). In contrast, females on the small islands (< 16 ha) exhibited a variety of spot frequencies which tended to fall into three groups: unimodal at 0-spots, bimodal at 0- and 2-spots and unimodal at 2-spots (Dowdeswell *et al.* 1960; Creed *et al.* 1964).

The other major series of studies by Ford's group has been concerned with the so-called 'boundary phenomenon' which occurs along the south-west peninsula of England. Females in populations in Cornwall to the west and extending some distance into Devon are more highly spotted than in populations further east and extending throughout southern England (Fig. 9.3). In some years the transition between the western populations tending to be bimodal, with peaks at 0- and 2-spots, to those unimodal at 0-spots in the east has been an abrupt or a sharp one. For example, in 1956 when first discovered, the boundary along the east–west study

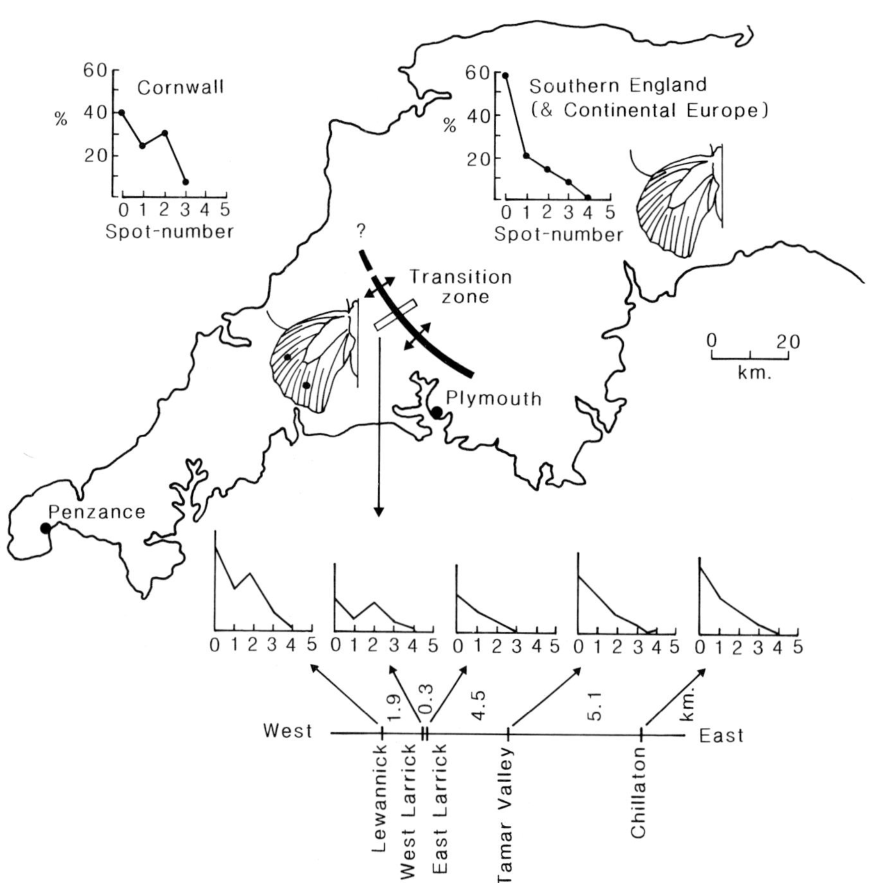

Fig. 9.3 The 'boundary phenomenon' for the hindwing spot-number in females of the meadow brown *Maniola jurtina* across the south-west peninsula of England. The forms of spot-frequency distributions found by Ford and his co-workers to be characteristic of populations in southern England and in Cornwall are illustrated. The zone of the transition between these forms occurred, in some years, along the line indicated. The results of surveying spot-number variation in populations in 1956 along a transect bisecting the transition zone are also shown (inset). (After material in Ford 1975; Berry 1977; and Parkin 1979.)

transect was associated with two adjoining fields separated by a hedge (Creed *et al.* 1959; see Fig. 9.3). In other years the change-over is much less abrupt with a more or less steep clinal change over some tens of kilometres between regions characterized by the western and eastern types of spot frequencies. Male spot-numbers, other spot characters (see below), and the allele frequencies for several polymorphic enzymes (Handford 1973; Brakefield and Macnair, unpublished data) are also associated with fluctuations or clines across the boundary region. These observations have again led to some controversy involving alternative explanations based on either divergence during a period of past isolation (allopatric differentiation) or on present-day spatial changes in natural selection (sympatric evolution).

A study of museum material was also used to examine variability in spotting throughout the range of the species. Dowdeswell and McWhirter (1967) interpreted their data as demonstrating a large, more or less central region extending into southern England characterized by a rather uniform spotting with other, more peripheral regions, including the Isles of Scilly and Cornwall, showing different forms of spot frequency. Such regions were described as stabilization areas and transitions between them were considered to be sharp as in the boundary region in southwest England. However, when a further study of this type of Frazer and Wilcox (1975) and the field surveys within particular regions by other workers are considered, it is unclear how precise the distinctions between stabilization areas are. Certainly enclaves of populations with differing spot frequencies, such as the higher spotted populations of the Grampian Mountains in Scotland, can occur within these areas (see Brakefield 1984).

9.1.2 The inheritance of wing-spotting variation

The evolutionary significance of such survey results can only be assessed with any confidence when the basis of the phenotypic variation in terms of genetic and environmental effects has been rigorously examined. To take an extreme situation, if differences in spot-number depend solely on the environment during development, as for example if temperature alone accounts for the amount of spot pigment synthesized, then the field data will merely reflect the developmental conditions within populations. They will then be of no evolutionary significance. In practice, the control of continuous variation will usually be intermediate between this extreme and variation arising solely from genetic influences. Geneticists assume that variation in a quantitative character results from a combination of genetic and environmental differences (see Falconer 1981). Their initial aim is to divide the total or phenotypic variation (V_P) into its components, the genetic variance resulting from additive effects of the polygenes influencing the character (V_A) and the environmental variance resulting from external effects (V_E). The heritability (h^2) of a character is then the proportion of the total variance which is additive:

$$h^2 = V_A / V_P.$$

The heritability of a character can be estimated from the degree of resemblance between relatives. For example, if a graph of the mean values for a character of the offspring of different families is plotted against the corresponding values for the parents (i.e. mid-parent values), then the line fitted through the points represents the degree of resemblance between offspring and parents (see Fig. 9.4). The slope of this, so-called, parent–offspring regression line estimates the heritability of the character and can vary between 0 and 1. A value close to 1 indicates that offspring are closely similar to their parents and that substantial genetic variation exists with little environmental influence. The measurement of the resemblance between relatives gives heritability one of its most important properties, namely as a predictor of the response to directional selection. For example, say that the character of wing-size has a high heritability in a particular species, then if only the largest butterflies in a population provide the parents of the next generation, this will have a higher mean value than the original unselected population and one more similar to that of the selected parents. Thus a rapid change will be apparent in the population mean for the character. This type of response is also evident when artificial selection is applied on highly heritable variation in plant breeding or animal husbandry programmes. A heritability of zero, as occurs in the absence of additive genetic variance, precludes a response to any form of selection.

Reliable estimates for heritability in *Maniola jurtina* were obtained by breeding 16 families,

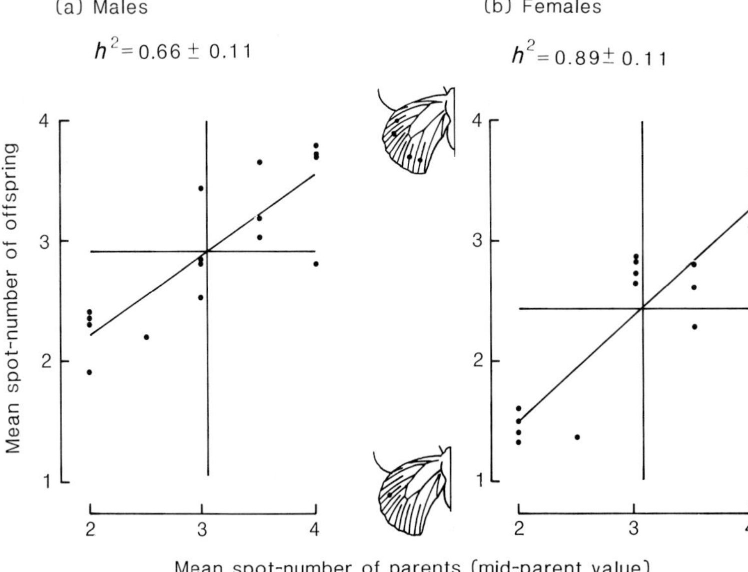

Fig. 9.4 The genetics of variation in the number of spots on the ventral surface of the hindwings of the meadow brown *Maniola jurtina*. The graphs illustrate the family resemblance between offspring of each sex and their parents. The estimates of heritability and their standard errors as given by the indicated regression lines are shown. (After Brakefield 1984; courtesy of Academic Press.)

comprising a total of 1340 butterflies, from a Dutch stock (Brakefield and van Noordwijk 1985). The families and their parents were all reared on growing grasses at similar temperatures and humidities in an unheated laboratory. Parent–offspring regression analyses yielded estimates of heritability (Fig. 9.4). These indicate that there is a fairly high heritability and substantial additive genetic variance for spot-number. However, some caution is necessary in interpreting these estimates since they can only be considered precise for the stock used and the particular conditions in which the butterflies were reared. Indeed it is recognized that such estimates tend to overestimate these parameters in natural populations.

Some additional experiments were performed with one of the families of *Maniola jurtina* to examine further the possibility of direct effects of temperature on spotting during the period of pattern determination in the early pupal stage (see section 8.3). The background to these experiments includes much early work which demonstrated that wing patterns could be altered by subjecting the pupae to extremes of temperature (see Kühn 1926; Köhler and Feldetto 1935). Every pattern element was found to have its particular sensitive period. Lorković (1938, 1943) and Høegh-Guldberg (1971*a*, 1974) found that prolonged cooling of the pupae of two lycaenids led to effects on the underside spot pattern. Furthermore, Høegh-Guldberg and Hansen (1977) found that a lower spot-number was sometimes produced in the northern brown argus *Aricia artaxerxes* by subjecting insects, just before or after pupation, to one of more periods of cooling at between 2 and 5 °C for 9–12 hours. Their experimental pupae also yielded some rare forms which may provide an explanation for such aberrations in nature (see chapter 8). The similar experiments to those of Høegh-Guldberg and Hansen on the family of *M. jurtina* failed to detect an environmental effect on hindwing spotting (Brakefield and van Noordwijk 1985). However, this does not preclude the possibility of some more indirect, so far undetected, effect occurring at some earlier stage of development.

The genetical and breeding experiments with *Maniola jurtina* suggest that differences in spot frequency distributions between populations are likely to reflect genetic differentiation, whether arising from random effects or the influences of natural selection. The same is likely to apply to changes which have been detected within populations over a number of generations. Some of these changes (Dowdeswell and Ford 1955; Dowdeswell *et al.* 1957) have been associated with an altered habitat as for example on Tean in the Isles of Scilly where removal of cattle led to more luxuriant vegetation cover and increased spotting. Other examples have been associated with exceptional droughts or warm and dry summers on islands (Dowdeswell *et al.* 1960; Bengtson 1978) or with climatic changes in England and Italy (Creed *et al.* 1959, 1962; Scali 1972). At lower altitudes in some Mediterranean regions *M. jurtina* shows a modified life cycle with an earlier adult emergence followed by a mid-summer aestivation of females prior to oviposition in late summer (Scali 1971; Masetti and Scali 1972; Scali and Masetti 1975). Scali and his colleagues found that the female spot frequency changed from a 'flat-topped' form to one unimodal at 0 spots over the period of aestivation, indicating a 65–70 per cent selection against 2–5 spotted specimens.

Detailed measurements of variation in the complete spot pattern of *Maniola jurtina* on the ventral surface of both wings revealed the existence of numerous phenotypic correlations. Butterflies of each sex which have more hindwing spots also tend to have larger apical forewing eyespots. These relationships also extend to populations (Brakefield 1984). The forewing eyespot is the most striking and well developed element of the spot pattern, with a white pupil and a large black area which contrast markedly with the background wing colour. It is about twice as large in females as in males and often bipupilled though females usually exhibit fewer and smaller hindwing spots. The eyespot is also extremely variable in size and often in shape within and between populations of each sex. Hindwing spots in females tend to be more often positioned or most heavily expressed towards the costal area of the wing which is closest to the forewing eyespot. These types of phenotypic correlations are probably characteristic of variable species of satyrines and other butterflies (see Frazer and Willcox 1975; Ehrlich and Mason 1966; Mason *et al.* 1967, 1968; Dennis *et al.* 1984) and in *M. jurtina* hindwing spot-number can be conveniently taken as a 'trend' character for the expression of the complete spot pattern. It will be interesting to determine by more detailed genetic analyses or selection experiments whether or not such phenotypic correlations are based on genetic correlations involving common effects of some genes on different spot characters. Studies on the positioning of the hindwing spots have, however, shown that this spot-placing variation is largely independent of spot-number (McWhirter and Creed 1971). This is emphasized in Scotland where a steep cline of increasing costality in the position of hindwing spots with altitude is not associated with any corresponding change in spot frequency (Brakefield 1984).

Analysis of reared material demonstrates a substantial genetic component in determination of forewing eyespot-size (h^2 = 0.59–0.80; Brakefield and van Noordwijk 1985). The families also revealed evidence for a component of polygenic control for bipupillation of the forewing eyespot (f. *bioculata*), for the shape of the eyespot in males, for the presence of additional spots on the forewing (f. *addenda*) and for the presence of white pupils in the hindwing spots (f. *infra-pupillata*) (Fig. 9.1). Similarly, additive genetic variance exists for the position of the hindwing spots (h^2 = 0.35–0.57). One feature of both hindwing spot-number and spot-placing is that those estimates of h^2 for offspring on their same-sex parent are higher than those on that of the opposite sex (Brakefield and van Noordwijk 1985). In other words, males apparently resemble their male parent more closely than their female parent, and females are more like their female parent. This seems to be due to sexual differences in the inheritance and expression of the characters. Certain of the individual hindwing spots can be characterized as typically female while others are typically male. This will cause a greater resemblance to the same-sex parent in respect of both spot number and spot position.

9.1.3 Wing-spotting and natural selection

This pattern of partial sex-dependence for expression of spotting is of particular interest because it is consistent with a model to account for variability in spot patterns in populations (see chapter 5, Fig. 5.3). This model provides an alternative to the assumption that the spots themselves are 'trivial' or unimportant to the individual. It predicts that visual selection by insectivorous birds and lizards operates differently for males and females of *Maniola jurtina*. These differences in selection are associated with differences in the resource requirements of the sexes and in adult behaviour (Brakefield 1982*a*). The marked sexual dimorphism in spot pattern is an important component of the model. The large eyespot is frequently hidden when females are at rest, but as in the grayling *Hipparchia semele* (Tinbergen 1958; chapter 5) it is sometimes exposed when disturbed. Overall, the model predicts that an even wing-spotting will be most advantageous in deflecting attacks by predators to the more active butterflies. Male *M. jurtina* are generally more active than females and have a more evenly expressed spotting.

Mark–release–recapture experiments performed in populations near Liverpool also suggested that the distances moved by butterflies of each sex within a habitat-area increased with spot-number (Brakefield 1984). A particularly clear example of this tendency is illustrated in Fig. 9.5. If such greater dispersal occurs more generally within populations of the species and if it reflects a general increase in activity and frequency of changing behaviour and shifting position, which tends to attract searching predators, then the relationship with spot expression is consistent with the model. Further capture–recapture experiments on other populations and in larger habitat areas would be most useful. Interestingly, there is some evidence for differences in behaviour between two hindwing spotting phenotypes in the speckled wood *Pararge aegeria* (Shreeve 1987). The selection against spotting, particularly on the hindwing, in inactive butterflies depends on their resting background and the effectiveness of their crypsis (see chapter 5). Circular spots may tend to be more conspicuous in grasslands dominated by linear shapes and thus attract the attention of searching predators (a comparable example is shown in Fig. 5.2).

Many of the components and details of the model require testing, although Bengtson (1978, 1981) working in Sweden found differences in the frequency of wing damage (see Fig. 5.1) between spotted and unspotted females consistent with differences in their activity and consequently in their exposure or 'apparency' to predators. However, the model is of value in the absence of alternative hypotheses about how precisely natural selection influences the phenotypic variation.

An important finding from the breeding programme was that in some larger families there were differences in the timing of adult emergence between the different hindwing spot phenotypes represented

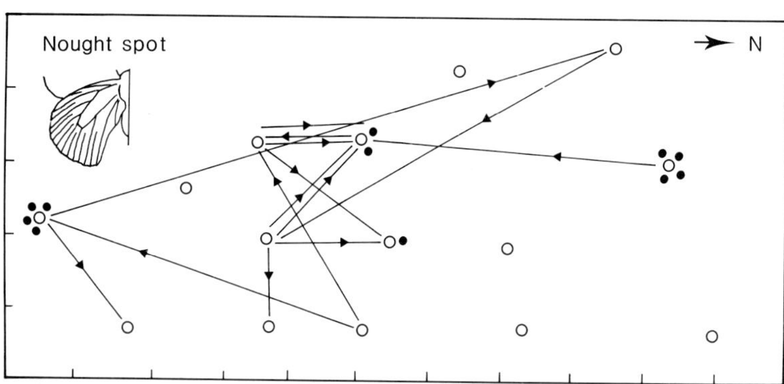

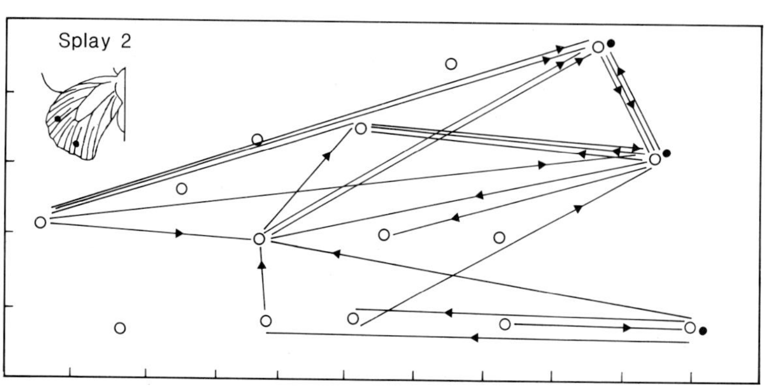

Fig. 9.5 The movements of marked females of two hindwing spot-types of the meadow brown *Maniola jurtina* at Hightown near Liverpool in 1976. Open circles show capture–release sites and closed circles show recaptures at the same site as release. Arrows indicate the direction of moves. Each map (a) and (b) shows 25 movements selected at random from the complete sample for a spot type. Divisions on the edge of each map are at intervals of 15 m. The illustrated patterns of movement are significantly different. (After Brakefield 1979.)

by particular combinations of the hindwing spots. Overall such differences led to declines in spotting during the emergence period. There may be differences in development rates between the phenotypes which result from other (pleiotropic) effects of the spotting genes than those on wing pattern. Such non-visual effects may be critical in the overall selection influencing spotting. These observations are consistent with numerous findings of intra-seasonal changes in spot-number within populations (examples in Creed *et al.* 1959; Dowdeswell 1962; Scali and Masetti 1975; see Fig. 9.6b). When they occur such changes are always in the direction of a declining spot average with time. The species is characterized by extreme geographical variation in the timing and length of emergence. There are consistent differences between populations with a trend towards a more synchronized emergence—beginning later and ending earlier—northwards through England and Wales (Brakefield 1987*d*). There is also a clear trend towards an earlier emergence in years when the month of June is comparatively warm. Figure 9.6a shows that differences in the timing of emergence may also occur in the field between spot types which differ in the positioning of the spots. In Scotland the cline of increasing anality of spotting with altitude may be related to such effects (Brakefield 1984). It is tempting to suggest that spotting is partly related to spatial and habitat-related patterns of variation in timing and length of emergence.

Earlier research on the larvae of *Maniola jurtina* by Dowdeswell (1961, 1962) which involved a comparison of the spotting in adults reared from wild-collected, late-instar larvae with that of flying adults from populations in Hampshire suggested that intra-seasonal changes in spotting could result from differential parasitism by the ichneumonid *Apanteles tetricus*. However, the causal nature of the observed relationship between level of parasitism and spotting has not been demonstrated (Brakefield 1984).

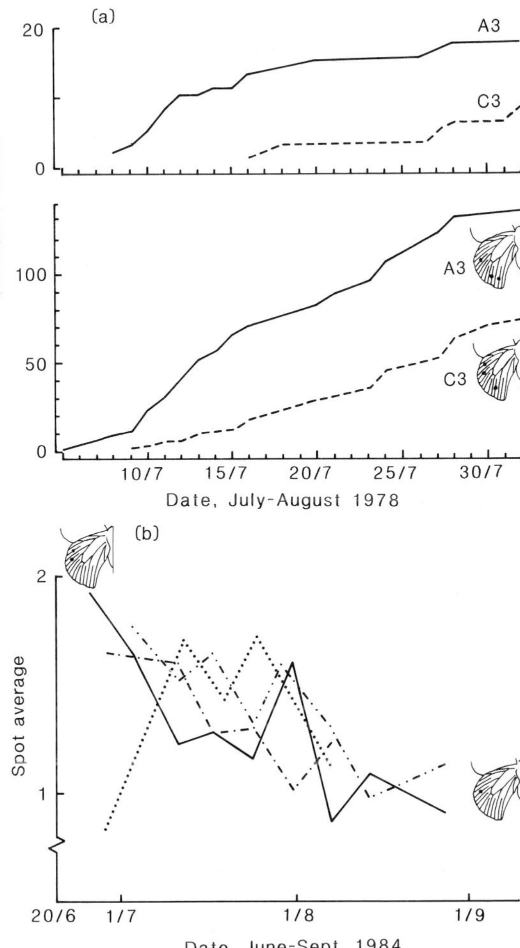

Fig. 9.6 Emergence patterns and hindwing spotting variation in the meadow brown *Maniola jurtina*. (a) Cumulative daily totals of males of two-spot types in fresh condition and captured for the first time in two populations near Liverpool. (b) The intra-seasonal change in female spot average for four populations in the boundary region. Each point is based on about 30 females. (After Brakefield 1979, 1984; Brakefield and Macnair, unpublished data.)

9.1.4 Interpreting geographical variation

With this background information in mind the results of the survey work on the Isles of Scilly and south-west England can be discussed with regard to alternative explanations for the patterns of population differentiation. The spotting of *Maniola jurtina* on the three large islands of the Isles of Scilly is similar whereas the small island populations are more variable tending towards one of three different spot frequencies (Fig. 9.2b). Ford and his co-workers believe that those populations on the large, and diverse islands result from natural selection producing a gene complex simultaneously adapted to a wide range of environments. In contrast, they suggest that the small islands are each characterized by one of a range of different environments and that consequently selection has favoured a more specific gene complex closely adapted to specialized conditions (Fig. 9.2c). McWhirter (1957) suggested that females unimodal at 0-spots are associated with areas of open, often exposed grassland whilst those in luxuriant habitats with patches of scrub tend to show high spot averages. The discontinuity in spotting found between populations on the two ends of White Island is also a clear example of this association (see Ford 1975). This difference is, therefore, consistent with the predictions of the model for visual selection described above and with selection for low dispersal rates in more exposed, open, and windswept habitat and higher ones in a more mixed, diverse habitat. The high spotted and low density populations of the Grampian Mountains in Scotland which occupy a mixed habitat of bracken scrub and poor grassland also fit this pattern (Brakefield 1982*a*; 1984). However, in many other regions populations in widely differing habitats show similar spotting. Dennis (unpublished data) found spotting patterns to be similar in 10 widely separated populations in North Wales on very different rock types and associated with different habitat types.

An alternative to an explanation based on differences in natural selection related to habitat is one in which random effects are paramount. Waddington (1957) considered that the differences between small islands resulted from periods of intermittent random genetic drift associated with periods of very small population size known as bottlenecks. A similar reasoning was developed by Dobzhansky and Pavlovsky (1957), who suggested that the small island populations were derived from small founder groups with differing gene frequencies from which relatively stable but different gene pools developed. Such effects do seem to be reflected in frequencies of melanic forms of the spittlebug *Philaenus spumarius* on the Isles of Scilly (Brakefield 1987*e*, 1990*b*). However, Ford and his colleagues have countered such hypotheses for *Maniola jurtina* with their observations of a population passing through an extreme bottleneck in size, with no subsequent disruption of spot frequency (Creed *et al.* 1964), and an example of a change in habitat due to the removal of cattle on Tean being associated with one in spotting. Unfortunately recent fieldwork on the Isles of Scilly found that a switch in land use on smaller, uninhabited islands away from grazing by cattle has led to the spread of bracken scrub through grassland and the extinction or decline of many colonies of *M. jurtina* (Brakefield 1987*e*).

The studies of the 'boundary phenomenon' in south-west England (Fig. 9.3) also led to alternative hypotheses to account for the population differentiation. Ford and his co-workers consider that their data demonstrate the existence of powerful natural selection which differs on each side of the boundary. Some laboratory experiments with Drosophila fruit flies have shown how such disruptive patterns of selection practised on artificial populations may sometimes lead to divergence and effective isolation (see Sheppard 1969). Ford (1975) discusses the boundary phenomenon in relation to sympatric evolution in which distinct races or local forms can arise without isolation past or present. Handford (1973) elaborates on this interpretation, suggesting that there is a switch-over between two co-adapted genetic systems at a critical point in an environmental gradient. Genes are said to be co-adapted if high fitness depends upon specific interactions between them. Oliver's (1972*a*,*b*; 1979) hybridization studies provide some evidence of such genetic systems in several species of butterfly. He found various forms of disruption in development and in the relative timing of emergence of males and females in hybrids between geographically well-separated populations of, for example, wall brown *Lasiommata megera*, and small pearl-bordered fritillary *Bolaria selene*.

Dennis (1977) discusses evidence, particularly from palaeobotanical studies of pollen, which shows that from about 9500 to 5000 years ago a disjunct or allopatric distribution of *Maniola jurtina* may have occurred in southern Britain. During the warm dry climate of the Boreal period (see Table 10.2), forest cover would have led to populations being restricted to the granite or sandstone upland areas and to the coastal fringe of Cornwall and Devon in the west and to the wide expanse of interconnected calcareous uplands in southern and south-east England. Dennis uses this reasoning to develop the alternative explanation for the boundary phenomenon based on divergence in allopatry originally suggested by Clarke (1970). The boundary is then considered to represent the zone to which two groups of populations of *M. jurtina* have expanded their range after a period of isolation and divergence. The discreteness of the main population groups may be maintained by some form of selection against hybrids. Alternatively, the boundary zone may be in the process of decay involving a progressive change to a shallow cline, and eventual uniformity and mixing of gene pools. Such a breakdown may be slowed by some present-day changes in selection from east to west and the comparatively low dispersal rate of the species. In practice it is extremely difficult to distinguish between differentiation evolving with or without allopatry, especially without a detailed knowledge of the geological and biological history of the region across which a present day hybrid zone or steep cline occurs (Endler 1977).

The most difficult feature of the boundary phenomenon to account for is the frequent shift in the geographical position of the boundary itself. This may be up to 60 km east or west between generations (Creed *et al.* 1970). Clarke (1970) in suggesting that the boundary region may be a zone of hybridization also postulated that the shifts could result from individuals within this zone being particularly prone to developmental instability. Ford (1975) argues that such instability would lead to a mosaic of populations with differing spot frequencies, which is not found. Recent work by P. Brakefield and M. Macnair (unpublished data) has suggested that progressive declines in spotting during adult emergence (see Fig. 9.6b) in combination with variation between years in the timing of the flight period and/or sampling dates could contribute to such (apparent) shifts in the position of the boundary.

Brakefield and Macnair's research was based on analysis of morphometric data from a grid of over seventy populations covering the whole boundary region. This reveals a more complex pattern of geographical differentiation than the results of Ford and his co-workers obtained by using one or two transects (Fig. 9.7). The morphometric data for each spot-character in males and females showed a series of more or less coincident clines of varying steepness from east to west along the peninsula between 1982 and 1984. The width of the clines was of the order of a few tens of kilometres rather than in single figures. No simple boundary was detected running from north to south across the peninsula. Various types of multivariate analysis of the morphometric data for all spot-characters failed to separate the populations into two discrete clusters corresponding to a western and an eastern grouping. Rather, they indicated a single cluster with western populations tending to one side and eastern to the other. Overall these results are more consistent with a pattern of fairly

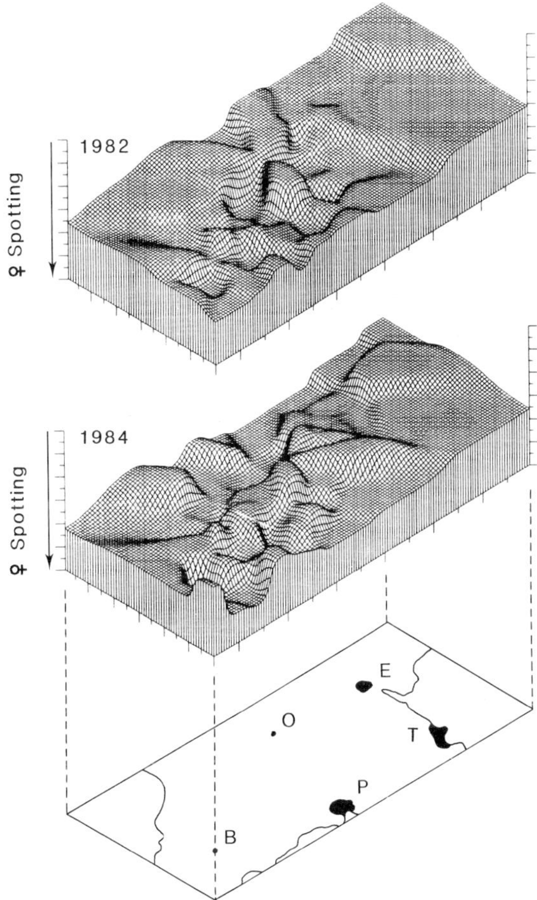

Fig. 9.7 Results of a multivariate analysis of spotting variables in the meadow brown *Maniola jurtina* for a grid of populations covering the boundary region in the south-west peninsula of England in 1982 and 1984. (See also Fig. 9.3.) Pseudo three-dimensional plots for the first principal component over the boundary region. The component is a linear combination of variables and measures the overall extent of development of wing-spotting in females. Note: plot ignores sea areas. Major towns: E, Exeter; T, Torquay; P, Plymouth; O, Okehampton; B, Bodmin. (After Brakefield and Macnair, unpublished data.)

steep clines resulting from some form of present day spatial change in selection in response to an environmental gradient rather than with a hybrid zone following extensive genetic differentiation in allopatry. Such an environmental gradient might, for example, involve a change in climate and the effects of spot genes on timing of emergence. However, such environmental gradients have been in existence throughout the Holocene (Dennis 1977; Table 10.2) and at this stage one can only speculate about the nature of selective influences across the boundary region.

9.1.5 Variation in genitalia

Thomson (1973, 1975, 1976) has made some interesting studies of another aspect of the phenotypic variability of *Maniola jurtina*. He investigated the distribution of different morphological forms of the genitalia (Fig. 9.8). This geographical variability does not appear to parallel that in spot pattern. It is most marked in the form of the valve of males but is also evident in the shape of the paired gnathos. Although such characters can be difficult to measure, two main types—the 'eastern' and the 'western'—are distinguished. These differ particularly in the shape of the dorsal process. The change-over between these types apparently occurs along a more or less broad zone running southwards from Sweden, The Netherlands, Belgium and north-east France to south-east France, and the Mediterranean. This region is characterized by a transitional type of valve and is much narrower in the south (Fig. 9.8). Unfortunately the inheritance of the variation in genitalia has not been examined. However, Thomson (1987) has found that changes in allele frequencies at 10 polymorphic enzyme loci (see chapter 8) are associated with the change-over in types of genitalia. This distribution of genitalia types and enzyme variation may reflect some form of hybrid zone or secondary contact between two races or similar entities which diverged in allopatry. Such isolation was probably associated with changing spatial patterns of 'refugia' of favourable habitat occurring in the period of gross

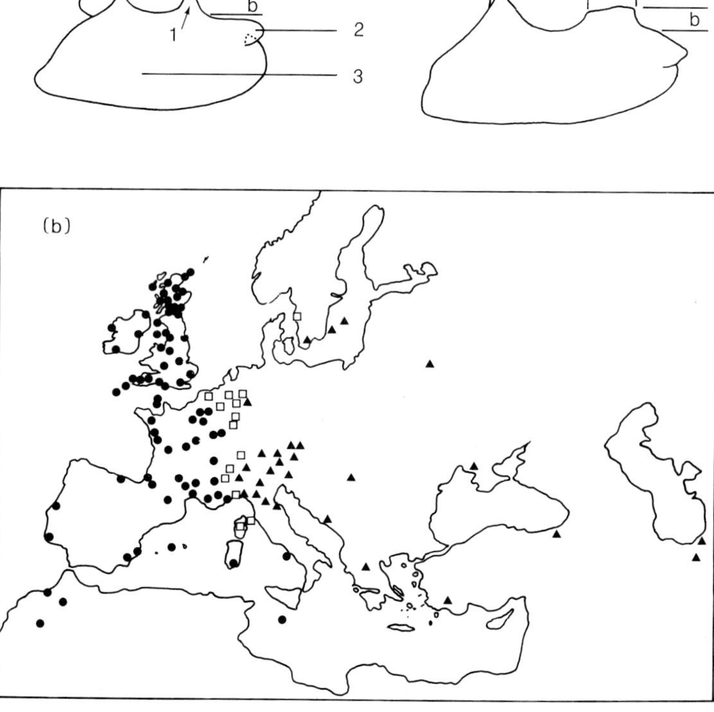

Fig. 9.8 Geographical variation in male meadow brown *Maniola jurtina* genitalia forms over Europe. (After Thomson 1973, 1975, 1987; courtesy of G. Thomson.) (a) The two principal forms of the valve of the male genitalia: left, 'western' form; right, 'eastern' form. 1, dorsal process; 2, distal process; 3, valve. The simple coefficient b–a gives positive values when populations are 'western' and negative when values are 'eastern'. (b) The distribution of ▲, 'eastern', ●, 'western', and ◻, transitional male genitalia forms.

climatic fluctuations of the glacial advances and retreats of the Pleistocene (see section 10.5). Thomson considers that both the eastern and western types of valve originated from ancestral populations in western Asia (Iran). These populations are represented today by a so-called 'primitive' valve structure. The two main types of valve may, however, have originated independently in the eastern and western regions of the Mediterranean (see Dennis 1977; Dennis *et al.* 1991). Within Britain there is a diversity of valve forms, transitional types tending to be restricted to dry calcareous landscapes (Thomson 1973, 1975). This diversity occurs at individual sites and there may also be an element of seasonality in valve shape (Shreeve 1989).

9.2 The large heath: evolution of races

Apart from the variation in genital morphology, the study of *Maniola jurtina* has examined processes involved in microevolution and the adaptation of local populations to particular environmental conditions. The differences examined between populations have been rather limited in extent, often concerning gradual or clinal changes in the relative frequency of different spotting types with any pair of populations differing in mean number, size, or position of spots. Research on another satyrine butterfly, the large heath *Coenonympha tullia*, has begun to examine more substantial differences in wing pattern which are probably associated with the evolution of races in different geographical regions.

Coenonympha tullia is a northern or alpine species and together with the small mountain ringlet *Erebia epiphron* probably represents the most ancient of our butterfly fauna (see section 10.3). Colonies of *C. tullia* are confined to peat mosses, lowland raised bogs, damp acid moorland, and upland blanket bog from sea-level up to at least 800 m. The species has a disjunct distribution in Britain, although this is likely to be partly due to major losses of lowland raised bogs. Several major races or subspecies are recognized by taxonomists. The use of these terms is somewhat imprecise but they implicate more distinctive and far-reaching differences between populations inhabiting different regions than those involving hindwing spot pattern in *Maniola jurtina*. The races of *Coenonympha tullia* are distinguished principally on the basis of development of the submarginal rings of eyespots especially on the ventral surface of both wings (Fig. 9.9). These are most striking in populations inhabiting some of the peat mosses of England from lowland Cumbria to Shropshire. This phenotype, which is also characterized by a comparatively dark ground colour, is known as *davus* (an important synonym is *philoxenus*). Populations of the race *polydama* with intermediate expression of these spots are found in the central lowlands of Scotland southwards to Cumbria and North Wales in the west and Yorkshire and Lincolnshire in the east. The eyespots are greatly reduced both in number and size in populations of the race *scotica* in the Scottish Highlands and Orkney. The butterflies are also comparatively pale in ground colour.

Some indication of more extensive genetic differentiation than that associated with the phenomena studied in *Maniola jurtina* is given by some limited crossing experiments made by Ford (1949). He reared some 'intersex' specimens from a mating between material from Merioneth in Wales and Caithness in northern Scotland, that is *polydama* × *scotica* (none were observed for a cross of Merioneth and Carlisle in northern England). This form of disruption in the more distant cross indicates substantial disruption of the mechanism of sex determination. Oliver's (1972*a*,*b*; 1979) experiments, which included crosses of *Lasiommata megera* from England and France, quantified in a more complete manner the disruption of features such as the sex ratio and the emergence pattern of males and females commonly found in hybrids between distantly spaced populations. Such effects indicate substantial disruption of development and hybrid inviability, probably as a result of the breakdown of coadapted gene complexes.

A detailed morphometric analysis of material from British collections of *Coenonympha tullia* has been done by Dennis and Porter over the last decade or so. Their latest results (Dennis *et al.* 1986) describe a multivariate analysis of records of spot presence and measurements of spot size and wing area in samples from thirty-one localities in the British Isles

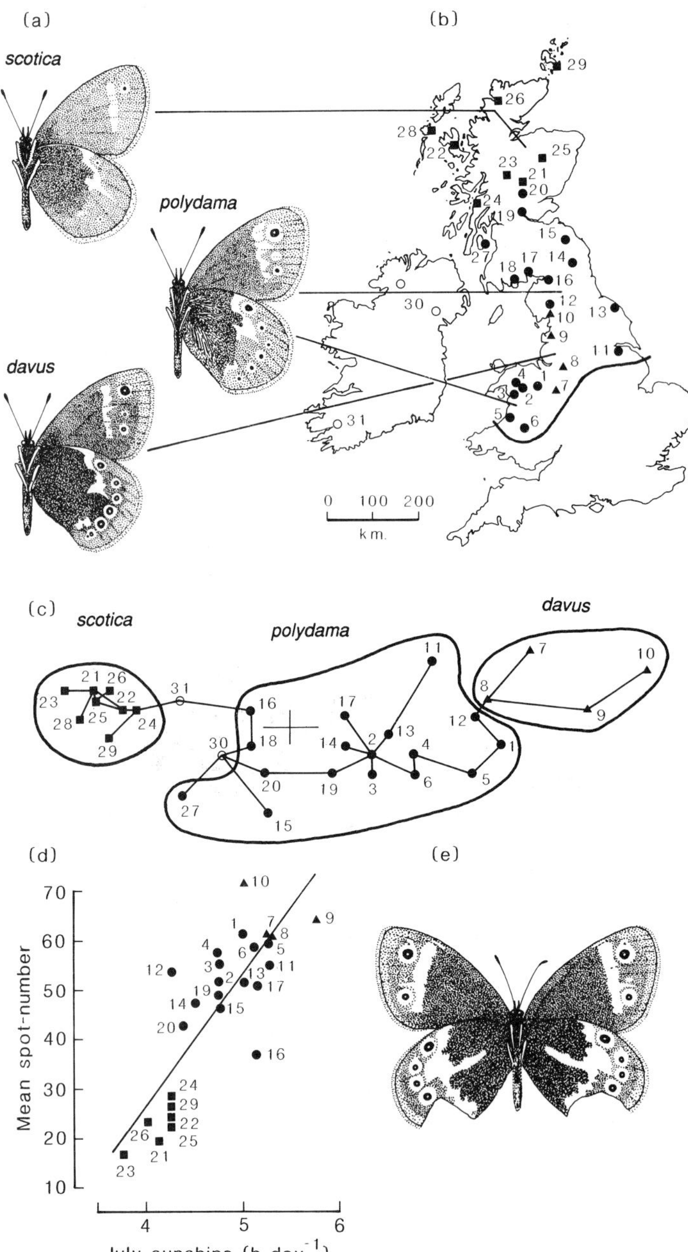

Fig. 9.9 Variation in wing-spotting among British populations of the large heath *Coenonympha tullia*. (Redrawn from sources in Dennis *et al*. 1986; courtesy of *Entomologist's Gazette*.) (a) Examples of the ventral spotting to illustrate the degree of variation from north to south in Britain. (b) Sites from which population samples were obtained for morphometric analysis. (c) A non-metric scaling plot of the male populations for some fifty spot variables; most of the geographical variation can be accounted for in a single dimension. Lines join nearest neighbour populations from single linkage cluster analysis of the same data. Clusters specify traditional subspecies. (d) Relationship between th mean spot-number (index of spot development) for the samples of males and the mean daily hours of sunshine ir July at the sites. The fitted regression line is shown. (e) A male of the race *davus* collected at Whixhall Moss, Shropshire, shows symmetrical damage to the hindwings likely to have resulted from the partial evasion of an attack by a bird or lizard. Forty per cent of the males sampled in 15 of the populations given in (b) had symmetrical wing damage to the hindwings.

(Fig. 9.9b). A non-metric two dimensional scaling plot (see p. 6) produces three clusters corresponding to populations of the three phenotypes associated with the described subspecies (Fig. 9.9c). The clusters to the left and right of Fig. 9.9c include samples from the north of Scotland and from Shropshire, Cheshire, and Cumbria, respectively. Th larger, central envelope comprises the geographic ally more heterogeneous group covering the regio from the central lowlands of Scotland to norther England with North Wales. The two Irish collection appear to be somewhat intermediate between *scotic* and *polydama*. It is interesting that the position within the *polydama* envelope tends to reflect the geographical order either southwards from Scotland or westward into North Wales. This relationship can be seen more clearly within the array for the Scottish Highlands and lowland English mosses.

These findings suggest several possible evolutionary explanations. For example, the local disjunct distribution of *Coenonympha tullia*, throughout much of its range, may tend to mask what is essentially a pattern of equivalent clines in spotting extending both northwards and westwards from the lowland English mosses in response to an environmental gradient. At the other extreme, the three phenotypes may correspond to three (or four) groups of populations which have diverged genetically from each other during a past period of allopatry and isolation. If so, why is there evidence of clinal change within each race? This could represent a common response to similar environmental gradients within each region. The occurrence of intersexes when distant populations are crossed is consistent with this type of explanation. One can then ask whether the divergence in spotting patterns reflects:

(1) a by-product of more general genetic differentiation unrelated to adaptation processes;
(2) pleiotropic effects of selected differences in genes controlling spotting unrelated to the spotting phenotypes *per se*; or
(3) an integral component of adaptation to differences in the environment inhabited by each race.

A further possible complication is that the heterogeneous grouping of populations with intermediate phenotypes could represent a zone of mixing or introgression between the races with extreme phenotypes produced by a spread and mixing of populations resulting from recent human activities.

Unfortunately, as Endler (1977) has emphasized, it is very difficult to distinguish between hypotheses involving past allopatry and differentiation with secondary contact and those concerning responses to present-day selection regimes and environmental gradients (see preceding section). This applies to *Coenonympha tullia* since it is difficult to speculate about the likely historical changes in distribution of the population groupings in relation to both ecological and climatic constraints. If future work demonstrates more profound ecological differences, and confirms the major genetic differentiation between populations of the lowland mosses and those of the Scottish Highlands, then the hypothesis involving secondary introgression becomes more attractive. The often patchy distribution of colonies of *C. tullia* even in regions where it is relatively well distributed makes it rather difficult to distinguish real discontinuities in distribution and gene flow between populations. The northern and southern races of the brown argus, which are now usually given the taxonomic status of species (*Aricia agestis* and *A. artaxerxes*), provide a particularly interesting example of well separated population groupings associated with both genetically controlled differences in wing pattern and other morphological characters and with ecological differences especially in larval hostplants (section 10.5).

The model described in the previous section relating variation in spotting between meadow brown butterflies to differences in adult activity patterns and exposure to vertebrate predators can be applied to *Coenonympha tullia* (Brakefield 1979, 1984; Dennis *et al.* 1984, 1986). The basic components suggest that the combination of a habitat with mixed vegetation and high sunshine experienced by the lowland Shropshire populations will favour the evolution of eyespots for deflection of predator attacks. Figure 9.9e illustrates a specimen from Whixhall Moss in Shropshire with symmetrical damage to the hindwing consistent with deflection by the eyespots of an attack by a bird or lizard away from the body of the insect (see also Fig. 5.1). At the other extreme it can be argued that in northern Scotland a comparatively homogeneous moorland habitat and a cooler climate which reduces adult activity will favour smaller eyespots which enhance crypsis. Analogous to the scenario for *Maniola jurtina*, large eyespots in these conditions might attract predators to resting insects which were for the most part incapable of escape. It is also noteworthy that when other species of heaths on the continent are considered, for example, *Coenonympha arcania*, *C. doris C. gardetta*, and *C. vaucheri*, populations at lower altitudes show a more pronounced spotting than those of more mountainous regions.

Dennis *et al.* (1986) found a close positive association between spotting in *Coenonympha tullia* and the duration of bright sunshine in the adult flight period (Fig. 9.9d). Although there are statistical associations with other climatic variables, that with sunshine is the highest and most consistent with geographical changes in levels of adult activity. An increase in activity with more frequent changes in behaviour and position could in turn shift the balance of selection on wing pattern towards an emphasis on more highly developed spotting functioning primarily in the deflection of attacks by predators. Interestingly, in the *scotica* phenotype the pupils of small spots are silver in colour which may tend to increase their conspicuousness in sunshine by reflective effects. Significant positive correlations between spotting and sunshine have also been found for *Maniola jurtina* in Britain (Brakefield 1979) suggesting that this may be a more general relationship. Dennis *et al.* (1986) summarize the incidence of wing damage in their material of *C. tullia* which was consistent with failed attacks by predators (see chapter 5). There was evidence of a high level of predation by birds, with 43 per cent of males and 58 per cent of females bearing at least one damage mark. They had noted earlier (Dennis *et al.* 1984) that the pattern in the positioning of such damage was consistent with the hypothesis that larger wing spots are the best decoys.

Dennis *et al.* point out that certain populations of *Coenonympha tullia* in Britain, including some with well expressed spots, are associated with grassland habitats at the margins of woodlands or where substantial invasion by birch scrub has occurred. These are habitats where the abundance of foraging birds and heterogeneity of resting backgrounds are greater than in uniform grassland. Interestingly an additional spot frequently occurs at the anal border of the hindwing in populations near Witherslack, Cumbria which is the most wooded locality. This is consistent with the influence of the strongest directional selection for spotting. Shapiro and Cardé (1970) described an increase of spotting in some woodland satyrines of the genus *Lethe* in an area of New York State USA. Brakefield (1979, 1984) suggests that there may be a general tendency for woodland satyrines to be more heavily spotted, possibly because their resting backgrounds tend to be more heterogeneous and less dominated by the linear shapes of grasslands. However, this does not seem to apply to the dry season forms of tropical species which are inactive and rely on a resemblance to dead leaves for survival (see Fig. 5.2). Dennis *et al.* (1986) suggest that more numerous and larger eyespots are favoured in conditions of changing light and shade because butterflies may be less able to monitor, and therefore avoid, approaches by predators and that the low light levels necessitate more contrasting spots to divert predator attention.

Although some interesting explanations of geographical variability in wing pattern have been given to some butterflies such as large heaths, it must be emphasized that tests for the entire chain of causal mechanisms are still required. Evidence that birds can select for small differences in wing pattern or details such as eyespots is discussed in chapter 5. In common with most research on variation in wing patterns where visual selection by predators is implicated, we need to know much more about the precise incidence and mechanics of feeding by predators on living Lepidoptera in the wild. In *Coenonympha tullia* we also need to know much more about its ecology before we can be more confident about our adaptive explanation for variation in wing pattern. There are no quantitative data describing the postulated differences in activity levels between populations of different regions, although they could be readily obtained.

The other major area for which empirical data must be obtained before we can be more definite about the role of natural selection in the observed spatial differentiation of wing pattern is the mode of control of the phenotypic variation. There is no information from breeding work about the inheritance of spotting variation in *Coenonympha tullia*. Indeed Turner (1963) suggested that the distribution of the races of *C. tullia* is to some extent correlated with temperature: for example, on Islay in the Western Isles, Scotland, which has a mean annual temperature comparable to that of Shropshire, a small proportion of individuals approach the *davus* phenotype, with many more of intermediate spotting development. Dennis *et al.* (1986) confirmed the close statistical correlations between spotting and temperature variables for July. Although such effects are quite consistent with the model of selection developed above, there is a need to establish that the development of spotting is insensitive to such direct environmental effects (see previous section and chapter 8).

9.3 The speckled wood: geographic variation in Europe

The studies of *Maniola jurtina* illustrate clinal variation and microevolution, and those of *Coenonympha tullia*, evolution and differentiation in isolation. More substantial variation, in wing pattern and seasonal polyphenism (see section 8.5) is exhibited by the speckled wood *Pararge aegeria* in Europe. This is discussed below in relation to the evolution and maintenance of distinct forms within a continuous distribution.

Within its whole range (Fig. 9.10) *Pararge aegeria*, occupies a variety of habitats, including forest, grassland, and wasteland. Some geographic regions are characterized by a gradual change of the wing phenotype, but there are also more abrupt changes despite an apparently continuous distribution. There are also differences in the life histories of populations from different areas (Thompson 1952; Robertson 1980; Wiklund *et al.* 1983; Shreeve 1985; Nylin *et al.* 1989; E. Lees, personal communication) which may not coincide with changes in the adult phenotype. This whole pattern of variation is not easily explained by either clinal variation in response to present-day environments or by adaptation to past isolation but rather by some combination of the two.

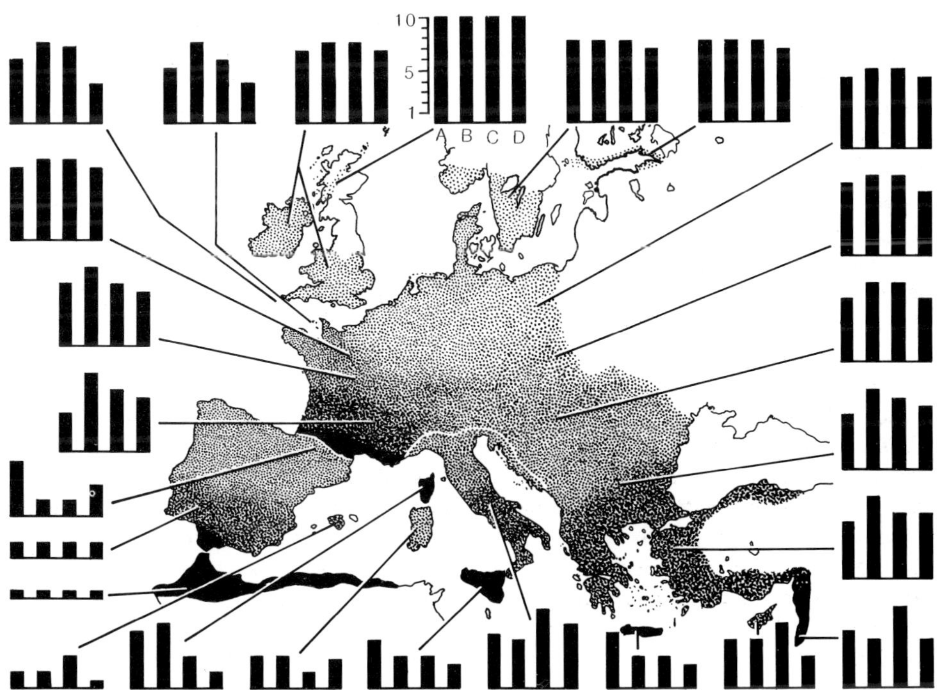

Fig. 9.10 Geographic variation in the phenotype of the speckled wood *Pararge aegeria* in Europe. The three main forms, *P. aegeria aegeria* in Iberia and North Africa, *P. aegeria tircis* over most of Europe, and *P. aegeria italica* restricted to the Italian peninsula, are clinally variable. Most Mediterranean island forms resemble adjacent mainland forms, but those from Sicily, Crete, and Cyprus are probably unique. Size, wing shape, pattern, and colour of the various forms are represented on an arbitrary scale from 1 to 10 by histograms. A. Size: 1, smallest (Africa), 10, largest (Scotland); B. wing shape (forewing angularity): 1, most indented (Africa), 10, least indented (northern Europe); C. pattern (dorsal pale patches): 1, most extensive (Africa), 10, least extensive (Scotland); D. dorsal patch colour: 1, most orange (Africa), 10, pale cream colour (Scotland). Subspecies *tircis* is the most seasonably variable. Distribution in eastern Europe is not well documented but may extend to the Ural mountains and Pripyat marshes.

9.3.1 Species, subspecies, and geographic variation

The three species within the genus *Pararge* (see Higgins and Riley 1983) are distinguished by their wing shape and colour. *P. aegeria*, which occurs throughout mainland Europe, western Asia, North Africa, and Madeira is the smallest and most variable species. *P. xiphia*, is the largest and darkest species and is restricted to Madeira. *P. xiphioides*, which occurs on the Canary Islands is intermediate in size and colour between *P. xiphia* and *P. aegeria*. Numerous races of *P. aegeria* have been described by various authors as varieties or subspecies (e.g. Verity 1916; Gaede 1931), but two subspecies are universally recognized, *P. aegeria aegeria* from southern Europe and *P. aegeria tircis* from northern Europe. They are distinguished by the colour of the pale areas on the upper wing surfaces; in the former these are reddish orange and in the latter creamy yellow. The southern European form also has more scalloped margins on the forewing.

Within the range of the southern *aegeria* form there is obvious clinal variation, specifically an increase in wing expanse and a decrease in the intensity and size of the reddish orange patches on the upper wing surface northwards and with altitude. The wing shape appears constant and there is little indication of seasonal polyphenism. In southern Spain and in Africa the species is continuously brooded but is least abundant in the dry summer months. Within this area individuals from the higher parts of the Pyrenees are strikingly different, being much larger, with more hair-scales and darker wing coloration. Presumably it is a response to thermoregulatory needs in the cooler environment (see section 2.2).

The northern *tircis* form is more uniform in wing shape and pattern throughout its range. It is usually seasonally polyphenic, spring adults having larger and more extensive pale wing areas than those flying in summer. This variation is more extreme in the north. However, populations in eastern Finland, central Sweden, and probably north-western Russia have only one generation a year (Wiklund *et al.* 1983), and may therefore be less variable (see also section 8.5). Although coloration is one of the features which distinguishes *tircis* from the southern European *aegeria*, this coloration is variable. At low altitude in southern parts of its range it is similar in colour, but not pattern, to the southern form. At northern latitudes and at high altitude in southern areas the phenotype of *tircis* is very similar to that found in southern Britain.

Individuals from peninsular Italy are different from those of all other mainland European areas, both in wing shape and pattern, though this phenotype is also located on certain Aegean islands, in Syria, and Lebanon. This similarity of Italian and Middle-Eastern races was first commented on by Verity (1916) but Larsen (1974) disputes this view. He points out that current geographic distribution patterns cannot support any inferred similarity between Italian and Middle-Eastern races. The phenotype does not extend into the Alps but does occur along the northern Dalmatian Coast.

On each of the Mediterranean islands the phenotype is unique but on most it has a close affinity to the southern *aegeria* form. Interestingly, those located on Corsica and Sardinia, which are only 25 km apart, are very different, resembling *tircis* and *aegeria*, respectively. Neither resemble those from Italy, the nearest mainland area.

9.3.2 Adaptations and evolutionary arguments

The complex pattern of geographic variation does not always follow geographic boundaries and is not easily explained, particularly the abrupt boundaries in southern France, and in northern Italy (see Dennis *et al.* 1991).

However, clinal variation in the wing coloration within each of the northern and southern European subspecies follows a consistent pattern which, together with seasonal polyphenism, can be explained by adaptations to present-day environmental gradients. In each, the pale wing areas become more orange southwards and with lower altitude, a change which can be related to background matching. In a woodland habitat in Britain more than one half of individuals are damaged in a manner which is consistent with having been attacked by birds (Shreeve 1985). An examination of populations in 25 different sites in France reveals a similar incidence (Shreeve, unpublished data). Therefore, an appropriate form of crypsis may substantially enhance survival (see chapter 5). When settled in a potentially active state, the butterfly basks with open wings and when the general wing pattern matches that of the background the butterfly is difficult to detect. Although the effectiveness of the various colour forms on different backgrounds needs to be tested, orange coloration probably affords better crypsis in drier southerly, low-lying, habitats. A reduction in the contrast between dark and light wing areas of *aegeria* may decrease conspicuousness in more open habitats where the contrast between light and shade is less extreme than in densely wooded areas. This may also account for the development of orange coloration on the upper surfaces of individuals from the Isles of Scilly, described by Howarth (1971) as subspecies *insula*. Observations of the species in France (Shreeve, unpublished data) demonstrate that in southern areas individuals spend more time resting on bare earth and dead leaf litter than on vegetation, but in more northern areas the main resting background is vegetation, usually in woodland. Reduction of the wing area occupied by the pale patches with increasing latitude and altitude can be directly related to thermoregulatory constraints (see chapter 2), the most melanized individuals occurring in the coolest and cloudiest regions, such as the higher Pyrenees.

The geographic pattern of seasonal polyphenism can be related to patterns of habitat use and seasonal changes in these habitats. In southern regions, such as the southern part of the Rhône valley, the most frequently used habitats are open evergreen forest and ditches, and water-courses in very open agricultural areas. Unlike most northern habitats, which are chiefly dense deciduous and coniferous woodland, these are probably less seasonally variable in both colour and light intensity. Therefore, seasonal polyphenism, which may enhance background matching, is likely to be generally favoured in the north but not in the south.

Individuals from three Mediterranean regions—Spain, Italy, and Greece—differ, though they are in similar climatic zones. Three explanations may be offered to account for these differences. Firstly, they may all originate from a single source in the southern range of the species, variation between these geographic areas being related to local adaptation to differences in climate, predators, and habitat. Secondly, the present distribution of forms may be the result of colonization by distinct forms from two or more areas since the last major ice retreat some 15 000 years ago, each adjusting to current conditions. Thirdly, variation may be the consequence of phenotypic plasticity with different environmental and/or climatic conditions producing different phenotypes (see section 8.5).

At present it is impossible to provide a definitive explanation for the pattern of variation throughout Europe or to provide an accurate post-Pleistocene framework of range expansion (see Dennis *et al.* 1991; section 10.5). However, it is possible that the species has spread from more than one geographic area. Examination of museum material and field work in July 1987 and 1988 reveals an apparent absence of phenotypes intermediate between *Pararge aegeria aegeria* and *P. aegeria tircis* in their main contact zone in southern France. In the Rhône valley the former subspecies is located in agricultural areas at low elevation and the latter at higher elevation in both agricultural and woodland habitats, the two forms occurring some 1–2 km apart. Similar patterns of distribution are recorded for other species and subspecies in southern Europe (see Dennis *et al.* 1991). This lack of intermediates may be explained by abrupt habitat differences but a comparison between chromosome numbers of *P. aegeria aegeria* ($N = 27$) and *P. aegeria tircis* ($N = 28$) (Robinson 1971) points to genetic dissimilarity between the two subspecies, perhaps indicative of different geographic origins. The lack of intermediates may then indicate hybrid inviability or the existence of pre-mating barriers. However, the degree of genetic differentiation of these two forms is not quantified. It is possible that the phenomenon may be a more extreme example of the genital valve boundary phenomenon described for *Maniola jurtina*. The distance between *aegeria* and *tircis* is not great and may be traversed by flying individuals. Furthermore, these observations were made at the driest time of year when population density was low and damp areas for egg-laying scarce. At other times, particularly in spring, population density may be higher and with more abundant and widespread suitable larval resources, the distance between *aegeria* and *tircis* may be less.

The distribution of the various forms of *Pararge aegeria* is currently changing. In southern France there is some evidence that the southern European form is expanding its range northwards (Robertson, personal communication) perhaps in response to changes of landuse. In Sweden there are two geographically separate races of different origin, a univoltine race from Finland and a bivoltine race from Denmark, the latter being recent colonists (Nordstrom 1955).

Pararge aegeria aegeria is a recent colonizer of Madeira, being first found on the island in 1967 by Høegh-Goldberg (Oehmig 1980). The impact of this colonizer on the endemic *P. xiphia* is of considerable interest. Lace and Jones (1984) suggest that it may replace the endemic species, citing the disappearance of *P. xiphia* from low-lying areas in the south of the island as tentative evidence of the 'taxon cycle' (Wilson 1961; Ricklefs and Cox 1972). In this cycle a recent colonist is supposed to be competitively superior to a resident species because the invading organism has escaped from its natural enemies whereas the resident has a suite of enemies. This effect is presumably thought to outweigh the closer adaptation of the resident to the environment.

More recent work (Owen *et al.* 1986; Jones *et al.* 1987; Shreeve and Smith, in press; Lace and Jones, in press) provides evidence of limited habitat partitioning and overlap and of potential competitive interactions between the two species. Because *Pararge xiphia* is more active in cool, and *P. aegeria* in warm temperatures (Shreeve and Smith, in press), the distribution of the two species changes with the season. *P. xiphia* tends to be most abundant in cool, high altitude laurel forest in summer but more dispersed into more open, lower pine forest and agricultural areas in winter. *P. aegeria* tends to be more common in low altitude warm habitats. Although there may be interactions between individuals of the two species in relation to mate-location (Lace and Jones, in press), competitive displacement of adults is unlikely as the larger endemic species tends to occupy canopy layers and the colonizer ground layers while engaging in patrolling and perching (see chapter 2). There is a possibility that interactions during development, perhaps via shared natural enemies have an effect on the dynamics of any interaction between the two species: larvae of both species can occur on the same foodplant (*Brachypodium sylvaticum*) when the two occur together. Whilst the exact nature of the interaction between these two species remains in doubt, changes in agricultural practices may well have caused the loss of *P. xiphia* from low-lying areas before colonization of the island by *P. aegeria aegeria*.

The time of origin and establishment of *Pararge xiphia* on Madeira is not known. It is probable that this species, and *P. xiphioides*, are derived from *P. aegeria* or *P. aegeria*-like ancestors possibly from North Africa or southern Europe, the isolated populations then evolving into distinct, reproductively isolated species. The marked dissimilarity of these species from *P. aegeria* imply a relatively long period of isolation. However, the strength of past and present selection on the adult phenotype is not known; intense selection can cause rapid change and phenetic differentiation in isolation can be rapid (Dennis 1977). If change has been rapid it is possible that divergence of wing pattern of the different island species has occurred over a few thousand generations, though differences of other features of both island species, such as genitalia (Higgins 1975), larval morphology (Shreeve, personal observation), behaviour, and thermoregulation (Shreeve and Smith, in press; A. Smith, T. Shreeve, and M. Z. Baez, unpublished data) suggest a longer period of isolation (see section 10.5). Some of the Mediterranean island races are probably of more recent origin than the Atlantic island species. They may be the result of either recent colonization events, as perhaps in the case of the Balearics, or originate from earlier Pleistocene invasions.

Pararge is not unique in its complex pattern of variation and probable diverse origin of geographic races. Other equally complex patterns occur, for example in species of *Melanargia* (Higgins 1969; Descimon and Renon 1975; Wagener 1982; Mazel 1986; Tilley 1986), *Erebia* (Lorković and de Lesse 1954), *Lysandra* (de Lesse 1960, 1969, 1970), and *Hipparchia* (de Lesse 1951). In *Erebia* there are identifiable hybrid zones.

In this chapter studies of three species have been examined in detail to illustrate the processes of divergence, differentiation, and the formation of reproductively isolated forms of species. These issues are put into the wider historical perspective in chapter 10.

10

An evolutionary history of British butterflies

Roger L. H. Dennis

Reconstructing butterfly history is a speculative process, as butterfly fossils are rare, but none the less it is rewarding. First, each butterfly makes a unique set of demands on resources and these taken together with geographical opportunity, account for its distribution. Secondly, a wide array of geological and palaeontological techniques and an abundance of other fossil life-forms have provided a detailed historical template of past environments for butterfly evolution. Thus, data on butterfly biology and geography can be matched against environmental conditions that have changed markedly in time and space (Dennis 1977, in preparation).

No organism developed all its attributes overnight at the same moment in the past. Nor, despite the fact that evolutionary processes such as genetic drift and selection are ongoing, is the rate of change constant, either for different species or for different attributes in a single species. Butterflies reveal some features which are ancient as well as others which are constantly developing. Isolation of fauna, given time, is understood to result in evolutionary distinctions, ultimately new species. As such, features held in common by island and mainland butterflies probable pre-date their separation, whereas features unique to island faunas have usually evolved in isolation. These premises are not absolute (see p. 242) but, as no butterfly species is unique to Britain (unlike species on many Mediterranean islands), it is highly improbable that any British species, nor aspects of their basic biology, such as physical pattern and colour, behaviour, their hostplant family affinities, and basic elements of population structure, evolved here. On the other hand, aspects of their distributions in Britain, subspeciation and related physical modifications, unique genetic markers, and any associated hostplant, behavioural and physiological specificities probably have evolved in Britain. This useful distinction allows us to examine the development of British butterflies in two sections, namely that before and that after their latest arrival. Isolation and elimination of the British fauna has occurred frequently during the past two million years. We know it to be the result of Pleistocene glaciations, a series of repeated extraordinary climatic events and landscape transformations.

10.1 Evolution before glaciers

Butterflies are of great antiquity, although this is perhaps not the impression that might be gained from works on the butterfly fauna of large regions (Smart 1976; Higgins and Riley 1983) where the many species condense into far fewer pattern types. By the Eocene (54 million years BP; see Table 10.1) at least three of the modern major families had developed (Papilionidae, Nymphalidae, Lycaenidae) and by the Lower Oligocene (36 million years BP) all of them had evolved, long before man's hominid ancestors had emerged. Indeed, most butterfly families and subfamilies probably predate the Cretaceous–Tertiary (K–T) boundary. Rich locations for fossils have been the fine-grained sediments at Florissant in Colorado, Aix-en-Provence in France, and Radoboj in Yugoslavia (Leestmans 1983; Durden and Rose 1978). Moths date back much further, probably to the Permian (Shields 1976); one from the Lias clay of Dorset dates to 180 million years ago (Whalley 1986). The earliest butterflies were originally thought to have evolved from a pre-Tertiary Castnioid line (Brock 1971; Tindale 1985) but this is

Table 10.1 Generalized geological time scales with environmental and evolutionary features. (a) Geological time scale for the last 4000 million years including some relevant evolutionary events (Simmons 1979; Brown 1987). (b) Traditional terminology for glacial and interglacial stages for the British Quaternary (Godwin, 1975).

Analyses of: oxygen isotope ratios, CO_2 concentrations and microfaunal changes from deep cores through marine sediments and similar chemical spectra from cores through ice sheets have revealed some 46 glacial–interglacial cycles over the past 2.5 million years (Bartlein and Prentice 1989; Cronin and Schneider 1990). The quasi-cyclical growth and decay of ice sheets is explained by the Milankovitch theory of climatic change. Periodicity in three elements of the earth's orbit and axial inclination allow for the calculation of past insolation levels which correspond closely with inferred climatic changes in marine basin and ice stratigraphy. Asynchrony between elements of the Milankovitch model, feedbacks and phase lags between insolation, climatic changes and ice accumulation prevent a strict repetition of glacial–interglacial conditions.

(a)

Era	Duration (million years)	Period/Epoch	Life forms
	0.01	Holocene	
Quaternary	2	Pleistocene	Man, modern butterfly subspecies
Cenozoic	3	Pliocene	Modern butterfly species
	20	Miocene	Mammals, first hominids
Tertiary	13	Oligocene	Modern butterfly genera
	17	Eocene	All butterfly families
	12	Palaeocene	
		67 million years ago	
			Flowering plants dominant
	77	Cretaceous	First butterflies
Mesozoic	69	Jurassic	Reptiles, first flowering plants
	35	Triassic	Moth families developing
			First cycads, conifers dominant
		248 million years ago	
	38	Permian	Radiation of reptiles
Palaeozoic	74	Carboniferous	Insects, land plants
	48	Devonian	First amphibia
	30	Silurian	First land plants, fishes
	67	Ordovician	Marine jawless armoured vertebrates
	85	Cambrian	Algae, marine invertebrates
		590 million years ago	
	4000	Precambrian	Traces of bacteria, fungi, algae, back to 3000 million years

(b)

Stage	Climate
Upper Pleistocene	
Flandrian	Temperate
Devensian	Cold, glacial
Ipswichian	Temperate
Wolstonian	Cold, glacial
Hoxnian	Temperate
Middle Pleistocene	
Anglian	Cold, glacial
Cromerian	Temperate
Beestonian	Cold
Pastonian	Temperate
Baventian	Cold
Lower Pleistocene	
Antian	Temperate
Thurnian	Cold
Ludhamian	Temperate
Waltonian	Cold

contested by Scott (1986) who associates butterfly evolution with some branch of the Macrolepidoptera (see Brock 1990; Scott and Wright 1990; Scoble 1991).

Many of the pre-Pleistocene butterflies that have been found bear a striking resemblance to modern forms. *Vanessa karaganica* and *Pyrameis fossilis* (Upper Miocene, N. Caucasus), show close similarity to present day nymphalids such as the small tortoiseshell *Aglais urticae* and the painted lady *Cynthia cardui* (Nekrutenko 1965). Apparently, fossil species can be assigned to modern day biotic provinces (Scudder 1875; Comstock 1961; Zeuner 1962), additional clues being provided by the associated flora. Butterfly evolution is thus characterized by several features; great antiquity, early radiation of families, and affinity of modern to ancestral forms. Some interesting problems are waiting to be solved. Firstly, despite the early radiation of butterfly families (Miller 1968; Smart 1976), their diffusion throughout the ancient landmasses and continents has not occurred equally. What prevented this? Secondly, when did the temperate species evolve and why is there a distinct reduction in the number of species from the tropics to high latitudes? Thirdly, what factors underlie restricted 'pattern' development especially amongst temperate species?

These issues occurred well before the return of butterflies to the British islands after the last glacial episode (18 000 years BP). Instead, they occurred amidst massive biogeographical disruption on the planet, from the Cretaceous–Tertiary (K–T) event and the extinction of the dinosaurs 66.5 million years ago to mid-latitude temperate glaciations during the past two million years. The existence of indigenous taxa on various continents has its explanation in plate tectonics and thus in the post-Cretaceous separation of continents and the consequent transformations in climate and ecology (Smith *et al.* 1981).

It is not yet possible to date the evolution of modern temperate species but their origin too may be linked closely to plate tectonics. What is certain is that cooler climatic regimes in the northern landmasses date from the Eocene epoch (Barghoorn 1965; West 1968; Spicer and Chapman 1990). Such environments developed with mountain-building during the alpine orogeny; but lower temperatures at sea-level could relate to the polar drift of land-plates or to changes in solar and other external influences. The change in climate provided the environmental setting for new adaptations (Dennis, in preparation).

Numerous elegant explanations have been put forward for the paucity of temperate fauna and flora, not just butterflies. Generally, they divide into four groups of hypotheses: structural—the tropical environment contains more niches (Scriber 1984*a*); dynamic—levels of interaction between biota (competition and predators) induce higher rates of speciation or make more niches available; energetic—greater stability and high production in the tropics permits greater specialization; and historical—climatic changes associated with glacial cycles regularly effect extinctions in high latitudes (see Simmons 1979; Turner *et al.* 1987). Fundamental to all of these, directly or indirectly, is Wright's (1983) 'species energy' hypothesis. This relates species diversity in terrestrial habitats directly to the amount of solar energy available (see chapter 1 and Dennis, in preparation) and is relevant at the scale of the British Isles (Dennis 1977, 1985*a*; Dennis and Williams 1986; Turner 1986; Turner *et al.* 1987) and for even smaller areas.

One of the most interesting aspects of butterfly biology is the conservatism of pattern, form, ecology, and behaviour within butterfly families. Part of the reason for this stability and restricted pattern of development in butterflies may be inherent, that is they are due to the existence of universal genetic structures among different taxa (see chapter 8). The fact that Batesian mimics can be produced at all implies the existence of underlying structures which can be modified for the assemblage of convergent patterns. However, part of the explanation may be external to the butterflies and dependent on wider ramifications of insect–hostplant 'coevolution' than has hitherto been expressed. As we have seen, butterflies use few plant taxa, with related groups of butterflies tending to feed on related groups of flowering plants (see section 7.4.4). Whatever the historical scenario associated with 'coevolution' (Jermy 1976, 1984; Edwards and Wratten 1980; Strong *et al.* 1984), butterfly dependence on hostplants is absolute. Not only has the need to overcome chemical defences led to hostplant family specialization, but a number of taxa are now known to use and rely on these chemicals for their own defence against predators and parasitoids, as kairomones for hostplant location and oviposition and as pheromones in mate-location and courtship.

But the evolution of insect–host plant associations probably involves much more than the extension of a chemical arsenal. Hostplants differ in life-form (bryophytes, herbs, shrubs, and trees), ecological context (biotope, seral stage, and lifespan), and physical conditions (temperature, moisture, nutrient supply, pH, and light); butterflies, as plant specialists, are contextually constrained. Hostplant family specialization has led not just to chemical adaptation and application but also to ecological and behavioural evolution within the ecological confines of the hostplant group. Wiklund (1981) discusses 'finer tuning' specializations associated with monophagy. Butterflies feeding on different plants have their own choice of habitats restricted. Thus roost-sites, hibernation sites, nectar, predators, parasites, and potential symbionts are often determined in large part by the hostplant habitat. Aspects of oviposition, phenology, and mate-location will also be determined by the hostplant habitat. How eggs are laid (e.g. singly, large or small batches; edge sites) and how the ovipositing females behave will depend much on the geography, condition, and predictability of the plant (e.g. spacing, clumping, quantity, apparency, phenology, toughness, and nutrient status). The production of new growth by the plant, making available high levels of nitrogen, water, and allelochemics as opposed to low resin, tannin, and tough plant tissues, often dictates the period when early stadia grow (Slansky 1974; Scriber 1984*b*). The distinctions between different forms of mate-location, such as between invariant patrollers compared to localized perchers and territoriality, may depend more on the physical structure of the (ancestral) hostplant habitat, that is on topographic vantage points such as edge- and junction-sites, than on other aspects of the geography of larval resources and adult population structure (see Fig. 2.5; Dennis and Shreeve 1988).

Hostplant habitat carries with it indirect influences which may further bond it to butterfly biology. A butterfly's morphology, colour pattern, and physiology are determined by predators, parasitoids, symbionts (Beattie 1985), nectar, and the need to thermoregulate, each of which is in part contextural in hostplant habitat and chemistry (see chapter 5). The ramifications of cause, effect, and feedback are far-reaching and developed long before the final arrival of butterflies on British shores. Ultimately hostplants, through their predictability in space and time, influence butterfly dispersal and, in turn, population and genetic structure, all of which have considerable significance for their modern day survival. Although generalizations on butterfly biology still evade us, these various connections and interrelationships based on hostplant habitat structure promise greater understanding and may partly explain the conservatism of pattern among butterfly taxa.

10.2 Evolution with glaciation

The Pleistocene is characterized by abrupt and intense cyclic changes in climatic and environmental conditions. In Britain, invasion of much of the fauna and flora occurs during each warm phase (interglacials; interstadials) and extinction with the return to cold (glaciations; stadials) (West 1969; Table 10.1), but cold-tolerant fauna reacts in the opposite manner. The slate is never quite wiped clean during glaciations; techniques in the geological sciences allow us to determine whether or not particular species have survived from previous warm phases. Some obvious questions follow. Did our current butterflies survive the succession of glacial stages? In response to what pressures did unique adaptations in British butterflies evolve? Firstly, some insight is required into environmental history.

10.2.1 Reconstructing the past

Few butterfly fossils have been found in Britain. Those that exist such as *Nymphalites zeuneri* (Nymphalidae) and *Lithopsyche antiqua* (Lycaenidae) from the Bembridge Marls in the Isle of Wight, which date to the late Eocene or early Oligocene circa 35 million years BP, belong to sub-tropical biota and pre-date the Pleistocene (Figure 10.1; Jarzembowski 1980). Establishing the modern day origins of British Lepidoptera requires two things; first, a template describing changing conditions over Britain and Europe; second, critical thresholds for each butterfly species for climate, hostplant, and habitat. An historical reconstruction, then, depends on matching species' thresholds to past environments (Dennis 1977, and in preparation). The process is feasible, firstly because butterflies are sensitive to changes in physical conditions, especially to temperatures, and secondly, because environmental conditions have changed markedly, but generally in phase in different regions, over the past two million years.

Data on past environments are acquired mainly from minerogenic and biogenic sediments, analysed by sophisticated sampling, levelling, identification, and dating techniques (West 1968; Tooley 1981). Environmental reconstruction is not without difficulty. Past climates can be determined indirectly only from the distribution of present life-forms. This determination is based on the adaptive stability of life-forms. Migrational lag of fauna behind plants and changes in climate affects these estimates as does the influence of man and animals on vegetation. Radiometric dates carry various errors (Pearson *et al.* 1977; Evin 1990). Furthermore, due to adaptive differences among life-forms and human destruction of habitats, few present analogues of past environments exist for the interpretation of fossil assemblages (Huntley 1990).

10.2.2 Past environments

Numerous glaciations, interglacials, stadials, and interstadials have occurred over the past two million years (Bowen 1978; Denton and Hughes 1981; Bartlein and Prentice 1989), all of which must have made for substantial changes in butterfly biogeography. But we need concentrate only on the latter half of the last glacial stage (the Devensian) (Table 10.1, Fig. 10.2). During the last 45 000 years, ice has advanced twice from the mountainous regions in Britain; first around 18 000 years BP, when it covered northern Britain and extended to north Norfolk, South Wales and southern Ireland; subsequently for a short period between approximately 11 000 years and 10 200 years BP. At 18 000 years BP conditions were as severe as any previous glacial stage. During the shorter period, Zone III, ice advance was restricted to the Grampian highlands and to corrie glaciation in other upland areas. During both stages, ice areas and their margins were reduced to polar desert, and beyond the ice lay steppe and tundra, characterized by landscapes of permafrost, with pingos, ice wedge polygons, stone polygons, and involutions (Williams 1975; Watson 1977; Mitchell

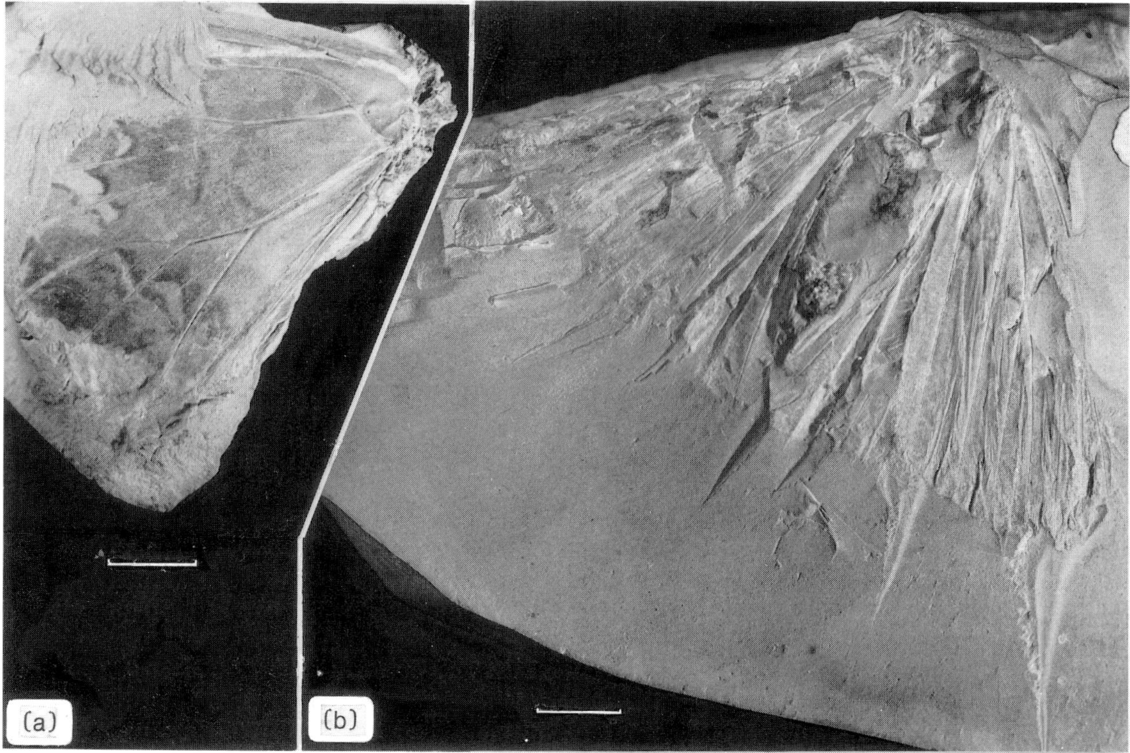

Fig. 10.1 Fossil butterflies from the Bembridge Marls of the Isle of Wight, southern England, dating to the Late Eocene or Early Oligocene, *c*.35 million years BP. Scale line = 5 mm. (a) *Nymphalites zeuneri* Jarzembowski (Nymphalidae); (b) *Lithopsyche antiqua* Butler (Lycaenidae). (Courtesy of E. A. Jarzembowski, and the Natural History Museum.)

1977). Short summers were dominated by meltwater, mudflows, braided streams, and solifluction, caused by volumes of meltwater unable to penetrate the perched water-table (Fig. 10.2).

During glacial and arctic phases, plants were typically either hardy, low-growing, woody, and desiccation-resistant perennials (e.g. *Calluna*; ericales) or short-lived annual herbs, grasses, obligate halophytes, and bryophytes. Ponds and marshes were dominated by rushes, sedges, and *Sphagnum* communities and the only trees—birch and willow (*Betula nana; Salix herbacea*)—were dwarf species adopting prostrate life-forms. A return of temperate flora, shrubs, and trees occurred after 13 000 BP, but these were forced to retreat after 11 500 years BP only to complete colonization in the Flandrian (Pennington 1969; Godwin 1975).

Three main influences dominated Flandrian Britain; immigration of plants and animals, vegetation succession, and humans (Table 10.2). The first two resulted in a sequence of 'communities' enveloping the landscape in a climatic-climax regionally diverse forest (7000–5000 years BP) and limited open grazing for ungulates to higher altitudes (Bennett 1989). Humans, developing fast in technology (stone to iron), economy (nomadic hunting and gathering, transhumance and pastoralism, and settled farming), and in population from the Mesolithic, reversed the process. Their influence on wildlife has been little short of devastating. The tally is virtual clearance of our natural forest cover, impoverishment of habitats, particularly those of uplands in northern and western Britain, and extinctions and introductions of the fauna and flora. Interference began early with firing of upland and other habitats by humans in the Mesolithic period to

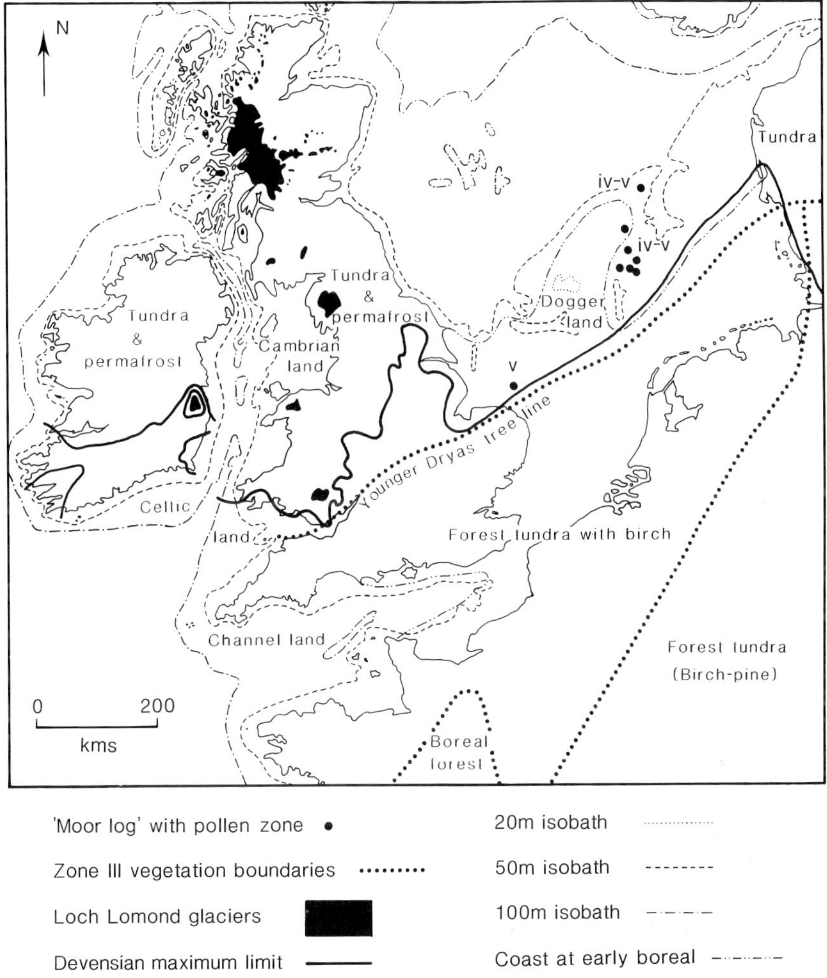

Fig. 10.2 The Devensian maximum limit and British environments during the Younger (Upper) Dryas (Loch Lomond Readvance). *c*.11,000 to 10,000 years BP. (After Grichuk 1973 in Starkel, 1977; Godwin 1975; Dennis 1977; Mitchell 1977; Watson 1977; Frenzel 1979.)

improve hunting, but prerequisites for wholesale clearance were iron technology, the plough, and changes in emphasis in domesticated animals from cattle and pigs to sheep. By the end of the Iron Age, in many areas, the essential lineaments of modern landscapes had emerged (Spratt and Simmons 1976; Table 10.2). Extinctions of large mammals and birds in the late Devensian and Flandrian (woolly rhino, mammoth, reindeer, elk, and auroch) continued through medieval times (brown bear, wolf, and beaver) to the modern day (great auk). Closely affiliated is the contraction in the distribution of many species of mammals (e.g. wildcat, pine martin, and polecat) and of birds (e.g. red kite), which used to occur in southern Britain (Simmons and Tooley 1981).

Corresponding with these transformations in habitat were substantial changes in climate. During the Devensian maximum and Zone III, mean summer temperatures were reduced to less than 8 °C in the Midlands; ice wedges point to mean annual temperatures less than −6 °C with January means well below −20 °C. One supposed 'refuge' area was south-west Ireland (Beirne 1947) but even during

Table 10.2 Late Glacial and Holocene environments (terminology, climate, human cultures, vegetation changes, hydrology, and sea levels) and biogeographical and evolutionary events in the British butterfly fauna

Radio-carbon (years BP)		Blytt–Sernander units	Godwin zones	Cultures	Climate: Generalized curve for average summer temperatures	Main vegetation features: Ireland	England/Wales	N Sco
					8 10 12 14 16 18 °C			
0	Holocene (Flandrian Interglacial)	Sub-Atlantic	VIII	Industrial & Agricultural Revolutions	Greenhouse scenario warming	Destruction of wildscape and wildl		
					'Little Ice Age' cooler and wetter	Clearance of woodland for cultivat		
				Middle-Ages Norman	Warm dry period			
				Anglo-Saxon	Warmer and drier	Alder–birch–oak	Alder–birch–oak–(beech)	L w
				Romano British				
2000				Iron Age	Cool and wet			
		Sub-Boreal	VIIb	Bronze Age	Little evidence of substantial changes, perhaps slightly cooler and wetter in north and west. Evidence or dry episodes in lowland Britain	Alder–oak	Alder–oak–lime	
4000				Neolithic		Beginning of large scale clearance (elm decline)		
6000		Atlantic	VIIa	Mesolithic	Warmer summers than today by 2 or 3 °C Winter temperatures 2 °C above present. Rainfall 10 per cent higher than today	Hazel–alder–oak elm–pine	Alder–oak–elm–lime	
		Boreal	VIc		Increasing oceanicity	Hazel–pine	Pine–hazel (mixed oak forest)	
8000			VI b a	Proto Maglemosian	Cooler winters than today rainfall 5 to 7 per cent less than today			
			V			Hazel–birch	Hazel–birch–pine	J en
10 000		Pre-Boreal	IV			Birch	Birch–(pine)	
	Late Devensian	Younger Dryas	III	Upper Palaeolithic	Arctic conditions	Open ground with forbs, grass, and dw	Birch in south-east	
12 000		(Windermere interstadial) Allerød	II			Open ground (forbs, grasses, sc with birch 'parkland'		
						Tundra vegetation with grasses sedges, rushes, forbs, woody perennials, and scrub		
14 000		Older Dryas	I		January temperatures less than −20 °C			
					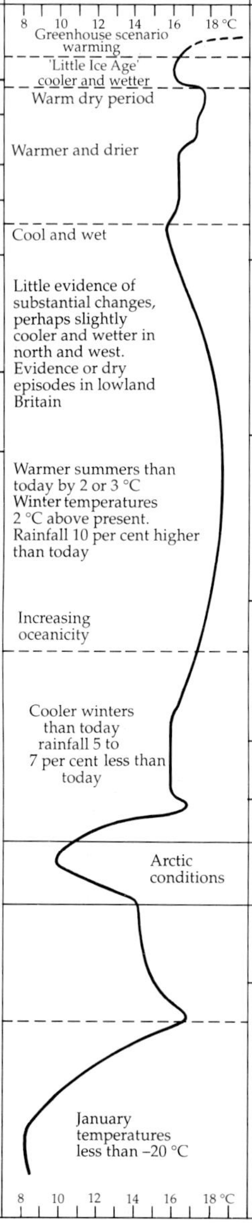 8 10 12 14 16 18 °C			

Status of butterflies for different periods given in the order of: climate/habitats; +, beneficial; −, detrimental; Δ, changing to conditions indicated.

After Pennington (1969); Evans (1975); Godwin (1975); Dennis (1977), and in preparation; Simmons and Tooley (1981); Musk (1985); Rackham (1986); Atkinson *et al.* (1987); Bennet (1989); Dennis and Shreeve (1989); J. A. Thomas, personal communication.

...centage ...st cover (50 – 100)	Coasts, lakes and mires	Status of butterflies typical of: tundra and montane habitats	Status of butterflies typical of: open habitats, scrub and clearings	Status of butterflies typical of: mature forest habitants	Evolution and adaptations in British butterflies	Radio-carbon (years BP)
Plagioseres (heath, short turf, cultivated landscapes) created by humans		−/−	+/Δ+	+/−	Greenhouse conditions: shift to cooler, shaded, damper habitats	
	Granlund's recurrence surfaces	Δ−/Δ− Contraction under intensive land use	+/− Losses	Δ+/−	butterfly extinctions	0
	I	Δ+/Δ− Δ−/Δ−	−/+ Expansion in dry climate and with coppice and open fields	−/− Possible extinction of high forest species +/Δ−		
	II Sea-level oscillations around OD III IV Fenland marine transgressions V	Δ+/Δ+ Expansion of cold-tolerant species but not of alpine species Δ+/Δ+	+/+ Δ−/Δ+ Withdrawal in north and west. Expansion with deforestation and early woodland management in lowlands	Δ+/Δ− Significant decline as deforestation rate increases Δ−/Δ− Beginning of habitat decline	Increased dependence of species on warmer habitats: south-facing slopes and plagioseres maintained by humans (J.A.T.). Clinal adaptations in voltinism, physiology and wing patterns to cooler, damp conditions in north-west.	2000 4000
	Wide extension of ombrogenous mires	−/− Minimal range of British alpine, and cold-tolerant butterflies	+/Δ+ Expansion in areas of hunting activity. Beginning of withdrawal from uplands +/−	+/+ +/+ Gradual withdrawal from north and west Britain	Beginning of withdrawal of species from uplands in highland Britain	6000
	Severance of Britain from Europe Rapid-rise in sea-level Low lake levels North Sea dry	−/− Extinction of remaining arctic species	Increased dependence on habitats created by humans +/− Forest encroachment limiting habitats +/Δ−	Withdrawal from uplands Optimal period for warmth-demanding butterflies of climax forest +/+	Evolutionary changes with forest closure in species dependent on open habitats: genectic divergence in isolated populations. Specialization on fewer larval hostplants	8000
		Δ−/Δ−	+/+ Rapid return	+/Δ+ Arrival	Beginning of isolation and evolutionary changes in cold adapted species	10 000
	Solifluction and permafrost	+/+	−/Δ− Loss of all warmth demanding species	−/−	Base-line for entry of most temperate species	
Natural open landscapes	Celtic and Cambrian land extensive Permafrost conditions	Δ+/Δ+ Expansion of ranges Losses from south and lowlands Δ−/+ +/+ Maximum range of arctic butterflies	Δ−/+ Retreat +/+ Post-Devensian glaciation arrival of temperature species of open habitats Δ+/Δ+ −/−	−/− +/− −/−	Extinction of temperate species Spread of temperate species from Mediterranean latitudes	12 000 14 000

Zone III conditions there were too severe for the Lusitanian element (Fig. 10.3a) (Mitchell 1970, 1977; Mitchell and Watts 1970; Godwin 1975; Ruddiman and McIntyre 1981; Coope 1986). The climate improved to reach a peak in the late Devensian prematurely at 12 500 years BP (Bishop and Coope 1977; Coope 1981; Atkinson *et al.* 1987) and recovered fully by 9600 years BP, after the Loch Lomond readvance, to exceed modern day summer temperatures by 8500 years BP. Climatic oscillations continued through the Flandrian to the modern day (Table 10.2; Manley 1974; Lamb 1977).

For several reasons, previous attempts to determine the arrival of British butterflies have made much of the need for land-bridges and nearby refugia during glacial episodes, both of which depend on changing sea-levels. Firstly, it was thought that butterflies could not cross sea-barriers. Secondly, plants and animals having a west-coast distribution, particularly the Lusitanian element (see

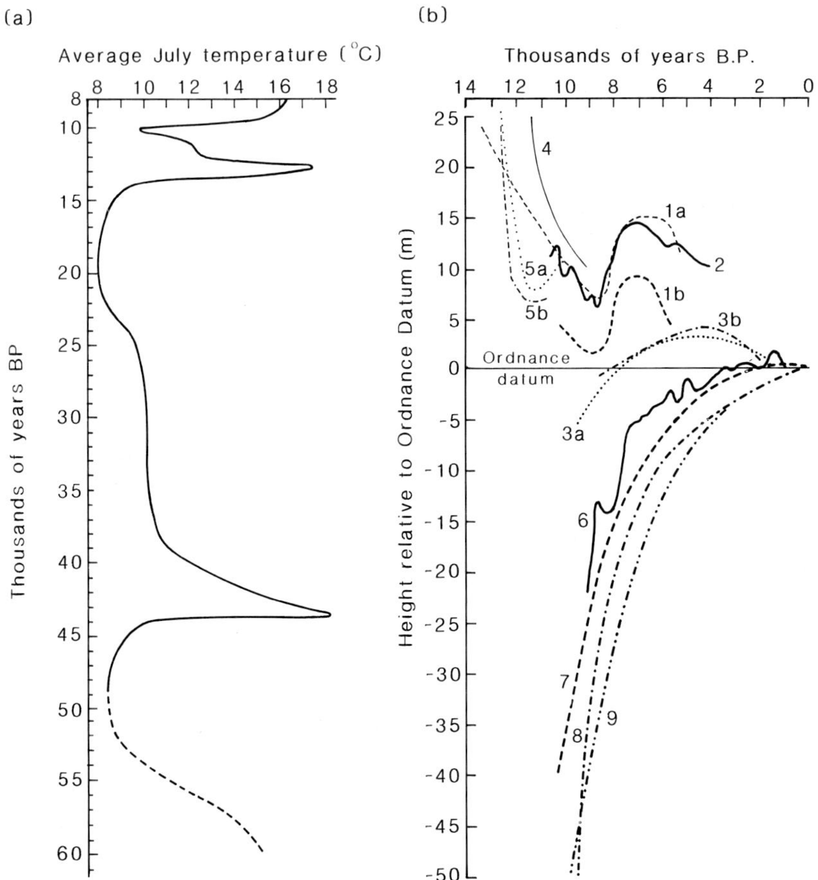

Fig. 10.3 Climate and sea-levels for Devensian and Holocene Britain. (a) July mean temperatures since the early Devensian glaciation inferred from assemblages of fossil Coleoptera. (Courtesy of Professor G. R. Coope; see Dennis 1977.) (b) Relative sea-levels for Late and Post Glacial Britain. 1a, 1b. Central Scotland and 'marginal' areas of Scotland (Donner 1970); 2. Forth valley, Scotland (Sissons and Brooks 1971); 3a, 3b. Northern seaboard of Solway Firth (Jardine 1975); 4. Oban-Lorne area, Scotland (Synge 1977); 5a, 5b. Ardyne and Lochgilphead areas, Scotland (Peacock *et al.* 1978); 6. North-west England (Tooley 1974); 7. South-east England (Devoy 1982); 8. South-west Britain (Hawkins 1971); 9. South-west Ireland (Stillman 1968 in Synge 1977).

Godwin 1975) and certain British butterfly subspecies (Ford 1945, 1975 Fontana edition; Beirne 1947; Heslop-Harrison 1947), were thought to have been confined to British environments during the Devensian. Warm refugia close to the North Atlantic drift were surmised. However, slow evolution of racial forms under periglacial conditions is now understood to be unnecessary as rapid adaptations can be related to Flandrian habitat changes and environmental gradients (Dennis 1977; Dennis and Shreeve 1989), and conditions were too cold in south-west Ireland and its continental shelf extension for the survival of many Lusitanian plants (see Ruddiman and McIntyre 1976, 1981; Price 1983; Preece *et al.* 1986).

If it is assumed that land-bridges are necessary for butterfly colonization, the problem is not merely of their existence but of their timing. Sea-levels were affected by two main factors which varied in strength according to regional provenance. Around the British coastline, sea-levels depended much on the conflict between eustacy and isostacy. World-wide, the sea-level falls when the hydrological cycle is interrupted during a glaciation, and snow builds up on land as ice instead of running back as river discharge to ocean areas. At the same time, areas under ice are depressed and areas relieved of ocean rise. The release of meltwater with deglaciation reverses the process. Such was the weight of ice that Scotland and northern Ireland had higher, not lower, sea-levels during glaciation (Evans 1975; Gray and Lowe 1977; Greensmith and Tooley 1982). Thus no land-bridges occurred there that could usefully allow the colonization of Ireland from Scotland (Fig. 10.3b). Moreover, at the same time that amelioration in climate triggered northward migration in plants and animals, sea-levels were rising in southern Britain and land-connections narrowed down (Fig. 10.2). Early Flandrian crossings were available for the British mainland as well as for nearby islands (Wight, Anglesey, Man, and perhaps Orkney) but not for the Outer Hebrides from the British mainland. Whether connections occurred for Ireland and the Isles of Scilly is still unknown. Estimates of eustacy 18 000 years BP (−120 m to −130 m OD) (Curray 1965; Milliman and Emery 1968; Jelgersma 1979; Street 1980) ensure late Devensian connections. Unknowns, such as uplift resulting from abstraction of overlying seawater and erosion of unconsolidated deposits from the floor of the Irish Sea, may just have extended connections into the Flandrian. But dependence of butterflies on land-bridges is probably a great deal less than was originally thought. Despite the apparent difficulty some butterflies have in colonizing new ground (e.g. silver-studded blue *Plebejus argus*, Dennis 1972*b*; black hairstreak *Satyrium pruni*, Thomas 1974), small butterflies have reached lighthouses and tiny islands (e.g. common blue *Polyommatus icarus* and small heath *Coenonympha pamphilus*; see chapters 1 and 6). Moreover, Mesolithic and Neolithic man were also responsible for introducing a variety of ungulates to Britain and Ireland (red deer, pigs, sheep, auroch) (Stuart 1974; Simmons and Tooley 1981). Many insect species may have been introduced in fodder or bedding discarded by humans on British shores in much the same way as plant species are introduced by migrating birds.

10.3 The pattern of butterfly arrivals

There are two reasons why we cannot expect the return of butterflies after the Devensian maximum to occur simply in a single invasion. First, the Loch Lomond readvance made a 'false start' for the Flandrian, delaying the colonization process. Secondly, although British butterflies lack for varied faunistic elements, they do not fall into a single homogeneous group. There are distinctions in their ability to withstand different conditions of climate, habitat, and hostplants, and the latter vary in life-form and tolerance of conditions. Thus environmental transformations resulted in varied responses by different groups of butterflies. For simplicity, periods of the Late Devensian are used here as a chronological guide to the early sequence.

Response by butterflies to changes is assessed by matching their present-day environmental tolerance in temperatures, hostplants, and habitats to the conditions of each stage. Climatic tolerance has been determined by comparing the minimum July mean temperatures at the edge of their Eurasian range with the best estimate of mean July temperatures during Late Devensian stages. Summer more than winter temperatures are believed to be critical (chapter 1; Dennis 1977, 1985*a*; but see Dennis, in preparation). The widest hostplant choice is used, but pollen records need to be interpreted with caution and greater reliance should be placed on macroscopic remains. Matching habitats is more difficult because of the lack of modern equivalents to those of Late Devensian and Early Flandrian age. Nevertheless, basic distinctions in habitat such as between open conditions of heath, grassland, and wetlands compared to closed woodland, often follow specific hostplants and permit comparison.

Successive colonization and extinctions of butterflies occurred in the Late Devensian. Conditions causing extinction in one group allowed others to take up residence and created large-scale movements. With the return of warmer temperatures, not only did butterflies colonize new areas freed of permafrost and ice but often, simultaneously, they must have become extinct in some of their glacial refuges. The changes in climate which triggered plant migration and extended their habitats northwards also transformed communities through vegetation succession in their 'refuges'. Over the millenia since 18 000 years BP, species moved hundreds of kilometres across an increasingly recognizable map of Europe. Survival for butterflies at that time, depended as it does now, on their ability to colonize new habitats, an ability frequently underestimated, based nowadays on dispersal from small residual colonies amid stable or deteriorating habitats. For reasons mentioned above, it is unlikely that anything other than extensive sea areas formed effective barriers to movement. Some species undoubtedly spread across land connections together with their appropriate habitats; others simply flew across widening sea straits. A number of butterflies may have been conveyed as eggs, larvae, or pupae in fodder with human populations and their domesticated animals (see McNeil and Duchesne 1977). Developing warmth, the spread of habitats and vegetation succession must have resulted in large populations and increasing numbers of insects seeking out new resources (Dennis 1977, p. 129).

10.3.1 Glacial *tabula rasa* and refugia in southern Europe

During the extreme conditions of the Devensian maximum, it is unlikely that any present-day British butterfly occupied the British islands, particularly as summers would have been cloudier and cooler than on the mainland. There were times when the beetle fauna is believed to have been totally eliminated (Coope 1987); even for arctic butterflies, the Devensian maximum in Britain may then have been a glacial *tabula rasa*. Instead, they were forced to low ground between and beyond the advancing ice sheets centred on Scandinavia, Britain, and other mountainous regions. As the timberline (July 10 °C isotherm) has been pushed as far south as northern Spain and along the Mediterranean shore to the Black Sea, all but cold tolerant butterflies of montane and arctic habitats retreated to Mediterranean latitudes. Erebias and some of our hardy species (e.g. green-veined white *Pieris napi*, green hairstreak *Callophrys rubi*, small copper *Lycaena phlaeas*, *Polyommatus icarus*, dark green fritillary *Argynnis aglaja*, large heath *Coenonympha tullia*, and northern brown argus *Aricia artaxerxes*) probably found suitable habitats in south-west France along the coastal fringe. Earlier authors have argued that some species with arctic–alpine distributions survived between the Scandinavian and alpine ice sheets on the north German plain (Warnecke 1958). Athough this may have been feasible (see Frenzel 1979) it is not the only possibility, as this region could have been steadily colonized by these species from the south-west and the south-east in the early stages of deglaciation before colonization of the Alps and Scandinavia. Species requiring warmth, especially those dependent on mature deciduous woodland hostplants, could have found suitable habitats only in isolated stands compressed along sheltered river valleys and at mid-elevational bands on southern mountains from Iberia to Armenia, as much of the Mediterranean landmass was covered in dry steppe (Beug 1975; Starkel 1977; Bowen 1978; Brice 1978; Frenzel 1979; Blondel 1987; Bennett *et al.* 1991).

10.3.2 Butterflies of the Windermere interstadial

Despite evidence for a rapid rise in temperatures before 15 000 years BP (Emiliani 1966), substantial warming of the climate did not occur until 13 000 years BP (Atkinson *et al.* 1987) and response from the fauna was relatively slow. Many butterflies had been forced more than 1200 km from Britain and some could recolonize northern landscapes only at the dispersal rate of their resources (Dennis, in preparation); this depended not just on the dispersal rate of plants (seed dispersal), but on the retarded interaction between biota and mineral deposits, soil development, and vegetation succession. Bryophytes, forbs, and grasses are the first colonizers and in turn shrubs and trees increasingly demanding of warm conditions. By 13 000 years BP montane and arctic butterflies which had remained closer to the ice-front had retreated to north and north-east Europe and to mountain slopes. Probably many, if not all, of Britain's present-day temperate butterflies of open landscapes were then more plentiful and covered a wider area than during the past two centuries. The more hardy element now found in the Outer Hebrides and Orkneys would have entered first and may even have occupied these northern islands. Narrow sea-barriers were probably ineffective in the face of mass immigration and large population turnovers. The last to arrive (*c.* 12 150 years BP) were the most warmth demanding such as the Lulworth skipper *Thymelicus acteon* and the Glanville fritillary *Melitaea cinxia*, and with them probably many species now finding a limit in northern France (see Dennis 1977; Table 1.2). Certainly, beetles of southern latitudes, such as *Bembidion callosum*, occurred that now have ranges further south (Bishop and Coope 1977).

Butterflies of mature deciduous woodland, made slower progress and failed to arrive in the Late Glacial. As light *Betula–Salix–Pinus* woods spread across the country *c.* 12 500 years BP, summer temperatures suddenly deteriorated. 'The survivors of this catastrophe were almost all the widespread species (beetles) whose climatic tolerance is seemingly broad-based' (Coope 1981, p. 218). Thus butterflies such as the purple hairstreak *Quercusia quercus*, the white-letter hairstreak *Satyrium w-album*, the large tortoiseshell *Nymphalis polychloros*, and holly blue *Celastrina argiolus*, dependent on mature tree hostplants, probably did not arrive in the Late Glacial (see Godwin 1975), nor perhaps those known to require habitats associated with mixed oak woodlands (e.g. purple emperor *Apatura iris* and white admiral *Ladoga camilla*). Blackthorn *Prunus spinosa*, purging buckthorn *Rhamnus catharticus*, and alder buckthorn *Frangula alnus* are not recorded in Late Glacial deposits and therefore it is unlikely that the brown hairstreak *Thecla betulae*, *S. pruni* and the brimstone *Gonepteryx rhamni* were resident. But other 'woodland' butterflies may have been found among the *Betula* 'parkland' before 12 000 years BP (e.g. chequered skipper *Carterocerphalus palaemon*, wood white *Leptidea sinapis*, speckled wood *Pararge aegeria*, high brown fritillary *Argynnis adippe*, silver-washed fritillary *Argynnis paphia* and comma *Polygonia c-album*), and, if not, none were distributed far to the south. As *Argynnis paphia* is known to lay its eggs readily on the trunks of pine trees (Kudrna 1986), it may well have been present in southern Britain.

10.3.3 Retreat during the Loch Lomond stadial (Younger Dryas)

Less than a quarter of our resident butterflies survived in Britain with the reimposition of arctic conditions at 10 800 years BP (Dennis, in press). Butterflies of arctic tundra wastes occupied the Midlands and coastal areas of northern Britain. Possible candidates were the northern clouded yellow *Colias hecla*, the pale clouded yellow *Colias nastes*, the cranberry blue *Vaccinia optilete*, and a number of *Boloria* species (e.g. mountain fritillary *B. napaea* and cranberry fritillary *B. aquilonaris*); *Clossiana* species (e.g. Titania fritillary *C. titania*, arctic fritillary *C. chariclea*, and polar fritillary *C. polaris*, etc.); *Oeneis* species (e.g. Norse grayling *O. norna* and arctic grayling *O. bore*); *Erebia* species (e.g. Arran brown *E. ligea*, Lapland ringlet *E. embla*, and arctic woodland ringlet *E. polaris*) and other butterflies occupying similar habitats today. The more northerly British fauna at this time would have included some of the current hardy residents such as the small mountain ringlet *Erebia epiphron*, which can extend its life cycle over a two-year period (Wheeler 1982) and *Coenonympha tullia*, with larvae that can survive being frozen in standing water (Melling 1984*b*; Table 10.3). As summer conditions were colder and cloudier to the north and west, these regions probably received fewer species (Walker and Lowe 1990). The present-day arctic insect fauna is severely depleted, comprising less than five per cent

Table 10.3 The post-Devensian arrival sequence for British butterflies into southern Britain and Ireland; periods from which species have probably been resident to the present day

Period	Sub-period	Species	Species
Flandrian	Boreal	*Satyrium pruni*	*Apatura iris*
		Quercusia quercus	*Satyrium w-album*
		Aporia crataegi	*Thecla betulae*
	Pre-Boreal	*Aphantopus hyperantus*	*Coenonympha pamphilus*
		Pyronia tithonus	*Maniola jurtina*
		Melanargia galathea	*Hipparchia semele*
		Pararge aegeria	*Erebia aethiops*
		Melitaea cinxia	*Lasiommata megera*
		Argynnis paphia	*Eurodryas aurinia*
		Argynnis adippe	*Argynnis aglaja*
		Inachis io	*Polygonia c-album*
		Ladoga camilla	*Nymphalis polychloros*
		Maculinea arion	*Hamearis lucina*
		Cyaniris semiargus	*Celastrina argiolus*
		Lysandra coridon	*Lysandra bellargus*
		Plebejus argus	*Aricia agestis*
		Lycaena dispar	*Cupido minimus*
		Pieris rapae	*Anthocharis cardamines*
		Gonepteryx rhamni	*Pieris brassicae*
		Erynnis tages	*Pyrgus malvae*
		Thymelicus acteon	*Ochlodes venata*
		Thymelicus sylvestris	*Thymelicus lineola*
		Carterocephalus palaemon	
Loch Lomond Stadial or Younger Dryas			
Late Glacial		*Mellicta athalia*	*Coenonympha tullia*
		Boloria selene	*Boloria euphrosyne*
		Polyommatus icarus	*Aglais urticae*
		Lycaena phlaeas	*Aricia artaxerxes*
		Leptidea sinapis	*Callophrys rubi*
		Papilio machaon	*Pieris napi*
		Hesperia comma	
Devensian glaciation		*Erebia epiphron*	

Several migrant species, *Colias* spp., *Vanessa atalanta*, and *Cynthia cardui* have been excluded. The reconstruction is based on summer screen temperatures of the most cold tolerant forms for each species providing a liberal estimate for arrival, but does not account for local climatic influences, such as slopes for southerly aspects.
From Dennis, in preparation; cf. Dennis (1977).

of the species of corresponding temperate areas (Downes 1965). Other cold-tolerant species, found today in the taiga fringe of Scandinavia, would probably have occurred in southern England occupying favourable habitats on south-facing scarps and in sheltered 'dry' valleys along the belt of limestones and old cliffs marking earlier coastlines. Populations would have been small and scattered. More warmth-demanding species and those of mature deciduous woods were arrested across the 'channel land' in Aquitaine and northern Spain.

10.3.4 Post Glacial butterfly invasion

The Younger Dryas differs from the Devensian maximum in one important respect. Despite the cold and its effect on the fauna, much of the flora, including many butterfly hostplants, remained intact. Conditions for interglacial invasion were merely suspended, not destroyed. Coope (1981) remarks on the speed at which arctic–alpine species were exterminated and replaced by warmth-demanding Coleoptera to as far as south-west Scotland (*c.* 9640 years BP). The reaction of most British butterflies—opportunistic, benefiting from herb-rich carpets on base-rich soils and triggered by sudden warmth—would have been the same. Instantaneous incursion, massive population build-up, gene-flow, panmixis, and dispersal to outlying islands undoubtedly followed for all but those species dependent on more advanced communities with succession to woodland (Table 10.3). Even then, there is evidence of a rapid spread of deciduous habitats up western Britain along the Irish Sea (Taylor 1975). Some plants apparently did not spread extensively over northern Britain until the late Boreal, in which case the last phase of diffusion was that of *Quercusia quercus*, *Thecla betulae*, *Celastrina argiolus*, and *Gonepteryx rhamni* to Ireland and Scotland. Many warm temperature species, now resident in France, probably occurred here but have since been lost (see Table 1.2).

10.4 Butterfly adaptations to Britain's Post Glacial environments

Earlier work on butterfly origins attributed little significance to Flandrian environmental changes (Ford 1945, 1975 Fontana edition; Beirne 1947; Heslop-Harrison 1947), but the influx of butterflies to Britain was but the first stage in their establishment. Thereafter, they had to adapt to fluctuations in Flandrian climate and to environmental changes. Depending on their tolerance of conditions, butterflies have responded in three different ways:

(1) changes in their distributions;
(2) modifications in their ecology, phenology, and behaviour;
(3) modifications in their phenotypes and genetics.

Interpretation and dating of these adaptations is based on their date of entry, the nature of adaptations compared to conspecific populations abroad, and the sequence of environmental changes, uniquely those imposing isolation (sea barriers or unsuitable habitats) and likely to enforce specific adaptations. Reconstruction of evolutionary events is made easier where unrelated species assume similar patterns of adaptation—in identity of distribution, of hostplant or of alterations in phenotype. Thresholds of tolerance in different species have been exceeded and general environmental controls are implicated.

Environmental changes often impinge simultaneously on distributions, ecology, and genetics. Divergence of populations, resulting from behavioural and ecological adaptations and genetic changes, has its most orthodox explanation in the geographical separation of a once, more or less, continuous distribution, the parts of which then are subject to different environmental conditions, habitats, and hostplants. Subspecies and species are potential end products. As a chronological narrative would involve some repetition, the following two sections are devoted to distinct themes. The first investigates the impact of Flandrian environments on distributions, the speed of changes, disjunctions, and the abandonment by butterflies of upland Britain. Some of our butterflies reveal unusual and 'inflexible' ecologies; how and when these adaptations became 'fixed' is the subject of the second section. Our butterflies are also characterized by 'subspecies', phenetic gradients, and local races, aspects of differentiation at different scales. In section 10.5 these are placed in an historical context and related to long-term environmental changes which also affected the continent of Europe.

10.4.1 The development of Post Glacial distribution patterns

Fundamental changes have occurred in butterfly distributions since the early Flandrian. The end-result has mainly been one of contraction: typically range reductions, distributional disjunctions, scattered and isolated populations, and the vacation of mountain and moorland. The processes leading to these patterns were more complex. Different butterflies were affected by different conditions. Environmental changes were either beneficial or deleterious to species depending on the tolerance of different butterflies, the nature of changes, and whether different aspects of environmental change operated in unison or opposition (Table 10.2).

Temperate species of open habitats expanded in two phases: (i) rapidly in the Pre-Boreal and early Boreal, only to be severely curtailed by woodland succession in the late Boreal and Atlantic, and then (ii) from the Sub-Atlantic onwards as forest clearance proceeded with increased vigour under Iron Age and later cultures. Woodland butterflies, strictly those requiring mature tree hostplants, diffused widely in the Boreal and Atlantic climax forests and regressed under deforestation, gradually to the twelfth century and then dramatically to the eighteenth century. Both groups must have been adversely influenced by the climatic downturns of the Sub-Atlantic and of the Little Ice Age, a double disadvantage for strictly woodland butterflies. Range retractions of butterflies of open habitats from the north and uplands occurred then but, as their habitats in the south and lowlands were abundant and 'untreated', populations there were probably large enough to absorb the impact. Despite substantial increases in the human population by the seventeenth and eighteenth centuries and pressure on the land, weeds and 'pests' were as yet not controlled by artificial chemicals and their removal was dependent on manual labour. Fallow ground was extensive and pasture still burgeoning with native grasses and herbs (Mitchell 1965). Enclosure, well under way in the Midlands, removed the vagaries of management on adjacent open-field strips but created new semi-natural habitats in hedgerows. In short, the butterfly scene must have been healthy.

The heyday of British butterflies was probably the early medieval period. Not only would they have benefited from the warm dry summers of the time but more substantially from the developing mosaic of habitats created by ubiquitous and contiguous village communities (Roberts 1977) with their arable or fallow land, common pasture, and woodland. There must have been a profusion of butterflies of open habitats as well as woodland butterflies. Most woodland species depend on early or mid-successional hostplants (Appendix 1) and woodlands managed as standards and coppice, and used for feeding pigs, created space and light and prodigious ground flora, maintaining an effective plagioclimax for such butterflies. Many species, typically less common in the natural context of earlier interglacials, were made indicators of agricultural settlement. Nymphalids, for example *Aglais urticae* and the peacock *Inachis io*, must have proliferated as early as the Neolithic period, since nettles dominate open ground on nitrogen-enriched human and animal waste. Thistle invasions of prehistoric and medieval fields, and crops of brassicas dating from the Roman period (Godwin 1975) must have encouraged temporary abundances of the vanessid *Cynthia cardui*, permanent residence of the pierids the large white *Pieris brassicae* and the small white *Pieris rapae*, and population explosions of these pierids as their cultivation expanded with the agricultural revolution. Ribwort plantain *Plantago lanceolata*, was a characteristic marker of cultivated land. Did *Melitaea cinxia* make use of this and was this butterfly found widely on the British mainland? As the heath fritillary *Mellicta athalia*, a close relative, has been known to thrive in abandoned strawberry fields in Cornwall (Warren *et al.* 1984), there is every chance that it did.

Restricted distributions and disjunctions imply confinement of populations. For much of the British butterfly fauna—those of open habitats and of the exposed woodland floor—such pressure was induced by extensive and thick forest cover. Forest conditions dating to 7000–5000 years BP, contrast strikingly with the managed and medieval landscapes. Although many species spread with forest clearance and climatic change, expanding arable and pastureland failed to provide the correct habitats for others. The confinement of *Erebia epiphron* to mountain-tops dates from the Boreal. Similarly, the scotch argus *Erebia aethiops* and *Aricia artaxerxes* became increasingly restricted to northern scree slopes, crags, thin mountain limestone parkland, and coastal cliffs and *Coenonympha tullia* to upland blanket peat and isolated, lowland acid mosses. During the development of climatic climax forest most butterflies were squeezed out to stable subseres, to scattered fragile habitats (i.e. coastal fringe, uplands, escarpments, and porous sands) and temporary open spaces associated with wind-throw, the activity of animals such as beavers, fires, and regeneration. The large copper *Lycaena dispar* and the swallowtail *Papilio machaon* became restricted to fenland, the Lulworth skipper *Thymelicus acteon* (if it was not introduced later in historical times) to mature tor grass *Brachypodium pinnatum* on steep Dorset cliffs and limestone scarps, and *Melitaea cinxia* largely to undercliffs abounding in young *Plantago lanceolata*. Some butterflies remained widespread but never regained their Pre-Boreal status (e.g. *Plebejus argus*, small blue *Cupido minimus*, and large blue *Maculinea arion*). Disjunct segments of many woodland floor butterflies (e.g. *Mellicta athalia*, *Carterocephalus palaemon*, and *Leptidea sinapis*) reflect the same process, surviving to the twentieth century where regular coppicing and patchy woodland clearance occurred, a process which dates back to the Neolithic period (Chippendale 1985; Coles and Coles 1986).

The distribution of *Satyrium pruni*, centred on the Rockingham and Bernwood forest complex, is a puzzling case. Prior to Iron Age agriculture, it was probably confined to open wood formations where lower storey *Prunus spinosa* could abound and mature. Oxford clay and till over Oolite are unlikely to have provided the 'ancestral' habitat; and the butterfly, with its hostplant, probably moved off higher scrubby ground on limestones to lower ground as the latter opened up and the escarpments were more intensively grazed. Artificially established colonies do well in other parts of southern England and there is no reason why its distribution in medieval times and before should not have been more extensive than today. The butterfly has poor dispersal capability and fragmentation of woodland would have led to local extinction. It is maintained by an ancient long-term coppice cycle, over a period of 20 years, of blackthorn (Thomas 1974; see chapter 11)—an atypical regime in that hazel, willow and other shrubs are generally preferred and cut more frequently. Not all disjunctions and distributional restrictions date back to the Atlantic period. The distributions of many butterflies, both of open habitats (e.g. wall brown *Lasiommata megera*, small skipper *Thymelicus sylvestris*, orange-tip *Anthocharis cardamines*, *Polyommatus icarus*, and *Lycaena phlaeas*) and of woodland (e.g. *Pararge aegeria*, *Leptidea sinapis*, *Ladoga camilla*, *Polygonia c-album*) have expanded and contracted substantially in more recent times (see chapter 11) as may also have *Thymelicus acteon* and *Melitaea cinxia*.

It may seem remarkable that so few species occupy our mountains and open moors when it is to just these areas that many of our butterflies would have been confined during the climatic climax, and when numbers of species are compared with those at similar altitudes in continental Europe. Although pockets of butterflies are found at higher elevations (e.g. large skipper *Ochlodes venata*, *Lasiommata megera*, *Callophrys rubi*, and meadow brown *Maniola jurtina* at 420 m at Greenhow Quarries in Nidderdale on the Pennines; see Barnham and Foggitt 1987), the species 'lapse rate' for the Pennines, Cumbrian, and Welsh mountains is extremely steep with but a fraction of those British butterflies occupying the Welsh coast and the Cheshire Plain extending much above 200 m. *Coenonympha tullia* and *Erebia epiphron* are rather restricted but attain the greatest elevations (Pollard *et al.* 1986). *Coenonympha pamphilus* and *Pieris napi* are habitual residents; pockets of *Polyommatus icarus* and *Aglais urticae* also occur, and occasionally the small pearl-bordered fritillary *Boloria selene* and *Argynnis aglaia*. The latter depend on *Viola riviniana*, which has survived deforestation under heather on the Durham fells and in western Scotland (Dunn, personal communication). Acid moorland, thicknesses of peat, and cold, wet, and windy microclimates are directly responsible for the lack of butterflies and belie the presence of rich habitats in the Pre-Boreal and Boreal, known from macroscopic remains and stratigraphy.

Upland habitat degradation—retrogressive succession—is typical of interglacial phases but is usually restricted to the latter (telocratic) part of the interglacial (Iversen 1958), when excess moisture from cooler, wetter conditions increases the downward translocation of minerals and nutrients, and so in turn leads to podsolization and impoverishment of vegetation communities (Dimbleby 1962). In the Flandrian, however, the onset of these processes has been much earlier, resulting from Mesolithic man's firing of scrub and light woodland to improve hunting and Neolithic and later clearances for cultivation by human populations and long-term intensive grazing (Fig. 10.4). Water supply increased, and with the collapse of soil structures and formation of durable pans, infiltration capacity of soils was substantially

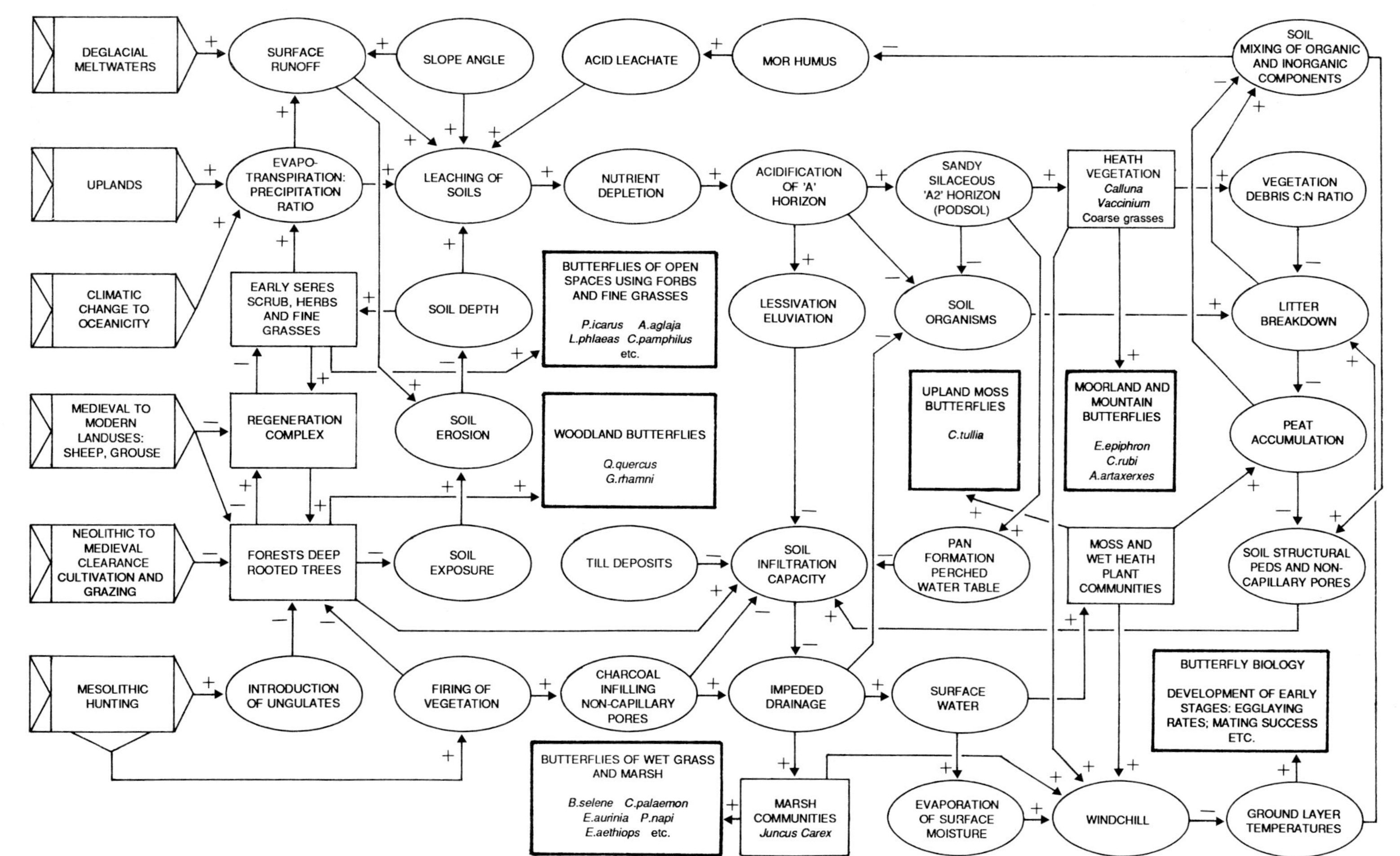

Fig. 10.4 Climatic and human factors leading to the impoverishment of the British upland environments and their influence on the butterfly fauna. Boxes, vegetation; highlighted boxes, butterflies; ellipses, conditions and factors influencing butterflies. For interpretation of signs, refer to **process-response models** in glossary. This diagram excludes more recent influences of industrial pollution which have seriously affected the ecology of some uplands (e.g. Pennines).

reduced. The consequence has been the permanent replacement of deciduous woodland associations and subseres by species of impoverished but specialized acid heath, where summer free-drainage still occurs, as well as anaerobic-tolerant bog communities where waterlogging is dominant. Microclimates became cooler and wetter on the exposed moorland. All but a handful of specialized or hardy butterflies retreated to lower ground in step with Bronze Age and Iron Age clearances.

10.4.2 Butterflies and Post Glacial ecological 'bottlenecks'

The British butterflies mainly belong to a wide-ranging group of Eurasiatic species which in continental Europe are tolerant of varied hostplants and habitats (Kostrowicki 1969; Dennis *et al.* 1991). This observation is interesting because, although many species still accept a wide range of hosts from a single plant family (oligophages) in any one locality, a number in Britain have become restricted to single hostplants and habitats. Such regional monophages are *Papilio machaon*, *Lycaena dispar*, the Adonis blue *Lysandra bellargus*, the chalk hill blue *Lysandra coridon*, and *Cupido minimus*. Other butterflies, perhaps as in Europe, feed on a number of hostplants but are possibly predisposed to being locally monophagous (e.g. probably *Callophrys rubi*, *Mellicta athalia*, *Erebia aethiops*, brown argus *Aricia agestis*, *Polyommatus icarus*, *Leptidea sinapis*, and at one time *Carterocephalus palaemon*) or may be more specialized in Britain (e.g. *Polygonia c-album*; Nylin 1988). Some obvious questions are: how unusual are these hostplant habitat restrictions and how have these different adaptations been produced under the Flandrian historical setting? First, it is necessary to look at the basis of these adaptations.

Individuals in butterfly populations generally vary in feeding preferences, both in the order of plants chosen and in their thresholds of acceptance (Wiklund 1981; Thomas, C. 1985*a*). Ecological specialists (monophages) and generalists occur and are at an advantage in different circumstances. Specialization on one hostplant enables physiological and behavioural adaptations and should lead to increased reproductive success. Should the hostplant become unpredictable, then fecundity is reduced and generalists are at an advantage as raised fecundity will offset lower offspring-survival on less favourable plants. As Wiklund (1981) points out in a discussion of Swedish *Papilio machaon* (see section 3.4.1), it is possible for both strategies to be maintained in a population if the lower fecundity and higher offspring survival associated with a specialist strategy is balanced by the higher fecundity and lower offspring survival associated with a generalist strategy. This latter situation must occur in populations of many British butterflies where a range of hostplants is available, for instance among satyrines (e.g. *Maniola jurtina* and *Pararge aegeria*) and pierids (e.g. *Pieris napi* and *Anthocharis cardamines*) (Lees and Archer 1974, 1980; Courtney and Duggan 1983; Shreeve 1984).

Local and regional monophagy could, however, imply two very different historical processes. First, one hostplant may have been chosen in a situation where larval survival on it has outstripped that on others equally abundant and predictable in the same vicinity. Alternatively, the butterfly may have had no choice because only a single hostplant species occurred; in that case a generalist strategy is redundant. At this point it is relevant to ask how unusual are the regional monophages. Although they differ from their European counterparts, they are not unique. A fenland equivalent of *Lycaena dispar* occurs in Friesland, Holland (Bink 1986) and the butterfly feeds locally on water dock *Rumex hydrolapathum* in West Germany and in Italy (Blab and Kudrna 1982; Kudrna 1986). Similarly *Papilio machaon* occurs on inland fens in Sweden feeding preferentially on milk parsley *Peucedanum palustre*, but also occupies other habitats there (seashores and moist meadows) and feeds on other hostplants (Wiklund 1981). In Europe, *Lysandra coridon*, *L. bellargus*, and *Cupido minimus* inhabit both calcareous and base-rich environments, but select other Leguminosae in addition to horseshoe vetch *Hippocrepis comosa* and kidney vetch *Anthyllis vulneraria*. Thus, British populations of these species are distinct only in that alternative population specialists and generalists have been eliminated. There is some evidence that the second explanation above fits the British fenland butterflies in that *Papilio machaon* and *Lycaena dispar* failed to disperse when alternative hostplants became available with disforestation. *Papilio machaon* has not taken to ubiquitous angelica *Angelica sylvestris* and neither has *Lycaena dispar* taken to other docks (*Rumex* spp.) and bistorts (*Polygonum* spp.); this may imply that these plants were scarce in the Fenland community and its immediate surrounds. In the case of three lycaenids, *Cupido minimus*, *Lysandra bellargus*, and *L. coridon*, their habitat preference and hostplant choice in Britain is less different from that observed in continental Europe and the first explanation seems more probable. Scattered and isolated colonies of *C. minimus* all use *Anthyllis vulneraria* despite the widespread availability of other Leguminosae in Britain and Europe. Obligate monophagy probably occurs locally in *Callophrys rubi*, whereas the varied choice of habitats and hostplants in *Plebejus argus* (limestone grassland, acid heath, and acid bogs) is explained by their greater dependence on ant attendance (Fig. 10.5). In these two cases, particular hostplant species are less important than the seasonal availability of young plant growth.

All these adaptations date, along with distributional disjunctions, to Boreal and Atlantic forest confinement. Species which were widely distributed and which occupied different settings along the coast and inland, have retained a generalist strategy. Much hostplant and habitat specialization obviously pre-dates the return of butterflies to Britain in the Flandrian (see section 10. 1). But the entrenchment of the two fenland species and of southern limestone butterflies dates to the Flandrian. Fenland habitats have been maintained in balance with the water-table. Higher water-table levels associated with rising sea-level resulted in peat accumulation, whereas lower water-table levels during falling sea-level exposed the peat to aerobic conditions. When exposed, peat decomposes and shrinks to the lower water-table level (Shennan 1982). *Lycaena dispar* and *Papilio machaon* adopted this habitat, the former species on either side of the North Sea, as Doggerland was inundated before 7800 years BP (Fig. 10.2; Kolp 1976). As forest encroached, both butterflies also became isolated from populations of the same species inland.

The southern limestones are unusual and are important habitats. Not only are they strongholds for *Lysandra coridon* and *L. bellargus* (Fig. 10.6), they also form major habitats for other butterflies, particularly for such species as the silver-spotted skipper *Hesperia comma*, and also for *Aricia agestis*, the Duke of Burgundy fritillary *Hamearis lucina*, and the marbled white *Melanargia galathea*. The exceptional chemistry of these habitats—virtually pure calcium carbonate—arrested the development of mature forest cover but also later offset serious environmental impoverishment, as occurred elsewhere on high ground. As tree-growth was retarded—a result of thin soils due to steep slopes, Devensian solifluction, and a fractional residue from weathering—Mesolithic man was able to maintain open spaces. High base-status of soils attracted Neolithic and later farmers. It has also partly offset leaching and has encouraged a rich flora. Limestones, however, differ in another respect. They are permeable; the soil dries out rapidly and warms up quickly. Chalks are highly porous and dry out faster still. When the climate became wetter and cooler in the Sub-Atlantic and the 'Little Ice Age', many species probably retracted from marginal habitats to these warm, dry grasslands. Similar conditions occur for butterflies in the Irish Burren region (Carboniferous limestones). Only the higher limestones of the northern Pennines (Alston and Askrigg blocks; e.g. Malham area), which are colder, under incessant rainfall, and intensively grazed, have failed to retain healthy butterfly populations (Michaelis 1963; Nelson 1971).

10.5 The evolution of subspecies, races, and character gradients

Britain may lack unique butterfly species but it abounds in subspecies named for different islands and local races. Earlier authors (Beirne 1947; Ford 1945, 1975 Fontana edition) accredited great significance to these taxa and argued unsuccessfully for glacial maximum survival of subspecies in British refugia (Dennis 1977). Unfortunately, too much status was attributed to the category 'subspecies' and perhaps too much enthusiasm was expended on naming populations of the restricted British fauna, with too little attention being paid to the nature of variation, which is not great. Even now, little work has been done to determine the biochemical and genetic differences between populations (see Geiger

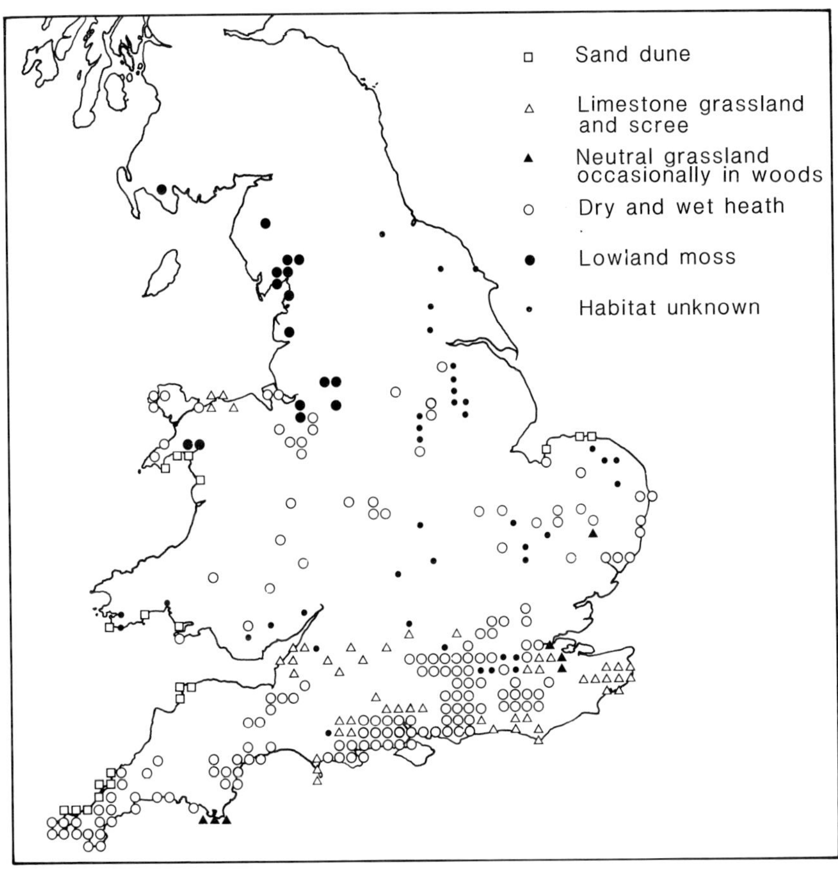

Fig. 10.5 The distribution of habitats for the British silver-studded blue *Plebejus argus* adopted during the Holocene. Recent introductions in North Wales have been omitted. Habitat details for sites in South Wales are poorly known; those in Pembrokeshire overlie limestone-solid geology but can be on other superficial deposits such as calcareous wind-blown sands, as near Castle Martin. Dune sites can be either acid or alkaline in soil pH reaction, depending on the admixture of shell fragments. Most heaths in the west of Britain (e.g. Dorset) are probably humid heaths with cross-leaved heath *Erica tetralix* as a common component, much as on the lowland mosses further north. Typical hostplants: sand-dunes, birdsfoot trefoil *Lotus corniculatus*; limestone grassland and scree, common rockrose *Helianthemum chamaecistus*, hoary rockrose *H. canum*, *L. corniculatus*, gorse *Ulex europaeus* and wild thyme *Thymus praecox*; dry and wet heath, ling *Calluna vulgaris*, *E. tetralix* and *U. europaeus*; lowland moss, *C. vulgaris* and *E. tetralix*. (Site data courtesy of the Biological Records Centre, Institute of Terrestrial Ecology and Harding, P. T., Habitat data from Chalmers-Hunt 1960–61; L. T. Colley; R. L. H. Dennis; J. W. Donovan; D. Elias; Mendel and Piotrowski 1986; I. K. Morgan; C. D. Thomas 1985*a*; J. A. Thomas; and M. S. Warren.)

1980, 1982; Courtney 1980; Morton 1985; Thomson 1987).

Much of the physical variation named for subspecies is also clinal, beginning on the British mainland and developing in intensity towards Ireland, the Hebrides, and Orkney islands (Dennis 1977; Majerus 1979; Lorimer 1983; Brakefield 1984; Fig. 10.7). It was not considered that other aspects of biological variation coincide with this trend in phenetic variation. Species become fewer further north, as do the number of annual broods, which are also shorter. Physiological clines exist as in the case of *Pieris napi* (Lees and Archer 1980). There is evidence that population structure differs. Populations become

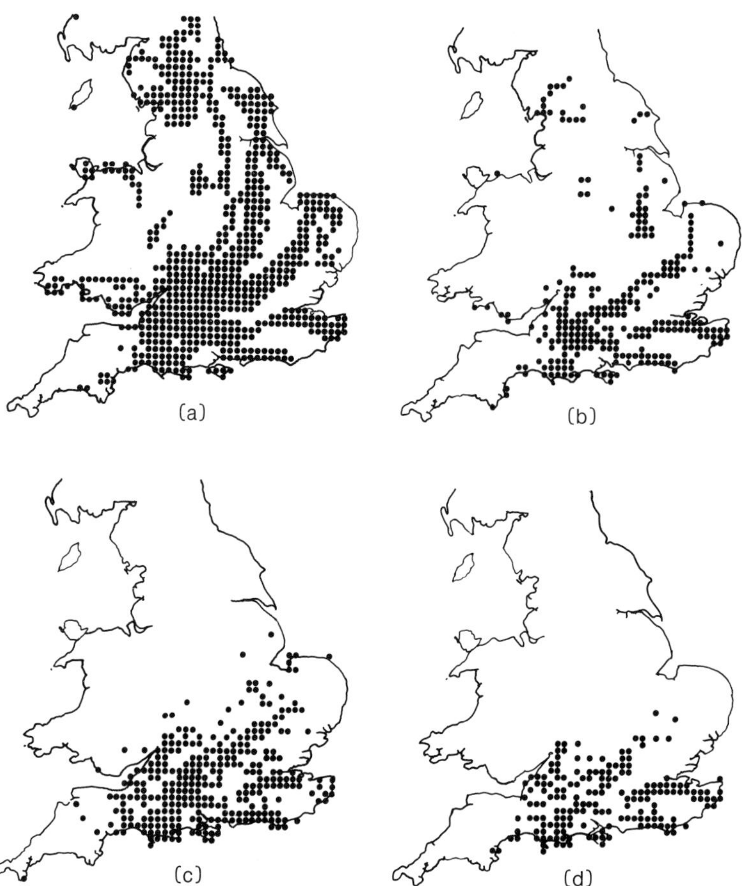

Fig. 10.6 The distribution of two lycaenid butterflies associated with limestones in southern Britain. (a) Limestone outcrops (where these exceed 5 per cent of the area of each 10 km square); (b) horseshoe vetch *Hippocrepis comosa*, all records; (c) chalk hill blue *Lysandra coridon*, all records; (d) Adonis blue *Lysandra bellargus*, all records. (Data courtesy of the Biological Records Centre, Institute of Terrestrial Ecology, and P. T. Harding.)

more isolated; average population size is smaller and dispersal is less (Thomson 1980; Brakefield 1982*a*; Courtney and Duggan 1983; Dennis, pers. obs.). Population fluctuations also become more dramatic (Pollard *et al.* 1986; see Fig. 4.7). Insufficient research has been done on the ecology of Scottish butterflies and the extent to which they differ from their English counterparts is not known, but a number of differences in habitat and hostplant have been identified (Dennis 1977). All in all, the butterflies appear to be adapting to marginal conditions at the edge of their ranges.

10.5.1 Seeking purpose for adult wing pattern variation

What is the basis for the various changes? One explanation is to visualize species as allocating resources (energy) among metabolic compartments (storage, growth, reproduction, defence, foraging, etc.) to their various subunits and to different life-stages, so as to maximize reproductive success (Townsend 1987). The balance of resource allocation changes as the environment changes and marked environmental gradients are crossed in Britain. Britain's climate is notably oceanic (wet, mild winters, and cool summers) but becomes more so to the north-west; physical variation in the butterflies correlates highly with this (Fig. 10.7; Dennis 1977, p. 201).

The reasons for some of the trends (i.e. number of species and broods, population size, and hostplants) were discussed earlier (see chapter 1). But as the form of physical variation in adult butterflies northwards contrasts for different taxa, an explanation for it is more elusive. A model has been

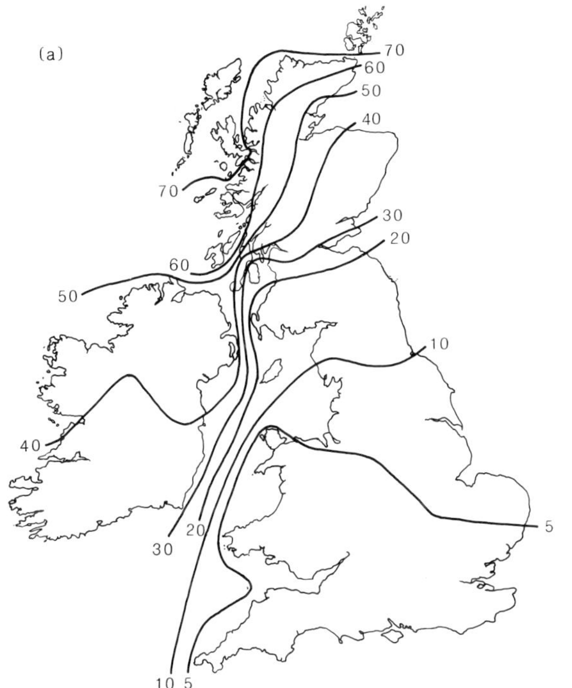

Fig. 10.7 Phenetic variation in British butterflies. (a) Gradients in phenetic variation (percentage of total fauna from a base of zero in southern England) among British butterflies plotted for 100 km grid square intersections. (b) Some of the potential factors influencing the wing-morphology in northern common blues *Polyommatus icarus*, a reflectance basker, during the present Flandrian interglacial. (From Dennis and Shreeve 1989; courtesy of the Linnaean Society and Academic Press.) Octagons, zonal factors; rectangles, butterfly morphology; ellipses, factors affecting butterfly morphology. For interpretation of signs, refer to **process-response models** in glossary.

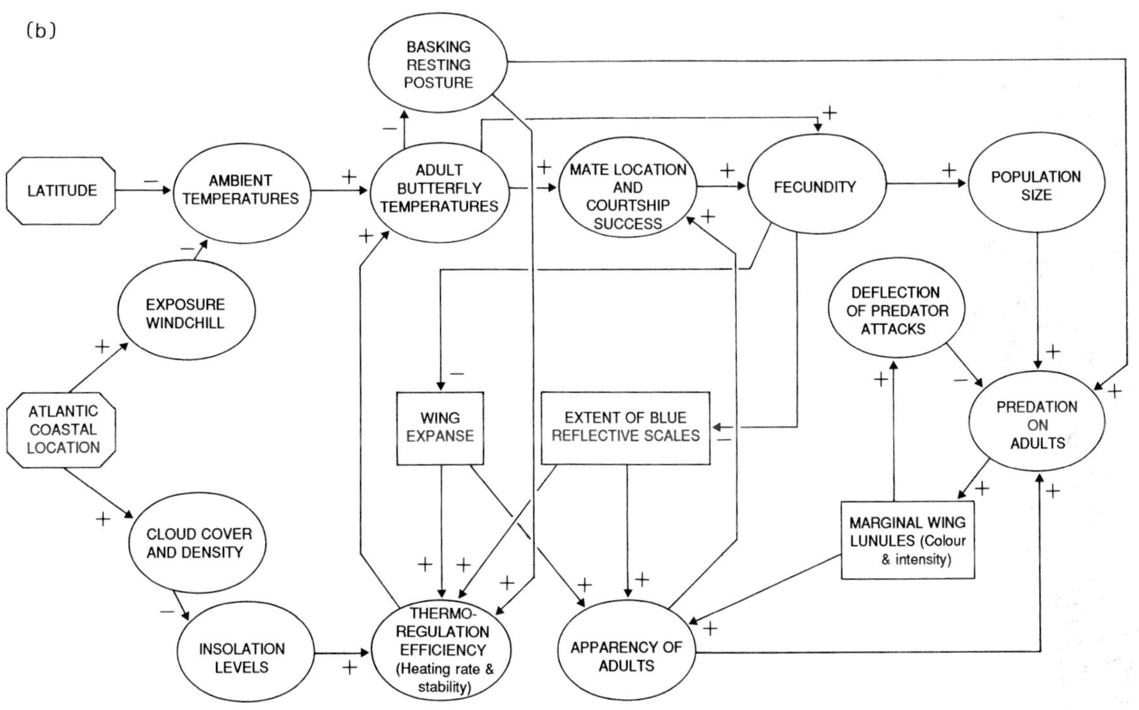

developed which accounts for the transformation in wing-phenotypes in relation to climate, especially ambient temperatures and radiation levels (Dennis and Shreeve 1989; Dennis, in preparation). This suggests that in the cooler less predictable summer conditions in the north and west of the British Isles, selection has favoured modifications in adult phenotypes that maintain efficiency in three functions: thermoregulation, mate advertisement, and predator escape. The form that wing modifications take depends mainly on constraints imposed by basking-posture (see chapter 2), which determines the allocation and interaction of functions on different wing surfaces, and by hostplant habitat structure, which influences insect thermal stability and the milieu of selection pressures. Increased basal melanism (e.g. *Argynnis aglaja*, *Pieris napi*, marsh fritillary *Eurodryas aurinia*, and *Boloria selene*; Dennis 1977; Thomson 1980) probably improves thermoregulation in dorsal and lateral baskers (see Tables 2.2 and 2.3 for details), whereas the pale colours of *Coenonympha pamphilus* and *C. tullia* may ensure concealment of these insects from predators during inclement weather (see section 9.2). Brighter colours also occur in some northern butterflies which are dorsal and lateral baskers (e.g. *M. jurtina*, grayling *Hipparchia semele*, the underside of *A. aglaia*; Dennis 1977; Thomson 1980), usually on wing surfaces exposed when the insects are active; these may enhance mate-location and courtship, or perhaps secondary defence.

The evolution of wing-morphs can be complex (see section 8.3). Some morphs may have dual functions; others are the outcome of interactions between different functions. The ventral silver pupillation and maculation in northern *Coenonympha tullia* and *Argynnis aglaja* is relatively unapparent in dull conditions but distinctly evident in sunlight (see section 9.2); thus, the spotting is cryptic when the insects are inert but has potential to deflect predators or attract mates when they are active. The way in which different functions could plausibly interact can be illustrated with reference to the striking changes in phenotype that make up female *Polyommatus icarus mariscolore* from the Orkney islands (Fig. 10.7b). The large wing-expanse and increased area of reflective bright blue scales may enable the insect to warm up more rapidly and to maintain thermal stability better than the smaller duller phenotype in a northern cool, cloudy climate; thus to find mates and to lay eggs more successfully. As populations tend to be small, more apparent females with brighter wings may also have been subject to sexual selection (Lundgren 1977). On the other hand, striking red lunules may then be required to deflect predator attacks away from vital body organs to the wing-margin, as in the case of satyrine marginal ocellation. The culmination of wing-morph modifications occurs among arctic endemic species which are invariably small, dark, or pale insects. In conditions where both ambient temperatures and radiation intensity levels are low in summer, transformations for thermoregulation, that is small size and melanized phenotypes (see section 2.2) are in phase with the dominant requirement for primary defence, that is crypsis, during extended periods of inertia when the weather is unsuitable for activity.

10.5.2 Dating variation in butterfly populations

Not all this variation need have evolved at the same time and it is difficult, if not impossible, to determine precisely periods of origin from any relationship between populations on the basis of the physical appearance of adults and early stages. It is equally difficult to identify 'ancestral' stock from among related species. Assuming that precursor populations survive in much the same geographical region, which is by no means certain, it is perhaps a little too much too expect them to be 'frozen', physically and biochemically, in evolutionary time. Phenetic similarities between isolated populations could imply close relationship. On the other hand, because we do not know the biological basis for much of the variation in wing-patterning and other attributes, despite recent advances (see chapters 8 and 9), similarities in phenotype could also reflect on adaptations to similar selection pressures (convergence) in different regions or simply on the impact of environmental influences (Thomson 1973). Conversely, identity in wing pattern can conceal significant genetic differences associated with evolution during isolation under other pressures (Oliver 1972*a*, 1977).

Important advances are promised by electrophoretic techniques, and by carefully designed cross-breeding experiments, but much of this work is still in its infancy. Electrophoretic techniques, more closely than aspects of morphology, reflect direct gene products, their loci, and alleles, and provide measures of genetic distance between populations. The alleles of these chemical variants—allozymes—are believed by some (see Throckmorton 1978; Korey 1981; Thorpe 1982) to 'evolve' at approximately constant rates. They argue that this, in conjunction with 'genetic' distances between organisms, allows some determination of their period of origin, much as radioactive isotopes permit the dating of rock strata and fossils. Although an accurate molecular clock is unlikely, and absolute values of such 'distances' may prove difficult to translate between different life-forms and into geological time, the relative 'genetic' distances obtained between populations could provide an improvement on the results of phenetic studies, especially as many of the latter have been based on inappropriate sampling designs and inadequate numerical techniques. Nevertheless, caution is advised; it is important to remember that when allozymes are being compared between populations only a small sample of biochemical variation within each organism is being assessed. It still remains necessary to demonstrate the genetic basis of the variation and to understand the function of biochemical differences and similarities (see Courtney 1982*b*; Watt *et al.* 1983).

Most of the physical variation in British butterflies is not great; much of it is unique to these islands and coincides with present-day gradients or specific ecological settings. As a result there is a temptation to date it to the post-Devensian maximum entry, following the Late Glacial and Flandrian incursions. The smaller the unit of differentiation and the more closely it can be related to a physical or ecological island, the more probable is *in situ* development. A number of interesting local races have been described, good examples of which occur on small islands such as *Maniola jurtina cassiteridum* and *Pararge aegeria insula* on the Isles of Scilly. Included in this category are isolated populations of *Plebejus argus* (de Worms 1949), *Maculinea arion* (Thomas 1976) and inland northern colonies of *Hipparchia semele*. Other examples are provided by *H. semele clarensis* and the dingy skipper *Erynnis tages baynesi* on the Burren limestone, *Pararge aegeria drumensis* in the Welsh Carneddau, and the ringlet *Aphantopus hyperantus* from Kerry. This scale of variation, as described for the Fenland butterflies, *Lycaena dispar dispar* and *Papilio machaon britannicus*, typically coincides with distinctive environments and rarely requires consideration beyond *in situ* development. It frequently occurs for populations at the edge of a species' range. In the case of *Coenonympha tullia* and *Erebia aethiops*, populations at the southern limit are distinct (Dennis 1977; Dennis *et al.* 1986).

On the Great Orme in North Wales there are two unrelated species—*Hipparchia semele* and *Plebejus argus*—both of which comprise small adults which emerge earlier than conspecifics elsewhere and which have parallel distributions on the west and south scree slopes (Fig. 10.8). *H. semele*, although larger on the adjoining sand-dunes at Llandudno and Conway Morfa, is also significantly smaller than it should be in that habitat (Dennis 1972*a*, 1977). Other butterflies (e.g. *Maniola jurtina*) are apparently unaffected (Dennis, unpublished data). Both local races may well represent adaptations to water-deficiency and desiccated hostplants on unstable screes since isolation by surrounding tree-cover in the Boreal. The steep slopes and permeable limestones of the Great Orme receive very little precipitation (<700 mm per annum) and experience very high evaporation from prevailing winds and direct afternoon sunshine. Higher sea-levels in the Atlantic period isolated the populations further, but when the sandy tombolo linked the Great Orme to the mainland once more, after 4000 years BP, *H. semele* spread to the adjoining sand-dunes. The north and east sides of the Great Orme are much damper, and the dense turf makes an unsuitable habitat for both these butterflies and for the hostplants (mainly *Helianthemum* spp.) and ants on which *P. argus* depends, but these areas and the rest of the headland would have been occupied by other butterflies with the incursion of Neolithic man.

Perhaps the most likely explanation for the early emergence and the smaller size of adults of *Hipparchia semele* and *Plebejus argus* on the Great Orme, is that selection eliminates individuals which are slow to develop in early summer. This is the period when the larvae of both species grow most rapidly. Although development may start in the early spring, because of premature plant growth under the influence of the mild climate along the North Wales coastline, the larvae depend on hostplants which are maturing rapidly as the limestone screes dry out. Larvae which develop slowly will have to contend

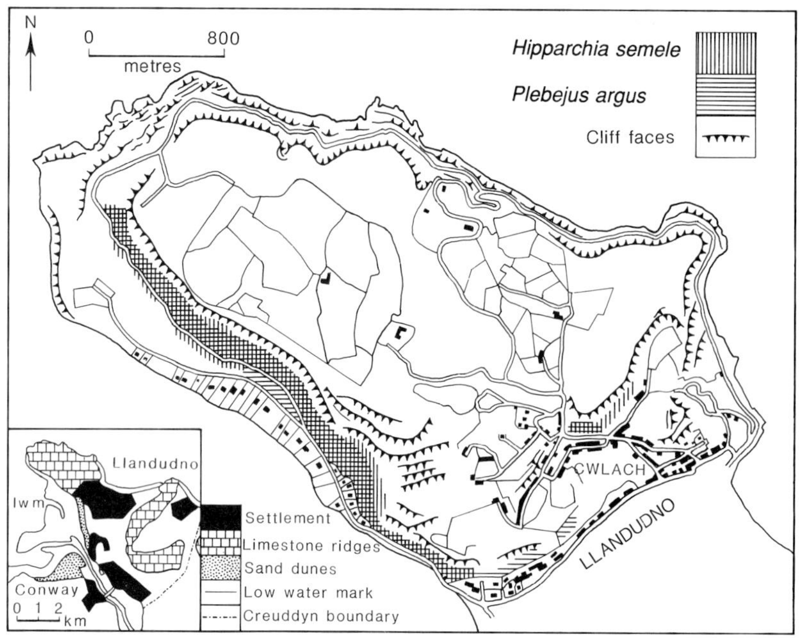

Fig. 10.8 The distribution of local races of the grayling *Hipparchia semele* and the silver-studded blue *Plebejus argus* on the Great Ormes Head, North Wales. (After Dennis, 1972*a*,*b*; 1977.)

with plant material which is perhaps tougher and contains less nitrogen and water; their growth will be further retarded and they will then be exposed for longer periods to parasitoids and predators. The influence of mild springs and early summer desiccation would have been more dramatic in the Boreal when *Hipparchia semele* and *Plebejus argus* became isolated on the headland. Selection is less likely to impinge on the two butterflies once the adults have eclosed in mid-summer, as *P. argus* remains in diapause in the egg-stage until spring, and *H. semele* larvae develop slowly and intermittently through the first two instars on tiny amounts of new growth on sheep's fescue *Festuca ovina*. For the larger part of the summer they aestivate on dead grass blades and remain well-camouflaged (Shreeve, personal communication). There are other populations (e.g. on the Malvern Hills) of *H. semele* which comprise small adults, but it must not be assumed that the same combination of factors, whatever these are, are responsible for small wing expanse in different locations.

Clinal variation through northern Britain, culminating in western Scotland, is more difficult to date. It is worth exploring the reasons in some detail as a cautionary tale for those intent on speculating on evolutionary origins.

First, subspecific terms used to describe the phenetic changes northwards may be unique to Britain (see Dennis 1977) and may embrace variation differing in magnitude and biological significance. Neither the designation of subspecies nor the associated physical markers can be used unequivocally as measures of antiquity or degree of mixing among sibling taxa. The attributes may simply be subject to selection and the genes for them common to the collective gene pool.

Second, the physical variation lacks quantification and genetic analysis. Thus it is difficult to determine any relationships with similar trends in variation described for population units on the European mainland north into Scandinavia (see Petersen 1947). The physical variation as understood provides an inadequate relational system for determining evolution pre- or post-isolation of the British landmass (Dennis 1977; Dennis and Shreeve 1989).

Finally, even when variation is shown to be unique to part of Britain, it is not proof that it evolved here. The possibility of evolution in allopatry on the European landmass and biased survival to particular regions after the Devensian maximum remains.

In formulating evolutionary scenarios for organisms without fossil material there can be no substitute for understanding the genetics and functional significance of attributes used and basing reconstructions on appropriate tolerance criteria and detailed geological templates. Much of the infraspecific variation northwards (Fig. 10.7) may represent adaptation to Holocene climatic gradients as the high correlations with contemporary climate and the model for phenetic differentiation, outlined briefly above, may suggest. But, this could be a gross misrepresentation. The species differ in type of physical modifications, distributions, adaptations, and almost certainly in degree of differentiation. This is not evidence in itself that the variation differs in age, but neither can this be dismissed. There are also precedents among other organisms, in different species, for the co-variation of different types of variation, that is, coincidental examples of primary and secondary intergradation, the classic example of the latter being the tension zone between *Corvus corone corone* and *C. corone cornix* (Aves). There is no reason why the marked environmental gradients into Scotland, exaggerated by altitude, should not only be a cause of phenetic differentiation but also provide opportunities for the stabilization of tension zones between closely related taxa which have evolved elsewhere in allopatry. Furthermore, similar environmental gradients, differing in location and shifting with climatic change, have undoubtedly existed throughout the Pleistocene in Europe, allowing opportunity for evolution in allopatry or parapatry during any one of numerous glacial–interglacial cycles or earlier.

These observations suggest that it is inadvisable to attempt evolutionary reconstructions of the British fauna in isolation of patterns and events on the European mainland. Although at this stage it is not possible to determine the timing and detailed pathways for the evolution of phenetic patterns and taxa in Europe, multivariate analysis of the distributions of European species and their subunits reveals broad geographical units (faunal units and elements) which suggest cyclical evolutionary events involving extinctions, isolation, separate development of populations in refugia, and subsequent migrations and colonization, corresponding to Pleistocene polyglacialism. A model has been developed based on long-term survival of core populations in southern Europe during Pleistocene glaciations that accounts not only for the faunal patterns but also for a bias in endemism in southern Europe, the maintenance of tension zones and widespread infraspecific variation in northern Europe, and for evolutionary rates (speciation rates) potentially far more rapid than indicated for fossil carnivorous Coleoptera (Coope 1965, 1970, 1987), in spite of the absence of long-term geographic isolation (>100 000 years) (Dennis *et al.* 1991). However, in detail, the evolutionary history of European butterflies is extremely complicated (see Lorković 1958). Species differ in their requirements for hostplants, habitats, and environmental conditions; as the climate of one glacial-interglacial failed to reproduce that of any other, this must have made for differences in response—adaptation, susceptibility to extinction, colonization potential, refuge locations, and the positioning of tension zones (Dennis *et al.* 1991). The most suitable opportunities for evolutionary analysis are provided by sibling taxa (e.g. *Pieris napi*, and *P. bryoniae*; Lorković 1962, 1986; Geiger 1980; Courtney 1982*b*; Bowden 1985) particularly at 'tension' zones between them (e.g. bath whites *Pontia daplidice* and *P. edusa*; Geiger *et al.* 1988; Wagener 1988). However, the plain fact remains that insufficient data currently exist, especially of the functional significance of genetic attributes studied, for the purpose of detailed evolutionary reconstructions (see Guillaumin and Descimon 1976).

The *Aricia agestis-artaxerxes* complex in Britain presents a particularly interesting issue, as it comprises three population groups falling into two pairs of taxa. Their evolution may be attributed to either side of the critical temporal divide—pre- or post the latest arrival date into Britain—and therefore to different locational settings. The group includes two sibling species, southern *Aricia agestis* and northern *A. artaxerxes* and two races of *Aricia artaxerxes*, *salmacis* and *artaxerxes* (Fig. 10.9). Much painstaking work has been done on this group of butterflies by Høegh-Guldberg and Jarvis (Høegh-Guldberg 1966, 1968, 1974, 1979, 1985; Høegh-Guldberg and Hansen 1977; Høegh-Guldberg and Jarvis 1970, 1974; Jarvis 1962, 1966, 1968, 1969, 1976); despite this, the taxonomy of the whole group is still very unclear. The sibling-species pair seems to belong to a larger assemblage including possibly two bivoltine (*cramera* and *agestis*) and two univoltine (*montensis* and *artaxerxes*) species in two western (*cramera* and *montensis*) and two eastern (*agestis* and *artaxerxes*)

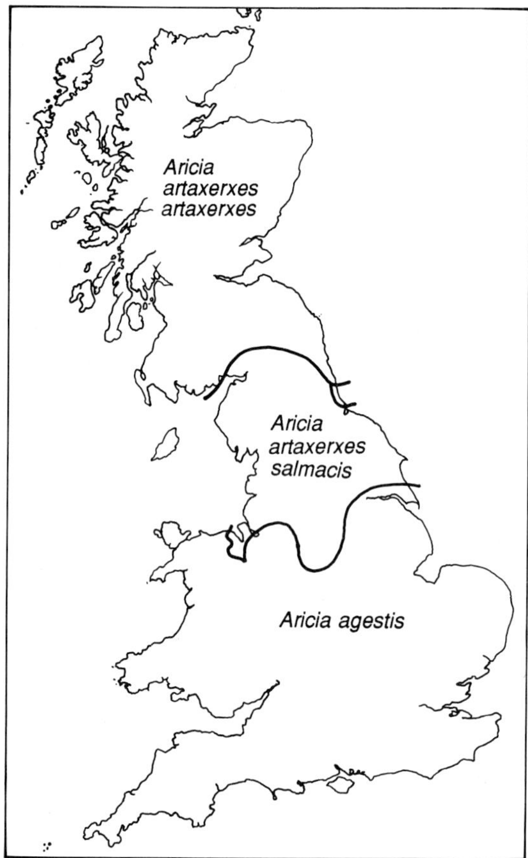

Fig. 10.9 The generalized distribution of brown argus *Aricia agestis* and northern brown argus *A. artaxerxes* in Britain. Conventional boundaries are illustrated separating *A. agestis* and *A. artaxerxes* on the basis of voltinism and *A. artaxerxes salmacis* and *A. artaxerxes artaxerxes* on the frequency of the fully expressed dorsal white discal spotting (cf. Dennis 1977). On these criteria areas of overlap may occur in north-east Wales and Derbyshire (R. W. Whitehead, personal communication) and along the Durham coast where the *artaxerxes* gene declines from fixation in Scottish populations to low frequencies (recessive homozygotes <10 per cent).

groups (Høegh-Guldberg, personal communication). *Aricia agestis* and *A. artaxerxes* differ morphologically, genetically, biochemically, and ecologically, especially in thermal environment (Frydenberg and Høegh-Guldberg 1966; Jelnes 1975*a,b,c*; Appendices 1 and 2) but overlap in all criteria with perhaps the exception of voltinism which, apparently, is the sole absolute species-specific character. Both occur in eastern Denmark and southern Sweden where differences in voltinism ensure temporal isolation in a zone of overlap. Some genetic disturbance occurs when these two species are crossed, but they are capable of producing fertile offspring. There is some evidence that the two species may overlap in North Wales and Derbyshire, in which case, variable weather conditions could result in introgression from time to time.

The coexistence of these species, both of which occupy Britain and Europe, points to their evolutionary divergence prior to their Late Glacial and Holocene entry into Britain. Research on fossil Coleoptera would suggest that insect species are of great antiquity. Coope (1965, 1970, 1987) found ample evidence in the Carabidae for long-term conservatism in morphology (> 200 000 years), for low extinction rates and for the ability of beetles to track climatic changes over thousands of kilometres. He argued that conditions have only been sufficiently stable for the evolution of new beetle species on southern Atlantic islands, small isolated refuges equatorward of the region influenced by movements of the polar front. Although it is not possible to date the origin of the *Aricia* species units, for several reasons it would be unwise to assume that these too are necessarily of great antiquity and that more recent glacial–interglacial cycles are incapable of triggering speciation events. Firstly, butterflies are phytophagous and are more dependent on specific habits than carnivorous Coleoptera; thus they are more easily isolated and their movements dictated by these resources. Secondly, long-term isolation of core populations in disjunct parts of southern Europe is maintained during the glacial periods as populations in regions of confluence and admixture in northern Europe are erased. Each interglacial allows colonization of northern territories and testing of genetic differences at tension zones (Barton and Hewitt 1989; Dennis *et al.* 1991). Thirdly, many of the physical differences in *Aricia* and other good biological species, would not make sound morphological species and would be indistinguishable if found in the fossil record. Fourthly, larger physical and genetic differences, as between *Aricia cramera* and the remaining species are not evidence, *ipso facto*, of antiquity; differences in genitalia do not necessarily imply large genetic divergence nor necessarily inability to mate (Shapiro 1978; Porter and Shapiro 1990). Finally, there is some evidence that significant genetic differences can accumulate rapidly in areas that can only have been colonized by genetically uniform populations in the Holocene (Dennis 1977; Thomson 1987).

Aricia artaxerxes salmacis and *A. artaxerxes artaxerxes* are primarily distinguished by the white discal spots on the dorsal wing surfaces in the latter; this feature behaves fundamentally as an autosomal recessive gene and is allegedly unique to Britain (Jarvis 1966, 1968, 1969). Cross-breeding experiments between English *A. agestis* and Scottish *A. artaxerxes artaxerxes* produced greater disparity than between *A. agestis* and *A. artaxerxes* in southern Scandinavia, including all female broods, F1 proterogyny and several simple genetic differences in morphology (Frydenberg and Høegh-Guldberg 1966). Differences between *A. artaxerxes artaxerxes* and the varied *A. artaxerxes salmacis* are less substantial; they overlap on the Durham coast where the recessive gene for the white discal spotting occurs but at a much reduced frequency. As the differences between *salmacis* and *artaxerxes* are infraspecific, the question revolves around the evolution and maintenance of the recessive gene. *A. artaxerxes* could have colonized Britain at least 13 000 years ago and more likely than not survived the Younger Dryas, although environmental conditions would have been marginal for it over much of Britain, but especially in the north (Dennis 1977; in preparation). The simplest scenario that fits the facts would envisage the fixation of the autosomal recessive in an isolated population in Britain during the Younger Dryas, perhaps on a south facing limestone escarpment. In the immediate Post Glacial (*c.* 10 000 years BP), this group would have colonized northern Britain, followed by additional incursions of *A. artaxerxes* (*salmacis*) and, lagging behind this, *A. agestis* from further south. These movements would not necessarily have involved extensive introgression between *A. agestis* and *A. artaxerxes* despite an overlap in thermal tolerance. Although this cannot be ruled out (see Smyllie 1992), it cannot be determined from the distribution of wing morphs (e.g. wing marginal lunulation and spot pupillation) especially those which have a polygenic basis, are common to the gene pool of the two species and could be influenced by development time and selection (see p. 211; Endler 1977). Following colonization, they would have been separated largely by phenology coupled with differences in thermal tolerance and temperature gradients, and then isolated by forest development and subsequent land exploitation from the Boreal onwards. Disjunction and relative stabilization of populations by Flandrian afforestation makes it less likely that the autosomal recessive evolved later than the Pre Boreal. All this, of course, is highly speculative. It has yet to be confirmed that the *artaxerxes* gene is absent from other parts of Europe (see Henriksen and Kreutzer 1982).

If developing an evolutionary scenario for genetic variation and ecological adaptation has its pitfalls, then there are many more uncertainties to palaeogeographic reconstructions. Probably the most that can be attained when there is no fossil record are parsimonious 'best-fit' solutions that make biological sense and account for all the available facts. Caution is required, especially as butterfly taxonomy is in a state of flux and uncertainty (cf. Higgins 1975 with Kudrna 1986). Far better retrodictive modelling is required if the perpetuation of 'just so' narratives is to be avoided (Dennis, in preparation). At the very least, as geological events have become clearer and advances have been made in butterfly biology, so the mystery of our butterfly origins lessens. Their incursion is far more recent than was thought 20 years ago. If British butterflies are at all distinct from their European counterparts, then most of these differences have probably emerged during the last 15 000 years. Throughout this time, the role of human populations has become increasingly crucial and the paramount question is now, how do we hold on to them.

11

The conservation of British butterflies

Martin S. Warren

11.1 Changing butterfly populations

Britain's butterflies have been documented since the mid-nineteenth century enabling the construction of detailed maps of their past and present distributions (Heath *et al.* 1984; Emmet and Heath 1989). Over this period many species have undergone considerable changes, with the majority contracting their ranges following local extinctions, and few species expanding their ranges (see summaries in Dennis 1977; Thomson 1980; Heath *et al.* 1984). However, there are major problems with interpreting the pattern of change recorded by the presence or absence data given in standard distribution maps. Firstly, the data-base itself has a very uneven geographic coverage and the declines recorded are linked, to some extent, with the thoroughness of past and present recording. Thus, fewer declines have been detected in south-west England, Wales, and Scotland, partly because they are more remote, and so have been poorly recorded in the past. Secondly, the declines shown on distribution maps may be related to the original abundance of a species, or the number of colonies within each recording square. Thus, if the rate of decline was even across the country, extinctions would be more apparent in regions where species were sparsely represented to begin with. For many species, these sparse regions occur in eastern Britain, and far higher densities have always been present in the south and west. There is ample evidence that declines have been very severe on a more local scale in traditionally rich butterfly counties such as Dorset (Thomas and Webb 1984), but these have not yet been reflected in obvious changes on distribution maps, especially when plotted at a scale of 10 km squares.

Despite these provisos, the decline of Britain's butterflies has been extremely severe, probably more so than in any other well known group of plants or animals (Thomas 1991). Nearly half of the 59 resident species have experienced major contractions in range, and four have become extinct (Table 11.1). The first recorded extinction was of the large copper *Lycaena dispar* in 1851 (see section 11.3.1); the second was the mazarine blue *Cyaniris semiargus* which was once recorded from 22 counties but became extinct in 1877 (Fig. 11.1); the third was the black-veined white *Aporia crataegi* which formerly bred in 24 counties but died out around 1925 after becoming extremely scarce from about 1900 (Fig. 11.1). The most recent extinction was of the large blue *Maculinea arion*, which survived in numerous scattered colonies until the 1950s when it suddenly declined becoming extinct in 1979 (Fig. 11.2). Its recent history and re-introduction are described in section 11.4.2.

At least 18 other species are now far more restricted than formerly, but the timing of their declines are varied. Some, like the wood white *Leptidea sinapis* and purple emperor *Apatura iris* mainly declined in the late nineteenth century, although the former has experienced a recent revival (Warren 1984). The heath fritillary *Mellicta athalia* also began to decline in the nineteenth century, but this decline has continued and the species is now extremely rare (Figs 11.3 and 11.7). In contrast, many species have experienced major range contractions only recently. For example, the woodland fritillaries (*Boloria* and *Argynnis* species) disappeared from much of England during the 1950s and 1960s, following an earlier, slower, decline (Fig. 11.7). They are still declining rapidly. The chequered skipper *Carterocephalus palaemon* also declined at this time

Table 11.1 Major changes in the range of British butterflies over the past 150 years

Extinctions (date of extinction)	Major contraction of range	Contraction and partial re-expansion	Contraction and equal or greater re-expansion	Major expansion of range
Lycaena dispar (1851)	*Carterocephalus palaemon*	*Polygonia c-album*	*Anthocharis cardamines*	—
Cyaniris semiargus (1887)	*Hesperia comma*	*Pararge aegeria*	*Ladoga camilla*	
Aporia crataegi (*c*.1925)	*Leptidea sinapis*	*Lasiommata megera*	*Inachis io*	
Maculinea arion (1979)	*Thecla betulae*	*Thymelicus sylvestris*		
	Plebejus argus			
	Cupido minimus			
	Lysandra bellargus			
	Hamearis lucina			
	Apatura iris			
	Nymphalis polychloros			
	Boloria selene			
	Boloria euphrosyne			
	Argynnis adippe			
	Argynnis aglaja			
	Argynnis paphia			
	Eurodryas aurinia			
	Mellicta athalia			
	Melanargia galathea			

These major changes concern 29 species. Thirty-three further species including three migrants which do not usually overwinter here, have remained essentially stable over the period, although many of these species have declined in abundance within their ranges.
Updated from Heath *et al.* (1984).

and was lost from England in 1975 (Fig. 11.7), though numerous colonies remain in Scotland (Collier 1986; Ravenscroft 1991).

Butterflies with a more southerly distribution have tended to decline most severely in the northern and eastern parts of their ranges. This is most pronounced in the *Boloria* and *Argynnis* species, and marsh fritillary *Eurodryas aurinia*, which are now virtually restricted to the more westerly parts of their ranges. In contrast, those species with a northern distribution, such as the small mountain ringlet *Erebia epiphron*, Scotch argus *Erebia aethiops* and northern brown argus *Aricia artaxerxes* have declined northwards and westwards. There are exceptions to this general trend, for instance *Aporia crataegi* declined eastwards before its extinction (Fig. 11.1).

Twenty-two British species show no major range changes over the last 150 years when their distributions are mapped to 10 km squares, but many are undoubtedly less abundant within their ranges than formerly (Table 11.1). Most have experienced numerous local extinctions which are only apparent when their distributions are mapped at a much smaller scale (e.g. the grizzled skipper *Pyrgus malvae* in Suffolk, Mendel and Piotrowski 1986).

Not all range contractions have been permanent. The white admiral *Ladoga camilla* was widespread in southern England in the early nineteenth century (Pollard 1979*b*), but was restricted to an area around Hampshire at the start of the twentieth century. During the 1930s and 1940s the range expanded and it now occurs in most of southern England and continues to colonize new localities (Fig. 11.4). The comma *Polygonia c-album* (see Fig. 11.5), speckled wood *Pararge aegeria* and wall brown *Lasiommata megera* all contracted in range at the turn of the twentieth century, but have since re-colonized most of their former areas. *P. aegeria* is still expanding its

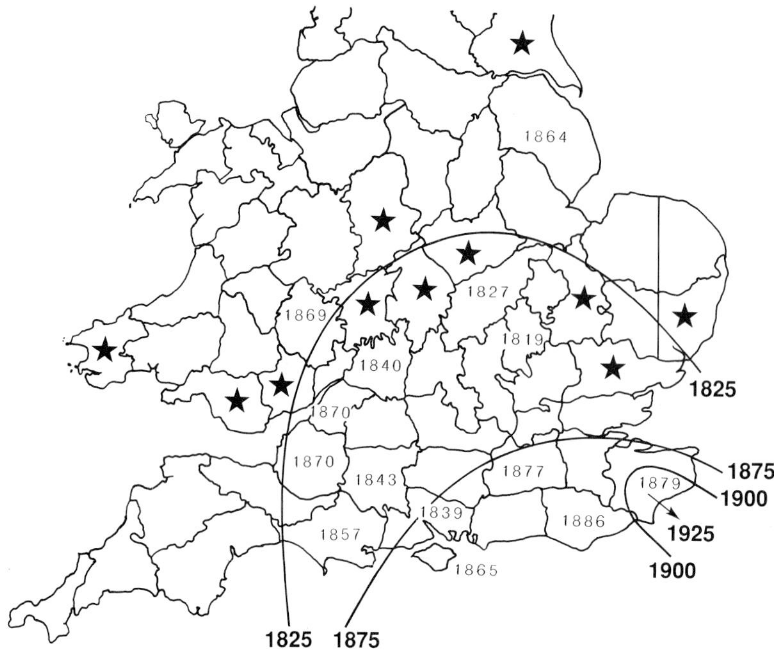

Fig. 11.1 The decline and extinction of the mazarine blue *Cyaniris semiargus* and the black-veined white *Aporia crataegi*. Dates within each vice-county indicate the last record *C. semiargus*, whereas a pre-1900 record with no accurate date is marked with an asterisk. (Compiled from Biological Records Centre data; courtesy of the Institute of Terrestrial Ecology, and P. T. Harding.) Dates on isolines illustrate the contraction in the range of *A. crataegi* in the 100 years prior to its extinction in 1925. (After Heath 1974, see Dennis 1977.)

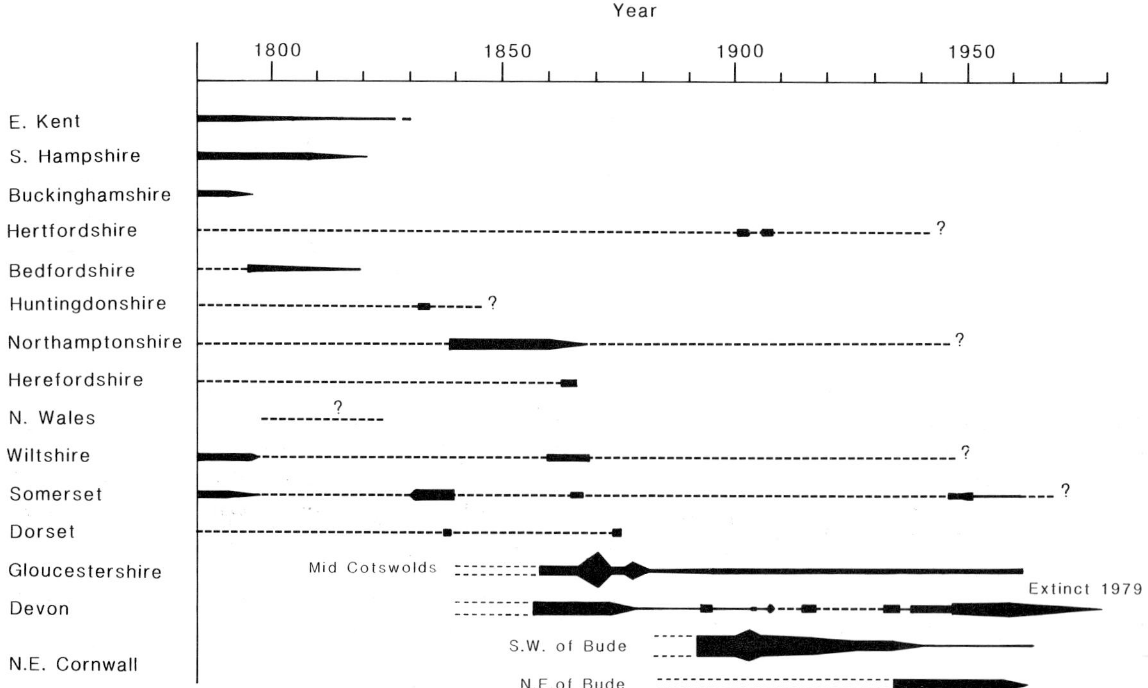

Fig. 11.2 The decline and extinction of the large blue *Maculinea arion*. Thickness of lines indicates approximate abundance. Note, *M. arion* was reintroduced to a site in south-west England in 1984. (Updated from Spooner 1963; courtesy of The Royal Entomological Society.)

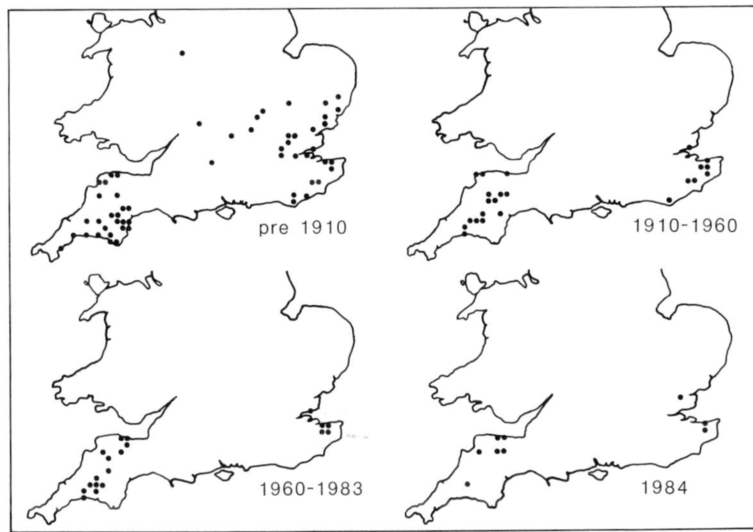

Fig. 11.3 Changes in the distribution of the heath fritillary *Mellicta athalia* plotted to 10 km grid squares. (After Warren *et al*. 1984; courtesy of *Biological Conservation* and Elsevier Applied Sciences.)

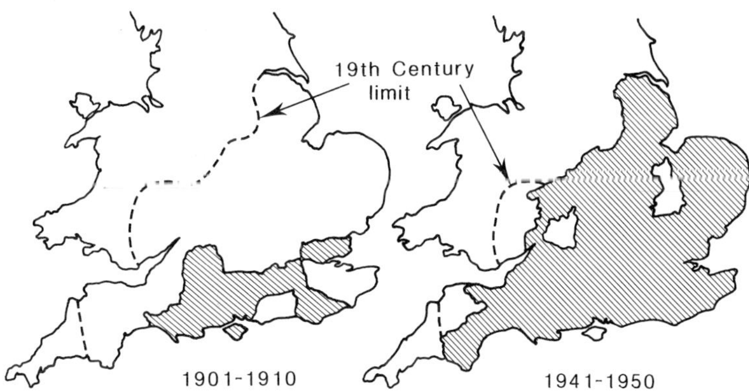

Fig. 11.4 Changes in the range of the white admiral *Ladoga camilla*. Data plotted for vice-counties from details in the entomological journals, 1901–10 and 1941–45. (After Pollard 1979*b*; courtesy of The Royal Entomological Society.)

range in East Anglia and Scotland (Emmet *et al.* 1985; Barbour 1986*b*; Mendel and Piotrowski 1986), but has not re-colonized north-east England. The peacock *Inachis io* and orange-tip *Anthocharis cardamines* have also increased their ranges following declines in Scotland and northern England (Long 1979–81; Thomson 1980).

More recently, declines of a few other species have been reversed at a local level. The marbled white *Melanargia galathea* and *Aphantopus hyperantus* have extended their ranges in northern England during the 1980s (Rafe and Jefferson 1983; Barnham and Foggitt 1987), and the Adonis blue *Lysandra bellargus* in Dorset and Wiltshire (Thomas and Webb 1984; Fuller, M. personal communication). Other species, such as the Lulworth skipper *Thymelicus acteon* are more abundant within their ranges than previously but have not increased in range (Thomas 1983*b*).

11.2 Causes of decline of British butterflies

11.2.1 The impact of butterfly collectors

Since the nineteenth century collecting has been frequently cited as a major cause of butterfly declines but in most cases, there is little evidence to support this. Moreover, the vast majority of local extinctions in Britain have occurred on sites where there has been little or no collecting. Most sites are either too remote, or poorly known, and most species are simply not collected in sufficient quantity for this to be the cause of the enormous number of local extinctions. Conversely, many populations of local species have survived lengthy periods of regular collecting.

Collectors impose additional predation on the adult stage which is largely selective of the freshest individuals. Thus, collecting tends to be directed against individuals with the greatest reproductive potential. The only populations at risk are those that occupy small areas in accessible sites. For most species the maximum proportion of the population that can be caught is too low to influence the chance of survival in the future (Thomas 1983*a*). The only substantive claims that collectors may have led to extinction have concerned small, isolated colonies of *Lycaena dispar* and *Mellicta athalia* (Duffey 1968; Warren *et al.* 1984). These colonies had already reached dangerously low levels through other causes and extinction was probably inevitable, but collectors apparently helped to speed their final demise. The Joint Committee for the Conservation of British Insects (JCCBI) code on insect collecting is reproduced in Appendix 3a.

11.2.2 Recent climatic changes

The influence of climate on butterfly populations is still poorly understood. Annual fluctuations of abundance are often the result of the action of weather at critical, short-lived, stages (see chapter 4). Examination of average climatic data may therefore mask important weather effects. Butterflies are also highly selective in their use of microclimates, especially during egg-laying (see chapter 3). Changes in climate may therefore have subtle effects on the populations of certain species. Although weather variation may determine the timing of extinctions of small populations, initially other factors are involved in reducing them to such critically low levels.

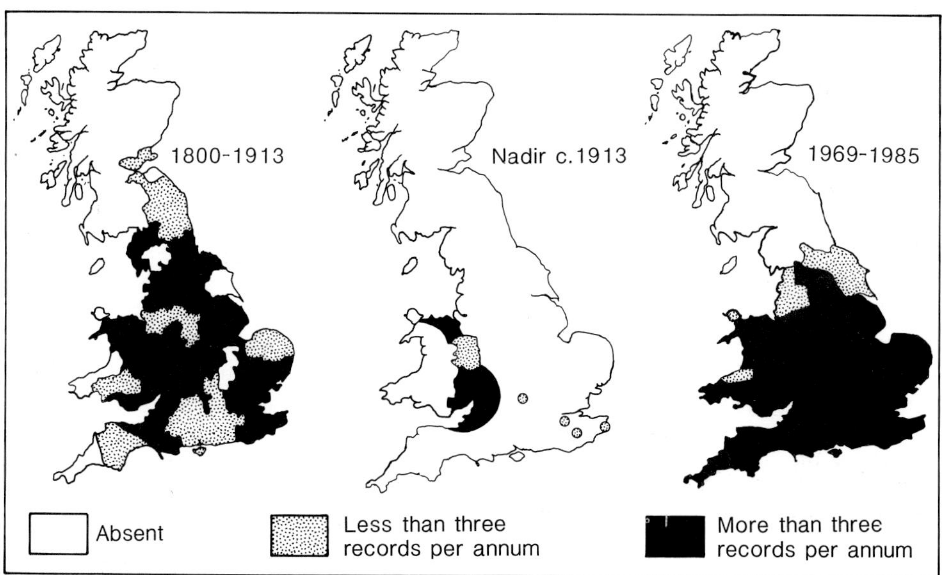

Fig. 11.5 The change in the range of the comma *Polygonia c-album* 1800–1985. (After Pratt 1986; courtesy of *The Entomologist's Record*.)

Climate undoubtedly has a major impact on the ecology of British butterflies, by causing annual fluctuations of abundance and by influencing the distributional limits of many species (see chapters 1 and 4). Favourable weather can facilitate population increase, thereby increasing the chances of colonizing new areas. This can lead to range expansion if there are suitable habitats for colonization. The expansion of the range of *Ladoga camilla* (Fig. 11.4) since the 1930s, which coincided with a period of warm weather, was probably assisted by the abandonment of coppicing which increased the suitability of woodland areas for the species (Pollard 1979*b*). Similarly, Pratt (1986) showed that major contractions and expansions in the range of *Polygonia c-album* (Fig. 11.5) were partly correlated with changes in winter temperatures, although habitat changes were also implicated. Warm summers may facilitate expansion in *Lycaena phlaeas* (Dempster 1971*b*) and *Eurodryas aurinia* (K. Paine, personal communication; M. S. Warren, personal observation) but cool summers may result in the extinction of populations in marginal habitats.

Beirne (1955) suggested that changes in the abundance of butterflies could be correlated with changing climatic conditions (Fig. 11.6). Thus, contractions of the ranges of *Polygonia c-album*, *Pararge aegeria*, and *Lasiommata megera* and the increasing scarcity of *Ladoga camilla* and *Leptidea sinapis* at the end of the nineteenth century coincided with cooler periods in the 1880s, though the warmer 1890s appear to have had little effect. The cool summers of the 1920s are thought to have been responsible for the drastic decline of *Ladoga camilla* and *Polygonia c-album* whereas warmer periods in the 1930s and 1940s may have caused expansions in their ranges (Pollard 1979*b*; Pratt 1986). However, cooler weather during the most recent 30-year period has not caused a contraction in the range of either species; indicating that other factors, such as habitat quality and quantity are also involved. For the majority of species there were no substantial range changes during these periods, indicating that they are less strongly influenced by climate, either because they occupy more stable habitats or because they are less directly affected by temperature. Thus, although the decline, or even local extinction, of a species may occasionally be due to climatic changes, these are unlikely to have been responsible for the large-scale decline of many British butterflies. The climatic shift expected as a result of global warming could, however, have a major impact on butterflies over the next few decades and is discussed by Dennis and Shreeve (1991).

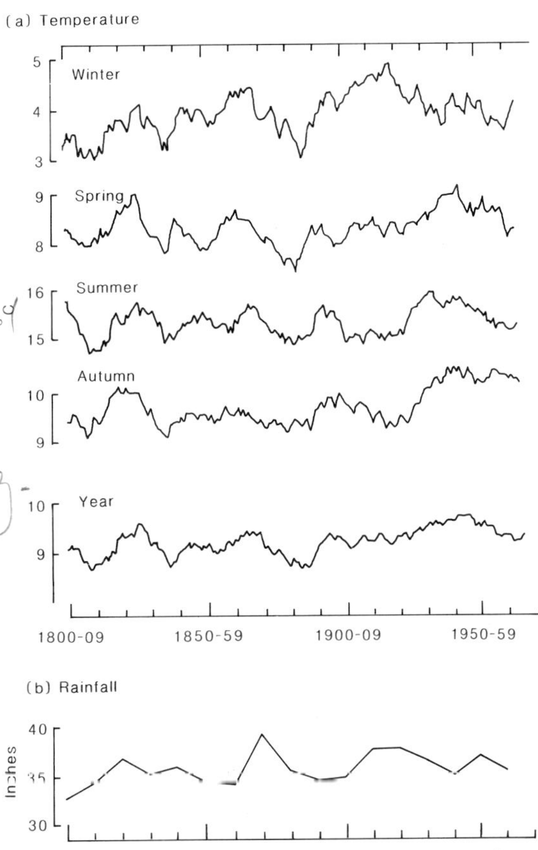

Fig. 11.6 Climatic changes in Britain since 1800. (a) Average temperature in central England by seasons and for the whole year: 10-year running means. (After Manley 1974; courtesy of the Royal Meteorological Society.) (b) Rainfall in England and Wales: 10-year running means. (After Wales-Smith 1971; courtesy of the Meteorological Office with permission of the Controller of HMSO.)

11.2.3 Chemicals and butterflies

During the last two centuries the use of chemicals—particularly artificial fertilizers, pesticides, and herbicides—in the British countryside has increased enormously. Certain persistent pesticides have caused the decline of birds of prey and certain mammals (Mellanby 1970; Moore 1983), but with the exception of the small white *Pieris rapae* (Dempster 1967, 1968) little is known of their impact on butterflies. Both the large white *P. brassicae* and *P. rapae* are a common pest of brassica crops which are subjected to regular insecticide spraying. Dempster (1967) found that most larvae of *P. rapae* are killed when the cabbages are sprayed, but a new generation of eggs is laid quickly afterwards. The survival of the new eggs and larvae was higher than in unsprayed areas though, because populations of their insect enemies, which were killed during the initial spraying, were slower to recover than those of this highly mobile butterfly.

Most British butterflies are not deliberately sprayed with pesticides and are only likely to receive low doses from sprays applied to nearby farmland. Research on the small tortoiseshell *Aglais urticae*, which regularly breeds on nettles growing in arable regions, discovered that spray drifts of two persistent chemicals (DDT and Dieldrin) rarely killed the butterfly or its larvae (Moriarty 1969*a*,*b*). In addition, wild caught adults contained pesticide concentrations that were less than one hundredth the level known to cause harmful effects in laboratory studies. However, when occasional reports of butterfly larval deaths from spray drift have been made, they nearly always refer to species such as *Inachis io* and *A. urticae* which are highly mobile and breed over wide areas. These are likely to be of minor importance compared to the major effects of parasites, or even cutting of the nettle hosts at a critical stage. Recent work has demonstrated that adult numbers of some of Britain's common butterflies tend to increase in 'conservation' field headlands where spraying with herbicides or pesticides is reduced (Rands and Sotherton 1986; Dover 1987, 1991), but it is not known if this influences breeding success. The headlands themselves are ploughed each year along with the rest of the field and chiefly provide a short-term source of nectar, with very few hostplants. They may, however, encourage some species to stay within the fields and breed more regularly in the neighbouring hedgerows.

The increased use of herbicides and fertilizers, particularly since 1945, has adversely affected most butterflies, mainly because it has contributed to habitat loss (see section 11.2.4). Both herbicides and fertilizers reduce plant diversity and destroy many important butterfly hostplants in permanent pastures, hay-meadows, and field edges. Intensively 'improved' pasture can support only one British species *Colias croceus*, which feeds on clover flowerheads, whereas 'unimproved' pasture is a major breeding habitat of 28 species (Thomas 1984*a*; Appendix 2). Perhaps the only butterfly hostplant that flourishes with heavy fertilizer use is stinging nettle, and this may have benefited nettle-feeders such as *Vanessa atalanta*, *Cynthia cardui*, *Inachis io*, and *Aglais urticae*.

Airborne pollutants, particularly acid depositions, have a major affect on wildlife (Fry and Cooke 1984), and there is some evidence that they are detrimental to certain European butterflies (Heath 1981). Barbour (1986*a*) demonstrated that in Britain the regions with the most extinctions since 1970, the north midlands and eastern England, are those with the most depleted epiphytic lichens which are indicators of air quality. However, this correlation need not imply a causal relationship between air pollution and butterfly depletions (see section 1.2; Warren 1989). Nevertheless, it remains a legitimate cause for concern, and experiments are needed urgently to determine its true effects.

11.2.4 Changing land use and habitat loss

Until the eighteenth century the net effect of human impact on the countryside was probably beneficial to butterflies (see chapter 10). Many British species are dependent on early successional habitats which were maintained by traditional agricultural and forestry practices. Since the end of the eighteenth century, land management practices have changed drastically and human impact has become increasingly unfavourable, particularly during the last 50 years (Fig. 11.7).

Since the eighteenth century, changes in agricultural practices resulted in the loss of a large amount of rough 'marginal' land. Wetlands, including the huge Fenlands of East Anglia, the stronghold of the large copper *Lycaena dispar* and swallowtail *Papilio machaon*, were drained and ploughed. Heathlands, the home of specialized butterflies like the silver-studded blue *Plebejus argus* and grayling *Hip-*

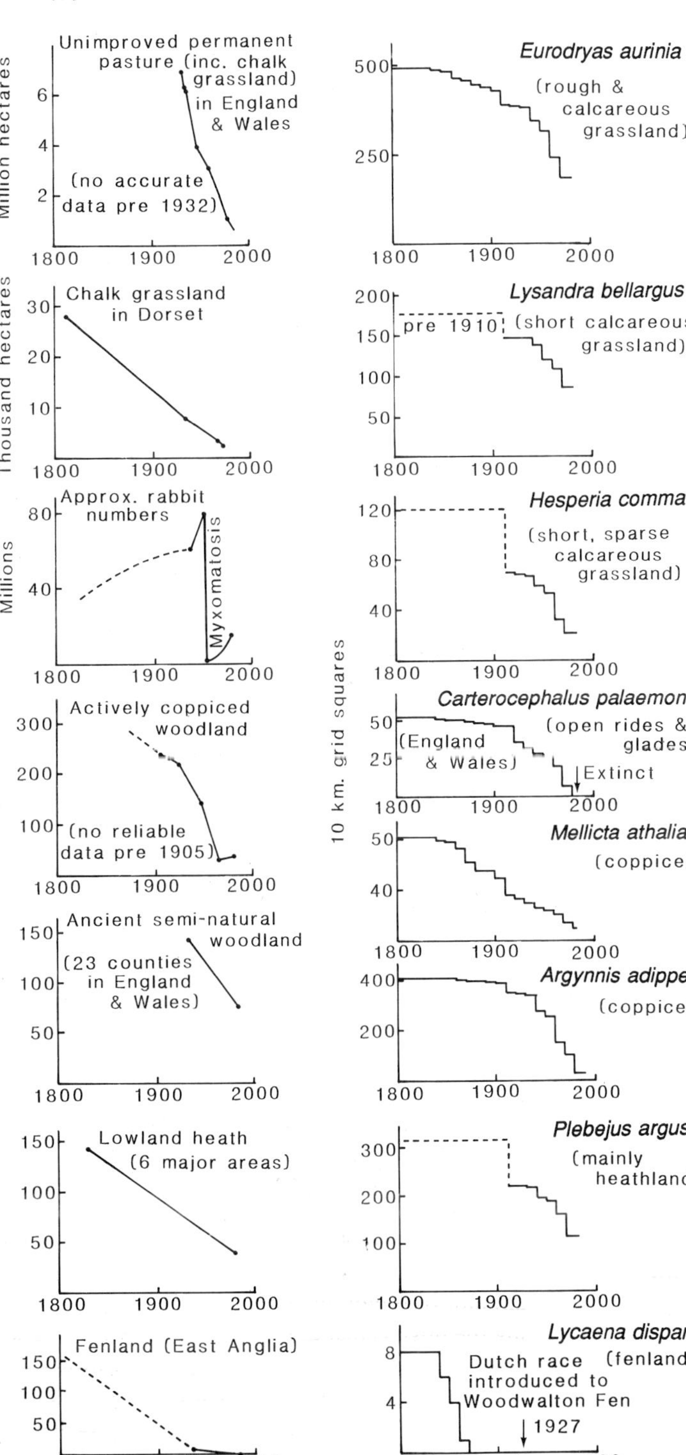

Fig. 11.7 Loss of major semi-natural habitats and the decline of selected butterflies since 1800. All data refer to Great Britain unless otherwise stated. Sources (a) (i) Fuller 1987; (ii) Jones 1973; (iii) Lloyd 1970, 1981; Sheail 1971; (iv) Peterken 1981; Forestry Commission 1984; (v) and (vi) Nature Conservancy Council 1984; (b) (i)–(vii) Biological Records Centre data. Note, butterfly declines are plotted as 'last date' records, assuming continuous prior residence in each 10 km grid square. The plots are not precise because of the limitations of the recording data, but give an indication of the timing and extent of the declines. Pre-1910 records have been summed for some species owing to the lack of precise dates.

parchia semele were reclaimed. Downland and flower-rich grasslands, which are important habitats for numerous butterflies, were radically changed either by fertilizer applications, drainage, ploughing, or reseeding. These agricultural improvements gathered pace during the early twentieth century but were sharply accelerated after 1939 in response to government policy. In more recent years, the entry of Britain into the EEC, and the price support system of the Common Agricultural Policy, has led to continued intensification (see reviews by Mabey 1980; Shoard 1980; Body 1982, 1987). In addition to these losses, a few butterflies have also suffered from natural disasters, such as Dutch-elm disease which has decimated the food plants of the white-letter hairstreak *Satyrium w-album*. Fortunately, this species appears to be more adaptable than at first thought, and many colonies have survived where there is abundant sucker regrowth of elms (Davies 1985).

Habitat loss since 1939 has been unprecedented (Rackham 1986). An estimated 95 per cent of lowland neutral grassland (which includes flower-rich hay-meadows), 80 per cent of chalk and limestone grassland, nearly 50 per cent of ancient broadleaved woodland, and 40 per cent of lowland heaths have been destroyed (Nature Conservancy Council 1984). The extent of habitat loss has not been even across Britain, and some regions, notably eastern England, have suffered more than others. This region contains a high proportion of flat and fertile land which is very productive agriculturally, and has been extensively drained and improved. It has therefore suffered relatively greater losses of major butterfly habitats such as unimproved grassland, ancient woodland and hedgerow. Because of this, the proportion of land under arable cultivation corresponds fairly closely with the main areas of butterfly loss (see chapter 1; Fig. 1.6). Many other factors such as climate, air pollution, and even pheasant *Phasianus colchicus* density have produced a similar pattern of loss (Barbour 1986*a*; Dennis and Williams, 1986; Turner 1986; Corke 1989) but this type of correlation cannot be taken as proof of cause and effect.

Not all declines have been caused solely by wholesale habitat destruction. In many cases, the loss of habitat has been far more subtle and has been caused by changes in management which affect habitat quality. For example, decreases in grazing of chalk downland led to an increase in vegetation height and a reduction of temperature within the sward (see chapter 3). This cooling resulted in the loss of the red ant host *Myrmica sabuleti* of *Maculinea arion* and loss of suitable egg-laying sites for the silver-spotted skipper *Hesperia comma* and Adonis blue *Lysandra bellargus* (Thomas 1980, 1983*a*; Thomas *et al.* 1986). Consequently all three species became extinct on many sites. Declines were particularly severe in the 1950s when myxomatosis reduced rabbit populations which had formerly helped to maintain short, sparse turf (Fig. 11.7).

Several butterflies rely on the early successional stages following clearance within woodland, and soon disappear if woodland is unmanaged (Warren and Key 1991). In the past, suitable conditions were created in woodlands throughout lowland Britain by the traditional management of coppicing. This created a diverse woodland age-structure and a variety of habitats from open sunny clearings to dense, shady woodland. Furthermore, coppicing was carried out on a fairly strict rotation, usually 5–15 years, so that open habitats were created continuously within each wood. During the Middle Ages, most woods in lowland Britain were managed as coppice (see section 10.4), but, at the start of the nineteenth century this practice began to decline when wood was replaced by coal as a major fuel. It has now disappeared from most parts of the country (Fig. 11.7; Rackham 1980, 1986; Peterken 1981). This major change in woodland management has had repercussions for many butterflies and has led to severe declines in many fritillaries, notably *Mellicta athalia*, *Boloria euphrosyne*, *B. selene*, and *Argynnis adippe* which often breed in newly cleared areas (Fig. 11.7). Further west in Britain, these butterflies tend to occur in more open grassland habitats, which may partly explain why they have declined most severely in the east.

Following the decline of coppicing, the extensive clearing and replanting of woods with conifers during the twentieth century has provided temporary breeding habitats. However, conifer plantations become unsuitable after a few years due to their deep shade, and thereafter remain unsuitable until felling, some 60–80 years after planting. One redeeming feature of many plantations are the grassy rides maintained for timber extraction. These can provide important breeding habitats for butterflies—but only when they are relatively open and sunny, and not cut annually across their full width (Warren and Fuller 1990). The wood white *Leptidea sinapis* is associated with lightly shaded woodland rides and over half its British colonies now occur in commercial forestry plantations (Warren 1984). However, this is a species which has declined severely due to the increased shading of many of its former coppiced sites and may decline further as the numerous new plantations mature.

Not all butterflies have suffered from these changes in woodland management and grazing levels, some have actually benefited. The decline of coppicing and general lack of management in many deciduous woods during the twentieth century may have benefited *Ladoga camilla* and *Pararge aegeria*, which breed in shaded conditions (Pollard 1979*b*; Shreeve 1986*a*). Also, the reduction in grazing pressure that led to declines in *Hesperia comma* and *Lysandra bellargus* during the 1950s and 1960s has allowed *Thymelicus acteon* to increase, as it prefers long turf (Thomas 1983*b*, 1984*b*). The Duke of Burgundy *Hamearis lucina* may have benefited temporarily from the widespread cessation of grazing on chalk downland, as it prefers longer grass and scrub but it has since died out on many sites which have become dominated by dense scrub (Oates, M. R., personal communication).

11.2.5 Habitat fragmentation and isolation

The enormous changes outlined above have not only destroyed the habitats once used by many butterfly species but have left the remaining habitats fragmented and isolated (see section 1.3). Habitat fragmentation has been cited as the cause of extinction of certain species because of:

(1) the inability of populations to survive in small habitat areas;
(2) genetic problems associated with inbreeding and reduced gene flow; and
(3) lack of recolonization following local extinctions.

The first hypothesis does not seem to be true of the majority of species which appear to be capable of sustaining viable populations in small, discrete areas, at least over 50–100 years (see section 11.4.3). For example, *Maculinea arion* is known to have survived for over 100 years on many habitat islands that were much smaller than those on which it had become extinct (Thomas 1980). The restriction of butterflies to very small habitat patches may, however, increase the likelihood of local extinction, and isolation may subsequently reduce the chances of natural recolonization. In fragmented habitats, the regional survival of species may thus depend on the functioning of a metapopulation which relies on a network of habitat patches rather than on individual small sites (Hanski and Gilpin 1991). Species that require much larger areas of breeding habitat are likely to be particularly vulnerable to habitat fragmentation and this may be implicated in some local extinctions of the brown hairstreak *Thecla betulae* and purple emperor *Apatura iris*, which breed in pockets of habitat spread over large areas. Restriction to small breeding areas may also lead to genetical changes in a population, such as the tendency for small wing-length observed in *Papilio machaon* at Wicken Fen, an isolated site in Cambridgeshire (Dempster *et al.* 1976). Similar effects have been recorded following the isolation of colonies of *M. arion* (Dempster 1991).

Possible consequences of habitat fragmentation include an increase in inbreeding and reduced gene flow caused by isolation and genetic drift associated with small breeding populations (see Frankel and Soulé 1981). Genetic problems resulting from inbreeding have been cited as the cause of decline in a few butterfly species, most notably *Maculinea arion* (Muggleton and Benham 1975), but this has been criticized on both practical and theoretical grounds (Morton 1979; Thomas 1984*a*). Populations are unlikely to suffer from inbreeding unless they go through 'bottlenecks' which reduce population size to fewer than ten adults for several generations (Nei *et al.* 1975). Loss of variability through genetic drift (see chapters 8 and 9) is also unlikely to occur unless the effective breeding population remains under about 50 adults for 20–30 generations (Franklin 1980; Soulé 1980). It is important to note that effective breeding numbers may be far smaller than total population size, as many males may not mate and many females may not lay eggs (see chapters 2, 3, and 4). Both genetical problems are overcome if a population receives about one immigrant every two to three generations (Nei *et al.* 1975). It is generally accepted therefore that if any population is consistently small enough to suffer from genetical difficulties, it probably has ecological problems that are far more severe (Berry 1977). Consistently small population size and lack of gene flow may, however, lead to the gradual loss of genetic diversity which could cause major problems over longer time scales (Brakefield 1991).

Once a butterfly becomes extinct on a site, for whatever reason, the probability of recolonization will depend upon several factors including mobility, the size and distance of the nearest surviving populations, and the nature of the intervening land. The interplay of these factors is poorly understood, but many butterflies have been unable to recolonize suitable habitats in the past, even when thriving colonies exist in fairly close proximity. The lack of natural recolonization in many regions is confirmed by the success of a large number of man-made re-introductions (see section 11.5.5). Clearly, the increased isolation of their colonies, because of habitat loss and fragmentation, reduces the ability of butterflies to recolonize new habitats if these become available. Also, it has important implications for the role of reintroduction in any overall conservation strategy.

Table 11.2 Causes of decline in British butterflies (conclusions of detailed studies only)

Species	Reference	Main cause of decline
Carterocephalus palaemon	Collier 1986	Not known (possibly decline in woodland management and conversion to conifers)
Hesperia comma	Thomas *et al*. 1986	Loss of chalk downland and reduction of grazing
Papilio machaon	Dempster *et al*. 1976	Drainage of fens
Leptidea sinapis	Warren 1985*a*	Decline in woodland management, especially coppicing
Aporia crataegi	Pratt 1983	Not known (possibly climate and disease)
Thecla betulae	Thomas 1974	Hedgerow removal, and regular hedge trimming
Satyrium w-album	Davies 1985	Loss of elms through Dutch elm disease.
Lycaena dispar	Duffey 1968	Drainage of fens
Cupido minimus	Morton 1985	Loss of calcareous grassland
Plebejus argus	Thomas, C. D. 1985*a*,*b*; Read 1985	Loss of heathland and reduction in grazing on calcareous grassland
Lysandra bellargus	Thomas 1983*a*	Loss of chalk downland and reduction of grazing
Maculinea arion	Thomas 1980	Reduction in grazing
Hamearis lucina	Oates, M. R., personal communication	Decline of coppicing and loss of chalk downland
*Ladoga camilla**	Pollard 1979*b*	Climate and habitat changes
*Polygonia c-album**	Pratt 1986	Possibly climate and decline in hop growing?
Boloria selene	Warren and Thomas (1992)	Decline of coppicing
Boloria euphrosyne	Warren and Thomas (1992)	Decline of coppicing
Argynnis adippe	Warren and Thomas (1992)	Decline of coppicing
Argynnis paphia	Warren and Thomas (1992)	Decline of coppicing
Eurodryas aurinia	Porter 1981	Improvement and drainage of damp grassland
Mellicta athalia	Warren 1987*c*	Decline of coppicing

* Decline and subsequent re-expansion.

11.2.6 Conclusions on the decline of British butterflies

Numerous environmental factors influence the inherent suitability of different regions for butterflies, making them more or less susceptible to local extinctions. Slight changes may be all that is needed to cause extinction of some species in one region, while the same changes may have little effect on others. The subject is therefore highly complex, and a different combination of factors may be involved for each species. Simplistic explanations of the pattern of butterfly declines can therefore be misleading and further research is needed to draw more definite conclusions about their real causes. However, nearly all detailed investigations into the declines of British butterflies have reached the same general conclusions—they are ultimately caused by habitat changes or loss (Table 11.2).

11.3 Early attempts at conservation

11.3.1 Maintaining the large copper

One of the earliest attempts at butterfly conservation concerned the British race *dispar* of *Lycaena dispar* (Anon 1929; Duffey 1968). The butterfly was reported to be abundant in East Anglia as late as the 1820s but declined rapidly in the 1840s and was last recorded in 1851. During the early 1900s there were a number of attempts to reintroduce *L. dispar* to the British Isles, initially using the small European race *rutilis*. Then, in 1915, the more similar Dutch race *batavus* was discovered and attempts were subsequently made to establish populations on at least four sites. Nearly all succeeded in establishing colonies which thrived for several years, however, all but one died out within 15 years (see Table 11.8).

The only colony to survive was at Woodwalton Fen, but this was assisted by the caging of young, wild larvae and regular supplementing of the wild population with cage-reared adults. The open fenland habitat also was managed extensively, not only by the regular planting of great water docks but also by scrub clearance, and from the 1950s onwards by shallow peat cutting. Few precise records were kept of the size of the wild population until 1955, but from then onwards the Woodwalton colony was studied in great detail by Duffey (1968, 1977). The wild population declined steadily during the 1960s and the colony would have become extinct in 1966 and again in 1970, had it not been for the release of cage-reared adults. The conclusion of Duffey's research was that the extent of the breeding habitat that could be created on the fen for *Lycaena dispar* (about 30 ha) was too small to support a viable population.

The attempts to re-establish *Lycaena dispar* were probably the earliest examples of practical butterfly conservation in Britain (and probably in the whole world). Woodwalton Fen was the first nature reserve to be established in Britain when it was purchased by Charles Rothschild in 1910. However, there are huge problems associated with its effective management. It stands 2 m above the surrounding farmland where constant cultivation has caused the peat soils to shrink from erosion and oxidation. The high water-table, necessary for the fen vegetation, has to be maintained by a complicated system of dykes and sluices. The reserve has maintained much of its fenland character and remains one of the last relicts of the vast East Anglian fens of which *L. dispar* was an integral part. The butterfly is now an endangered species in the rest of Europe—the Dutch race *batavus* in particular is under threat (Heath 1981). Important lessons have been learnt from the reintroduction of *L. dispar* back into Britain, especially the difficulty of conserving species on small reserves. However, it is clearly not desirable to maintain a species by such artificial means and a far more satisfactory solution is to ensure that correctly managed reserves are large enough to support populations of sufficient size to overcome natural mortalities without external assistance. The British Butterfly Conservation Society (BBCS) have recently taken over responsibility for monitoring the Woodwalton population and are actively searching for other, larger, sites, but it seems unlikely that any now exist.

11.3.2 Conservation on early nature reserves

Early attempts at conserving butterflies on nature reserves had a poor success rate, most noticeably with rare or local species. As an example, Monks

Table 11.3 Changes in the butterfly fauna of Monks Wood NNR and Castor hanglands NNR compared with the region as a whole

Species	Monks Wood	Castor Hanglands	Northants/ Cambridge
Extinctions (date of last record)			
Carterocephalus palaemon	1971	1971	1975
Erynnis tages	1969	1969	Still present 1987
Pyrgus malvae	Still present 1987	1968	Still present 1987
Leptidea sinapis	19th century	—	Still present 1987
Aporia crataegi	19th century	—	19th century
Callophrys rubi	Early 1970s	1977–80?	Still present 1987
Thecla betulae	1966	1961	1980?
Satyrium pruni	1914–25	Still present 1980	Still present 1987
Satyrium w-album	1980?	Still present 1980	Still present 1987
Aricia agestis	1969	1965	Still present 1987
Lysandra coridon	—	pre-1961	Still present 1987
Maculinea arion	19th century	—	19th century
Hamearis lucina	1957	1970	19th century
Ladoga camilla	—	1960	19th century
Apatura iris	1914–25	1925	19th century
Nymphalis polychloros	1939–46	1948	1974?
Polygonia c-album	—	*c*.1850	Still present 1987
Boloria selene	19th century	—	Early 1960s
Boloria euphrosyne	1963	1962	1963
Argynnis adippe	1962	1960	1962
Argynnis paphia	1969	1966	1978
Argynnis aglaja	late 1950s	1960	1976
Eurodryas aurinia	19th century	—	1947
Melanargia galathea	1976	1969	Still present 1987
Gains			
Carterocephalus palaemon	1939–46		
Leptidea sinapis	1985 (reintroduced)	—	
Satyrium pruni	1914–25 (reintroduced)		
Polygonia c-album	1934	1933	
Ladoga camilla	1939–46	1946	
Historical total number of species			
	40	38	
Total number of losses since establishment of NNRs in 1954			
	12	14	9

—, not recorded on site, or arrived since nineteenth century. NNR, National Nature Reserve.
Updated from Collier (1966), (1978); Flinders (1982); Peachey (1982); Thomas (1984*a*); D. Goddard, personal communication.

Table 11.4 The survival of selected rare butterflies on nature reserves in Britain between 1960 and 1982

(A) Species	(B) Date	No. of known colonies at date in (B)	Total No. of colonies	Colonies on reserves at date in (B)					Extinctions on reserves 1960–81	Colonies on reserves that became extinct 1960–81 (%)	Source
				All GB colonies (%)	NNR	County wildlife trust	Other (LNR, FNR, RSPB, private)	NT			
Lycaena dispar	1982	1	1	100	1	—	—	—	1*	100*	Duffey (1968, 1977)
Maculinea arion	1967–79	4	4	100	—	2	—	1	4	100	Thomas (1976, 1980)
Carterocephalus palaemon (England only)	1960–75	16	4	25	3	—	1	—	4	100	Farrell (1973), and unpublished data
Melitaea cinxia	1982	18	7	39	—	—	—	7	0	0	Simcox and Thomas (1979) updated
Mellicta athalia	1982	29	7	24	1	—	5	1	2	100	Warren, Thomas, and Thomas (1981), and personal communication)
Satyrium pruni	1982	30	11	25	2	4	5	—	0	0	Thomas (1975) updated
Hesperia comma	1982	52	33	63	7	5	2	19	9†	24	D. J. Simcox and C. D. Thomas, personal communication
Lysandra bellargus	1982	*c*.75	19	25	3	6	2	8	4**	27	Thomas (1983*a*) updated
Thymelicus acteon	1978	81	25	30	1	2	5	17	?	?	J. A. Thomas, unpublished data
Leptidea sinapis	1979	87	10	12	3	4	3	—	?	?	Warren (1981)
Erebia epiphron (England only)	1980	?	?	70	?	?	?	70% of all	?	?	J. Hemsley, personal communication

* Reintroduced 1970
** Excluding NT properties.
† Four reserves colonized during the same period.
NNR, National Nature Reserves; LNR, Local Nature Reserves; FNR, Forest Nature Reserves; RSPB, Royal Society for the Protection of Birds.
After Thomas (1984*a*).

Wood National Nature Reserve in Huntingdonshire contained 35 species of resident butterfly when it was established in 1953. Since then no fewer than 11 have become extinct, including several nationally rare species (Table 11.3). A nearby reserve, Castor Hanglands, has lost at least 13 species, since its establishment in 1954 and several other reserves demonstrate similar losses. It should be noted, however, that the reserves exhibiting most losses occur in East Anglia where many butterflies have shown severe declines and several have suffered regional extinction (Table 11.3). As far as butterflies are concerned the reserves have failed to fulfil their function largely because they were not sufficiently managed and surrounding habitats were destroyed leaving them increasingly isolated.

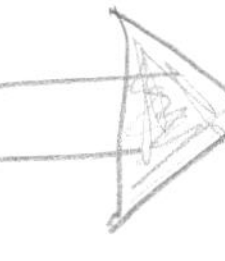

Reserves established primarily to conserve individual butterfly rarities have fared little better (see Table 11.4). Several reserves were set up during the 1950s and 1960s specifically to conserve *Maculinea*

Table 11.5 The representation of British butterflies on 131 statutory nature reserves (National, Local and Forest Nature Reserves) in 1964 and 100 National Nature Reserves in 1980[1]; and 384 County Trusts for Nature Conservation reserves and 56 Royal Society for the Protection of Birds reserves 1980–83[2]

Species	Number of reserves with current records			
	1964	1980	County Trust reserves 1980–83	RSPB reserves 1980–83
HESPERIIDAE				
Carterocephalus palaemon	4	1	0	0
Thymelicus sylvestris	33	44	—	27
Thymelicus lineola	3	12	45	11
Thymelicus acteon	0	0	2	0
Hesperia comma	2	3	4	0
Ochlodes venata	37	47	—	34
Erynnis tages	30	18	144	12
Pyrgus malvae	21	12	87	5
PAPILIONIDAE				
Papilio machaon	2	3	4	1
PIERIDAE				
Leptidea sinapis	4	2	31	2
Colias croceus†	20	8	—	28
Gonepteryx rhamni	37	44	—	28
Pieris brassicae	53	80	—	55
Pieris rapae	58	74	—	53
Pieris napi	58	91	—	55
Anthocharis cardamines	39	51	—	42
LYCAENIDAE				
Callophrys rubi	26	28	134	15
Thecla betulae	4	0	13	0
Quercusia quercus	10	18	107	16
Satyrium w-album	4	5	53	6
Satyrium pruni	4	2	8	0
Lycaena phlaeas	58	58	—	46
Lycaena dispar	1	1	0	0

Species	Number of reserves with current records			
	1964	1980	County Trust reserves 1980–83	RSPB reserves 1980–83
Cupido minimus	4	8	43	2
Plebejus argus	7	3	11	4
Aricia agestis	24	20	77	2
Aricia artaxerxes	0	6	11	2
Polyommatus icarus	53	76	—	47
Lysandra coridon	5	10	48	0
Lysandra bellargus	1	3	14	0
Celastrina argiolus	21	28	—	17
Maculinea arion	0	0	0	0
RIODININAE				
Hamearis lucina	9	4	37	1
NYMPHALIDAE				
Ladoga camilla	10	6	78	6
Apaura iris	2	0	17	0
Vanessa atalanta†	52	77	—	56
Cynthia cardui†	47	86	—	45
Aglais urticae	60	85	—	56
Nymphalis polychloros	3	0	0	0
Inachis io	49	66	—	47
Polygonia c-album	23	36	—	25
Boloria selene	22	26	67	17
Boloria euphrosyne	24	14	45	9
Argynnis adippe	9	3	18	5
Argynnis aglaja	30	35	77	12
Argynnis paphia	17	9	92	9
Eurodryas aurinia	6	6	38	2
Melitaea cinxia	0	0	0	0
Mellicta athalia	3	1	2	1
SATYRINAE				
Pararge aegeria	37	40	—	19
Lasiommata megera	43	47	—	39
Erebia epiphron	2	3	0	0
Erebia aethiops	1	11	1	6
Melanargia galathea	17	16	139	8
Hipparchia semele	30	29	49	13
Pyronia tithonus	33	51	—	31
Maniola jurtina	65	82	—	52
Aphantopus hyperantus	30	44	—	22
Coenonympha pamphilus	62	88	—	46
Coenonympha tullia	8	18	9	8

† Migrant species.

[1] After Morris (1967) and Peachey (1982).

[2] From Steel and Parsons (1985).

arion. These colonies died out one by one and in 1979 the butterfly was declared extinct in Britain. Two large populations of *Mellicta athalia* were protected on nature reserves over the same period, but by 1980 both had dwindled to a few individuals (see sections 11.4.2 and 11.4.4).

There have been two major reviews of the representation of butterflies on National Nature Reserves (NNRs) since 1945; the first was in the early 1960s and the second in the early 1980s (see Table 11.5). Several important species have declined in numbers severely over this period, even though the data covered a similar number of sites and were probably more complete for the later survey. For example, the number of NNRs with colonies of *Boloria euphrosyne* fell from 24 to 14, *Argynnis adippe* from nine to three, *Hamearis lucina* from nine to four, *Erynnis tages* from 30 to 18, and *Pyrgus malvae* from 21 to 12. This again provides evidence of the failure of reserves to conserve some of the more specialized butterflies, a problem mirrored on statutory sites of Special Scientific Interest (SSSI) (Warren, in press). However, the results should be seen in the context of wider conservation problems. Since nationally important wildlife sites were being rapidly destroyed in the post-war period, scarce resources were concentrated on reserve acquisition and little was then available for their proper management. With the benefit of hindsight, this policy has been far from satisfactory for butterflies and probably many other insects, although it may have been justifiable in a broader context.

11.4 The ecological approach to conservation

11.4.1 Historical background

In 1923 the Royal Entomological Society established its Lepidoptera Protection Committee (later renamed the Conservation Committee) specifically to address conservation matters. This committee was replaced in 1968 by the more broadly-based Joint Committee for the Conservation of British Insects (JCCBI). Most of the early attempts at butterfly conservation were aimed purely at the protection of sites, and were often unsuccessful when applied to the scarcer butterfly species. Several pioneering ecological studies on butterflies were carried out before 1940 (Ford and Ford 1930; Dowdeswell *et al.* 1940), but it was not until the 1960s that long-term ecological research began to examine all the stages of a butterfly's life cycle, often in the form of life table studies (see chapter 4). It is now apparent that many species require active habitat management and were lost from nature reserves because their requirements were not adequately known. The importance of applied ecological research is most easily appreciated by examining a few case histories.

11.4.2 The large blue

Maculinea arion always occurred in small, discrete colonies breeding on calcareous grassland where its hostplant, thyme *Thymus praecox*, was abundant. However, entomologists consistently failed to rear the larvae through to the adult stage in captivity until 1915 when Captain Purefoy discovered its parasitic life cycle (see chapter 7). He found that the larvae only feed on thyme flowers for a few weeks before they are carried underground by the red ant *Myrmica sabuleti*, where they feed to maturity on ant grubs.

The first serious attempts at conserving population of *Maculinea arion* began over 40 years ago, and in 1962 the Committee for the Conservation of the Large Blue was formed by the Royal Entomological Society's Insect Conservation Committee (now the JCCBI). Much of the early conservation effort went into the acquisition and protection of sites as many were being destroyed by activities such as ploughing. There was little research into the butterfly's ecological requirements, the main preoccupation of the Committee being to protect surviving colonies from collectors. Site protection proved just as ineffective and the butterfly continued to decline nationally until, by 1972, it was reduced to two small populations. It was only then that an intensive and detailed ecological study was commissioned to see if it could be rescued. This research led to vital discoveries about the butterfly's life history but, unfortunately the results came too late to save the British race of *M. arion* from extinction in 1979 (Thomas 1980).

Perhaps the most important discovery was that

Maculinea arion parasitizes the nests of only two species of red ant—*Myrmica sabuleti* and *M. scabrinodis*—and the survival rate of larvae was much higher in the former than in the latter (Thomas 1977). *M. sabuleti* was found to predominate only in very closely cropped grassland as this provides the warmest conditions at ground level. If grazing pressure was reduced even slightly, this ant was replaced by the less suitable *M. scabrinodis*. If grazing ceased altogether, both species of ant disappeared very rapidly and the habitat became unable to support a colony of the butterfly. In contrast to the ants, the initial larval hostplant, thyme, was found to persist for several years in ungrazed swards, and occurred in many grasslands which were unsuitable for either ant species.

Following this discovery, the habitat on all the former *Maculinea arion* sites was re-examined, but none was being grazed hard enough to support sufficiently high densities of *Myrmica sabuleti* for the butterfly's survival. This was partly due to the decline in rabbits through myxomatosis in the 1950s (section 11.2.4) and the reduction in stock grazing on poor, unimproved pasture. The only exception was the last remaining site which was still, by chance, heavily grazed. Unfortunately, the results of the detailed research emerged just too late and despite strenuous efforts, the last colony died out in 1979.

Following its extinction, plans were formulated to reintroduce *Maculinea arion* and the last site in the South West Peninsula was specifically maintained for such an event. When extinction had been fully confirmed, a trial reintroduction was attempted in 1983 using Swedish stock. This was highly successful, and a full-scale reintroduction was attempted in 1986 by placing nearly 300 larvae in the territories of red ants' nests. These have bred successfully over several years and there is every reason to believe that a viable population has again been established in Britain (Thomas 1989). Suitable conditions are now being restored on a few other former sites and further reintroductions are planned for the next decade. With luck, these will secure the butterfly's presence once more in Britain, an aim that has become more desirable since the species has declined drastically in the rest of Europe. The old British race, however, can never be restored.

11.4.3 The black hairstreak

Satyrium pruni has a very restricted distribution in Britain, being recorded in only a few sites in the east Midlands, though it feeds on blackthorn, which is both common and widespread. The first national survey of *S. pruni* was done in 1969 by Thomas (1974), who compiled a list of 59 former sites, 30 of which were still known to support colonies. However, the species was found to be so elusive in nearly every stage of its life cycle that detailed, quantitative research was virtually impossible. To overcome this difficulty, information was gathered on the approxiate size of each population, and the nature of the management of the breeding habitat.

Large populations of *Satyrium pruni* were nearly always associated with dense stands of blackthorn growing in sheltered, but generally unshaded, woodland situations. Most colonies occurred in small discrete areas within each wood, and all known cases of extinction resulted from the destruction of the whole wood, or a crucial part of it, by forestry operations. Colonies were unaffected by even major forestry operations in the rest of the wood (including clear-felling) providing that the breeding centre was left intact. The results also indicated that the butterfly is extremely sedentary and slow to colonize new habitats, even when these occur within the same wood. Although new habitats were being created elsewhere within its historical range, these were not colonized quickly enough to compensate for the destruction of existing sites by modern forestry. The study did not, however, completely solve the mystery of its restricted distribution, as suitable-looking habitats are by no means confined to the east Midlands. In fact an introduction of the butterfly during the 1950s to a wood in Surrey, well outside its historical range, has been highly successful and the colony still survives today (Thomas 1989). One possible explanation of its distribution is that traditional coppice cycles in the east Midlands were formerly longer than elsewhere, which allowed blackthorn to reach 20 or so years of age required for the butterfly to breed. In most other regions, coppice cycles were usually between 8 and 20 years, probably too short to support a colony.

The requirements and poor mobility of *Satyrium pruni* make it particularly vulnerable to habitat changes, but colonies can usually be safeguarded for some time by saving the breeding centres from destruction. This may explain why colonies on reserves have fared better than those of many other species. However, blackthorn stands eventually become moribund, and there is a risk of extinction without management. Ideally, the blackthorn should be cut on a long rotation (more than 20 years), in a piecemeal fashion, producing uneven-aged stands (Thomas 1975*a*). Up to a quarter of the blackthorn stand can be cut in any one year, but for good conservation this should not be repeated more than once every 5–10 years. A safer type of management (particularly useful with small colonies) is to leave the breeding centre intact and to encourage the regeneration of nearby blackthorn. On most of the heavy clay soils in the east Midlands, blackthorn generally responds well to such treatment.

The recommendations for the management of *Satyrium pruni* habitats have been put into effect on a number of nature reserves, with great success. By 1986, nearly half of the British sites had been incorporated in nature reserves or covered by conservation agreements. Many further sites have been protected within SSSIs, and relatively few unprotected colonies are now at risk from wholesale destruction. Thus, the future status of *S. pruni* rests firmly in the hands of conservationists and there is every sign that this will be successful.

11.4.4 The heath fritillary

For most of the twentieth century, the decline of *Mellicta athalia* has been a matter of concern and debate among entomologists. The first full national survey, organized by the JCCBI (Warren *et al.* 1984), confirmed that its decline was still continuing and that it was one of Britain's most endangered butterflies (Fig. 11.3). The survey located only six colonies in the west country and 25 in Kent, all in three large woodland blocks. Even more disturbingly three-quarters of the colonies were estimated to be fairly small (fewer than 200 adults at the peak flight period) and many of the remaining breeding sites were deteriorating rapidly following conifer planting. Also, the only two colonies protected on nature reserves had experienced severe declines to just a few individuals. The remaining sites were largely unprotected and the survey report recommended that conservation measures be taken as a matter of great urgency.

In 1982, the NCC commissioned a three-year ecological study to identify the butterfly's precise habitat requirements and discover the most appropriate forms of habitat management (Warren 1987*a,b,c*). *Mellicta athalia* was found to be an extremely sedentary species, with most adults having a range of less than 150 m, yet it often lives in ephemeral habitats. Three distinct types of habitat were identified—woodland, grassland, and heathland. In woodland, the butterfly had long been known to flourish in new clearings, where it uses common cow-wheat *Melampyrum pratense* as its main hostplant, but research showed that these remained suitable for just a few years after cutting (see Fig. 4.6). Over the years, the colony would move from one new clearing to another, earning the butterfly a local nickname as the 'woodman's follower'. Thus, in larger woods, with scattered patches of habitat, the species survives as a metapopulation within which the extinction and formation of local populations is a regular phenomenon. In the past, suitable conditions were provided under traditional coppice rotations, and the decline of *M. athalia* was thought to be attributed directly to the abandonment of this form of wood management (Fig. 11.7). The most beneficial form of management for this and several other butterflies that breed in new clearings is therefore to copy the traditional coppice rotations, ensuring that new clearings are cut each year. A network of rides connecting the coppice blocks is also highly valuable as it enables the free movement of adults and ensures the rapid colonization of the new clearings (Warren 1991*a*).

In grassland habitats, research demonstrated that some form of cutting is essential to prevent the formation of tall grassland and eventually scrub and woodland. Experimental mowing during the autumn, combined with some ground disturbance to encourage the larval hostplants (ribwort plantain *Plantago lanceolata* and germander speedwell *Veronica chamaedrys*) proved to be satisfactory on at least two sites in Cornwall. Subsequent monitoring revealed that the colonies managed deliberately in this way exhibited smaller fluctuations in population size compared to other unmanaged colonies, most of which soon became extinct.

The occurrence of *Mellicta athalia* in the third

type of habitat—heathland—was discovered very recently and its ecology there is still being unravelled. Only one heathland colony was discovered during the 1980 survey, but since then about 22 distinct colonies have been identified, all within one small region of Somerset (Warren 1991*a*). All the colonies breed in sheltered coombes where the main larval hostplant *Melampyrum pratense*, grows as scattered plants among bilberry *Vaccinium myrtillus*. Searches during the spring revealed that larger larvae can also feed on foxglove *Digitalis purpurea*, sometimes in large numbers, even though the plant is apparently ignored by egg-laying females. The importance of this secondary host to the survival of heathland populations and the effects of habitat management are the subjects of current research. However, the deliberate burning of small patches of heathland on a regular cycle, combined with moderate levels of grazing, is thought to be beneficial in the long-term.

Following the loss of *Maculinea arion*, many conservation organizations responded quickly when the threat to *Mellicta athalia* was publicized. Since 1980, several new reserves have been established primarily for this species, including two on Duchy of Cornwall land which were created at the express wish of HRH Prince Charles. The newly acquired knowledge of the butterfly's habitat requirements have been incorporated into the management plans of all the reserves, sometimes with spectacular results. On the Blean Woods NNR, numbers rose from just a few adults in 1980 to an estimated peak of 6000 in 1986 (Warren 1991*a*). Furthermore, most existing colonies are now protected either as nature reserves, or by conservation agreements that have been negotiated with the owners by the Nature Conservancy Council (NCC). The butterfly has also been re-established successfully since 1982 on two coppiced reserves in Essex (Warren 1991*a*). Like that of *Satyrium pruni*, the future of *M. athalia* will depend on the correct management of the protected areas by conservationists.

11.5 Strategies for conservation

11.5.1 Managing for butterflies

Many recent studies have addressed the subject of habitat management for butterflies and several publications are now available to give guidance to land managers (e.g. Nature Conservancy Council, 1981). In most cases, the general aim is to return to more traditional forms of land management, under which the species thrived in the recent past. However, traditional systems often need to be modified as conservationists are faced with concentrating all the former wildlife interest onto the small habitat fragments that still remain. This difficult feat is only possible following detailed ecological research into the requirements of each species. Some general principles are now beginning to emerge. A handbook entitled *Chalk Downland Management for Butterflies* (BUTT 1986) describes the habitat requirements and preferred sward heights for each downland species (see Fig. 7.2). Several grazing regimes are suggested to provide a range of sward heights, which will cater for a variety of species. One of the most valuable systems is rotational grazing which maintains a herb-rich sward, while providing a good range of microclimate and vegetation structure. This system is also known to be highly beneficial for many other insects (see Morris 1971). Rotational cutting or burning during the winter is also thought to be beneficial for the heathland habitats of *Plebejus argus* (Thomas 1985*b*; Read 1987; Ravenscroft 1990) and *Mellicta athalia* (Warren 1991*a*). Without some form of regular management, the heathland vegetation can become tall and unsuitable for these species, and there is increased risk of accidental summer fires which are catastrophic.

Within woodland, many butterflies require very open, grassy regions such as rides and clearings, while others prefer partially shaded rides (Figs 11.8 and 7.9). Warren (1985*a*) demonstrated how the width of a ride can be manipulated to vary shade levels, and suggests that a few rides in mature woodland need to be at least 20–30 m wide to cater for a full range of species. Several systems have been proposed for the management of ride vegetation, all involve zoning the ride to produce a gradation from short grass in the centre, with strips of tall grass and scrub either side (Steel and Khan 1987; Warren and

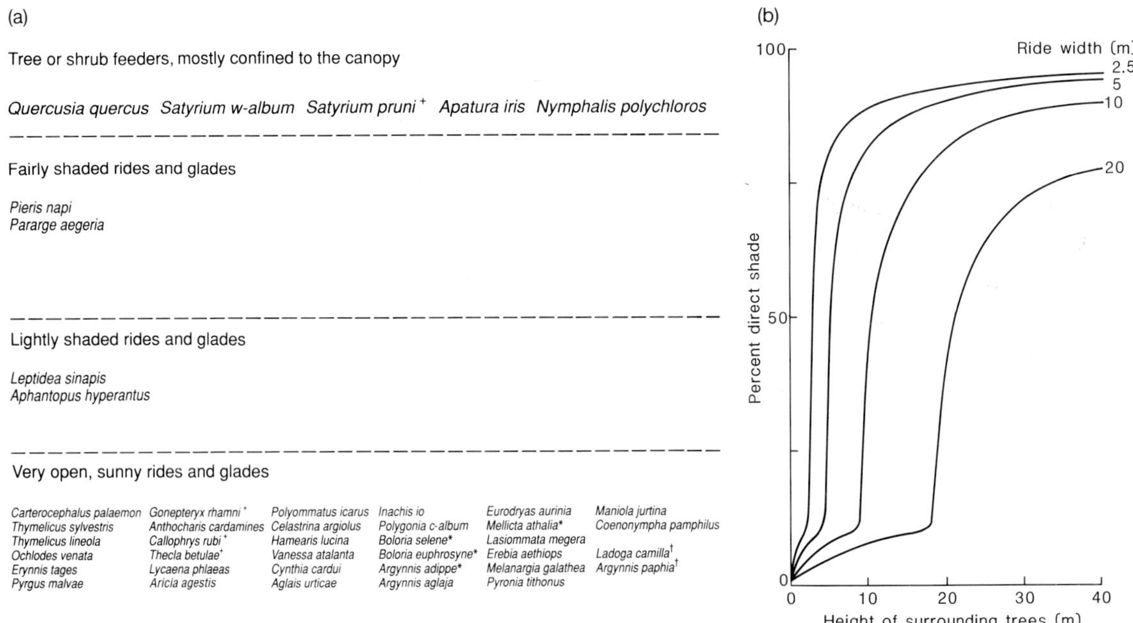

Fig. 11.8 The influence of shade on British butterflies: (a) The shade tolerance of butterflies breeding in woodland. +, shrubs needed; ★, prefer newly-cleared woodland; †, use sunny rides for nectar, but breed in adjacent deciduous woodland or ride edges. (b) The levels of shade predicted during June in model rides orientated east to west of different width and with different heights of surrounding trees. (After Warren 1985*a*; courtesy of *Biological Conservation* and Elsevier Applies Sciences.) Note, rides orientated north to south will have lower shade levels than east to west rides during the summer months.

Fuller 1990). Variations include the creation of a series of 'scallops' along the ride margin and the enlargement of ride intersections in the form of box junctions.

The active coppicing of woodland is also vital to many butterflies, notably the fritillaries (Warren 1987*b*, 1991*a*; Thomas and Snazell 1989). This management not only exposes the woodland ground flora, which includes many important foodplants such as violets, but also provides a particularly warm microclimate for the developing larvae. The effects of coppicing on butterflies has been reviewed by Warren and Thomas (1992), and recommendations for the management of coppice for butterflies and other wildlife are given by Fuller and Warren (1990).

Numerous surveys and conservation-oriented projects have been conducted recently using funds raised by the World Wildlife Fund during Butterfly Year 1981–82 (Table 11.6). Many butterfly enthusiasts are also now using the transect method to monitor populations on important sites, quite independent of the National Butterfly Monitoring Scheme (see chapter 4). This may be particularly important as butterflies are very sensitive to environmental change, including global warming (Dennis and Shreeve 1991), and may be valuable indicators of changes in other less noticeable animal groups. Although much has been learnt about the general principles of butterfly conservation, detailed management prescriptions have been worked out for only a few species. More specific and detailed practical information is greatly needed on managing habitats for butterflies.

11.5.2 Genetic considerations

In order to ensure the long-term viability of each species, conservation should aim at maintaining genetic variability throughout their geographic range. This includes the conservation of examples of

Table 11.6 Projects undertaken following butterfly year 1981–82 under sponsorship raised by the World Wildlife Fund

Project title	Project organizer or reference
National survey of *Apatura iris*	Willmott 1987
National survey of *Papilio machaon*	Hall, M. L.
National survey of *Carterocephalus palaemon*	Scottish Wildlife Trust
National survey of *Satyrium w-album*	Davies 1985
National survey of *Hamearis lucina*	Oates, M. R.
Survey of *Mellicta athalia* in Somerset	Jarman, R. 1984–86
Survey of *Coenonympha tullia* in Northumberland	Northumberland Trust for Nature Conservation
Survey of *Plebejus argus* in North Wales	Thomas, C. D. 1983
Survey of *Boloria euphrosyne* and *B. selene* in Berks, Bucks, and Oxon	Steel, C. and Shreeve, T. G.
Survey of *Erebia aethiops* in Cumbria	Cumbria Trust for Nature Conservation
Management programme for butterflies on a site in Leicestershire	Leicester & Rutland Trust for Nature Conservation
Butterfly survey of Cardigan	Fowles, A.
Butterfly survey of Suffolk	Mendel and Piotrowski 1986
Common butterfly survey in Avon	Bristol Environmental Records Centre
Study of administrative aspects of butterfly conservation	Torrie, P.
Importance of garden habitats to butterfly populations	Stephens and Warren, in press

the full range of different races and ecotypes (e.g. important butterfly races to conserve include the unique populations that occur on islands and headlands such as Great Ormes Head in North Wales). The long-term conservation of a species is also considered to be possible only in situations which permit continued evolution (Frankel 1974; Greig 1979). An effective breeding population of at least 50 has been suggested to prevent adverse effects of inbreeding, but a population of at least 500 may be needed to give evolution a chance of continuing (Frankel and Soulé 1981). If genetic diversity within a population is lost it will be less able to adapt to any environmental changes, and extinction will be more likely. It is therefore highly desirable to maximize the effective breeding size of the population to be conserved in the long-term (Brakefield 1991). The figure of 500 may not necessarily apply to butterflies which are likely to be less vulnerable to inbreeding depression owing to their high rate of reproduction. Many species, such as the small blue *Cupido minimus* and *Hamearis lucina* often occur naturally in far smaller populations without showing any obvious adverse effects in the short-term. Further research is urgently needed on this subject, but 500 is probably a useful guideline, particularly for isolated populations that are unlikely to receive any additional genes through immigration.

11.5.3 Nature reserves

In 1984, the total area of land covered by nature reserves was 266 000 hectares—two per cent of the land area of Britain. This is extremely small compared with many other forms of land use, but most reserves have been selected very carefully and represent the 'cream' of our remaining semi-natural vegetation. In addition, a large area of land

(214 000 ha) is owned by the National Trust. Although this land was not usually acquired specifically for its nature conservation value, this is often considerable and is increasingly recognized as a prime consideration of subsequent management.

Much has been published on the importance of the size and shape of nature reserves, but no clear consensus has emerged (Diamond and May 1976; Hooper 1980). Studies of both plant and animal distributions demonstrate that larger areas usually contain more species, but it is very difficult to use the results to predict how large a reserve should be to conserve any particular groups of wildlife (e.g. Boecklen and Gotelli 1984). In a British context, the theoretical arguments about size, shape, and contiguity of land to be designated as reserves are largely irrelevant (see Mason *et al.* 1984). Many British butterflies can occur in very small areas of habitat (section 4.3) and, potentially, can be conserved successfully on relatively small reserves, provided that the habitat is correctly managed. However, larger reserves or linked networks of reserves will be necessary for some species with populations which breed on patches of widely dispersed habitat, or to provide for species that breed in successional vegetation which necessitates rotational management (see section 11.4.4). A strategy for conserving butterflies in the long-term should also include better links between reserves to counteract the loss of gene flow and genetic diversity that follows isolation (e.g. Noss 1987).

Another problem that affects reserves is that their boundaries are usually determined by pre-existing land ownership and may not correspond to the natural distributions of the populations they are established to conserve. Some extinctions are therefore inevitable if peripheral habitats are destroyed. Thomas (1984*a*) suggested that this factor probably accounted for the extinction of *Thecla betulae* and *Argynnis aglaja* at Monks Wood. Reserves for butterflies therefore need to be selected with a thorough understanding of the ecology of the species concerned, and a good local knowledge of the potential sites. Subsequently, reserves need to be actively managed to maintain, and possibly enhance, the essential features on which each species depends. Adequate monitoring is also desirable to ensure that the management is successful, and that adverse trends are detected at an early stage.

Table 11.5 shows that most butterflies are well represented on designated nature reserves. National Trust land also covers numerous important butterfly localities, including a large proportion of rarities such as the Glanville fritillary *Melitaea cinxia* and *Thymelicus acteon* (Warren, in press). However, past experience has shown that the designation of reserves alone is rarely sufficient for successful conservation and reserves must be seen in the context of wider land use (see section 11.3.2).

11.5.4 Non-reserve land

Even though many important sites have now become nature reserves, the vast majority of butterfly populations still occur on privately-owned land (Warren, in press). The best wildlife areas on private land are being scheduled as SSSIs with the new protective powers of the 1981 Wildlife and Countryside Act (see reviews in Adams 1984). SSSIs are carefully selected to protect a representative selection of the best biological sites, ensuring a good geographical spread. The selection procedure places special emphasis on populations of butterflies that are either listed in the *Insect Red Data Book* or are classified as 'scarce' (Table 11.7). However, a recent study has shown that SSSIs are not effective at conserving these rarer butterflies (Warren, in press). Over 5000 sites, covering around 1.7 million hectares had been scheduled as SSSIs by 1990 (NCC 1990), and this figure is expected to rise to around two million hectares in the next few years. Although SSSI protection is essentially negative, the Act does allow for management agreements with the NCC which will cover payment for active conservation measures. Management agreements have far-reaching consequences for butterfly conservation as they can be tailored to the specific needs of the important resident species. However, by 1990, only 4.1 per cent of the total SSSI area was covered by agreements (NCC 1990) and the majority of payments were purely to prevent site damage rather than for positive measures.

Outside SSSIs, the countryside is influenced by general land use policies determined by governmental bodies such as the Ministry of Agriculture, Fisheries and Food (MAFF), the Forestry Commission, the Countryside Commission, or local authorities. Whilst nature conservation is undoubtedly

Table 11.7 Red Data Book or 'scarce' butterflies which are taken into account during the selection of Sites of Special Scientific Interest by the Nature Conservancy Council. (NB other species are also taken into account in regions where locally rare or declining)

Red Data Book species (Shirt 1987)		Scarce species (thought to occur in less than 100 ten-kilometre grid squares)
Category 1	Endangered	*Thymelicus acteon*
*Maculinea arion**†		*Leptidea sinapis*
Nymphalis polychloros		*Thecla betulae*
Category 2	Vulnerable	*Satyrium w-album*
*Papilio machaon**		*Plebejus argus*
Argynnis adippe		*Aricia artaxerxes*
*Mellicta athalia**		*Lysandra bellargus*
Category 3	Rare	*Hamearis lucina*
Hesperia comma		*Apatura iris*
Melitaea cinxia		*Boloria euphrosyne*
Category 4	Out of danger	*Eurodryas aurinia*
Carterocephalus palaemon		*Erebia epiphron*
Satyrium pruni		
Category 5	Endemic (races)	
Plebejus argus caernensis‡		
Hipparchia semele thyone‡		
Appendix	Believed extinct	
Aporia crataegi		
Lycaena dispar†		
Cyaniris semiargus		

* Species protected under section 5 of the 1981 Wildlife and Countryside Act.
† Believed extinct but reintroduced.
‡ Infraspecies—not threatened, on Great Ormes Head, North Wales.

being incorporated into these policies it is still a relatively minor concern. An encouraging development has been the designation of Environmentally Sensitive Areas but, the scheme is entirely voluntary and has been much criticized by conservationists. Overproduction of food within the Common Market has also led to discussions about the removal of land from agriculture—perhaps as much as four million hectares in Britain (see NCC 1987). This is undoubtedly the most exciting opportunity for conservationists for many years but it may not necessarily be advantageous to wildlife. It could, for example, be extremely detrimental if the less productive farmland, often of great value for wildlife, is converted to alternative forms of land use, such as conifer afforestation or building.

Whatever the outcome of these new developments, there is still a great need for advice to private landowners who are interested in the conservation of butterflies. The British Butterfly Conservation Society plays a valuable role in promoting this aspect of conservation through its numerous local branches. There is a huge potential for improving conditions for butterflies on non-reserve land, notably gardens, urban areas, parks, and farmland. For example, at least 25 species were recorded during a recent survey of the vegetation of railway verges, indicating the considerable value of this type of marginal habitat (Sargent 1984, and C. Sargent, personal communication). Hedgerows and road verges are also very important; at least 24 species use them for breeding in at least part of their British range (see Appendix 2). One species, *Thecla betulae*, relies almost entirely on hedgerows and, apart from hedgerow removal, the biggest threat to this species is annual hedge trimming. A switch to less regular hedge cutting, even cutting a half of total length each year, would be highly beneficial (J. A. Thomas, personal communication).

Thecla betulae, and several other species, tend to breed at low densities over large areas, and are dependent on a specific landscape mosaic. The aim of conservation must therefore be to maintain landscape mosaics which incorporate the vital small- and large-scale elements of butterfly habitats. The problem of habitat fragmentation and isolation also needs to be tackled by creating and enhancing habitat corridors and links between the remaining patches (Harris 1984; Noss 1987). Similarly, a far greater effort is needed to integrate the reserve and SSSI system with the conservation of the wider countryside if it is to maintain butterfly populations successfully in the longer term.

11.5.5 Species legislation

Three butterflies are protected by law in Britain under schedule 5 of the 1981 Wildlife and Countryside Act:

Mellicta athalia heath fritillary
Papilio machaon swallowtail
Maculinea arion large blue

A fourth species *Carterocephalus palaemon*, was included in the original list but was deleted in 1988 when a review considered that it was out of danger. The Act makes it illegal to kill or injure any individuals of these species, or to have them in your possession without a licence from the NCC. A further 22 species are now also listed on schedule 5 with respect to provision 9 (5) which covers trade only (Stubbs 1991). The main drawback with such legislation is that it is primarily concerned with protecting the species from collectors, and does not tackle more crucial problems such as habitat loss or incorrect management. A far more effective form of legislation is in force in the United States, where a Recovery Programme has to be written, and published, for each endangered species that is scheduled. The Recovery Programme lists critical sites that should be protected and the management that is required to maintain, or improve, the habitat. If the recommended measures are not carried out within a designated period, the landowners responsible are liable for prosecution. If the measures are carried out successfully, and the species is no longer considered to be under threat, its name is removed from the list. The American-style legislation, therefore, addresses the real problems faced by endangered species, and many conservationists are backing its introduction into Britain (Whitten 1991).

11.5.6 Man-made introductions

The deliberate introduction of butterflies to new sites, or reintroduction to former sites, has a long tradition in Britain, which began with *Lycaena dispar* in 1909 (see section 11.3.1). Most recorded examples have concerned the movement of resident species from site to site within the country, or the introduction of a foreign race of a British species (as in the case of *L. dispar* and *Maculinea arion*). A few attempts have also been made to introduce foreign species, for example the European map *Araschnia levana*. Such introductions of new species to Britain are widely held to be inadvisable for a number of reasons, and are now prohibited under the 1981 Wildlife and Countryside Act.

A recent survey discovered that a minimum of 300 butterfly introductions have been carried out throughout Britain, covering at least 43 species (Oates and Warren 1990). Some attempts were highly successful and established populations which survived for many years (Table 11.8). Various other introductions succeeded for short periods only before becoming extinct, while many others failed completely.

The success of so many introductions demonstrates a very interesting and important fact: that there are numerous suitable habitats for butterflies currently not being exploited. Many of our species are so sedentary that they are unlikely to colonize new sites by natural means unless the sites are close to large, existing colonies. In the landscape of modern Britain, butterfly habitats have become fragmented and isolated, and local extinctions of sedentary species are likely to be permanent. The expansions observed in many more mobile species have often been incomplete, most noticeably where potential habitats are now more highly fragmented and, thus, slower to be colonized naturally. Ideally, the situation should be overcome by encouraging butterfly movement along a system of habitat 'stepping-stones' or 'corridors' as advocated earlier (section 11.5.3). Such a system may be very difficult to achieve in practice and, for some species, a programme of reintroduction may be the only viable alternative.

Morton (1983) has recommended that special Captive Breeding Institutes should be created to maintain captive populations of endangered species which can be used for reintroductions. Similarly, Collins (1987) suggested that the numerous Butterfly Houses that have recently become popular in Britain could rear stock for reintroductions. The JCCBI has produced a code of conservation practice for insect re-establishments which attempts to overcome some of the problems and criticisms (see Appendix 3b). Perhaps the most crucial recommendations are that each introduction is considered carefully, with full consultation with the relevant conservation bodies (e.g. NCC, Country Trusts for Nature Conservation and local BBCS groups), and that the details of any releases should be fully documented.

In the past, conservationists have regarded the deliberate introduction of species as a dubious practice since it interferes with natural distributions and genetic varieties. However, there has been a gradual shift in attitudes and many now believe that introductions have a vital role to play in the conservation of some of our rarer and more sedentary species, but only as part of an overall conservation strategy (Oates and Warren 1990). It should be stressed that although re-establishment may be able to play a limited role in conserving threatened butterflies, it will be impossible to recreate whole communities of insects and other animals that are being lost at the same time.

11.5.7 New habitats for butterflies

Human influence on the countryside has extended beyond the modification of the semi-natural vegetation to the creation of new habitats for wildlife. Wherever these reproduce the vital aspects of their more usual habitats, butterflies have been quick to take advantage. Railway embankments and cuttings contain a wide range of habitats (see Sargent 1984), and linear habitats have been created along other transport routes such as roads and canals. Today, many man-made habitats provide important breeding areas for many common species, and often for certain rarities. Within chalk downland, the focal point of many butterfly colonies are often ramparts of ancient hill-forts, or old quarry workings (Warren, in press). Such features have thin, broken soil, and can provide warm south-facing slopes with abundant food plants for *Lysandra bellargus* and other downland species. Disused railway-lines provide suitable conditions for a wide variety of butterflies, notably *Leptidea sinapis* (see Heal 1965; Warren 1984). Derelict or waste-land can also provide valuable habitats for butterflies, particularly within urban areas (Goode 1986; Plant 1987), but usually accommodates only the more widespread species (Thomas 1983*d*).

Perhaps the most familiar of all the habitats used by butterflies are gardens. Nearly all published research suggests that gardens are used primarily for their abundant nectar supplies by the more mobile species (Killingbeck 1985; Oates 1985). There is little evidence that gardens provide major breeding habitats for more than a few common butterflies, although large, rural gardens may contain areas of semi-natural habitat suitable for some species (Stephens and Warren, in press). Only three species rely heavily on garden habitats for breeding, *Pieris brassicae* and *P. rapae*, which are destructive pests of cabbage crops, and the holly blue *Celastrina argiolus*. The latter is attracted to gardens by the presence of its two main foodplants, holly *Ilex aquifolium* and ivy *Hedera helix*, especially in mature sheltered gardens. Numerous other butterflies have been recorded as breeding in gardens, though normally at low densities (Owen 1976; Stephens and Warren, in press).

The beneficial effects of the majority of artificial habitats are generally accidental, but we now have the technical ability to deliberately create new wildlife habitats (Boorman 1983; Davis 1989). The introduction of wild flowers into rural settings requires a certain amount of caution, summarized in the

Table 11.8 Summary of the outcome of all recorded attempts to establish or reinforce butterfly populations in Britain and Ireland.

Native species	No. releases involving reinforcement of existing colonies	No. establishments poorly documented	No. establishments reported 1985–88 and outcome awaited	No. unsuccessful establishments (i.e. survived less than 3 years)	Successful establishments: number of years survived		Total no. releases recorded
					Colonies now extinct	Colonies still surviving (1985–88)	
Erynnis tages	—	—	—	1	—	—	1
Thymelicus acteon	—	1	—	—	—	—	1
Hesperia comma	—	—	—	1	—	—	1
Papilio machaon ssp. *britannicus*	—	6†	—	1	4	—	8†
Papilio machaon ssp. *gorganus*	—	3	—	1	—	—	4
Leptidea sinapis	—	1	—	3	5, 10	12, 14, 19	9
Aporia crataegi	—	2†	1	5	—	—	7†
Gonepteryx rhamni	—	1	2*	—	—	—	3*
Aricia agestis	—	1	—	—	—	—	1
Cupido minimus	1	—	1	8	—	1, 2, 2	13
Plebejus argus	—	2	3	2	—	3, 5, 6, 8, 36	12
Lysandra coridon	2	2	—	2	—	2, 10†	8
Lysandra bellargus	2	7	—	2	*c*.15	3, 3†, 7	15
Maculinea arion	—	—	—	3†	—	4	4†
Lycaena dispar ssp. *rutilus*	—	—	—	1	5, 22	—	3
Lycaena dispar ssp. *batavus*	—	3	—	4	11, 13	(61)	10
Satyrium pruni	—	3††	—	2	—	35, 66, *c*.88	8††
Thecla betulae	—	1	—	1	—	—	2
Callophrys rubi	—	1	—	—	—	—	2
Hamearis lucina	1	3	1	3	8	3	10

Boloria euphrosyne	—	—	2	1	—	—	3
Boloria selene	—	—	1	—	3†	1†	3
Argynnis adippe	1	1	—	—	3†		3
Argynnis aglaja	—	—	1	—	—	—	1
Argynnis paphia	1	—	2	8	—	2	12
Melitaea cinxia	1††	5	1	7	3, 8, 22	3, 3	19††
Mellicta athalia	—	6†	—	10	5†, (20), 39	1, 4	21†
Eurodryas aurinia	6	8	6	23	3, 3, 3, 4, 4†, 5†	2, 2†, 2†, 3†, 10, 21, 26	56
Nymphalis polychloros	—	4	3	3	—	—	10
Polygonia c-album	1	—	2	—	—	—	3
Ladoga camilla	—	1	1	2	—	—	4
Apatura iris	7	3	—	1	—	—	11
Pararge aegeria	—	2	—	—	3, 5†	4	5
Erebia aethiops	—	1	—	5	3†	—	7
Melanargia galathea	—	2	2	1	—	1, 1, *c*.5, 22	9
Hipparchia semele	—	—	1	2	—	—	3
Pyronia tithonus	—	1	—	—	—	4	2
Aphantopus hyperantus	2	1	—	—	—	30	4
Coenonympha tullia	—	1	—	—	—	—	1
Non-native species							
Papilio bianor	—	—	—	2†	—	—	2†
Iphiclides podalirius	—	1	—	—	—	—	1
Gonepteryx cleopatra	—	—	—	1	—	—	1
Nymphalis antiopa	—	10††	—	2	—	—	12††
Araschnia levana	—	1	1	4	4††	—	7
Boloria dia	—	—	1	—	—	—	1
Grand totals	25††	85††	32	112	27	42	323††

Parenthesis indicate colonies surviving through regular reinforcement; †, more than indicated; †⁻, many more suspected; asterisk indicates a single attempt involving at least 70 separate releases. After Oates and Warren (1990.)

pioneering reports of Wells *et al.* (1981, 1986), but it undoubtedly has an important role to play in man-made landscapes. The commoner butterflies, with less demanding habitat requirements, are usually the first to colonize newly-created areas, but in time it may be possible to cater for some of the more fastidious species (Stephens 1988).

11.6 Future prospects

Butterflies are renowned for their large fluctuations in abundance from year to year, and their distributions have undoubtedly been in a state of flux for much of their recorded history. However, since 1945, the rapid changes in our countryside have caused unprecedented declines in a large number of species. The decline of butterflies in Britain has been paralleled by the decline of many other forms of wildlife which has spawned a large number of voluntary conservation bodies. Also the environment is increasingly becoming a political issue. Despite some recent beneficial changes of government policy, many land use decisions continue to be made, both locally and nationally, with little or no conservation input. Consequently, wildlife habitats are still being destroyed and the decline of many butterfly species is likely to continue. In the future, hopefully we will develop a more rational use of our countryside resources, with our natural heritage duly catered for.

The main support of the conservation movement in Britain has been the acquisition and management of nature reserves, and this is likely to remain vitally important in the future. Effective butterfly conservation also requires far more effective conservation of habitats in the wider countryside as well as sound reserve management. Conservation-oriented research into the requirements of our threatened species and into the most beneficial methods of managing habitats for butterflies is therefore essential. Of course, not all protected areas can be managed primarily for butterflies but, in nearly every case, measures taken to conserve this group will also benefit a great variety of wildlife.

In conclusion, the prospects for butterflies do not look completely bleak, and there is room for some optimism. Much has been achieved within the last decade to place butterfly conservation on a firm and practical footing. To a great extent, their future is now up to us—how much does society value butterflies?

Appendix 1

A Check list of British Butterflies and their Hostplants

Superfamily Hesperioidea

FAMILY HESPERIIDAE

Subfamily Hesperiinae

Carterocephalus palaemon Pallas. Chequered skipper. (England, extinct; Scotland). Main: *Brachypodium sylvaticum* (Huds.) Beauv. (slender or wood false-brome), *Molinia caerulea* (L.) Moench (purple moor grass), (Scotland). GRAMINEAE. Tall grasses. A range of tall grasses probably used (Warren 1991*b*).

Thymelicus sylvestris Poda. Smaller skipper. Main: *Holcus lanatus* L. (Yorkshire fog). Other: *Phleum pratense* L. (timothy or cat's-tail), *Brachypodium sylvaticum* (Huds.) Beauv. (wood false-brome) GRAMINEAE. Tall grasses.

Thymelicus lineola Ochsenheimer. Essex skipper. Main: *Dactylis glomerata* L. (cocksfoot), *Holcus mollis* L. (creeping soft-grass). Other: *Phleum pratense* L. (timothy), *Brachypodium sylvaticum* (Huds.) Beauv. (wood false-brome), *B. pinnatum* (L.) Beauv. (chalk false-brome or tor grass). Possibly *Bromus mollis* L. (soft brome or lop grass). GRAMINEAE. Tall grasses.

Thymelicus acteon Rottemburg. Lulworth skipper. Main: *Brachypodium pinnatum* (L.) Beauv. (tor grass). GRAMINEAE. Tall grass.

Hesperia comma Linnaeus. Silver-spotted skipper. Main: *Festuca ovina* L. (sheep's fescue). GRAMINEAE. Short grass.

Ochlodes venata Bremer & Grey. Large skipper. Main: *Dactylis glomerata* L. (cocksfoot). Other: *Brachypodium sylvaticum* (Huds.) Beauv. (wood false-brome), *Molinia caerulea* (L.) Moench. (purple moor grass). GRAMINEAE. Tall grasses.

Subfamily Pyrginae

Erynnis tages Linnaeus. Dingy skipper. Main: *Lotus corniculatus* L. (birdsfoot trefoil). Other: *Hippocrepis comosa* L. (horseshoe vetch), *Lotus uliginosus* Schkuhr. (greater birdsfoot trefoil). LEGUMINOSAE. Short herbs.

Pyrgus malvae Linnaeus. Grizzled skipper. Main: *Fragaria vesca* L. (wild strawberry), *Potentilla reptans* L. (creeping cinquefoil). Other: *Potentilla erecta* (L.) Räuschel (tormentil), *Agrimonia eupatoria* L. (agrimony), *Rubus idaeus* L. (raspberry), *R. fruticosus* L. (bramble). ROSACEAE. Short herbs and small shrubs.

Superfamily Papilionoidea

FAMILY PAPILIONIDAE

Subfamily Papilioninae

Papilio machaon Linnaeus. British subsp. *britannicus* Seitz Swallowtail. Main: *Peucedanum palustre* (L.) Moench. (milk parsley). UMBELLIFERAE. Tall herb.

FAMILY PIERIDAE

Subfamily Dismorphiinae

Leptidea sinapis Linnaeus. Wood white. Main: *Lathyrus pratensis* L. (meadow vetchling), *L. montanus* Bernh. (bitter vetchling), *Vicia cracca* L. (tufted vetch), *Lotus corniculatus* L. (birdsfoot trefoil). Other: *Lotus uliginosus* Schkuhr. (greater birdsfoot trefoil). LEGUMINOSAE. Short herbs.

Subfamily Coliadinae

Colias croceus Geoffroy. Clouded yellow. Main: *Trifolium repens* L. (white clover), *Lotus corniculatus* L. (birdsfoot trefoil). Other: several LEGUMINOSEA. Short herbs.

Gonepteryx rhamni Linnaeus. Brimstone. Main: *Rhamnus catharticus* L. (buckthorn), *Frangula alnus* Mill. (alder buckthorn). Larvae have been found to fail on other shrubs. CELASTRACEAE (see BUTT 1986). Shrubs.

Subfamily Pierinae

Aporia crataegi Linnaeus. Black-veined white. (Extinct.) Main: *Crataegus monogyna* Jacq. (hawthorn), *Prunus spinosa* L. (blackthorn). Other: cultivated damsons, plums, and apples. ROSACEAE. Shrubs and trees.

Pieris brassicae Linnaeus. Large white. Main: *Brassica* spp. such as *B. oleracea* L. (wild cabbage), *Tropaeolum* spp. (nasturtium) (Introduced TROPAEOLACEAE). Other: several Cruciferae and Resedaceae reported, such as *Reseda lutea* L. (wild mignonette), but significance not established. CRUCIFERAE. Small to moderate sized herbs.

Pieris rapae Linnaeus. Small white. Main: *Brassica* spp. (wild cabbage), *Tropaeoleum* spp. (nasturtium) (Introduced TROPAEOLACEAE). Other: *Alliaria petiolata* (Bieb.) (garlic mustard), *Sisymbrium officinale* (L.) Scop. (hedge mustard), *Reseda lutea* L. (wild mignonette), *Cardaria draba* (L.) Desv. (hoary cress) (Introduced, Kent). CRUCIFERAE. Small to moderate sized herbs.

Pieris napi Linnaeus. Green-veined white. Main: *Cardamine pratensis* L. (cuckoo flower or lady's smock), *Alliaria petiolata* (Bieb.) (garlic mustard), *Sisymbrium officinale* (L.) Scop. (hedge mustard), *Nasturtium officinale* R. Br. & *N. microphyllum* (Boenn.) Reichenb. (watercress). Other: a wide range of CRUCIFERAE. Small to moderate sized herbs.

Anthocharis cardamines Linnaeus. Orange-tip. Main: *Cardamine pratensis* L. (cuckoo flower), *Alliaria petiolata* (Bieb.) (garlic mustard). Other: numerous CRUCIFERAE and possibly two RESEDACEAE, *Reseda lutea* L. (wild mignonette) & *R. luteola* L. (weld). (See Courtney and Duggan 1983.) Small to moderate sized herbs.

FAMILY LYCAENIDAE

Subfamily Theclinae

Callophrys rubi Linnaeus. Green hairstreak. Main: *Vaccinium myrtillus* L. (bilberry), ERICACEAE; *Genista tinctoria* L. (dyer's greenweed), *Lotus corniculatus* L. (birdsfoot trefoil), *Ulex europaeus* L. (gorse), LEGUMINOSAE; *Helianthemum chamaecistus* Mill. (common rockrose), CISTACEAE; *Thelycrania sanguinea* (L.) Fourr. (dogwood), CORNACEAE. Other: *Rhamnus catharticus* L. (buckthorn), CELASTRACEAE; *Vaccinium oxycoccos* L. (cranberry), ERICACEAE; other *Ulex* spp., *Sarothamnus scoparius* (L.) Wimmer ex Koch (broom), LEGUMINOSAE and rarely bramble ROSACEAE. Herbs and shrubs.

Thecla betulae Linnaeus. Brown hairstreak. Main: *Prunus spinosa* L. (blackthorn). Other: *Prunus* spp., ROSACEAE. Shrubs.

Quercusia quercus Linnaeus. Purple hairstreak. Main: *Quercus robur* L. (pedunculate oak), *Q. petraea* (Mattuschka) Liebl. (sessile oak). Other: *Quercus* spp., FAGACEAE. Trees.

Satyrium w-album Knoch. White-letter hairstreak. Main: *Ulmus glabra* Huds. (wych elm), *Ulmus procera* Salisb. (English elm), URTICACEAE. Trees.

Satyrium pruni Linnaeus. Black hairstreak. Main: *Prunus spinosa* L. (blackthorn). Other: *Prunus* spp., ROSACEAE. Tall shrubs.

Subfamily Lycaeninae

Lycaena phlaeas Linnaeus. Small copper. Main: *Rumex acetosa* L. (common sorrel), *R. acetosella* L. (sheep's sorrel). Other: occasionally larger leaved docks such as *Rumex obtusifolius* L. (broad-leaved dock). POLYGONACEAE. Small herbs.

Lycaena dispar Haworth. The large copper. (extinct; introduced Dutch subsp. *batavus* Oberthür managed). Main: *Rumex hydrolapathum* Huds. (water dock), POLYGONACEAE. Herb.

Cupido minimus Fuessly. Small blue. Main: *Anthyllis vulneraria* L. (kidney vetch), LUGUMINOSAE. Small herb.

Plebejus argus Linnaeus. Silver-studded blue. Main: *Calluna vulgaris* (L.) Hull (ling or heather), *Erica tetralix* L. (cross-leaved heath), *E. cinerea* L. (bell heather), ERICACEAE; *Ulex europaeus* L. (gorse), *Lotus corniculatus* L. (birdsfoot trefoil), LEGUMINOSAE; *Helianthemum chamaecistus* Mill. (common rockrose), *H. canum* (L.) Baumg. (hoary rockrose),

CISTACEAE. Other: *Thymus praecox* Opiz. (wild thyme), LABIATAE; other *Ulex* spp. (see Thomas, C. 1985*a*). Small herbs and shrubs.

Aricia agestis Denis & Schiffermüller. Brown argus. Main: *Helianthemum chamaecistus* Mill. (common rockrose), CISTACEAE; *Erodium cicutarium* (L.) L'Hérit. (common storksbill), *Geranium molle* L. (dovesfoot cranesbill), GERANIACEAE. Small herbs.

Aricia artaxerxes Fabricius. Northern brown argus. Main: *Helianthemum chamaecistus* Mill. (common rockrose), CISTACEAE. Other: *Geranium sangiuneum* L. (bloody cranesbill), GERANIACEAE. Small herbs.

Polyommatus icarus Rottemburg. Common blue. Main: *Lotus corniculatus* L. (birdsfoot trefoil), *L. uliginosus* Schkuhr. (greater birdsfoot trefoil). Other: *Trifolium dubium* Sibth. (lesser trefoil), *Medicago lupulina* L. (black medick), *Trifolium repens* L. (white clover) (Pennines; Lees 1969), *Ononis* spp. (rest-harrow). LEGUMINOSAE. Small herbs.

Lysandra coridon Poda. Chalk hill blue. Main: *Hippocrepis comosa* L. (horseshoe vetch), LEGUMINOSAE. Small herb.

Lysandra bellargus Rottemburg. Adonis blue. Main: *Hippocrepis comosa* L. (horseshoe vetch), LEGUMINOSAE. Small herb.

Cyaniris semiargus Rottemburg. Mazarine blue (extinct). Main: *Trifolium pratense* L. (red clover). Other: LEGUMINOSAE. Small herb.

Celastrina argiolus Linnaeus. Holly blue. Main: *Ilex aquifolium* L. (holly), AQUIFOLIACEAE: *Hedera helix* L. (ivy), ARALIACEAE. Other: *Thelycrania sanguinea* (L.) Fourr. (dogwood), CORNACEAE; *Euonymus europaeus* L. (spindle-tree), *Frangula alnus* Mill. (alder buckthorn), CELASTRACEAE; *Ulex* spp. (gorse), LEGUMINOSAE. Trees, shrubs, and climbers.

Maculinea arion Linnaeus. Large blue (extinct, introduced Swedish race managed). Main: *Thymus praecox* Opiz. (wild thyme), LABIATAE. Eggs, larvae, or prepupae of ant *Myrmica sabuleti* Meinert in Britain. Small herb and ant brood.

Subfamily Riodininae

Hamearis lucina Linnaeus. Duke of Burgundy fritillary. Main: *Primula veris* L. (cowslip), *P. vulgaris* Huds. (primrose), PRIMULACEAE. Small herb.

FAMILY NYMPHALIDAE

Subfamily Limenitinae

Ladoga camilla Linnaeus. White admiral. Main: *Lonicera periclymenum* L. (honeysuckle), CAPRIFOLIACEAE. Shrub and climber.

Subfamily Apaturinae

Apatura iris Linnaeus. Purple emperor. Main: *Salix caprea* L. (goat willow or great sallow), *S. cinerea* L. (grey willow or grey sallow), SALICACEAE. Shrubs.

Subfamily Nymphalinae

Vanessa atalanta Linnaeus. Red admiral. Main: *Urtica dioica* L. (stinging nettle), *U. urens* L. (annual nettle), URTICACEAE. Others: *Humulus lupulus* L. (hop), CANNABACEAE. Tall herbs and climber.

Cynthia cardui Linnaeus. Painted lady. Main: *Carduus nutans* L. (musk thistle). Other: *Cirsium vulgare* (Savi) Ten. (spear thistle), COMPOSITAE; *Urtica dioica* L. (stinging nettle), *U. urens* L. (annual nettle), URTICACEAE; *Malva* spp. (Mallows), MALVACEAE (see Warren 1986); *Plantago lanceolata* L. (ribwort plantain), PLANTAGINACEAE (see Hall 1984). Tall and short herbs.

Aglais urticae Linnaeus. Small tortoiseshell. Main: *Urtica dioica* L. (stinging nettle), *U. urens* L. (annual nettle), URTICACEAE. Tall herb.

Nymphalis polychloros Linnaeus. Large tortoiseshell. Main: *Ulmus glabra* Huds. (wych elm), *U. procera* Salisb. (English elm), URTICACEAE. Other: *Salix* spp. (sallows); *Populus tremula* L. (aspen), SALICACEAE; *Betula* spp. (birches), BETULACEAE. Other: *Populus* spp.; also *Pyrus pyraster* Burgsd. (cultivated pear), *Prunus* spp., ROSACEAE and other trees. Trees.

Inachis io Linnaeus. Peacock. Main: *Urtica dioica* L. (stinging nettle), URTICACEAE. Tall herb.

Polygonia c-album Linnaeus. Comma. Main: *Urtica dioica* L. (stinging nettle), *Ulmus procera* Salisb.

(English elm), *U. glabra* Huds. (wych elm), URTICACEAE; *Humulus lupulus* L. (hop), CANNABACEAE. Other: *Ribes rubrum* L. (red currant), *R. nigrum* L. (black current), GROSSULARIACEAE (see Pratt 1986). Trees, shrubs, tall herbs, and climber.

Subfamily Argynninae

Boloria selene Denis & Schiffermüller. Small pearl-bordered fritillary. Main: *Viola riviniana* Reichb. (common dog violet) (southern Britain mainly), *V. palustris* L. (marsh violet) (northern Britain mainly). Other: *Viola* spp., VIOLACEAE. Small herbs.

Boloria euphrosyne Linnaeus. Pearl-bordered fritillary. Main: *Viola riviniana* Reichb. (common dog violet) (southern Britain mainly), *V. palustris* L. (marsh violet) (northern Britain mainly). Other: *Viola* spp., VIOLACEAE. Small herbs.

Argynnis adippe Linnaeus. High brown fritillary. Main: *Viola riviniana* Reichb. (common dog violet), *V. hirta* L. (hairy violet). Other: *Viola* spp., VIOLACEAE. Small herbs.

Argynnis aglaja Linnaeus. Dark green fritillary. Main: *Viola hirta* L. (hairy violet), *V. palustris* L. (marsh violet), *V. riviniana* Reichb. (common dog violet). Other: *Viola* spp., VIOLACEAE. Small herbs.

Argynnis paphia Linnaeus. Silver-washed fritillary. Main: *Viola riviniana* Reichb. (common dog violet). Other: *Viola* spp., VIOLACEAE. Small herbs.

Subfamily Melitaeinae

Eurodryas aurinia Rottemburg. Marsh fritillary. Main: *Succisa pratensis* Moench. (devilsbit scabious). Other: *Knautia arvensis* (L.) Coulter (field scabious), *Scabiosa columbaria* L. (small scabious), DIPSACACEAE (BUTT 1986); *Lonicera periclymenum* L. (honeysuckle), CAPRIFOLIACEAE. Moderate sized to tall herb and shrub.

Melitaea cinxia Linnaeus. Glanville fritillary. Main: *Plantago lanceolata* L. (ribwort plantain). Other: *P. coronopus* L. (buckshorn plantain), PLANTAGINACEAE. Small herbs.

Mellicta athalia Rottemburg. Heath fritillary. Main: *Plantago lanceolata* L. (ribwort plantain), PLANTAGINACEAE; *Melampyrum pratense* L. (common cow-wheat), *Veronica chamaedrys* L. (germander speedwell). Other: *Digitalis purpurea* L. (foxglove), SCROPHULARIACEAE (see Warren 1987*a*). Small herbs mainly.

Subfamily Satyrinae

Pararge aegeria Linnaeus. Speckled wood. Main: *Brachypodium sylvaticum* (Huds.) Beauv. (wood false-brome), *Dactylis glomerata* L. (cocksfoot), *Poa trivialis* L. (rough meadow-grass), *Holcus lanatus* L. (Yorkshire fog), *Festuca rubra* L. (red fescue), *Bromus ramosus* Huds. (hairy or wood brome), *Agrostis stolonifera* L. (creeping bent). Others: a wide range of grass species (see Shreeve 1986*a*), GRAMINEAE. Grasses.

Lasiommata megera Linnaeus. Wall brown. Main: *Brachypodium pinnatum* (L.) Beauv. (tor grass), *B. sylvaticum* (Huds.) Beauv. (wood false-brome), *Dactylis glomerata* L. (cocksfoot), *Holcus lanatus* L. (Yorkshire fog), *Deschampsia flexuosa* (L.) Trin. (wavy hair-grass). Other: a wide range of grass species, GRAMINEAE. Grasses.

Erebia epiphron Knoch. Small mountain ringlet. Main: *Nardus stricta* L. (mat grass), GRAMINEAE. Grasses.

Erebia aethiops Esper. Scotch argus. Main: *Molinia caerulea* (L.) Moench. (purple moor-grass) (Scotland), *Sesleria caerulea* (L.) Ard. (blue sesleria) in limestone habitats in northern England. Other: probably several grass species used, GRAMINEAE. Grasses.

Melanargia galathea Linnaeus. Marbled white. Main: *Festuca rubra* L. (red fescue). Other: a wide range of grass species may be used to supplement diet including, *Festuca ovina* L. (sheep's fescue), *Phleum pratense* L. (timothy), *Dactylis glomerata* L. (cocksfoot), *Brachypodium pinnatum* (L.) Beauv. (tor grass), *Arrhenatherum elatius* (L.) Beauv ex J. & C. Presl. (tall or false oat-grass) (see Wilson 1985). GRAMINEAE. Grasses.

Hipparchia semele Linnaeus. Grayling. Main: *Festuca ovina* L. (sheep's fescue), *Agrostis setacea* Curt. (bristle leaved-bent), *Elymus repens* (L.) Gould (couch or twitch), *Festuca rubra* L. (red fescue). Other: *Ammophila arenaria* (L.) Link (marram grass) in need of confirmation; probably a range of fine grass species, GRAMINEAE. Grasses.

Pyronia tithonus Linnaeus. Gatekeeper or hedge brown. Main: *Agrostis* spp. (bents), *Festuca* spp. (fescues) such as *Festuca rubra* L. (red fescue), *Poa* spp. (meadow grasses), *Elymus repens* (L.) Gould (couch), *Anthoxanthum oderatum* L. (sweet vernal-grass) (BUTT 1986). Other: a range of fine grass species, GRAMINEAE. Grasses.

Maniola jurtina Linnaeus. Meadow brown. Main: *Poa* spp. (meadow grasses), such as *Poa pratensis* L. (smooth meadow grass), *Holcus lanatus* L. (Yorkshire fog), *Agrostis* spp. (bents). Other: a wide range of fine grass species, GRAMINEAE. Grasses.

Aphantopus hyperantus Linnaeus. Ringlet. Main: *Elymus repens* (L.) Gould (couch), *Poa pratensis* L. (smooth meadow grass). Other: various grasses, but species unknown. *Deschampsia cespitosa* (L.) Beauv. (tufted hair-grass) is quoted as a hostplant but this is a very coarse grass, GRAMINEAE. Grasses.

Coenonympha pamphilus Linnaeus. Small heath. Main: *Festuca ovina* L. (sheep's fescue), *F. rubra* L. (red fescue), *Poa pratensis* L. (smooth meadow grass), *Agrostis* spp. (bents). Other: small fine grass species, GRAMINEAE. Grasses.

Coenonympha tullia Müller. Large heath. Main: *Eriophorum vaginatum* L. (common cotton-grass), CYPERACEAE. Other: *Molinia caerulea* (L.) Moench. (purple moor-grass), GRAMINEAE and *Rhyncho-spora alba* (L.) Vahl (white beaked-sedge), CYPERA-CEAE suspected; (see Melling 1984a).

After Dennis (1977); Heath *et al.* (1984); Thomas (1986); BUTT (1986); Emmet and Heath (1989). Additional references noted in the body of the Appendix. See Emmet and Heath (1989), for a listing of infrequent and rare immigrants, deliberate and accidental introductions, and adventives. Plant nomenclature from Clapham *et al.* (1987).

43
63

Appendix 2

Traditional Classification of Butterfly Breeding Biotopes in Britain

Species	*Carterocephalus palaemon* (E)	*Thymelicus sylvestris*	*Thymelicus lineola*	*Thymelicus acteon*	*Hesperia comma*	*Ochlodes venata*	*Erynnis tages*	*Pyrgus malvae*	*Papilio machaon*	*Leptidea sinapis*	*Colias croceus*	*Gonepteryx rhamni*	*Aporia crataegi* E	*Pieris brassicae*	*Pieris rapae*	*Pieris napi*	*Anthocharis cardamines*	*Callophrys rubi*	*Thecla betulae*	*Quercusia quercus*	*Satyrium w-album*
Family	HESPERIIDAE								PAPILIONIDAE	PIERIDAE								LYCAENIDAE			
Seral stages	3	3	3	3	1–2	3	2	2	3	3–5	2	4	4–5	2–3	2–3	2–3	2–3	3–4	4–5	6	6
Ubiquists		R				R					M	R		M/R	M/R	R	R	R			
Skiophils (partial shade)										○				○		○					
Heliophils (sunny)	○	○	○	○	○	○	○	○	○	○	○	○	○	○	○	○	○	○	○	○	○
Cool habitats																					
Wide temperature tolerance	○					○	○			○		○		○	○	○	○	○		○	
Warm habitats		○	○	○	○			○	○		○		○						○		○
Hygrophils (wet)	○					○			○			○				○	○	○			
Mesophils (moist)	○	○	○			○	○	○		○	○	○	○	○	○	○	○	○	○	○	○
Xerophils (dry)		○	○	○	○	○	○	○				○						○			
Embankment and cuttings Railway; motorway		○	○	○		○	○	○		●		○				○	○	●	○	○	○
Road verge, field edge and banks		●	●			●					○		○		○	●	●				
Hedgerow		●	●			●						●	○			●	●		●	○	●
Gardens, allotments, market gardens fields of *Brassica* crops		○				○						○	○	●	●	○	○			○	
Arable and improved pasture											●			●	●	○					
Sand-dune and dune slacks		○				○	●	○													
Salt marsh			●																		
Cliff, scree and rock outcrops walls, quarries, and derelict land		○	○	○		○	●	○		○	○			○	○	●	○	○			
River bank, and lake sides levees; cutoffs		○	●			●									○	●	●				
Basic flush and fen	○		●						●			●			○	●	●				
Marshland bog and valley mires; ditches	●	●				●									○	●	●				
Blanket bog and raised bogs																		●			
Moorland upland heath and coarse grass																		●			
Wet lowland heath						●											○				
Dry lowland heath			○			○	○	○										●			
Calcareous grassland short turf; sparse taller herbs/grasses		○	○		●		●	●			●							●			
Calcareous grassland long grass and tall herb; dense cover		●	●	●		●												●			
Acid and neutral grassland short turf; sparse taller herbs/grasses		○	○				○	○			○						○				
Acid and neutral grassland long grass and tall herb; dense cover	○	●	●			●											○	○			
Shrubs and scrub scrub, grasses, and herbs; bracken	●	●	●			●		○		●		●	○					●	●		
Conifer plantations rides; early growth stages	○	●	●			●	○	○		●		○			○	○	○	○			
Broadleaved woodland grass rides; permanent clearings	●	●	●			●	●	●		●		●			○	●	●	●			
Broadleaved woodland understorey component; coppice	○	○				○		○		●		●					●				○
Broadleaved woodland dependence on climax component																			○	●	●

Biotopes

Appendix 2 (cont.)

Species	*Satyrium pruni*	*Lycaena phlaeas*	*Lycaena dispar* (E)	*Cupido minimus*	*Plebejus argus*	*Aricia agestis*	*Aricia artaxerxes*	*Polyommatus icarus*	*Lysandra coridon*	*Lysandra bellargus*	*Cyaniris semiargus* E	*Celastrina argiolus*	*Maculinea arion* (E)	*Hamearis lucina*	NYMPHALIDAE *Ladoga camilla*	*Apatura iris*	*Vanessa atalanta*	*Cynthia cardui*	*Aglais urticae*	*Nymphalis polychloros*	*Inachis io*
Seral stages	5	2–3	3	2	2	2	2	2	2	2	2	5–6	1–2	3–4	5–6	6	3–4	2–3	3	6	3
Ubiquists		R						R									M	M	R/M	R/M	R/M
Skiophils (partial shade)		○										○			○	○					○
Heliophils (sunny)	○	○	○	○	○	○	○	○	○	○	○	○	○	○	○	○	○	○	○	○	○
Cool habitats							○														
Wide temperature tolerance		○		○	○			○				○							○		○
Warm habitats	○		○			○			○	○	○		○	○	○	○	○	○		○	
Hygrophils (wet)		○	○		○			○											○		○
Mesophils (moist)	○	○		○	○	○	○	○			○	○		○	○	○	○	○	○	○	○
Xerophils (dry)		○		○	○	○	○	○	○	○			○						○		○
Biotopes																					
Embankment and cuttings Railway; motorway	○	●		●		●		●	○	○		○					○	○	●		○
Road verge, field edge and banks		○				○		○			○						○	○	●		○
Hedgerow	●	○						○				○					○	●	●		○
Gardens, allotments, market gardens fields of *Brassica* crops		○										●					●	●	●		○
Arable and improved pasture		○															○	○	●		○
Sand-dune and dune slacks		●		○	●	●		●										●	●		○
Salt marsh																					
Cliff, scree and rock outcrops walls, quarries, and derelict land		●		●	●	●	●	○	○	○		○	○				○	○	○		
River bank, and lake sides levees; cutoffs		●				○		○									○	○	●		○
Basic flush and fen			●																		
Marshland bog and valley mires; ditches		●						●									○	●	●		○
Blanket bog and raised bogs					○																
Moorland upland heath and coarse grass		●						○											○		
Wet lowland heath					●																
Dry lowland heath		●			●			○					○						○		
Calcareous grassland short turf; sparse taller herbs/grasses		●		●	●	●	●	●	●	●	○		○	●				●			
Calcareous grassland long grass and tall herb; dense cover		●		○			●	○						●			○	●	●		
Acid and neutral grassland short turf; sparse taller herbs/grasses		●				○		●			○		○	○							
Acid and neutral grassland long grass and tall herb; dense cover		●						●						○			○	○	●		
Shrubs and scrub scrub, grasses, and herbs; bracken	●	●				○		○				●		●			○	○	●		○
Conifer plantations rides; early growth stages		○				○		○						○	○	○	●	●	●		●
Broadleaved woodland grass rides; permanent clearings		○			○	○		○						●	●	●	●	●	●		●
Broadleaved woodland understorey component; coppice	●	○				○		○				●		●	●	●	○	○	○		●
Broadleaved woodland dependence on climax component	○											●			●	●				●	

Appendix 2 (cont.)

Species	Broadleaved woodland dependence on climax component	Broadleaved woodland understorey component; coppice	Broadleaved woodland grass rides; permanent clearings	Conifer plantations rides; early growth stages	Shrubs and scrub scrub, grasses, and herbs; bracken	Acid and neutral grassland long grass and tall herb; dense cover	Acid and neutral grassland short turf; sparse taller herbs/grasses	Calcareous grassland long grass and tall herb; dense cover	Calcareous grassland short turf; sparse taller herbs/grasses	Dry lowland heath	Wet lowland heath	Moorland upland heath and coarse grass	Blanket bog and raised bogs	Marshland bog and valley mires; ditches	Basic flush and fen	River bank, and lake sides levees; cutoffs	Cliff, scree and rock outcrops walls, quarries, and derelict land	Salt marsh	Sand-dune and dune slacks	Arable and improved pasture	Gardens, allotments, market gardens fields of *Brassica* crops	Hedgerow	Road verge, field edge and banks	Embankment and cuttings Railway; motorway	Xerophils (dry)	Mesophils (moist)	Hygrophils (wet)	Warm habitats	Wide temperature tolerance	Cool habitats	Heliophils (sunny)	Skiophils (partial shade)	Ubiquists	Seral stages
Polygonia c-album		●	●	●	●																●	●	○	○		○		○			○	○	R	3–5
Boloria selene		●	●	●	●		●		○		●	●		●	○		●									○	○		○		○			2–4
Boloria euphrosyne		●	●	●	●				○								○								○	○		○			○			2–4
Argynnis adippe		●	○	○	●		○																		○	○		○			○			2–4
Argynnis aglaja		○	○	○	●	○	○	○	●	○		●					●		●						○	○	○		○		○			2–4
Argynnis paphia	●	●	●																			○				○			○		○	○		5–6
Eurodryas aurinia			○	○		●	●	○	●		○			●	○	○			○					○	○	○	○		○		○			3
Melitaea cinxia																	●								○	○		○			○			1–2
Mellicta athalia		●	○	●			○			●		○														○		○			○			3–4
Pararge aegeria		●	●	●	●												○				○	○		○		○			○		○	○	R	4–6
Lasiommata megera		○	○	○			●		●	●				○		●	●		●	○	○	●	●	●	○	○			○		○		R	1–2
Erebia epiphron												●														○				○	○			2
Erebia aethiops			●	○	●	●		●				●		●										○	○	○	○		○		○			3
Melanargia galathea			●	○	●	●		●									○		○			○	○	○	○	○		○			○			3
Hipparchia semele				○			●		●	●							●		●					○	○				○		○			1–2
Pyronia tithonus		○	●	○	●	●		○		●							○	○	○		●	●	○	●		○			○		○	○		3
Maniola jurtina		○	●	●	●	●		●		○				●		●	○	○	●		○	●	●	●	○	○	○		○		○		R	3
Aphantopus hyperantus		●	●	●	●	●		●						●				○	○			●	●	●		○	○		○		○	○		3
Coenonympha pamphilus		○	●		●		●		●	●	●	●					○		●		○	○	○	●	○	○	○		○		○			2
Coenonympha tullia												●	●														○		○		○			3

At best, all biotopes are now semi-natural and most are thoroughly artificial. A broad grouping is made of these biotopes on the basis of the tolerance of butterflies to moisture, temperature (based on the egg-laying and larval feeding sites during daylight hours in summer), and light. Ubiquists, cosmopolitan species found almost anywhere, are also distinguished as to whether they are regular migrants (M) or residents (R). Seral stages typified by hostplant habitats: 1, bare ground with sparse herb cover; 2, short herbs and grasses; 3, shrubs, tall herbs and grasses; 4, trees and shrubs; 5, pre-climax forest; 6, climax forest with regeneration patches. Black circles, habitats in which the butterflies are most frequently found; open circles, other habitats which can be regionally or locally important. E, species extinct; (E), species extinct but re-introduced or surviving in reduced habitat.

Notes: (i) Broadleaved woodlands – *Thecla betulae* and *Celastrina argiolus* depend on this habitat for mate-location; (ii) most biotopes provide a wide range of habitats, especially gardens, embankments and cuttings, which include the full range of seres to complete woodland cover, ground water levels and pH conditions.

Data from Dennis (1977); Thomson (1980); BUTT (1986); County texts referenced in Emmet and Heath (1989); I.K. Morgan personal communication; J.A. Thomas and M.S. Warren, personal communication.

Appendix 3

(a) A Code for Insect Collecting

Joint Committee for the Conservation of British Insects

This Committee believes that with the ever-increasing loss of habitats resulting from forestry, agriculture, and industrial, urban, and recreational development, the point has been reached where a code for collecting should be considered in the interests of conservation of the British insect fauna, particularly macrolepidoptera. The Committee considers that in many areas this loss has gone so far that collecting, which at one time would have had a trivial effect, could now affect the survival in them of one or more species if continued without restraint.

The Committee also believes that by subscribing to a code of collecting, entomologists will show themselves to be a concerned and responsible body of naturalists who have a positive contribution to make to the cause of conservation. It asks all entomologists to accept the following Code in principle and try to observe it in practice.

1. Collecting—general

1.1 No more specimens than are strictly required for any purpose should be killed.

1.2 Readily identified insects should not be killed if the object is to 'look them over' for aberrations or other purposes; insects should be examined while alive and then released where they were captured.

1.3 The same species should not be taken in numbers year after year from the same locality.

1.4 Supposed or actual predators and parasites of insects should not be destroyed.

1.5 When collecting leaf-mines, galls and seed heads never collect all that can be found; leave as many as possible to allow the population to recover.

1.6 Consideration should be given to photography as an alternative to collecting, particularly in the case of butterflies.

1.7 Specimens for exchange, or disposal to other collectors, should be taken sparingly or not at all.

1.8 For commercial purposes insects should be either bred or obtained from old collections. Insect specimens should not be used for the manufacture of 'jewellery'.

2. Collecting—rare and endangered species

2.1 Specimens of macrolepidoptera listed by this Committee (and published in the entomological journals) should be collected with the greatest restraint. As a guide, the Committee suggests that a pair of specimens is sufficient, but that those species in the greatest danger should not be collected at all. The list may be amended from time to time if this proves to be necessary.

2.2 Specimens of distinct local forms of Macrolepidoptera, particularly butterflies, should likewise be collected with restraint.

2.3 Collectors should attempt to break new ground rather than collect a local or rare species from a well-known and perhaps over-worked locality.

2.4 Previously unknown localities for rare species should be brought to the attention of this Committee, which undertakes to inform other organizations as appropriate and only in the interests of conservation.

3. Collecting—lights and light-traps

3.1 The 'catch' at light, particularly in a trap, should not be killed casually for subsequent examination.

3.2 Live trapping, for instance in traps filled with egg-tray material, is the preferred method of collecting. Anaesthetics are harmful and should not be used.

3.3 After examination of the catch the insects should be kept in cool, shady conditions and released away from the trap site at dusk. If this is not possible the insect should be released in long grass or other cover and not on lawns or bare surfaces.

3.4 Unwanted insects should not be fed to fish or insectivorous birds and mammals.

3.5 If a trap used for scientific purposes is found to be catching rare or local species unnecessarily it should be re-sited.

3.6 Traps and lights should be sited with care so as not to annoy neighbours or cause confusion.

4. Collecting—permission and conditions

4.1 Always seek permission from landowner or occupier when collecting on private land.

4.2 Always comply with any conditions laid down by the granting of permission to collect.

4.3 When collecting on nature reserves, or sites of known interest to conservationists, supply a list of species collected to the appropriate authority.

4.4 When collecting on nature reserves it is particularly important to observe the code suggested in section 5.

5. Collecting—damage to the environment

5.1 Do as little damage to the environment as possible. Remember the interests of other naturalists; be careful of nesting birds and vegetation, particularly rare plants.

5.2 When 'beating' for lepidopterous larvae or other insects never thrash trees and bushes so that foliage and twigs are removed. A sharp jarring of branches is both less damaging and more effective.

5.3 Coleopterists and others working dead timber should replace removed bark and worked material to the best of their ability. Not all the dead wood in a locality should be worked.

5.4 Overturned stones and logs should be replaced in their original positions.

5.5 Water weed and moss which has been worked for insects should be replaced in its appropriate habitat. Plant material in litter heaps should be replaced and not scattered about.

5.6 Twigs, small branches and foliage required as foodplants or because they are galled, e.g. by clearwings, should be removed neatly with secateurs or scissors and not broken off.

5.7 'Sugar' should not be applied so that it renders tree-trunks and other vegetation unnecessarily unsightly.

5.8 Exercise particular care when working for rare species, e.g. by searching for larvae rather than beating for them.

5.9 Remember the Country Code!

6. Breeding

6.1 Breeding from a fertilized female or pairing in captivity is preferable to taking a series of specimens in the field.

6.2 Never collect more larvae or other livestock than can be supported by the available supply of foodplant.

6.3 Unwanted insects that have been reared should be released in the original locality, not just anywhere.

6.4 Before attempting to establish new populations or 'reinforce' existing ones please consult this Committee.

(b) Insect re-establishment—a code of conservation practice (summary of main recommendations)

Joint Committee for the Conservation of British Insects

1. Consult widely before deciding to attempt any re-establishment.
2. Every re-establishment should have a clear objective.
3. The ecology of the species to be re-established should be known.
4. Permission should be obtained to use both the receiving site and the source of material for re-establishment.
5. The receiving site should be appropriately managed.
6. Specific parasites should be included in the re-establishment.
7. The number of insects released should be large enough to secure re-establishment.
8. Details of the release should be meticulously recorded.
9. The success of re-establishment should be continually assessed and adequately recorded.
10. All re-establishments should be reported to the Biological Records Centre and this committee.

See Stubbs (1988) and *Antenna* **10**(10): 13–18 (1986) for the entire code.

Appendix 4

Useful Addresses

Societies

The Amateur Entomologist Society, Registrar, 22 Salisbury Road, Feltham, Middlesex TW13 5DP. (Journal: *Bulletin AES*; Editor, B. O. C. Gardiner, F.L.S., F.R.E.S., 2 Highfield Avenue, Cambridge, CB4 2AL.)

The British Butterfly Conservation Society, Head Office, P.O. Box 222, Dedham, Colchester, Essex CO7 6EY. (Journal: *NEWS BBCS*; Editor, South View, Sedlescombe, Battle, East Sussex TN33 OPE.)

The British Entomological and Natural History Society, c/o The Royal Entomological Society, 41 Queen's Gate, London SW7 5HU. (Journal: *British Journal of Entomology and Natural History*; Editor, R. A. Jones, 13 Bellwood Road, Nunhead, London SE15 3DE.)

The Royal Entomological Society of London, 41 Queen's Gate, London, SW7 5HU. (Journals: *The Entomologist*; *Ecological Entomology*; *Systematic Entomology*; *Physiological Entomology*; *Medical and Veterinary Entomology*.)

Societas Europaea Lepidopterologica, Hemdener Weg 19, D-4290 Bocholt (Westf.). (Journal: *Nota Lepidopterologica*; Editor, S. E. Whitebread, Maispracher strasse 51, CH-4312, Magden, Switzerland.)

Journals

Entomologist's Gazette, Gem Publishing Company, Brightwood, Brightwell cum Sotwell, Wallingford, Oxfordshire OX10 0QD. Editor, W. G. Tremewan, C.Biol., M.I. Biol, Pentreath, 6 Carlyon Road, Playing Place, Truro, Cornwall TR3 6EU.

The Entomologist's Record and *Journal of Variation*, Editor, P. A. Sokoloff, M.Sc., C. Biol., M.I. Biol., F.R.E.S., 4 Steep Close, Orpington, Kent BR6 6DS.

Books, equipment & livestock

E. W. Classey Ltd., P.O. 93, Faringdon, OXON. SN7 7DR. Tel: 0367 820399 (Books).

Bio-Science Supplies, 4 Long Mill North, Wednesfield, Wolverhampton, West Midlands WV11 1JD (Equipment).

Butterfly Farm (Christchurch), 14 Grove Road East, Christchurch, Dorset BH23 2DQ (Livestock and Equipment).

Entomological Livestock Supplies, Unit 3, Beaver Park, Hayseech Road, Halesowen, West Midlands B63 3PD (Livestock).

Harley Books, Martins, Great Horkesley, Colchester, Essex CO6 4AH (Books).

The London Butterfly House, Syon Park, Brentford, Middlesex. (Associated butterfly houses at Edinburgh, Stratford-upon-Avon and Weymouth; these display live butterflies but do not sell livestock.)

Marris House Nets, 54 Richmond Park Avenue, Bournemouth BH8 9DR (Equipment).

Watkins & Doncaster Ltd., Four Throws, Hawkhurst, Kent, TN18 5ED (Equipment).

Worldwide Butterflies, Compton House, Over Compton, Sherborne, Dorset DT9 4QU (Livestock).

Scientific institutions

The Biological Records Centre, Dr P. T. Harding, Institute of Terrestrial Ecology, Monks Wood Experimental Station, Abbots Ripton, Huntingdon, PE17 2LS.

The Butterfly Monitoring Scheme, Mrs M. L. Hall, Institute of Terrestrial Ecology, Monks Wood Experimental Station, Abbots Ripton, Huntingdon, PE17 2LS.

English Nature, Northminster House, Peterborough PE1 1UA.

Glossary

aberration a distinctive, usually rare, form of a species or subspecies. Frequently the result of a genetic mutation but may be environmentally induced or arise from a developmental disorder.

adaptive zone the evolutionary divergence of sub-units of a single phyletic line (species) into a series of different niches.

advergence the gradual convergence of the pattern of a Batesian mimic on its model despite the selection for the model to evolve away from the pattern of the mimic.

aeropyle a narrow tube connecting the air spaces within the cuticle of an egg with the atmosphere.

aestivation a period of dormancy in the summer months similar to hibernation in winter.

allele alternative forms of a gene which, on account of their corresponding position on homologous (paired) chromosomes, are subject to Mendelian inheritance.

allelochemics or 'secondary' plant substances, noxious phytochemicals that are repulsive, unpalatable or poisonous, or that interfere with food assimilation by certain herbivores. Some species can circumvent these compounds which allegedly are not often involved in primary metabolic functions of the plants.

allomones chemical agents of adaptive value to the organism producing them.

allopatry species or populations originating in or occurring in different geographical regions.

allozymes different electrophoretic forms of the same enzyme which are products of alternative alleles segregating at a gene within a species (not to be confused with isozymes, which are alternative forms of an enzyme produced by different genes).

amensalism state in which one species is adversely affected by a second species but with no reciprocal reduction in fitness for the latter species.

anachoresis hiding in holes.

anal angle the corner of the hindwing nearest the posterior end of the body.

androconia scales specialized for distributing male pheromones to attract the female.

apex (and apical) the tip of the fore- or hindwings at the end of the costa which is farthest from the body.

aposematism warning coloration in those insects which are also known, or believed, to be distasteful to their predators.

apostatic selection frequency dependent selection of prey items which occur in different forms, such that predators select the commonest form and survival is inversely related to frequency. It is one mechanism whereby polymorphism, related to visual appearance in edible prey, is maintained.

apparency describes the ease with which individuals may be located by other organisms. Apparency may be conferred by brightness, individuals contrasting with their background, and being visually prominent, by behaviour, by scent especially in relation to mate-locating behaviour, and also by numerical abundance whereby individuals are frequently encountered despite other aspects of their morphology rendering them cryptic.

assortative mating non-random pairing within the mating portion of a population; usually used to describe a tendency for like individuals to pair.

assortment the random segregation of pairs of genes in the gametes.

autecology the study of an individual organism or species including its life history and behaviour.

backcross crossing of members of the F1 generation to either of the parental genotypes, usually to the parent carrying the recessive alleles.

base-status pertains to a soil, one containing much lime or potash (metal cations).

Batesian mimics a palatable species which comes to resemble a distasteful model and is protected to some extent against a predator which is wary of the distasteful species. The greater the relative number of mimics resembling the model, the less successfully can they shelter behind the model's defences.

BBCS British Butterfly Conservation Society.

biotic provinces or biome, is a major terrestrial community; it is the largest land community which can be conveniently recognized (e.g. tundra).

biotopes human perceived approximation to habitat, recognized by distinctive vegetation types or land-uses (e.g. oak woodland or acid heathland; see Appendix 2 and habitat).

blanket bog an area of peat which is not confined to a basin, but covers ground of moderate slope.

bottleneck a phase in the evolution of a species or a population when it is reduced to very low numbers.

boundary layer the thin layer of air immediately adjacent to a leaf or other surface; specific microclimatic conditions are associated with this layer.

BP before the present day, usually taken as 1950, in radiometric (^{14}C) years. It requires a correction of *circa* +720 years at 10 000 years BP.

Bronze Age in Britain, the infiltration of highly organized celtic folk from continental Europe (beaker, urn, and food vessel people) who brought with them the knowledge of copper metallurgy.

brown earth the characteristic soil type under deciduous forest in north-west Europe, in which the humus is of the mull type and well mixed with the topmost mineral soil.

bursal duct the tube connecting the bursa copulatrix to the outer area of the female abdomen; a spermatophore is passed along this duct during mating.

calcareous rock and soil substances containing a significant quantity of calcium carbonate; the rocks are known as limestones and chalk.

calcicole plants which require free calcium ions for successful growth and reproduction. They are usually restricted to chalk and limestone soils.

Castnioidea a superfamily of day-flying moths including more than 200 living species; three families are recognized within this group of which the Castniidae occur mainly in Australia and South and Central America.

character displacement the modification of the morphological form of a species as a result of the presence of interspecific competitors.

chromosome a microscopic thread-like structure of two chromatids, each consisting largely of DNA and proteins, numbers of which occur in the nucleus of every plant or animal cell. The chromatids are held together at the centromere. All nuclei of a given species have the same number and type of chromosomes. In butterflies, chromosomes exist in pairs (diploid number), one of which is derived from each parent.

climatic climax the final stage of the successional sequence of vegetation development which consists of a relatively stable ecosystem, said to be in equilibrium with its abiotic environment, especially climate.

cline a geographic gradient in a measurable character, or gradient in a gene, genotype, or phenotype frequency.

closed population one that breeds within a discrete 'home area' within which the majority of individuals hatch and die (see chapter 4.3.1).

C:N ratio the ratio of carbon to nitrogen in plant matter. The greater the amount of carbon compared to nitrogen the slower the breakdown of organic matter.

cocoon a case enclosing and protecting a pupa. It is made of silk, often strengthened with other materials.

codominant alleles which are both expressed in the heterozygote.

coefficient of variation is an expression of variation in a data set by converting the standard deviation to a percentage of the mean. See standard error.

coevolution reciprocal evolutionary change in interacting species (e.g. between phytophagous insects and higher plants) (Ehrlich and Raven 1965; see Thompson 1989).

commensalism state in which prequisite conditions for the existence of one species are maintained or provided by a second, but in which there is no associated reduction in fitness for the second species.

common oviduct the final tube in the ovary where all eight ovarioles end and pass their mature eggs for fertilization and laying.

communities groups of species that interact, or have the potential for interacting, with one another.

competition when more than one individual of the same or different species shares a common resource; the fitness of one or both may be reduced compared to that if the resource were unshared.

competitive exclusion when two species share a resource and the fitness of one of the pair is lowered to such an extent that it cannot coexist.

convergence morphological similarity in unrelated or distantly related forms.

correlation coefficient a measure of association (Pearson r, Spearman r_s) between two variables. Values typically vary from −1 (perfect inverse correlation), through 0 (no correlation) to +1 (perfect positive correlation); $r^2 \times 100$ gives an estimate of the amount of variance in one variable that can be accounted for by another.

costa (and costal) the front edge of the forewing and hindwing, bordered by the costal vein.

counter-shading the presence of dark dorsa and light ventra on animals, often supposed to ensure crypsis, but the adaptive effect by shape-obliteration is contested (see Kiltie 1988).

crypsis a pattern is cryptic if it resembles a random sample of the background perceived by the predator at the time and age, and in the microhabitat where the prey is most vulnerable to visually-hunting predators.

cultigen (or cultivar) a cultivated plant genotype.

deme a spatially discrete breeding unit; an effectively panmictic aggregate of organisms lasting for at least one breeding season and connected by gene flow with the neighbouring demes.

density-dependence a factor (i.e. mortality) that has an influence on a population which is related to the population density of an organism.

density effect with reference to egg-laying; a bias in the distribution of eggs for hostplants which occur at low density in excess of that expected by chance.

density-independence a factor that has an influence on a population which is unrelated to the population density of an organism.

Devensian the last glaciation in Britain extending from some 100 000 years ago to approximately 15 000 years ago. Maximum ice advance was *circa* 18 000 years ago.

diapause is a genetically determined state of suppressed development, the expression of which may be controlled by environmental factors.

diploid a cell or individual that has two copies of each chromosome (one from each gamete) in its genome.

directional selection selection favouring one extreme of the distribution of inherited variability.

disassortative mating tendency of unlike individuals to pair.

disjunction part of the geographical range of a species is separate (allopatric) from the rest of the species.

dispersal Movements made by individuals as a result of their daily activities which may allow transfer between habitat patches. The sum of such movements appears to be random and non-directional.

disruptive selection selection which favours the extreme members of a population at the expense of average members.

dominant in genetics, when a pair of chromosomes contain two different alleles for the same character (e.g. wing colour), then the dominant allele is the one that is expressed. In ecology, the organism, which by its size or other key role exerts a controlling influence over other living components of the ecosystem.

dorsal nectary organ a large slit-like depression on the larval seventh abdominal segment, which produces honeydew.

DNA for most organisms genetic information is encoded by molecules of deoxyribonucleic acid which is a large polymer of individual units called nucleotides. Each DNA nucleotide consists of a nitrogen-containing base, a dioxyribose sugar and a phosphate group.

drift *see* random genetic drift.

ecdysis the moulting of the chitinous cuticle of an insect.

eclosion the emergence of an adult from a pupa.

ecology the study of the relationship of plants and animals to each other and to their environment.

ecosystem a unit of space-time containing living organisms interacting with each other and with their abiotic environment by the interchange of energy and materials.

ecotone a sudden spatial change in the environment.

edaphic soil conditions including chemistry, texture, structure, depth, moisture status, etc.

edge effect with reference to egg-laying; a bias in the distribution of eggs to the margins of hostplant patches which are in excess of that expected by chance.

effective population size number of reproducing individuals in a population.

electrophoresis technique for separating molecules, based on their different mobility rates in an electric field.

eluviation a general term for the washing-out or removal of any material from the topsoil.

encapsulation a process whereby a parasitoid egg or larva is surrounded by host haemocyte cells and so depriving it of oxygen and nutrients.

endemism restriction of a particular species or other taxonomic group to a particular locality, due to factors such as soil, climate, or physical barriers.

epistasis interference of expression of one or more genes by another not allelic to them.

eustacy world-wide change in sea level associated with changes in ocean water volume.

evapotranspiration net transfer of water molecules into the air by the dual process of evaporation (conversion of a liquid to vapour) of water from lakes and rivers and from transpiration (water loss from plants).

evolution occurs when there is a change in gene frequency.

expressivity the degree of phenotypic expression of a fully penetrant gene; *see* penetrance.

fat-body an internal organ of insects consisting of an aggregation of cells enclosed within a membraneous sheath; it serves as a food storage and release organ.

fecundity the reproductive output of an organism; the number of eggs laid by a female, or the average laid by all females in a population, usually in a single generation.

fen a swampy area, where the ground water-level reaches the surface, composed of a non-acidic peat.

F1 generation first filial generation, the first generation of progeny from a parental cross. Symbol is F1.

fitness (synonym, biological fitness) the relative contribution of an individual or genotype to a subsequent generation(s) compared to those of other individuals or genotypes; fitness of an individual is measured by the number of his or her offspring reaching reproductive age. It depends on both individual success and that of other members of the same population.

fixation stage at which alleles become homozygous in a population.

Flandrian the present interglacial in Britain dating from 10 000 years BP.

follicle cells which surround and nourish the developing egg inside the ovariole.

Foraminifera a group of unicellular organisms (Protozoa) of the marine plankton, having calcareous shells.

forb a herb other than grass.

founder effect genetic drift due to the founding of a population by a small number of individuals.

frequency dependent selection natural selection the effects of which are determined by the relative frequencies of genotypes or phenotypes in the population.

gamete reproductive cell. Fusion of two gametes produces the zygote.

gene the functional unit of heredity.

gene flow is the process whereby the frequencies of alleles within a population is altered by the movement of reproductive individuals with different alleles or allele frequencies into the population.

gene frequency the proportion of an allele of a pair or series of alleles present in a population.

gene pool the total genetic information in a population.

genetic drift *see* random genetic drift.

genome all the genes present in the chromosome or set of chromosomes in a cell of an individual.

genotype the genetic constitution of an individual at one or more loci; usually written formally in gene symbols.

genus a category for a taxon including one species or a group of species, of common phylogenetic origin, which is separated from related similar units by a decided gap.

germinal bud a group of cells which exist through larval stages in insects eventually to form various organs of the adult in the pupal stage.

glacial an extensive period of time when colder conditions return to temperate latitudes dominated by periglacial processes, permafrost and tundra.

gleying soil forming processes occurring in waterlogged soils where alternative oxidizing and reducing conditions are present according to the level of water in the soil.

gnathos part of the chitinous male genitalia. It is a paired structure which descends from the terminal dorsal structure and forms the floor of the anal compartment. It is absent in some species (e.g. all European satyrinae), where it is replaced by other structures (i.e. brachia). When present, its morphology is often species-specific (see Higgins 1975).

guild a group of species that interact strongly because of their similar life-styles, for example share the same hostplants.

gynandromorph an individual displaying both male and female characteristics.

habitat the place where a plant or animal normally lives, characterized by its biotic or physical attributes. A subset of biotopes, perceived at a finer level by insects. Some habitats occur in more than one biotope.

haemolymph a clear fluid which is the insect equivalent to blood and lymph; contains food materials and specialized cells.

hair scales are unmodified cuticular scales of chitin. They have insulating properties and are often pigmented with melanin. Modified and flattened scales, sometimes containing complex pigments, cover the wings of butterflies.

halophytes salt-tolerant plant species.

haploid the number of chromosomes in the gametes.

Hardy–Weinberg rule unless disturbed by external influences (e.g. natural selection) the proportion of the various genotypes at a gene locus in the population remains the same in each successive generation provided that mating is at random.

heritability is a parameter h^2 indicating the proportion of the total phenotypic variance of a quantitative trait or character which is due to the segregation of genes (polygenes) as against environmental effects. The term is usually restricted to the additive effects of genes; those which do not involve dominance or epistasis.

heterogametic an individual that produces gametes differing in sex chromosome content.

heterozygote (and heterozygous) possessing two different alleles at the corresponding gene locus on a pair of chromosomes.

hibernation a quiescent state of dormancy in which some animals spend the winter.

homogametic an individual that does not produce gametes of different sex chromosome content.

homologous chromosomes chromosomes which occur in pairs and where members of pairs contain identical sets of gene loci although these may occur as different alleles.

homozygote (and homozygous) possessing similar alleles at the two corresponding pair of chromosomes.

honeydew partly digested sugars obtained from plant phloems; sometimes used to describe sugars obtained by ants from butterfly larvae.

hybrid an individual resulting from a cross between genetically dissimilar parents.

hybrid zone narrow belts (clines) with greatly increased variability in fitness and morphology compared to that expected from random mixing, separating distinct groups of relatively uniform sets of populations.

hyperparasites insect 'parasites' that infect and ultimately kill their host, the host being another parasitoid of a third insect.

ice wedge polygons surface patterns associated with deep wedges caused by the cracking of the ground and their infilling with ice and debris under severe frost action with temperatures below −20 °C.

inbreeding the crossing of closely related individuals, usually leading to increased homozygosity in a population or family.

inbreeding depression reduction in fitness or vigour as a consequence of inbreeding.

infiltration capacity the rate at which rainwater drains through the soil. This depends on the texture of the soil, its thickness, and degree of compaction.

inner margin the posterior edge of the forewings and hindwings.

instars the form assumed by an insect between each ecdysis; the imago, the pupa, and the phase between each larval moult.

interglacial a long interval between two glaciations during which the temperature rises at least as high as during the present period at temperate latitudes.

intergradation character gradients between groups of populations. Often refers to two or more clines for different characters in the same organism, and going in the same geographical direction.

intersex an individual whose reproductive organs are a mixture of the sexes.

interstadial a period of climatic fluctuation within a glacial period, of smaller magnitude and shorter duration than an interglacial.

intraspecific communication refers to the detection and recognition of individuals of the same species. It may involve more than one sensory mode (sight, sound, scent organs, touch) and includes male–male, male–female, and female–female communication.

involutions contorted structures in unconsolidated deposits associated with freeze–thaw cycles under periglacial conditions in which fine materials (clays/silts) are injected into coarser materials.

Iron Age in Britain, thought to begin in the sixth century BC with the coming of the Halstatt culture and iron metallurgy.

isostacy relative uplift or depression of a land area due to abstraction or addition of load, as in the case of ice caps in mountain regions.

ITE Institute of Terrestrial Ecology.

JCCBI Joint Committee for the Conservation of British Insects.

kairomones chemical agents of adaptive advantage to the receiving organism.

karyotype the artificial arrangement of the chromosome set so as to allow comparison of their morphology and subsequent analysis. Each species has a distinctive number of pairs of chromosomes.

key-factor method a method of analysing life table data in order to identify the 'key-factor' or 'factors' which determine the annual fluctuations in abundance. It also allows an examination of the effects of variations in the egg-laying rate and mortality rates on the population, and the recognition of any density-dependent relationships that exist.

***K*-selected strategy** a life cycle strategy where more importance has been placed on maintaining the highest possible density of individuals for most of the year (K) rather than a high rate of increase (r). K-selected species are characterized by a lower rate of increase (and death), slower development, larger body size, longer life, and reduced mobility.

K–T event the relatively short episode of mass extinction of the dinosaurs at the Cretaceous–Tertiary boundary at 66.5 million years BP leading to the radiation of mammals.

Late Glacial in Britain, the period from the end of the Devensian maximum ice advance some 15 000 years BP until the Post Glacial 10 000 years BP. It is normally divided into three time zones, characterized by changing conditions.

leachate organic solutions, comprising humic acids, which drain through permeable soils, and have the capacity to leach soils.

leaching solution and downwashing of soluble material in soils by rainwater.

lessivation downwashing of clays in soils.

linkage Genes situated on the same chromosomes are said to be linked. Except when crossing-over (*see* recombination) occurs they are inherited together and, therefore, do not assort or segregate independently. Crossing-over occurs less frequently the nearer together the genes are situated.

local climate the meteorological conditions (e.g. temperatures, radiation, humidity, precipitation) of different parts of habitats, such as sections of river valleys, hillsides, lake surrounds, woodlands, and urban areas.

locus the site on a chromosome occupied by a particular gene.

Lusitanian element an oceanic group of species extending from the Mediterranean to south-west Ireland up the western seaboard of Europe.

macroevolution evolution at and above the species level.

MAFF Ministry of Agriculture, Fisheries, and Food.

melanins dark pigments usually giving a black or brownish colour. Melanins are complex polymers which may be derived from the amino acid tyrosin, itself synthesized during insect metabolism.

Mendelian ratios the relative frequency of allele states in the progeny of a parental cross, corresponding with Mendel's first (principle of segregation) and second (law of independent assortment) laws.

Mesolithic period of middle Stone Age culture form the latter part of the Late Glacial to 5000 years ago when man, in Britain, still followed a hunter–gatherer economy.

metapopulation set of populations that interact via individuals moving among populations (Hanski and Gilpin 1991).

microclimates the meteorological conditions (e.g. temperatures, humidity, radiation) of very small parts of the environment. Microclimatic differences contrast conditions a short distance apart, for instance opposite sides of leaf, or the top and base of a grass sward.

microevolution evolution below the species level.

micropyle the minute opening in the chitinous shell of the egg, through which the sperm enters.

migration the relatively predictable long distance movements made by large numbers of individuals in approximately the same direction at approximately the same time; it is usually followed by a return migration; cf. dispersal. Immigration and emigration refer to the movement of individuals into and from a spatial unit or population.

mimicry occurs when an organism or group of organisms (the mimic) closely resembles a second living organism (the model), such that the mimic is able to take some advantage of the regular response of a sensitive signal-receiver (the operator) towards the model, through mistaken identity of the mimic for the model; *see* Vane-Wright (1976).

modifier gene a gene which alters the phenotypic expression of another gene not allelic to it.

monophagy restriction of hostplant use to one plant species.

monotypic a species having a single subspecies or form.

mor an acid soil humus that accumulates at the soil surface, and is too acid for the presence of earthworms.

morph a distinctive form, usually genetic, co-existing with the wild type and freely interbreeding with it.

morphogen a chemical substance which plays a role in the determination of the fate of cells during development. Most commonly a gradient in the concentration of the morphogen provides information about the position of cells during development.

MRR mark, release, recapture techniques for the study of population size and movement. These usually involve the capture and recapture of individuals but in recent studies marking and recording has occasionally been achieved without disturbance.

Müllerian mimics mimicry based on warning coloration indicating that all the species involved may be harmful (nauseous flavour, poisonous qualities) to predators.

multivariate analysis the simultaneous use of a large number of attributes to describe or compare organisms and the populations to which they belong (Sneath and Sokal 1973).

mutant gene an allele that differs from the recognized wild type gene.

mutation changes in the hereditary material. Gene or point mutations are those that change only one or a few DNA nucleotides in a gene. Chromosomal mutations are those changing the number of chromosomes or the number or arrangement of genes in chromosomes.

natural selection the process by which changes occur in the proportions of genetic types within a population due to differences in their fitness in the existing environment.

NCC Nature Conservancy Council.

Neolithic period of new Stone Age culture from 5200 to 4400 years ago when man introduced an agricultural economy to Britain including the domestication of livestock and cultivation of plants.

Newcomer's organ or honey gland; a large slit-like depression across the top of a larva's seventh abdominal segment, below which lies a sac containing honeydew that is secreted from four glands; it is used to appease ants.

niches (or ecological niche) is the range of environmental conditions, described by Hutchinson (1957) as an '*n*-dimensional hypervolume' within which fitness is positive. The fundamental niche describes the niche a species could potentially occupy, whereas the realized niche is some portion of the fundamental niche occupied by a species due to competition of other species or habitat limitations (see Begon and Mortimer 1981).

NNR National Nature Reserve.

non-metric scaling a computer technique that iteratively moves points (representing units) about until their relative positions in space equate as well as is possible with their affinities as measured on one or more variables (Sneath and Sokal 1973).

nurse cell specialized cells in the ovarioles which provide nutrients to the early developmental stages of the egg.

OD ordnance datum; mean sea-level measured at Newlyn, Cornwall.

oligophagy restriction of hostplant use to a single plant family; often a small number of species within a plant family.

ombrogenous peat that which lies above the ground water-level, so that plants on its surface depend on rainwater for their supply of minerals.

open population a population that does not breed within a discrete 'home area' and where individuals range widely over the countryside, regularly dispersing from one breeding area to another (see section 4.3.1).

outbreeding the mating of individuals in which the genes are not related by descent.

ovarioles the individual tubular organs which together comprise the ovary; butterflies have four ovarioles in each ovary, and two ovaries in their reproductive system.

pan indurated (often impervious) mineral horizon in soils. In Britain, pans are typically of iron oxides, resulting from the leaching of minerals (podsolization) from upper soil horizons and their precipitation and accumulation in lower horizons.

panmixis a random breeding population, in which any two individuals have an equal chance of mating.

parapatric two or more subspecies, incipient species, or species that are in contact over a very narrow zone.

parasitoids organisms which do not complete all their life cycle within the host, and usually have one free-living stage (typically the adult).

peds aggregates or structures in soil, usually as a result of cohesive properties of clay and humus, having a profound influence on soil properties.

penetrance frequency of phenotypic expression of a gene among individuals which carry it; cf. expressivity.

permafrost permanently frozen ground often as much as 600 m thick. Its cover is continuous where the annual average temperature does not exceed −5 °C and discontinuous where this is approximately between −5 and −1.5 °C. The surface metre of ground (active layer) thaws out in the brief arctic summer, becoming saturated and mobile.

pharate a stage in which a given instar is fully formed within the old cuticle of the previous instar, prior to 'moulting' of the old skin; it can persist for several weeks in pupae, or even months in eggs.

phenetic variation variation in populations in the expression of phenotypes.

phenocopy non-hereditary variation which mimics the expression of a gene.

phenology the seasonal occurrence of different life stages of a species.

phenotype the observable characteristic of an organism, resulting from the interplay of the genotype and the environment in which development takes place.

pheromone a chemical substance produced by one organism that influences the behaviour or development of another.

philopatry returning to the birthplace to reproduce, in spite of long-distance travel.

photoperiod the length of a period of light needed to stimulate a light regulated response, such as flowering in plants and reproduction in some animals.

phylogeny evolutionary (i.e. ancestor, descendant) relationships among organisms.

pingos large circular hills up to 300 m across and 60 m

high, some with a central crater containing a pond, caused by the segregation of underground ice in a developing cold climate.

plagioclimax a type of vegetation (sere) maintained by continual disturbance (heavy grazing, frequent burning).

plate tectonics the theory surrounding the knowledge of the earth's division into a number of large crustal plates. These move relative to one another and distinct activity (earthquakes, volcanoes, folding, and faulting) occurs at their margins.

pleiotropy multiple effects of the same gene.

Pleistocene the most recent part of geological time, the last two million years or so, including the recurrent glaciations in north-west Europe. *See* Table 10.1.

podsol soil type in which there is a surface layer of mor humus. Below this a severely leached grey sandy horizon is underlain by a lower dark horizon in which iron oxides have accumulated. Podsolization refers to the downward translocation of most of the soil solutes (iron and aluminium minerals) particularly by humic acids.

polygenic a character determined by many genes, each one having a small effect upon its expression in the phenotype.

polymorphism the occurrence within a freely interbreeding species of two or more differing inherited forms, the rarest of them being too common to be kept in existence by recurrent mutation; see Vane-Wright (1975).

polyphagy the use of hostplants from more than one plant family.

polyphenism discrete phenotypes which can be produced in genetically identical organisms in a switch-like manner by the influence of environmental controls on developmental pathways. The ability of organisms to produce such divergent forms under different environmental conditions is under genetic control.

polytopic a species having more than one subspecies or form.

population the total number of individuals of a particular species which are separated to some extent, either in space or time, from other groups of the same species (see section 4.1).

pore cupula microscopic pores on larvae and pupae which probably produce secretions to appease ants.

Post Glacial or Holocene, the relatively warm period in temperate latitudes from 10 000 years ago to the present day.

post-reproductive refers to adults which are incapable of further reproductive activity. Males are post-reproductive if they are no longer capable of successfully mating, and females when they cannot lay or produce viable eggs.

primary defence defences that operate regardless of whether or not there is a predator in the vicinity. The predator may not detect (anachoresis and crypsis) the prey at all, or it may be capable of detecting the prey but fail to recognize (aposematism and Batesian mimicry) it as something edible. (See Edmunds 1974*b*.)

primary intergradation intergradation between two geographical forms that have always been in contact.

process-response models natural systems drawn up in the form of a flow chart to indicate the links between variables and the nature of feedbacks (direct or indirect, positive or negative) between them. Logical connections (cause to effect) are determined by arrows, whereas the signs indicate the correlation between variables; +, an increase (decrease) in one variable results in an increase (decrease) in another variable; −, an increase (decrease) in one variable results in a decrease (increase) in another variable.

protandry the emergence from the same brood of males before the females.

proterogyny the emergence in the same brood of females before males; this is unusual and often indicates some form of developmental disruption (genetic differences) as when artificially crossing individuals from widely separated populations.

pteridine derived pigments that are white or yellow, and are synthesized from uric acid, itself produced during nucleic acid metabolism. Pteridine pigments are the most important 'white' pigments of the Pieridae. Leucopterin is a common white pigment which occurs in the Pieridae and xanthopterin gives the Coliadinae their yellow coloration. Other white or yellow pigments are most usually derivatives of flavones, obtained from plant material. Such pigments are responsible for the white coloration of *Leptidea sinapis* (Dismorphiinae) and *Melanargia galathea* (Satyrinae).

punctuated equilibrium term employed for the pattern of evolution consisting of long periods of stability interrupted occasionally by periods of relatively rapid evolutionary change, interpreted as true events of speciation; contrasts with a pattern of long-term gradual, progressive change.

qualitative variation characters that display discontinuities of expression, typically due to assortment of major genes with large effects upon the phenotype.

quantitative variation a character that shows continuous variation of expression and can be studied by measurement or, sometimes, by counting (*see* polygenic).

race a distinctive interbreeding, or potentially interbreeding, population or group of populations within a species or subspecies. Its taxonomic status may be approximately equal to, or lower than a subspecies.

radiation absorbing pigments that convert the electromagnetic energy of sunlight (photons) into heat energy (excited atoms).

radiometric dating dating artefacts by measuring the amount of different isotopes (forms) of elements occurring within them, based on a known decay rate (e.g. ^{14}Carbon returns to ^{14}Nitrogen emitting a beta particle. Organic material loses half its specific ^{14}C activity in 5730 years (half-life)).

raised bog a peat bog developed on a level substratum and having the form of a dome, the peat being thickest at the centre of the bog, so that the surface is convex.

random genetic drift genetic changes in populations caused by random phenomena rather than by natural selection.

range the maximum geographical extent of a species, usually determined by a line around all the locations where it is distributed.

recess effect reference to egg-laying; the bias in egg distribution to a recess in the hostplant patch or micro-landform (e.g. cattle hoof depression) or to where hostplants abut taller non-hostplant vegetation.

recessive a character is said to be recessive if it is only manifest in those individuals homozygous for the gene controlling it. It is not detectable (except sometimes by special tests) in the heterozygote.

recombination (or crossing-over) the exchange of genes between homologous chromosomes which takes place at meiosis (the processes leading to the production of the fertilized egg). It leads to a new association of genes in a recombinant individual.

recurrence surface a change in the type of peat in a bog profile, from dark highly humified and presumably slow-growing peat, to fresh unhumified and much faster growing peat; interpreted as indicating an increase in humidity.

regeneration complex for woodland it is the assemblage and sequence of plants colonizing an open space after tree fall. For bogs it is the assemblage of plants which form the newly wet bog surface and give rise to fresh peat immediately above a recurrence surface.

rendzina calcareous soils normally well drained and usually shallow, over shattered limestone, chalk, or soft, extremely calcareous (>40 per cent $CaCO_3$) material. They often have a dark surface horizon with a high content of organic material.

reproductive output refers to the number of offspring produced by an individual. In females this may be measured by the number of fertile eggs that are produced. In males it is related to the number of matings that are achieved, though this may not directly relate to the number of viable eggs that are produced by the females which are inseminated by a particular male since some females may not lay eggs.

reproductive strategy the pattern of life history (e.g. diapause, dormancy, dispersal, migration, phenology, voltinism, fecundity, size) adopted by individuals in a population under the influence of natural selection and believed to be most suited to maximize fitness in a specific environment. *See r*- and *K*-selected strategy.

residence time the time taken for half the adult population to have either died or emigrated. It is calculated from the residence rate obtained from the analysis of mark–recapture data, using the formula, $-1/\log_e$ residence rate (after Cook *et al.* 1967).

retrogressive succession the gradual devolution of plant communities from a climax state to lower status seres as a result of increased leaching and soil impoverishment.

***r*-selected strategy** a life cycle strategy where more importance has been placed on a high rate of increase (r) rather than maintaining the highest possible density of individuals for most of the year (K); r-selected species are characterized by a high rate of increase (and death), rapid development, smaller body size, shorter life, and greater mobility.

RSPB Royal Society for the Protection of Birds.

ruderal a plant characteristic of waste places.

seasonal polyphenism an annually repeating pattern of changing phenotypic ratios in successive generations under some kind of immediate environmental control. It involves phenotypic plasticity and contrasts with genetic polymorphism.

secondary compounds *see* allelochemics.

secondary contact *see* secondary intergradation.

secondary defence those defences of an animal which operate during an encounter with a predator. (See Edmunds 1974*b*.)

secondary intergradation intergradation between two geographic forms that at one time diverged in isolation.

second filial generation the generation descended from the first filial generation. Symbol F2.

selection *see* natural selection.

sere is any one of a number of stages in the successional development of a climax vegetation community. Early seral stages are usually dominated by fast growing and highly mobile species with good colonizing abilities; later seral stages contain a greater proportion of less mobile species which invest more energy into growth and defence than those of early seral stages.

sex chromosomes chromosomes which determine and differentiate the sexes.

sex-limited expression genes, the effects of which, in the phenotype, are only evident in one sex.

sex linkage genes borne by the sex chromosomes.

sibling species pairs or groups of closely related biological species which are reproductively isolated but morphologically identical or nearly so.

significance an observation having a low probability of occurring by chance alone. Common values for significance are 0.05 (1 in 20) and 0.01 (1 in 100); see Siegel (1956).

single linkage cluster analysis a multivariate statistical technique that orders the relationships between individuals and groups of individuals sequentially on the basis of their nearest neighbourhood affinities (Sneath and Sokal 1973).

solifluction the flowing or creeping of topsoil which occurs in permafrost conditions (arctic climate); spring melt mobilizes a saturated topsoil over permanently frozen ground.

species biological: groups of actually (or potentially) interbreeding natural populations which are reproductively isolated from other such groups (see Mayr 1963). Morphological: groups of populations which are entirely distinguishable from other such groups in the phenotype of each individual; it is also assumed that these different groups are reproductively isolated as with biological species (see Ferguson 1980).

species diversity is a measure of the number of species and the evenness with which individuals of these species occur in any area.

species richness the number of species in any area.

spermatheca an organ which receives and stores sperm prior to its being used to fertilize an egg in butterflies.

spermatophore a sac-like body containing sperm which is produced within the female by the male during copulation.

sphragis a plug secreted by males over the genital opening of females with which they have just mated; this physically prevents further matings with the female.

SSSI Sites of Special Scientific Interest.

stabilizing selection selection favouring the average individuals of a population and eliminating extreme variants.

stadials a short episode of glacial or arctic conditions, lasting a few thousand years at most, within a longer warm episode in temperate latitudes.

standard error describes the bounds around the sample mean within which the population mean is likely to occur with 68 per cent probability. It is calculated as, $SE = S/\sqrt{N}$, where S is the standard deviation of the sample calculated as the square root of the sum of the squared deviations of variates (x) from the mean ($\bar{x}$) divided by the sample size (N).

stone polygons sorted and unsorted collections of stones in various patterns depending on slope angle caused by differential movement of different sized particles in soils under a periglacial climate.

subspecies a geographically defined aggregate of local populations which differs taxonomically (usually with recognizable and consistent phenetic attributes) from other such subdivisions of the species. Such units are capable of interbreeding, if only to a limited extent.

succession the regular and progressive change in the components of an ecosystem from the initial colonization of an area to a stable state or mature ecosystem.

supergene term used to denote a series of genes which are thought to have become closely linked on the same chromosome because of the selective advantage of their being inherited as a unit.

symbiosis a condition where two dissimilar organisms are associated to the mutual benefit of both.

sympatric the occurrence of two or more actually or potentially interbreeding forms in the same place. The term is also used to describe species with overlapping distributions or ranges.

synecology the study of groups of organisms associated as a unit and their relationships to one another.

systematics the classification of living organisms into hierarchical groupings.

taiga the northern coniferous forest associated with a boreal climate.

taxon cycle the inferred cyclical evolution of species, from the ability to live in marginal habitats and disperse widely, to preference for more central, species-rich habitats with an associated loss of dispersal ability, and back again. (See MacArthur and Wilson 1967.)

taxonomy the classification of organisms in logical and natural groups along evolutionary lines. A particular group, that is a named species, is a taxon.

telocratic phase the terminal stage of an interglacial, retrogressive in a vegetational sense, with the soils becoming increasingly podsolized.

tentacle organs or retractile organs, are paired finger-like structures either side of the honey gland but on the eighth larval abdominal segment; they probably release scents to appease ants.

thanatosis death feigning; a predator escape mechanism.

thermoregulation the ability and processes whereby an organism can adjust its temperature to its surroundings.

threshold character a character which has an underlying polygenic continuity with a threshold which imposes a discontinuity on the visible expression in the phenotype.

till boulder clay, unsorted mineral debris conveyed by an ice sheet or glacier.

topogenous peat formed in drainage basins which receive groundwater from a land surface.

uniformitarianism 'the present is the key to the past', a principle enunciated by James Hutton in 1785; the same physical processes and laws that operate today have operated throughout geological time, although not necessarily always with the same intensity as now.

valence the range of environmental values that a given species or genotype can tolerate; also known as ecological amplitude.

valve paired lateral and moveable chitinous structures forming part of the male genitalia. During copulation the valves open to expose the penis and facilitate the joining of male and female genitalia. Valve morphology is extremely variable between the taxonomic groups and is often used for specific identification.

variance is a statistical measure describing the dispersion of values around the arithmetic mean (average) and is the mean squared deviation of each individual value from the mean value. It is calculated as the sum of squared deviations of variates (x) from the mean value ($\bar{x}$) divided by the sample size (N); see standard error.

voltinism the number of generations of a species per year; thus specifically univoltine, bivoltine, and multivoltine.

wild type the standard genotype with the normal or most frequently encountered phenotype in natural populations either with regard to the whole organism or to a particular gene locus.

zygote the product of fusion of male and female gametes.

Bibliography

ACKERY, P. R., 1988. Hostplants and classification: nymphalid butterflies. *Biological Journal of the Linnean Society*, 33: 95–203.

ADAMS, W. M., 1984. *Implementing the Act: A Study of Habitat Protection under Section II of the Wildlife and Countryside Act 1981*. Cambridge: British Association of Nature Conservationists.

ADLER, P. H. and PEARSON, D. L., 1982. Why do male butterflies visit mud puddles? *Canadian Journal of Zoology*, 60: 322–325.

AHMAD, S., 1983. *Herbivorous Insects. Host seeking behaviour and mechanisms*. London: Academic Press.

ALLEN, J. A., 1988. Frequency-dependent selection by predators. *Philosophical Transactions of the Royal Society of London B*, 319: 485–504.

ANON, 1929. Report of the Committee appointed by the Entomological Society of London for the Protection of British Lepidoptera. *Proceedings of the Entomological Society of London*, 4: 53–68.

APLIN, R. T., D'ARCY WARD, R., and ROTHSCHILD, M., 1975. Examination of the large white and small white butterflies (*Pieris* spp.) for the presence of mustard oils and mustard glycosides. *Journal of Entomology A*, 50: 73–78.

ARIKAWA, K., EGUCHI, E., YOSHIDA, A., and AOKI, K., 1980. Multiple extraocular photoreceptive areas in genitalia of butterfly *Papilio xuthus*. *Nature*, 288: 700-702.

ARMS, K., FEENY, P., and LEDERHOUSE, R. C., 1974. Sodium stimulus for puddling behavior by tiger swallowtail butterflies, *Papilio glaucus*. *Science*, 185: 372–374.

ATKINSON, T. C., BRIFFA, K. R., and COOPE, G. R., 1987. Seasonal temperatures in Britain during the past 22,000 years, reconstructed using beetle remains. *Nature*, 325: 587–592.

ATSATT, P. R., 1981a. Ant-dependent food selection by the mistletoe butterfly *Ogyris amaryllis* (Lycaenidae). *Oecologia*, 48: 60–63.

ATSATT, P. R., 1981b. Lycaenid butterflies and ants: selection for enemy-free space. *American Naturalist*, 118: 638–654.

BAKER, H. G. and BAKER, I., 1975. Studies of nectar-constitution and pollinator-plant coevolution. In L. E. Gilbert and P. H. Raven (Eds.), *Coevolution of animals and plants: Symposium V, First International Congress of Systematic and Evolutionary Biology*: 100-140. Austin, Texas: University of Texas Press.

BAKER, R. R., 1968a. A possible method of evolution of the migratory habit in butterflies. *Philosophical Transactions of the Royal Society of London*, 253: 309–341.

BAKER, R. R., 1968b. Sun orientation during migration in some British butterflies. *Proceedings of the Royal Entomological Society of London (A)*, 43: 89–95.

BAKER, R. R., 1969. The evolution of the migratory habit in butterflies. *Journal of Animal Ecology*, 38: 703–746.

BAKER, R. R., 1970. Bird predation as a selective pressure on the immature stages of the cabbage butterflies *Pieris rapae* and *P. brassicae*. *Journal of Zoology*, 162: 43–59.

BAKER, R. R., 1972a. Territorial behaviour of the nymphalid butterflies, *Aglais urticae* (L.) and *Inachis io* (L.). *Journal of Animal Ecology*, 41: 453–469.

BAKER, R. R., 1972b. The geographical origin of the British spring individuals of the butterflies *Vanessa atalanta* (L.) and *V. cardui* (L.). *Journal of Entomology A*, 46: 185–196.

BAKER, R. R., 1978. *The Evolutionary Ecology of Animal Migration*. London: Hodder and Stoughton.

BAKER, R. R., 1984. The dilemma: when and how to go or stay. In R. I. Vane-Wright and P. R. Ackery (Eds.), *The Biology of Butterflies*. Symposium of the Royal Entomological Society, Number 11: 279–296. London: Academic Press.

BALLETTO, E. and KUDRNA, O., 1985. Some aspects of the conservation of butterflies in Italy, with recommendations for a future strategy (Lepidoptera: Hesperiidae and Papilionidae). *Estratto dal Bollettino della Societa Entomologica Italiana*, 117: 39–59.

BARBOUR, D. A., 1986a. Why are there so few butterflies in Liverpool? An answer. *Antenna*, 10: 72–75.

BARBOUR, D. A., 1986b. Expansion of the range of the speckled wood butterfly, *Pararge aegeria* L. in north-east Scotland. *Entomologist's Record and Journal of Variation*, 98: 98–105.

BARD, J. B. L. and FRENCH, V., 1984. Butterfly wing patterns: how good a determining mechanism is the simple diffusion of a single morphogen? *Journal of Embryology and Experimental Morphology*, 84: 255–274.

BARGHOORN, E. S., 1965. Evidence of climatic change in the geological record of plant life. In H. Shapley (Ed.), *Climatic Change. Evidence, Causes and Effects*: 235–248. Cambridge, Massachusetts: Harvard University Press.

BARNHAM, M. and FOGGIT, G. T., 1987. *Butterflies in the Harrogate District*. Published by the authors. Harrogate: Harrogate and District Naturalists' Society.

BARRY, R. G. and CHORLEY, R. J., 1982. *Atmosphere, Weather and Climate*. London: Methuen.

BARTLEIN, P. J. and PRENTICE, I. C., 1989. Orbital variation, climate and paleoecology. *Trends in Ecology and Evolution*, 4: 195–199.

BARTON, N. H. and HEWITT, G. M., 1989. Adaptation, speciation and hybrid zones. *Nature*, 341: 497–503.

BAYLIS, M. and KITCHING, R. L., 1988. The myrmecophilous organs of the small blues, *Cupido minimus* (Lepidoptera; Lycaenidae). *Journal of Natural History*, 22: 861–864.

BAYLIS, M. and PIERCE, N. E., 1991. The effect of hostplant quality on the survival of larvae and oviposition by adults of an ant-tended lycaenid butterfly *Jalmenus evagoras*. *Ecological Entomology*, 16: 1–9.

BEATTIE, A. J., 1985. *The Evolutionary Ecology of Ant–Plant Mutualisms*. Cambridge: Cambridge University Press.

BEGON, M., 1979. *Investigating Animal Abundance: Capture-Recapture for Biologists*. London: Edward Arnold.

BEGON, M. and MORTIMER, M., 1981. *Population Ecology. A Unified Study of Animals and Plants*. Oxford: Blackwell Scientific Publications.

BEIRNE, B. P., 1947. The origin and history of the British macrolepidoptera. *Transactions of the Royal Entomological Society*, 98: 275–372.

BEIRNE, B. P., 1955. Natural fluctuations in the abundance of British Lepidoptera. *Entomologist's Gazette*, 6: 21–52.

BELL, B. G., 1982. *Biological monitoring in the Forth Valley*. I. T. E. Annual Report 1981: 63–65. Cambridge: Institute of Terrestrial Ecology.

BENGTSON, S. A., 1978. Spot-distribution in *Maniola jurtina* on small islands in southern Sweden. *Holarctic Ecology*, 1: 54–61.

BENGTSON, S. A., 1981. Does bird predation influence spot-number variation in *Maniola jurtina*? *Biological Journal of the Linnean Society*, 15: 23–27.

BENGTSON, S. A. and ENCKELL, P. H., 1983. Island ecology. *Oikos*, 41: 296–547.

BENNET, D. P. and HUMPHRIES, D. A., 1974. *Introduction to Field Biology*. London: Edward Arnold.

BENNET, K. D., 1989. A provisional map of vegetation for the British Isles 5000 years ago. *Journal of Quaternary Science*, 4: 141–144.

BENNET, K. D., TZEDAKIS, P. C., and WILLIS, K. J., 1991. Quaternary refugia for north European trees. *Journal of Biogeography*, 18: 103–115.

BERNAYS, E. A. and CHAPMAN, R. F. 1978. Plant chemistry and acridoid feeding behaviour. *Annual Proceedings of the Phytochemical Society of Europe*, 15: 99–141.

BERRY, R. J., 1977. *Inheritance and Natural History*. London: Collins.

BEUG, H. J., 1975. Changes in climatic and vegetation belts in the mountains of Mediterranean Europe during the Holocene. In Z. Pazdro *et al.* (Ed. 1), Palaeogeographical Changes of Valley Floors in the Holocene. *Biuletyn Geologicnzy*, 19: 101–110. Warszawa: Wydawnictwa University.

BIERHALS, E., 1976. Oekologische folgender der Vegetationsent-wicklung und des Wegfalls der Berwirtsschaftungs massnahmen. *Brackflachen in der Landschaft* KTBL-Schrift 195, Munster-Hiltrup, pp. 1–156.

BINK, F. A., 1972. Het onderzock naar de grote vuurvlinder (*Lycaena dispar batava* Öberthur) in Nederland (Lep. Lycaenidae). *Entomologische Berichten*, 32: 225–239.

BINK, F. A., 1985. Hostplant preference of some grass-feeding butterflies. *Proceedings of the 3rd Congress of European Lepidopterology, Cambridge 1982*: 23–29.

BINK, F. A., 1986 Acid stress in *Rumex hydralopathum* (Polygonaceae) and its influence on the phytophage *Lycaena dispar* (Lepidoptera; Lycaenidae). *Oecologia*, 70: 447–451.

BISHOP, W. W. and COOPE, G. R., 1977. Stratigraphical and faunal evidence for Late Glacial and early Flandrian environments in southwest Scotland. In J. M. Gray and J. J. Lowe (Eds.), *Studies in the Scottish Lateglacial Environment*: 61–88. Oxford: Pergamon Press.

BLAB, J. and KUDRNA, O., 1982. *Hilfsprogramm für Schmetterlinge. Ökologie und Schutz von Tagfaltern und Widderchen*. Naturschutz aktuell Nr. 6. Greven, W. Germany: Kilda-Verlag.

BLEST, A. D., 1957. The function of eyespot patterns in the Lepidoptera. *Behaviour*, 11: 209–256.

BLONDEL, J., 1987. From biogeography to life history

theory: a multithematic approach illustrated by the biogeography of vertebrates (birds). *Journal of Biogeography*, 14: 405–422.

BLOWER, J. G., COOK, L. M., and BISHOP, J. A., 1981. *Estimating the size of animal populations*. London: Allen and Unwin.

BODY, R., 1982. *Agriculture: the Triumph and the Shame*. London: Temple Smith.

BODY, R., 1987. *Red or Green for Farmers*. London: Broad Leys.

BOECKLEN, W. J. and GOTELLI, N. J., 1984. Island biogeography theory and conservation practice: species area or specious-area relationships? *Biological Conservation*, 29: 63–80.

BOGGS, C. L. and GILBERT, L. E., 1979. Male contribution to egg production in butterflies: evidence for transfer of nutrients at mating. *Science*, 206: 83–84.

BOGGS, C. L., SMILEY, J. T., and GILBERT, L. E., 1981. Patterns of pollen exploitation by *Heliconius* butterflies. *Oecologia*, 48: 248–259.

BOGGS, C. L. and WATT, W. B., 1981. Population structure of pierid butterflies. IV. Genetic and physiological investment in offspring by male *Colias*. *Oecologia*, 50: 320–324.

BOORMAN, L. A., 1983. *Studies on the Colonization of Planted Woodland by Herbaceous Plant Species*. I. T. E. Annual report 1982: 27–28. Cambridge: Institute of Terrestrial Ecology.

BOPPRÉ, M., 1984. Chemically mediated interactions between butterflies. In R. I. Vane-Wright and P. R. Ackery (Eds.), *The Biology of Butterflies*. Symposium of the Royal Entomological Society, Number 11: 259–275. London: Academic Press.

BOPPRÉ, M., 1986. Insects pharmacophagously utilizing defensive plant chemicals (pyrolizidine alkaloids). *Naturwissenschaften*, 73: 17–26.

BOURQUIN, F., 1953. Notas sobre la metamorfosis de *Hamearis susanne*, con oruga mirmecofila (Lepidoptera: Riodinidae). *Revista de la Sociedad Entomologica Argentina*, 16: 83–87.

BOWDEN, J., DIGBY, P. G. N., and SHERLOCK, P. L., 1984. Studies of elemental compostition as a biological marker in insects. I. The influence of soil type and hostplant on elemental composition of *Noctua pronuba* (L.) (Lepidoptera: Noctuidae). *Bulletin of Entomological Research*, 74: 207–225.

BOWDEN, S. R., 1985. Taxonomy for a variable butterfly? *Entomologist's Gazette*, 36: 83–90.

BOWDEN, S. R., 1987. Sex-ratio anomaly in an English pierid butterfly. *Entomologist's Gazette*, 37: 167–174.

BOWEN, D. Q., 1978. *Quaternary Geology. A Stratigraphic Framework for Multidisciplinary Workers*. Oxford: Pergamon Press.

BOWERS, M. D., 1979. *Unpalatability as a Defence Strategy of Checkerspot Butterflies with Special Reference to Euphydryas phaeton*. Ph.D. Thesis, University of Massachusetts, USA.

BOWERS, M. D., BROWN, I. L., and WHEYE, D., 1985. Bird predation as a selective agent in a butterfly population. *Evolution*, 39: 93–105.

BOWERS, M. D. and WIERNASZ, D. C., 1979. Avian predation on the palatable butterfly, *Cercyonis pegala*. *Ecological Entomology*, 4: 205–209.

BRAKEFIELD, P. M., 1979. *An Experimental Study of the Maintenance of Variation in Spot Pattern in* Maniola jurtina. Ph.D. Thesis, University of Liverpool.

BRAKEFIELD, P. M., 1982a. Ecological studies on the butterfly *Maniola jurtina* in Britain. I. Adult behaviour, microdistribution and dispersal. *Journal of Animal Ecology*, 51: 713–726.

BRAKEFIELD, P. M., 1982b. Ecological studies on the butterfly *Maniola jurtina* in Britain. II. Population dynamics: the present position. *Journal of Animal Ecology*, 51: 727–738.

BRAKEFIELD, P. M., 1984. The ecological genetics of quantitative characters of *Maniola jurtina* and other butterflies. In R. I. Vane-Wright and P. R. Ackery (Eds.), *The Biology of Butterflies*. Symposium of the Royal Entomological Society, Number 11: 167–190. London: Academic Press.

BRAKEFIELD, P. M., 1985. Polymorphic Müllerian mimicry and interactions with thermal melanism in ladybirds and a soldier beetle: a hypothesis. *Biological Journal of the Linnean Society*, 26: 243–267.

BRAKEFIELD, P. M., 1987a. Industrial melanism: do we have the answers. *Trends in Ecology and Evolution*, 2: 117–122.

BRAKEFIELD, P. M., 1987b. Tropical dry and wet seasonal polyphenism in the butterfly *Melanitis leda* (Satyrinae): phenotypic plasticity and climatic correlates. *Biological Journal of the Linnean Society*, 31: 175–191.

BRAKEFIELD, P. M., 1987c. Field studies of seasonal polyphenism in the butterfly *Melanitis leda*. *Bulletin of the British Ecological Society*, 18: 23–25.

BRAKEFIELD, P. M., 1987d. Geographical variability in, and temperature effects on, the phenology of *Maniola jurtina* and *Pyronia tithonus* (Lepidoptera, Satyrinae) in England and Wales. *Ecological Entomology*, 12: 139–148.

BRAKEFIELD, P. M., 1987e. Differentiation of island

populations and genetic drift in the Isles of Scilly. *Bulletin of the British Ecological Society*, 18: 238–241.

BRAKEFIELD, P. M., 1990a. Case studies in ecological genetics. In O. Kudrna (Ed.), *Butterflies of Europe*, Volume 2: 307–331. Weisbaden: Aula-Verlag.

BRAKEFIELD, P. M., 1990b. Genetic drift and patterns of diversity among colour polymorphic populations of the homopteran *Philaenus spumarius* in an island archipelago. *Biological Journal of the Linnean Society*, 39: 219–237.

BRAKEFIELD, P. M., 1991. Genetics and the conservation of invertebrates. In I. F. Spellerberg, F. B. Goldsmith, and M. G. Morris (Eds.) *The Scientific Management of Temperate Communities for Conservation*: 45–80. Oxford: Blackwell Scientific Publications

BRAKEFIELD, P. M. and DOWDESWELL, W. H., 1985. Comparison of two independent scoring techniques for spot variation in *Maniola jurtina* (L.) and the consequences of some differences. *Biological Journal of the Linnean Society*, 24: 329–345.

BRAKEFIELD, P. M. and LARSEN, T. B., 1984. The evolutionary significance of wet and dry season forms in some tropical butterflies. *Biological Journal of the Linnean Society*, 22: 1–12.

BRAKEFIELD, P. M. and LIEBERT, T. G., 1985. Studies of colour polymorphism in some marginal populations of the aposematic jersey tiger moth *Callimorpha quadripunctaria*. *Biological Journal of the Linnean Society*, 26: 225–241.

BRAKEFIELD, P. M. and NOORDWIJK, A. J. VAN, 1985. The genetics of spot pattern characters in the Meadow Brown butterfly *Maniola jurtina* (Lepidoptera: Satyridae). *Heredity*, 54: 275–284.

BRETHERTON, R. F., 1951. Our lost butterflies and moths. *Entomologist's Gazette*, 2: 211–240.

BRETHERTON, R. F. and CHALMERS-HUNT, J. M., 1981. The immigration of Lepidoptera to the British Isles in 1980. *Entomologist's Record and Journal of Variation*, 93: 47–54; 103–111.

BRETHERTON, R. F. and CHALMERS-HUNT, J. M., 1982. The immigration of Lepidoptera to the British Isles in 1981. *Entomologist's Record and Journal of Variation*, 94: 47–52; 81–87; 141–146.

BRETHERTON, R. F. and CHALMERS-HUNT, J. M., 1983. The immigration of Lepidoptera to the British Isles in 1982. *Entomologist's Record and Journal of Variation*, 95: 89–94; 141–152.

BRETHERTON, R. F. and CHALMERS-HUNT, J. M., 1984. The immigration of Lepidoptera to the British Isles in 1983. *Entomologist's Record and Journal of Variation*, 96: 85–91; 147–159; 196–201.

BRETHERTON, R. F. and CHALMERS-HUNT, J. M., 1985. The immigration of Lepidoptera to the British Isles in 1984. *Entomologist's Record and Journal of Variation*, 97: 140–145; 179–185; 224–228.

BRETHERTON, R. F. and CHALMERS-HUNT, J. M., 1986. The immigration of Lepidoptera to the British Isles in 1985. *Entomologist's Record and Journal of Variation*, 98: 159–163; 204–207; 223–230.

BRETHERTON, R. F. and CHALMERS-HUNT, J. M., 1987a. The immigration of Lepidoptera to the British Isles in 1982, 1983, 1984 and 1985 – a supplementary note. *Entomologist's Record and Journal of Variation*, 99: 147–152.

BRETHERTON, R. F. and CHALMERS-HUNT, J. M., 1987b. The immigration of Lepidoptera to the British Isles in 1986. *Entomologist's Record and Journal of Variation*, 99: 189–194; 245–250.

BRETHERTON, R. F. and CHALMERS-HUNT, J. M., 1988. The immigration of Lepidoptera to the British Isles in 1987. *Entomologist's Record and Journal of Variation*, 100: 175–180; 226–232.

BRETHERTON, R. F. and CHALMERS-HUNT, J. M., 1989a. Lepidoptera immigrant to the British Isles in 1985, 1986 and 1987 – a supplementary note. *Entomologist's Record and Journal of Variation*, 101: 131–135.

BRETHERTON, R. F. and CHALMERS-HUNT, J. M., 1989b. The immigration of Lepidoptera to the British Isles in 1988. *Entomologist's Record and Journal of Variation*, 101: 153–159; 225–230.

BRETHERTON, R. F. and CHALMERS-HUNT, J. M., 1990. The immigration of Lepidoptera to the British Isles in 1989. *Entomologist's Record and Journal of Variation*, 102: 153–159; 215–224.

BRIAN, M. V., 1975. Larval recognition by workers of the ant *Myrmica*. *Animal Behaviour*, 23: 745–756.

BRICE, W. C., 1978. *The Environmental History of the Near and Middle East Since the Last Ice Age*. London: Academic Press.

BRIGGS. D. and WALTERS, S. M., 1969. *Plant Variation and Evolution*. London: World University Press.

BROCK, J., 1971. A contribution toward an understanding of the morphology and phylogeny of ditrysian Lepidoptera. *Journal of Natural History*, 5: 29–102.

BROCK, J. P., 1990. Origins and phylogeny of butterflies. In O. Kudrna (Ed.) *Butterflies of Europe*, Volume 2 : 209–233. Weisbaden: Aula-Verlag.

BROCKIE, R. E., 1972. Evolutionary studies on *Maniola jurtina* in Sicily. *Heredity*, 28: 337–345.

BROOKS, M. and KNIGHT, C., 1982. *A Complete Guide to British Butterflies*. London: Jonathan Cape.

BROWER, L. P., 1984. Chemical defence in butterflies.

In R. I. Vane-Wright and P. R. Ackery (Eds.), *The Biology of Butterflies*. Symposium of the Royal Entomological Society, Number 11: 109–134. London: Academic Press.

BROWER, L. P. and BROWER, J. U. Z., 1972. Parallelism, convergence, divergence and the new concept of advergence in the coevolution of mimicry. *Transactions of the Connecticut Academy of Arts and Sciences*, 44: 57–67.

BROWER, L. P. and CALVERT, W. H., 1985. Foraging dynamics of bird predators on overwintering monarch butterflies in Mexico. *Evolution*, 39: 852–868.

BROWN, J. J. and CHIPPENDALE, G. M. 1974. Migration of the Monarch butterfly *Danaus plexippus*: energy sources. *Journal of Insect Physiology*, 20: 1117–1130.

BROWN, K. S., 1981. The biology of *Heliconius* and related genera. *Annual Review of Entomology*, 26: 42–456.

BROWN, K. S., 1987. Biogeography and evolution of Neotropical butterflies. In T. C. Whitmore and G. T. Prance (Eds.) *Biogeography and Quaternary History in Tropical America*: 66–104. Oxford: Clarendon Press.

BRUES, C. T., 1924. The specificity of food-plants in the evolution of phytophagous insects. *American Naturalist*, 58: 127–144.

BRUNTON, C. F. A., BAXTER, J. D., QUARTSON, J. A. S., and PANCHEN, A. L., 1990. The *valezina* morph of the butterfly *Argynnis paphia* (L.) in Corsica, *Entomologist's Record and Journal of Variation*, 102: 31–35.

BUTTERFLIES UNDER THREAT TEAM, (BUTT) 1986. *The Management of Chalk Downland for Butterflies*. Focus on Nature Conservation Series No. 17. Peterborough: Nature Conservancy Council.

CHAI, P., 1986. Field observations and feeding experiments on the responses of rufous-tailed jacamars (*Galbula ruficauda*) to free-flying butterflies in a tropical rainforest. *Biological Journal of the Linnean Society*, 29: 161–189.

CHALMERS-HUNT, J. M., 1960–61. The butterflies and moths of Kent: A critical account. Pt.1. Butterflies. *Entomologist's Record and Journal of Variation*, 72: 1–60; 73: 61–144.

CHALMERS-HUNT, J. M., 1977. The 1976 invasion of the Camberwell beauty (*Nymphalis antiopa* L.). *Entomologist's Record and Journal of Variation*, 89: 89–105.

CHAPLIN, S. B. and WELLS, P. H. 1982. Energy reserves and metabolic expenditures of Monarch butterflies overwintering in southern California. *Ecological Entomology*, 7: 249–256.

CHAPMAN, T. A., 1919. Contributions to the life history of *Lycaena euphemus* Hb. *Transactions of the Entomological Society of London*, 1919: 450–465.

CHEW, F. S., 1981. Coexistence and local extinction in two Pierid butterflies. *American Naturalist*, 18: 655–672.

CHEW, F. S. and ROBBINS, R. K., 1984. Egg-laying in butterflies. In R. I. Vane-Wright and P. R. Ackery (Eds.), *The Biology of Butterflies*. Symposium of the Royal Entomological Society, Number 11: 65–79. London: Academic Press.

CHEW, F. S. and RODMAN, J. E., 1979. Plant resources for chemical defence. In G. H. Rosenthal and D. H. Janzen (Eds.), *Herbivores: Their Interaction with Secondary Plant Metabolites*: 271–307. New York: Academic Press.

CHIPPENDALE, C., 1985. Flag Fen: new finds from the Bronze Age. *New Scientist*, 106: 36–39.

CLAPHAM, A. R., TUTIN, T. G. and MOORE, D. M., 1987. *Flora of the British Isles*. 3rd edition. Cambridge: Cambridge University Press.

CLARK, G. C. and DICKSON, C. G. C., 1956. The honey gland and tubercules of larvae of the Lycaenidae. *The Lepidopterist's News*, 10: 37–43.

CLARKE, B., 1970. Festschrift in biology. *Science*, 169: 1192.

CLARKE, C. A., CLARKE, F. M. M., COLLINS, S. C., GILL, A. C. L., and TURNER, J. R. G., 1985. Male-like females, mimicry and transvestism in butterflies (Lepidoptera: Papilionidae). *Systematic Entomology*, 10: 257–283.

CLARKE, C. A. and SHEPPARD, P. M., 1972. Genetic and environmental factors influencing pupal colour in the swallowtail butterflies *Battus philenor* and *Papilio polytes*. *Journal of Entomology A*, 46: 123–133.

CLENCH, H. K., 1966. Behavioural thermoregulation in butterflies. *Ecology*, 47: 1021–1034.

CLENCH, H. K., 1967. Temporal dissociation and population regulation in certain hesperiine butterflies. *Ecology*, 48: 1000–1006.

CODY, M. L., 1974. *Competition and the Structure of Bird Communities*. Princeton, New Jersey : Princeton University Press.

COLE, L. R., 1959. On the defences of lepidopterous pupae in relation to the oviposition behaviour of certain Ichneumonidae. *Journal of the Lepidopterists' Society*, 13: 1–10

COLE, L. R., 1967. A study of the life cycles and hosts of some Ichneumonidae attacking pupae of the green oak-leaf roller moth, *Tortrix viridiana* (L.) (Lepidoptera: Tortricidae) in England. *Transactions of the Royal Entomological Society of London*, 119: 267–281.

COLES, B. and COLES, J., 1986. *Sweet Track to Glastonbury*. London: Thames and Hudson.

COLLIER, A. E., 1959. A forgotten discard: the problems of redundancy. *Entomologist's Record and Journal of Variation*, 71: 118–119.

COLLIER, R. V., 1966. Status of butterflies on Castor Hanglands NNR 1961–1965 inclusive. *Journal of the Northamptonshire Natural History Society*, 35: 451–456.

COLLIER, R. V., 1978. *The status and decline of butterflies on Castor Hanglands NNR, 1919–1977*. Unpublished report. London: Nature Conservancy Council.

COLLIER, R. V., 1986. *The Conservation of the Chequered Skipper in Britain*. Focus on Nature Conservation Series No.16. Peterborough: Nature Conservancy Council.

COLLINS, N. M., 1987. *Butterfly Houses in Britain: The Conservation Implications*. Unpublished report. Cambridge: International Union for Conservation of Nature and Natural Resources.

COMSTOCK, W. P., 1961. *Butterflies of the American Tropics: the genus* Anaea, (*Lepidoptera Nymphalidae*). *A study of the species heretofore included in the genera* Anaea, Coenophebia, Hypna, Polygrapha, Protogonius, Siderone *and* Zanetis. New York: American Museum of Natural History.

CONNOR, E. F. and MCCOY, E. D., 1979. The statistics and biology of the species-area relationship. *American Naturalist*, 113: 791–833.

COOK, L. M., BROWER, L. P., and CROZE, H. J., 1967. The accuracy of a population estimation from multiple recapture data. *Journal of Animal Ecology*, 36: 57–60.

COOPE, G. R., 1965. Fossil insect faunas from the later Quaternary deposits in Britain. *Advancement of Science*, 21: 564–575.

COOPE, G. R., 1970. Quaternary insect fossils. *Annual Review of Entomology*, 15: 97–120.

COOPE, G. R., 1979. Late Cenozoic fossil coleoptera. Evolution, biogography and ecology. *Annual Review of Ecology and Systematics*, 10: 247–267.

COOPE, G. R., 1981. Episodes of local extinction of insect species during the Quaternary as indicators of climatic change. In J. Neale and J. Flenly (Eds.), *The Quaternary in Britain*: 216–221. Oxford: Pergamon Press.

COOPE, G. R., 1986. The invasion and colonisation of the north Atlantic islands: a palaeoecological solution to a biogeographical problem. *Philosophical Transactions of the Royal Society of London* (*B*), 314: 619–635.

COOPE, G. R., 1987. The response of Late Quaternary insect communities to sudden climatic changes. In J. H. R. Gee and P. S. Giller (Eds.), *Organisation of Communities, Past and Present*: 421–438. Oxford: Blackwell Scientific Publications.

COPPINGER, R. P., 1969. The effect of experience and novelty on avian feeding behaviour with reference to the evolution of warning colour in butterflies. Part I: reactions of wild-caught blue jays to novel insects. *Behaviour*, 35: 45–60.

COPPINGER, R. P., 1970. The effect of experience and novelty on avian feeding behaviour with reference to the evolution of warning colour in butterflies. Part II. Reactions of naive birds to novel insects. *American Naturalist*, 104: 323–335.

CORBET, S. A., 1978. Bee visits and the nectar of *Echium vulgare* L. and *Sinapis alba* L. *Ecological Entomology*, 3: 25–37

CORKE, D., 1989. Of pheasants and fritillaries: is predation by pheasants (*Phasianus colchicus*) a cause of the decline in some British butterfly species? *British Journal of Entomology and Natural History*, 2: 1–14.

COTT, H. B., 1940. *Adaptive Coloration in Animals*. London: Methuen and Co.

COTTRELL, C. B., 1984. Aphytophagy in butterflies: its relationship to myrmecophily. *Zoological Journal of the Linnean Society*, 79: 1–57.

COULTHARD, N., 1982. *An Investigation of the Habitat Requirements and Behavioural Ecology of the Small Blue Butterfly*, Cupido minimus *Fuessly, in Relation to its Distribution and Abundance in N.E. Scotland*. M. Sc. Thesis, University of Aberdeen.

COURTNEY, S. P., 1980. *Studies on the Biology of the Butterflies* Anthocharis cardamines *L. and* Pieris napi *L. in Relation to Speciation in Pierinae*. Ph. D. Thesis, University of Durham.

COURTNEY, S. P., 1981. Coevolution of pierid butterflies and their cruciferous foodplants. III. *Anthocharis cardamines* survival, development and oviposition on different hostplants. *Oecologia*, 51: 91–96.

COURTNEY, S. P., 1982a. Coevolution of Pierid butterflies and their cruciferous foodplants. IV. Crucifer apparency and *Anthocharis cardamines* (L.) oviposition. *Oecologia*, 52: 258–265.

COURTNEY, S. P. 1982b. Electrophoretic differentiation and *Pieris napi* (L.) (Lep. Pieridae). *Atalanta*, 13: 153–157.

COURTNEY, S. P., 1984. The evolution of egg clustering by butterflies and other insects. *American Naturalist*, 123: 276–281.

COURTNEY, S. P., 1986. The ecology of Pierid butterflies: dynamics and interactions. *Advances in Ecological Research*, 15: 51–116.

COURTNEY, S. P., 1988. Oviposition on peripheral hosts by dispersing *Pieris napi* (L.) (Pieridae). *Journal of Research on the Lepidoptera*, 26: 58–63.

COURTNEY, S. P. and CHEW, F. S., 1987. Coexistence and host use by a large community of Pierid butterflies: habitat as a templet. *Oecologia*, 71: 210–220.

COURTNEY, S. P. and COURTNEY, S., 1982. The edge-effect in butterfly oviposition: causality in *Anthocharis cardamines* and related species. *Ecological Entomology*, 7: 131–137.

COURTNEY, S. P. and DUGGAN, A. E., 1983. The population biology of the orange-tip butterfly *Anthocharis cardamines* in Britain. *Ecological Entomology*, 8: 271–281.

COURTNEY, S. P., HILL, C. P., and WESTERMAN, A., 1982. Pollen carried for long periods by butterflies. *Oikos*, 38: 260–263.

COX, C. B., HEALEY, I. N., and MOORE, P. D., 1973. *Biogeography, an Ecological and Evolutionary Approach*. Oxford: Blackwell Scientific Publications.

CREED, E. R., DOWDESWELL, W. H., FORD, E. B., and MCWHIRTER, K. G., 1959. Evolutionary studies on *Maniola jurtina*: the English mainland 1956–57. *Heredity*, 13: 363–391.

CREED, E. R., DOWDESWELL, W. H., FORD, E. B. and MCWHIRTER, K. G., 1962. Evolutionary studies on *Maniola jurtina*: the English mainland 1958–60. *Heredity*, 17: 237–265.

CREED, E. R., DOWDESWELL, W. H., FORD, E. B., and MCWHIRTER, K. G., 1970. Evolutionary studies on *Maniola jurtina*: the boundary phenomenon in southern England, 1961 to 1968. In M. K. Hecht and W. C. Steere (Eds.), *Essays in Evolution and Genetics in Honour of Theodosius Dobzhansky*: 263–287. New York: Appleton-Century-Croft.

CREED, E. R., FORD, E. B., and MCWHIRTER, K. G., 1964. Evolutionary studies on *Maniola jurtina*: the Isles of Scilly, 1958–59. *Heredity*, 19: 471–488.

CRIBB, P. W., 1986. Weaver's fritillary taken in Surrey. *The Bulletin of the Amateur Entomologists' Society*, 45: 193.

CRONIN, T. M. and SCHNEIDER, C. E., 1990. Climatic influences on species: evidence from the fossil record. *Trends in Ecology and Evolution*, 5: 275–279.

CURRAY, J. R., 1965. Late Quaternary history, continental shelves of the United States. In H. E. Wright and D. G. Frey (Eds.), *The Quaternary History of the United States*: 723–735. Princeton, New Jersey: Princeton University Press.

DAMMAN, H., 1986. The osmaterial glands of the swallowtail butterfly *Eurytides marcellus* as a defence against natural enemies. *Ecological Entomology*, 11: 261–265.

DAVEY, P. A., 1985. Notes on a remarkable immigration of Lepidoptera into the United Kingdom. *Entomologist's Record and Journal of Variation*, 97: 165–168.

DAVIES, G. H. N., FRAZER, J. F. D., and TYNAM, A. M., 1958. Population numbers in a colony of *Lysandra bellargus* Rott. (Lep. Lycaenidae). *Proceedings of the Royal Entomological Society of London*, 33: 31–36.

DAVIES, M., 1985. *White-letter haistreak project: annual report 1985*. Unpublished report. British Butterfly Conservation Society.

DAVIES, N. B., 1978. Territorial defence in the speckled wood butterfly, *Pararge aegeria*: the resident always wins. *Animal Behaviour*, 26: 138–147.

DAVIS, B. N. K., 1989. Habitat creation for butterflies in a landfill site. *The Entomologist*, 108: 109–122.

DAWKINS, R. and KREBS, J. R., 1979. Arms races between and within species. *Proceedings of the Royal Society of London B*, 205: 489–511.

DEBENHAM, F., 1962. *Map Making*. London: Blackie and Sons.

DEMPSTER, J. P., 1960. A quantitative study of the predators on the eggs and larvae of the broom beetle *Phytodecta olivacea* Forster, using the precipitin test. *Journal of Animal Ecology*, 27: 149–167.

DEMPSTER, J. P., 1967. The control of *Pieris rapae* with DDT. I. The natural mortality of the young stages of *Pieris*. *Journal of Applied Ecology*, 4: 485–500.

DEMPSTER, J. P., 1968. The control of *Pieris rapae* with DDT. II. Survival of the young stages of *Pieris* after spraying. *Journal of Applied Ecology*, 5: 451–462.

DEMPSTER, J. P., 1971a. The population ecology of the cinnabar moth, *Tyria jacobaeae* L. (Lepidoptera, Arctiidae). *Oecologia*, 7: 26–67.

DEMPSTER, J. P., 1971b. Some observations on a population of the small copper butterfly *Lycaena phlaeas* (L.) (Lep., Lycaenidae). *Entomologist's Gazette*, 22: 199–204.

DEMPSTER, J. P., 1975. *Animal Population Ecology*. London: Academic Press.

DEMPSTER, J. P., 1983. The natural control of populations of butterflies and moths. *Biological Review*, 58: 461–481.

DEMPSTER, J. P., 1984. The natural enemies of butterflies. In R. I. Vane-Wright and P. R. Ackery (Eds.), *The Biology of Butterflies*. Symposium of the Royal Entomological Society, Number 11: 97–104. London: Academic Press.

DEMPSTER, J. P., 1989. Insect introductions: natural dispersal and population persistence in insects. *The Entomologist*, 108: 5–13.

DEMPSTER, J. P., 1991. Fragmentation, isolation, and mobility of insect populations. In N. M. Collins and J. A. Thomas, *The Conservation of Insects and their Habitats*: 143–154. London: Academic Press.

DEMPSTER, J. P., KING, M. L., and LAKHANI, K. H. 1976. The status of the swallowtail butterfly in Britain. *Ecological Entomology*, 1: 71–84.

DEMPSTER, J. P., LAKHANI, K. H., and COWARD, P. A., 1986. The use of chemical composition as a population marker in insects: a study of the brimstone butterfly. *Ecological Entomology*, 11: 51–56.

DENNIS, R. L. H., 1972a. *Eumenis semele* (L.) *thyone* Thompson (Lep., Satyridae), a microgeographical race. *Entomologist's Record and Journal of Variation*, 84: 1–11; 38–44.

DENNIS, R. L. H., 1972b. *Plebejus argus* (L.) *caernensis* Thompson. A stenoecious geotype. *Entomologist's Record and Journal of Variation*, 84: 132–140.

DENNIS, R. L. H., 1977. *The British Butterflies. Their Origin and Establishment*. Faringdon, Oxon.: E. W. Classey.

DENNIS, R. L. H., 1982a. Observations on habitats and dispersion made from oviposition markers in north Cheshire *Anthocharis cardamines* (L.) (Lepidoptera: Pieridae). *Entomologist's Gazette*, 33: 151–159.

DENNIS, R. L. H. 1982b. Patrolling behaviour in orange-tip butterflies within the Bollin valley in north Cheshire, and a comparison with other pierids. *Vasculum*, 67: 17–25.

DENNIS, R. L. H., 1982c. *Erebia aethiops* (Esper) (Satyridae) on Arnside Knott. Confidential Survey for the Nature Conservancy Council.

DENNIS, R. L. H., 1982d. Mate locating strategies in the wall brown butterfly, *Lasiommata megera* (L.) (Lepidoptera: Satyridae): wait or seek? *Entomologist's Record and Journal of Variation*, 94: 209–214; 95: 7–10.

DENNIS, R. L. H., 1983a. Hierarchy and pattern in the spatial responses of ovipositing *Anthocharis cardamines* (Lep.). *Vasculum*, 68: 27–43.

DENNIS, R. L. H., 1983b. Spatial patterning in butterfly oviposition: density influences and the 'edge-effect' in the orange-tip butterfly (*Anthocharis cardamines* (L.) Lep; Pieridae). *The Bulletin of the Amateur Entomologists' Society*, 42: 55–60.

DENNIS, R. L. H., 1983c. Egg-laying cues in the wall brown butterfly, *Lasiommata megera* (L.) (Lepidoptera: Satyridae). *Entomologist's Gazette*, 34: 89–95.

DENNIS, R. L. H., 1984a. Brimstone butterflies (*Gonepteryx rhamni*) (L.) playing possum. *Entomologist's Gazette*, 35: 6–7.

DENNIS, R. L. H., 1984b. Egg-laying sites of the common blue butterfly, *Polyommatus icarus* (Rottemburg) (Lepidoptera: Lycaenidae). The edge effect and beyond the edge. *Entomologist's Gazette*, 35: 85–93.

DENNIS, R. L. H., 1984c. The edge effect in butterfly oviposition; batch siting in *Aglais urticae* (L) (Lepidoptera: Nymphalidae). *Entomologist's Gazette*, 35: 157–173.

DENNIS, R. L. H., 1985a. British butterfly distributions: space-time predictability. *Proceedings of the 3rd Congress of European Lepidopterology, Cambridge 1982*: 50–62.

DENNIS, R. L. H., 1985b. Small plants attract attention! Choice of egg-laying sites in the green-veined white butterfly (*Artogeia napi*) (L.) (Lep: Pieridae). *The Bulletin of the Amateur Entomologists' Society*, 44: 77–82.

DENNIS, R. L. H., 1985c. *Polyommatus icarus* (Rottemburg) (Lepidoptera: Lycaenidae) on Brereton Heath in Cheshire: voltinism and switches in resource exploitation. *Entomologist's Gazette*, 36: 175–179.

DENNIS, R. L. H., 1985d. Voltinism in British *Aglais urticae* (L.) (Lep.: Nymphalidae): variation in space and time. *Proceedings and Transactions of the British Entomological and Natural History Society*, 18: 51–61.

DENNIS, R. L. H., 1985e. The edge effect in butterfly oviposition: hostplant condition, edge-effect breakdown and opportunism. *Entomologist's Gazette*, 36: 285–291.

DENNIS, R. L. H., 1986a. Selection of roost sites by *Lasiommata megera* (L.) (Lepidoptera: Satyridae) on fencing at Brereton Heath Country Park, Cheshire. *Nota Lepidopterologica*, 9: 39–46.

DENNIS, R. L. H., 1986b. Motorways and cross-movements. An insect's 'mental map' of the M56 in Cheshire. *The Bulletin of the Amateur Entomologists' Society*, 45: 228–243.

DENNIS, R. L. H., 1987. Hilltopping as a mate-location strategy in a Mediterranean population of *Lasiommata megera* (L.) (Lep. Satyridae). *Nota Lepidopterologica*, 10: 65–70.

DENNIS, R. L. H. and BRAMLEY, M. J., 1985. The influence of man and climate on dispersion patterns within a population of adult *Lasiommata megera* (L.) (Satyridae) at Brereton Heath, Cheshire (U. K.). *Nota Lepidopterologica*, 8: 309–324.

DENNIS, R. L. H., PORTER, K., and WILLIAMS, W. R., 1984. Ocellation in *Coenonympha tullia* (Müller) (Lep. Satyridae): I Structures in correlation matrices. *Nota Lepidopterologica*, 7: 199–219.

DENNIS, R. L. H., PORTER, K., and WILLIAMS, W. R., 1986. Ocellation in *Coenonympha tullia* (Müller) (Lep. Satyridae): II Population differentiation and clinal variation in the context of climatically-induced anti-predator defence strategies. *Entomologist's Gazette*, 37: 133–172.

DENNIS, R. L. H. and RICHMAN, T., 1985. Egg-keel number in the small tortoiseshell butterfly. *Entomologist's Record and Journal of Variation*, 97: 162–164.

DENNIS, R. L. H. and SHREEVE, T. G., 1988. Host-plant-habitat structure and the evolution of butterfly mate-locating behaviour. *Zoological Journal of the Linnean Society*, 94: 301–318.

DENNIS, R. L. H. and SHREEVE, T. G., 1989. Butterfly morphology variation in the British Isles: the influence of climate, behavioural posture and the host-plant-habitat. *Biological Journal of the Linnean Society*, 38: 323–348.

DENNIS, R. L. H. and SHREEVE, T. G., 1991. Climatic change and the British butterfly fauna: opportunities and constraints. *Biological Conservation*, 5: 1–16.

DENNIS, R. L. H. and WILLIAMS, W. R., 1986. Butterfly 'diversity': regressing and a little latitude. *Antenna*, 10: 108–112.

DENNIS, R. L. H. and WILLIAMS, W. R., 1987. Mate locating behaviour of the large skipper butterfly *Ochlodes venata*: flexible strategies and spatial components. *Journal of the Lepidopterist's Society*, 41: 45–64.

DENNIS, R. L. H., WILLIAMS, W. R., and SHREEVE, T. G., 1991. A multivariate approach to the determination of faunal structures among European butterfly species (Lepidoptera, Rhopalocera). *Zoological Journal of the Linnean Society*, 101: 1–49.

DENTON, G. H. and HUGHES, T. J., 1981. *The Last Great Ice Sheets*. New York: Wiley.

DESCIMON, H. and RENON, C., 1975. Mélanisme et facteurs climatiques: II Corrélation entre melanisation et certains facteurs climatiques chez *Melanargia galathea* (L.) (Lepidoptera: Satyridae) en France. *Archives de Zoologie Experimentale et Génerale*, 116: 437–468.

DETHIER, V. G., 1970. Chemical interactions between plants and insects. In E. Sondheimer and J. B. Simeone (Eds.), *Chemical Ecology*: 83–102. New York: Academic Press.

DETHIER, V. G. and MACARTHUR, R. H., 1964. A field's capacity to support a butterfly population. *Nature*, 201: 728–729.

DEVOY, R. J., 1982. Analysis of the geological evidence for Holocene sea-level movements in south-east England. In J. T. Greensmith and M. J. Tooley (Eds.), I.G.C.P. Project 61, *Sea- Level Movements during the Last Deglacial Hemicycle*. Proceedings of the Geologists' Association, 93: 65–90.

DE VRIES, P. J., 1988. The ant associated larval organs of *Thisbe irenea* (Riodinidae) and their effects on attending ants. *Zoological Journal of the Linnean Society*, 94: 379–393.

DE VRIES, P. J., 1991a. Call production by myrmecophilous riodinid and lycaenid butterfly caterpillars (Lepidoptera): morphological, acoustic, functional and evolutionary patterns. *American Museum Novitates*, 3025: 1–23.

DE VRIES, P. J., 1991b. Evolutionary and ecological patterns in myrmecophilous riodinid butterflies. In R. Cutler and C. Huxley (Eds.) *Interactions between Ants and Plants*: 143–156. Oxford: Oxford University Press.

DE VRIES, P. J. and BAKER, I., 1989. Butterfly exploitation of an ant-plant mutualism: adding insult to herbivory. *Journal of the New York Entomological Society*, 97: 332–340.

DIAMOND, J. M. and MAY, R. M., 1976. Island biogeography and the design of natural reserves. In R. M. May (Ed.), *Theoretical Ecology: Principles and Applications*: 163–186. Oxford: Blackwell Scientific Publications.

DIGBY, P. S. B., 1955. Factors affecting the temperature excess of insects in the sun. *Journal of Experimental Biology*, 32: 279–289.

DIMBLEBY, G. W., 1962. *The Development of British Heathlands and their Soils*. Oxford Forestry Memorandum 23.

DIXON, A. F. G., 1973. *Biology of Aphids*. London: Edward Arnold.

DOBZHANSKY, T. and PAVLOVSKY, O., 1957. An experimental study of the interaction between genetic drift and natural selection. *Evolution*, 11: 311–319.

DOLINGER, P. M., EHRLICH, P. R., FITCH, W. L., and BREEDLOVE, D. E., 1973. Alkaloid and predation patterns in Colorado lupine populations. *Oecologia*, 13: 191–204.

DONISTHORPE, H., ST. J. K., 1927. *The Guests of British Ants*. London: Routledge.

DONNER, J. J., 1970. Land/sea-level changes in Scotland. In D. Walker and R. G. West (Eds.), *Studies in the Vegetational History of the British Isles*: 23–39. Cambridge: Cambridge University Press.

DOUGLAS, M. M. and GRULA, J. W., 1978. Thermoregulatory adaptations allowing ecological range expansion by the Pierid butterfly *Nathalis iole* Boisduval. *Evolution* 32: 776–783.

DOUWES, P., 1975. Territorial behaviour in *Hoedes virgaureae* (L.) (Lep., Lycaenidae) with particular reference to visual stimuli. *Norwegian Journal of Entomology*, 22: 143–153.

DOUWES, P., 1976. Distribution of a population of the butterfly *Heodes virgaureae*. *Oikos*, 26: 332–340.

DOVER, J., 1987. The benefits of conservation headlands to butterflies on farmland. *Game Conservancy Annual Review*, 18: 105–108.

DOVER, J., 1991. The conservation of insects on arable farmland. In N. M. Collins and J. A. Thomas (Eds.), *The Conservation of Insects and their Habitats*: 143–154. London: Academic Press.

DOWDESWELL, W. H., 1961. Experimental studies on natural selection in the butterfly, *Maniola jurtina*. *Heredity*, 16: 39–52.

DOWDESWELL, W. H., 1962. A further study of the butterfly *Maniola jurtina* in relation to natural selection by *Apanteles tetricus*. *Heredity*, 17: 513–523.

DOWDESWELL, W. H., 1981. *The Life of the Meadow Brown*. London: Heinemann Educational Books.

DOWDESWELL, W. H., FISHER, R. A., and FORD, E. B. 1940. The quantitative study of populations in the Lepidoptera I. *Polyommatus icarus* Rott. *Annals of Eugenics*, 10: 123–136.

DOWDESWELL, W. H., FISHER, R. A., and FORD, E. B. 1949. The quantitative study of populations in the Lepidoptera II. *Maniola jurtina* L. *Heredity*, 3: 67–84.

DOWDESWELL, W. H. and FORD, E. B., 1955. Ecological genetics of *Maniola jurtina* on the Isles of Scilly. *Heredity*, 9: 265–272.

DOWDESWELL, W. H., FORD, E. B., and MCWHIRTER, K. G., 1957. Further studies on isolation in the butterfly *Maniola jurtina*. *Heredity*, 11: 51–65.

DOWDESWELL, W. H., FORD, E. B., and MCWHIRTER, K. G., 1960. Further studies on the evolution of *Maniola jurtina* in the Isles of Scilly. *Heredity*, 14: 333–364.

DOWDESWELL, W. H. and MCWHIRTER, K. G., 1955. Stability of spot-distribution in *Maniola jurtina* throughout its range. *Heredity*, 22: 187–210.

DOWDESWELL, W. H. and MCWHIRTER, K. G., 1967. Stability of spot-distribution in *Maniola jurtina* throughout its range. *Heredity*, 22: 187–210.

DOWNES, J. A., 1948a. Lepidoptera feeding at puddle-margins, dung, and carrion. *Journal of the Lepidopterists' Society*, 27: 89–99.

DOWNES, J. A., 1948b. The history of the speckled wood butterfly (*Pararge aegeria*) in Scotland, with a discussion of range of other British butterflies. *Journal of Animal Ecology*, 17: 131–138.

DOWNES, J. A., 1965. Adaptations of insects in the Arctic. *Annual Review of Entomology*, 10: 257–274.

DOWNEY, J. C., 1966. Sound production in pupae of Lycaenidae. *Journal of the Lepidopterist's Society*, 20: 129–155.

DOWNEY, J. C. and ALLYN, A. C., 1973. Butterfly ultrastructure. 1. Sound production and associated abdominal structures of pupae of Lycaenidae and Riodinidae. *Bulletin of the Allyn Museum*, 14: 1–48.

DUFFEY, E., 1968. Ecological studies on the large copper *Lycaena dispar* (Haw.) *batavus* (Obth.) at Woodwalton Fen National Nature Reserve, Huntingdonshire. *Journal of Applied Ecology*, 5: 69–96.

DUFFEY, E., 1977. The reestablishment of the large copper butterfly *Lycaena dispar batava* Obth. on Woodwalton Fen National Nature Reserve, Cambridge, England, 1969–73. *Biological Conservation*, 12: 143–158.

DUNLAP-PIANKA, H. L., BOGGS, C. L., and GILBERT, L. E., 1977. Ovarian dynamics in *Heliconiine* butterflies: Programmed senescence versus eternal youth. *Science*, 197: 487–490.

DUNN, T. C., 1965. Lepidoptera seen on the Isle of Colonsay 1964. *The Entomologist*, 98: 243–247.

DURDEN, C. J. and ROSE, H., 1978. *Butterflies from the Middle Eocene: the earliest occurrence of fossil Papilionidae (Lep.)*. Pearce-Sellards Series, Texas Memorial Museum, 29: 1–25.

EDMUNDS, M. E., 1974a. Significance of beak marks on butterfly wings. *Oikos*, 25: 117–118.

EDMUNDS, M. E., 1974b. *Defence in Animals. A Survey of Anti-Predator Defences*. Harlow: Longman.

EDWARDS, P. J. and WRATTEN, S. D., 1980. *Ecology of Insect Plant Interactions*. Institute of Biology's Studies in Biology no. 121. London: Edward Arnold.

EGGELING, W. J., 1957. A list of butterflies and moths recorded for the Isle of May. *Scottish Naturalist*, 69: 75–83.

EHRLICH, P. R. 1984. The structure and dynamics of butterfly populations. In R. I. Vane-Wright and P. R. Ackery (Eds.), *The Biology of Butterflies*. Symposium of the Royal Entomological Society, Number 11: 25–40. London: Academic Press.

EHRLICH, P. R. and MASON, L. G., 1966. The population biology of the butterfly *Euphydryas editha*. III.

Selection and the phenetics of the Jasper Ridge colony. *Evolution*, 20: 165–173.

EHRLICH, P. R. and RAVEN, P. H., 1965. Butterflies and plants: a study in coevolution. *Evolution*, 18: 586–608.

EHRLICH, P. R. and RAVEN, P. H., 1967. Butterflies and plants. *Scientific American*, 216: 104–113.

EHRLICH, P. R., WHITE, R. R., SINGER, M. C., MCKECHNIE, S. W., and GILBERT, L. E., 1975. Checkerspot butterflies: a historical perspective. *Science*, 188: 221–228.

EISNER, T. and MEINWALD, Y. C. 1965. Defensive secretion of a caterpillar (*Papilio*). *Science*, 150: 1733–1735.

EISNER, T., SILBERGLIED, R. E., ANESHANSLEY, D., CARREL, J. E., and HOWLAND, H. C., 1969. Ultraviolet video viewing: the television camera as an insect eye. *Science*, 166: 1172–1174.

ELDRIDGE, N., 1987. *Life Pulse. Episodes from the Story of the Fossil Record*. Oxford: Facts on File Publications.

ELMES, G. W. and THOMAS, J. A., 1987. Die Gattung Maculinea. In W. Geiger (Ed.), *Tagfalter und ihr Lebensraume*: 354–368. Basel: Schweizerischer Bund für Naturschutz.

ELMES, G. W., THOMAS, J. A., and WARDLAW, J. C., 1991. Larvae of *Maculinea rebeli*, a large-blue buttefly, and their *Myrmica* host ants – wild adoption and behaviour in ant nests. *Journal of Zoology*, 223: 447–460.

ELMES, G. W., WARDLAW, J. C., and THOMAS, J. A., 1991. Larvae of *Maculinea rebeli*, a large-blue butterfly , and their *Myrmica* host ants – patterns of caterpillar growth and survival. *Journal of Zoology*, 224: 79–92.

EMILIANI, C., 1966. Palaeotemperature analysis of Caribbean cores P6304–8 and P6304–9 and a generalized temperature curve for the past 425,000 years. *Journal of Geology*, 74: 109–126.

EMMET, A. M., PYMAN, G. A., and CORKE, D., 1985. *The Larger Moths and Butterflies of Essex*. London: Essex Field Club, Pasmore-Edwards Museum.

EMMET, A. M. and HEATH, J., 1989. *Moths and Butterflies of Great Britain and Ireland*. Volume 7, Part 1. *Hesperiidae to Nymphalidae*. Colchester: Harley Books.

ENDLER, J. A., 1977. *Geographic Variation, Speciation, and Clines*. Princeton, New Jersey: Princeton University Press.

ENDLER, J. A., 1978. A predator's view of animal color patterns. *Evolutionary Biology*, 11: 319–364.

ENDLER, J. A., 1980. Natural selection on color patterns in *Poecilia reticulata*. *Evolution*, 34: 76–91.

ENDLER, J. A., 1981. An overview of the relationships between mimicry and crypsis. *Biological Journal of the Linnean Society*, 16: 25–31.

ENDLER, J. A., 1984. Progressive background matching in moths, and a quantitative measure of crypsis. *Biological Journal of the Linnean Society*, 22: 187–231.

ENDLER, J. A., 1986. *Natural Selection in the Wild*. Princeton, New Jersey: Princeton University Press.

ERHARDT, A., 1985. Diurnal Lepidoptera: sensitive indicators of cultivated and abandoned grassland. *Journal of Applied Ecology*, 22: 849–861.

EVANS, J. G., 1975. *The Environment of Early Man in the British Isles*. London: Paul Elek.

EVIN, J., 1990. Validity of radiocarbon dates beyond 35,000 years. *Palaeo*, 80: 71–78.

FAEGRI, K. and PIJL, L. VAN DER, 1979. *The Principles of Pollination Ecology*. London: Pergamon Press.

FALCONER, D. S., 1981. *Introduction to Quantitative Genetics*. 2nd ed. London: Longman.

FARRELL, L., 1973. *A Preliminary Report on the Status of the Chequered Skipper* (Carterocephalus palaemon) *in the British Isles*. Unpublished Report. London: Nature Conservancy Council.

FEENY, P., 1970. Oak tannins and caterpillars. *Ecology*, 51: 565–581.

FEENY, P., 1975. Biochemical coevolution between plants and their insect herbivores. In L. E. Raven and P. H. Raven (Eds.), *Coevolution of Animals and Plants*. Austin, Texas: University of Texas Press.

FEENY, P., 1977. Defensive ecology of the Cruciferae. *Annals of the Missouri Botanical Garden*, 64: 221–234.

FEENY, P., ROSENBERRY, L., and CARTER, M., 1983. Chemical aspects of oviposition behaviour in butterflies. In S. Ahmad (Ed.), *Herbivorous Insects. Host Seeking Behaviour and Mechanisms*: 27–76. London: Academic Press.

FELTWELL, J., 1981a. *Large White Butterfly*. The Biology, *Biochemistry and Physiology of* Pieris brassicae *(L.)*. Series Entomologia, Volume 18. The Hague: W. Junk.

FELTWELL, J., 1981b. Pupal spines of *Pieris Brassicae* (L.). *Entomologist's Record and Journal of Variation*, 93: 142–145.

FERGUSON, A., 1980 *Biochemical Systematics and Evolution*. London: Blackie.

FIEDLER, K., 1988. The preimaginal epidermal organs of *Lycaena tityrus* (Poda, 1761) and *Polyommatus coridon* (Poda, 1761) (Lepidoptera: Lycaenidae) – a comparison. *Nota Lepidopterologica*, 11: 100–116.

FIEDLER, K. 1990a. Effects of larval diet on myrmecophilous qualities of *Polyommatus icarus* caterpillars (Lepidoptera: Lycaenidae). *Oecologia*, 83: 284–287.

FIEDLER, K., 1990b. New information on the biology of *Maculinea nausithous* and *M.telius* (Lepidoptera: Lycaenidae). *Nota Lepidopterologica*, 12: 246–256.

FIEDLER, K. and MASCHWITZ, U., 1988a. Functional analysis of the myrmecophilous relationships between ants (Hymenoptera: Formicidae) and lycaenids (Lepidoptera: Lycaenidae) I. Release of food recruitment in ants by lycaenid larvae and pupae. *Ethology*, 80: 71–80.

FIEDLER, K. and MASCHWITZ, U., 1988b. Functional analysis of the myrmecophilous relationships between ants (Hymenoptera:Formicidae) and lycaenids (Lepidoptera: Lycaenidae) II. Lycaenid larvae as trophic partners of ants – a quantitative approach. *Oecologia*, 75: 204–206.

FIEDLER, K. and MASCHWITZ, U., 1989. The symbiosis between the weaver ant *Oecophylla smaragdina*, and *Anthene emolus*, an obligate myrmecophilous lycaenid butterfly. *Journal of Natural History*, 23: 833–846.

FINDLAY, R., YOUNG, M. R., and FINDLAY, J. A., 1983. Orientation behaviour in the grayling butterfly: thermoregulation or crypsis? *Ecological Entomology*, 8: 145–153.

FISHER, R. A. 1930. *The Genetical Theory of Natural Selection*. Oxford: Clarendon Press.

FISHER, R. A. and FORD, E. B., 1947. The spread of a gene in natural conditions in a colony of the moth *Panaxia dominula* L. *Heredity*, 1: 143–174.

FLINDERS, I., 1982. The status and distribution of butterflies in Northamptonshire 1976–81. *Journal of the Northamptonshire Natural History Society*, 264: 9–23.

FORD, E. B. 1940. Genetic research in the Lepidoptera. *Annals of Eugenics*, 10: 227–252.

FORD, E. B., 1945. *Butterflies*. London: Collins. (+ 1975 Fontana New Naturalist Series Edition).

FORD, E. B., 1949. Early stages in allopatric speciation. In G. L. Jepson, G. G. Simpson, and E. Mayr (Eds.), *Genetics, Palaeontology and Evolution*: 309–314. Princeton, New Jersey: Princeton University Press.

FORD, E. B., 1975. *Ecological Genetics*. 4th edition. London: Chapman and Hall.

FORD, H. D. and FORD, E. B., 1930. Fluctuations in numbers, and its influence on variation in *Melitaea aurinia*. *Transactions of the Entomological Society of London*, 78: 345–352.

FORESTRY COMMISSION, 1984. *Census of Woodlands and Trees, 1979–82: Great Britain*. Edinburgh: Forestry Commission.

FOWLER, S. U. and LAWTON, J. H., 1985. Rapidly induced defences and talking trees: the devil's advocates position. *American Naturalist*, 126: 181–195.

FRANK, J. H., 1967. The insect predators of the pupal stage of the winter moth *Operophtera brumata* (L.) (Lepidoptera: Hydriomenidae). *Journal of Animal Ecology*, 36: 375–389.

FRANKEL, O. H., 1974. Genetic conservation: our evolutionary responsibility. *Genetics*, 78, 53–65.

FRANKEL, O. H. and SOULÉ, M. E., 1981. *Conservation and Evolution*. Cambridge: Cambridge University Press.

FRANKLIN, I. R., 1980. Evolutionary change in small populations. In M. E. Soule and B. A. Wilcox (Eds.), *Conservation biology: An Evolutionary–Ecological Perspective*: 135–149. Sunderland, Massachusetts: Sinauer Associates.

FRAZER, J. F. D. and WILLCOX, H. N. A., 1975. Variation in spotting among the close relatives of the butterfly, *Maniola jurtina*. *Heredity*, 34: 305–322.

FRENZEL, B., 1979. Europe without forests. *Geographical Magazine*, 51: 756–761.

FROHAWK, F. W., 1924. *Natural History of British Butterflies* (2 volumes). London: Hutchinson.

FROHAWK, F. W. 1934. *The Complete Book of British Butterflies*. London: Ward Locke.

FRY, G. L. A. and COOKE, A. S., 1984. *Acid Deposition and its Implications for Nature Conservation in Britain*. Focus on Nature Conservation No.7. Peterborough: Nature Conservancy Council.

FRYDENBERG, O. and HØEGH-GULDBERG, O., 1966. The genetic differences between southern English *Aricia agestis* Schiff and Scottish *A.artaxerxes* (F.). *Hereditas*, 56: 145–158.

FULLER, R. J. and WARREN, M. S., 1990. *Coppiced woodland – its history, conservation, and management*. Peterborough: Nature Conservancy Council.

FULLER, R. M., 1987. The changing extent and conservation interest of lowland grasslands in England and Wales: a review of grassland surveys 1930–84. *Biological Conservation*, 40: 281–300.

FUTUYMA, D. J., 1983. Selective factors in the evolution of host choice by phytophagous insects. In S. Ahmad (Ed.), *Insects. Host-Seeking Behaviours and Mechanisms*: 227–244. New York: Academic Press.

FUTUYMA, D. J., 1986. *Evolutionary Biology*. 2nd edition. Sunderland, Massachusetts: Sinauer Associates.

GAEDE, M., 1931. Satyridae. In E. Strand (Ed.), *Lepidopterorum Catalogus*, 29 (43,46,48): 1–759. Berlin: W. Junk.

GALL, L. F., 1984. The effects of capturing and marking

on subsequent activity in *Boloria acrocnema* (Lepidoptera; Nymphalidae), with a comparison of different numerical models that estimate population size. *Biological Conservation*, 28: 139–154.

GAME, M., 1980. Best shape for nature reserves. *Nature*, 287: 630–632.

GAME, M. and PETERKEN, G. F., 1984. Nature reserve selection strategies in the woodlands of central Lincolnshire, England. *Biological Conservation*, 29: 157–181.

GARDINER, B. O. C., 1986. Some thoughts on red admirals in winter. *The Bulletin of the Amateur Entomologists' Society*, 45: 215–217.

GARLAND, S. P., 1981. *Butterflies of the Sheffield area*. Sheffield: Sorby Natural History Society and Sheffield Museums.

GEIGER, H. J., 1980. Enzyme Electrophoretic Studies on the Genetic Relationships of Pierid Butterflies (Lepidoptera: Pieridae) I. European Taxa. *Journal of Research on the Lepidoptera*, 19: 181–195.

GEIGER, H. J., 1982. Electrophoretic differentiation in *Pieris napi* L. *Atalanta*, 13: 234–236.

GEIGER, H. J., 1987.(Ed.) *Tagfalter und ihr Lebensraume*. Basel: Schweizerischer Bund für Naturschutz.

GEIGER, H. J., DESCIMON, H., and SCHOLL, A., 1988. Evidence for speciation within normal *Pontia daplidice* (Linneus 1758) in southern Europe (Lep: Pieridae). *Nota Lepidoterologica*, 11: 7–20.

GERAEDIS, W. H. J. M., 1986. *Voorlopige Atlas van de Nederlanse Dagvlinders Rhopalocera*. Landelijk Dagvlinder project L. H. Wageningen.

GERLACH, P., 1976. 40 Thesen zur problematik der Brechflachen. *Brechflachen in der Landschaft* KTBL-Schrift 195. Munster-Hiltrup.

GIBBS, R. D., 1974. *The Chemotaxonomy of the Flowering Plants* (4 volumes). Montreal: McGill-Queen University Press.

GIBO, D. L. and PALLET, M. J., 1979. Soaring flight of monarch butterflies, *Danaus plexippus*, during the late summer migration in southern Ontario. *Canadian Journal of Zoology*, 57: 1393–1401.

GIBSON, D. O., 1984. How is automimicry maintained? In R. I. Vane-Wright and P. R. Ackery (Eds.), *The Biology of Butterflies*. Symposium of the Royal Entomological Society, Number 11: 163–165. London: Academic Press.

GILBERT, L. E., 1975. Ecological consequences of coevolved mutualism between butterflies and plants. In L. E. Gilbert and P. R. Raven (Eds.), *Coevolution of Animals and Plants*: 210–239. Austin, Texas: University of Texas Press.

GILBERT, L. E. and SINGER, M. C., 1973. Dispersal and gene flow in a butterfly species. *American Naturalist*, 107: 58–72.

GILBERT, L. E. and SINGER, M. C., 1975. Butterfly ecology. *Annual Review of Ecology and Systematics*, 6: 365–397.

GILBERT, L. E., 1984. The biology of butterfly communities. In R. I. Vane-Wright and P. R. Ackery (Eds.), *The Biology of Butterflies*. Symposium of the Royal Entomological Society, Number 11: 41–54. London: Academic Press.

GILBERT, N., 1984a. Control of fecundity in *Pieris rapae*. I. The problem. *Journal of Animal Ecology*, 53: 581–588.

GILBERT, N., 1984b. Control of fecundity in *Pieris rapae*. II. Differential effects of temperature. *Journal of Animal Ecology*, 53: 589–597.

GILBERT, N., 1984c. Control of fecundity in *Pieris rapae*. III. Synthesis. *Journal of Animal Ecology*, 53: 599–609.

GILBERT, N., 1986. Control of fecundity in *Pieris rapae*. IV. Patterns of variation and their ecological consequences. *Journal of Animal Ecology*, 55: 317–329.

GILBERT, N., 1988. Control of fecundity in *Pieris rapae*. V. Comparisons between populations. *Journal of Animal Ecology*, 57: 395–410.

GILCHRIST, G. W. and RUTOWSKI, R. L., 1986. Adaptive and incidental consequences of the alba polymorphism in an agricultural population of *Colias* butterflies: female size, fecundity, and differential dispersion. *Oecologia*, 68: 235–240.

GILPIN, M. and HANSKI, I. (Eds.), 1991. *Metapopulation Dynamics: Empirical and Theoretical Investigations*. London: Academic Press.

GODWIN, H., 1975. *History of the British Flora*. Cambridge: Cambridge University Press.

GONSETH, Y., 1987. *Verbreitungratlas der Tagfalter der Schweiz (Lepidoptera; Rhopalocera)*. Documenta Faunistica Helvetica, 6: 1–242. Centre Suisse de Cartographie de la fauna Neuchâtel.

GOODE, D., 1986. *Wild in London*. London: M. Joseph.

GOSS, H., 1874. A new (?) foodplant for *Melitaea artemis*. *Entomologist*, 7: 203.

GOSSARD, T. W. and JONES, R. E., 1977. The effects of age and weather on egg-laying in *Pieris rapae*. *Journal of Applied Ecology*, 14: 65–71.

GRAHAM, S. M., WATT, W. B., and GALL, L. F., 1980. Metabolic resource allocation vs. mating attractiveness: adaptive pressures on the 'alba' polymorphism of *Colias* butterflies. *Proceedings of the National Academy of Sciences USA*, 77: 3615–3619.

GRAY, J. M. and LOWE, J. J., 1977. *Studies in the Late Glacial Environment*. Oxford: Pergamon Press.

GREENSMITH, J. T. and TOOLEY, M. J., 1982. I.G.C.P. Project 61. Sea-level movements during the last deglacial hemicycle (about 15,000 years). *Proceedings of the Geological Association*, 93: 1–125.

GREGORY, S., 1954. Accumulated temperature maps of the British Isles. *Transactions and Papers of the Institute of British Geographers*, 20: 59–73.

GREIG, J. C., 1979. Principles of genetic conservation in relation to wildlife management in southern Africa. *South African Journal of Wildlife Research*, 9: 57–78.

GUILLAUMIN, M. and DESCIMON, H., 1976. La notion d'espèce chez les lépidoptères. *Mémoires Société Zoologique de France*, 38: 129–201.

GUPPY, C. S., 1986. The adaptive significance of alpine melanism in the butterfly *Parnassius phoebus* F. (Lepidoptera: Papilionidae). *Oecologia*, 70: 205–213.

HALL, S., 1984. Unusual foodplant for painted lady. *The Bulletin of the Amateur Entomologists' Society*, 43: 137.

HANDFORD, P. T., 1973. Patterns of variation in a number of genetic systems in *Maniola jurtina*: the boundary region. *Proceedings of the Royal Society B*, 183: 265–284.

HANSKI, I. and GILPIN, M. 1991. Metapopulation dynamics: brief history and conceptual domain. *Biological Journal of the Linnean Society*, 42: 3–16.

HARCOURT, D. G., 1966. Major factors in survival of the immature stages of *Pieris rapae* (L.). *Canadian Entomologist*, 98: 653–662.

HARRIS, L. D., 1984. *The Fragmented Forest: Island Biogeography Theory and the Preservation of Biotic Diversity*. Chicago: University of Chicago Press.

HASSELL, M. P., 1978. *The Dynamics of Arthropod Predator-Prey Systems*. Princeton, New Jersey: Princeton University Press.

HAWKINS, A. B., 1971. Late Weichselian and Flandrian transgressions of south-western Britain. *Quaternaria*, 14: 115–130.

HAWKSWORTH, D. L. and ROSE, F. R., 1976. *Lichens as Pollution Monitors*. The Institute of Biology's Studies in Biology no. 66. London: Edward Arnold.

HAZEL, W. N., 1977. The genetic basis of pupal colour dimorphism and its maintenance by natural selection in *Papilio polyxenes* (Papilionidae: Lepidoptera). *Heredity*, 38: 227–236.

HAZEL, W. N., BRANDT, R., and GRANTHAM, T., 1987. Genetic variability and phenotypic plasticity in pupal colour and its adaptive significance in the swallowtail butterfly *Papilio polyxenes*. *Heredity*, 59: 449–455.

HAZEL, W. N. and WEST, D. A., 1979. Environmental control of pupal colour in swallowtail butterflies: *Battus philenor* and *Papilio polyxenes*. *Ecological Entomology*, 4: 393–400.

HAZEL, W. N. and WEST, D. A., 1982. Pupal colour dimorphism in swallowtail butterflies as a threshold trait: selection in *Eurytides marcellus* (Cramer). *Heredity*, 49: 295–301.

HAZEL, W. N. and WEST, D. A., 1983. The effect of larval photoperiod on pupal colour and diapause in swallowtail butterflies. *Ecological Entomology*, 8: 37–42.

HEADS, P. A. and LAWTON, J. H., 1984. Bracken, ants and extrafloral nectaries. II. The effect of ants on the herbivores of bracken. *Journal of Animal Ecology*, 53: 1015–1031.

HEAL, H. G., 1965. The wood white *Leptidea sinapis* L. and the railways. *Irish Naturalists' Journal*, 15: 8–13.

HEAL, J. R., 1982. Colour patterns of Syrphidae: IV. Mimicry and variation in natural populations of *Eristalis tenax*. *Heredity*, 49: 95–109.

HEATH, J., 1974. A century of change in the Lepidoptera. In D. L. Hawksworth (Ed.), *The Changing Flora and Fauna of Britain*: 275–292. London: Hawksworth.

HEATH, J., 1981. *Threatened Rhopalocera (Butterflies) in Europe*. Council of Europe, Nature and Environment Series No.23. Strasbourg.

HEATH, J., POLLARD, E., and THOMAS, J. A., 1984. *Atlas of Butterflies in Britain and Ireland*. Harmondsworth: Viking Press.

HEINRICH, B., 1972a. Energetics of temperature regulation and foraging in a bumblebee, *Bombus terricola* Kirby. *Journal of Comparative Physiology*, 77: 49–64.

HEINRICH, B., 1972b. Thoracic temperatures of butterflies in the field near the equator. *Comparative Biochemistry and Physiology*, 43: 459–467.

HEINRICH, B., 1985. Thermoregulation and flight activity of a satyrine, *Coenonympha inornata* (Lepidoptera: Satyridae). *Ecology*, 67: 593–597.

HENRIKSEN, H. J. and KREUTZER, I. S., 1982. *The Butterflies of Scandinavia in Nature*. Odense: Skandinavisk Bogforlag.

HESLOP-HARRISON, J. W. H., 1947. The Pleistocene races of certain British insects and distributional overlapping. *Entomologist's Record and Journal of Variation*, 59: 141–145.

HESLOP-HARRISON, J. W. H., 1953. Lepidoptera of

the lesser Skye Isles. *Entomologist's Record and Journal of Variation*, 67: 141–147; 169–177.

HIGGINS, L. G., 1969. Observations sur les *Melanargia* dans le midi de la France. *Alexanor*, 6: 85–90.

HIGGINS, L. G., 1975. *The classification of European butterflies*. London: Collins.

HIGGINS, L. G., and HARGREAVES, B., 1983. *The Butterflies of Europe*. London: Collins.

HIGGINS, L. G. and RILEY, N. D., 1983. *A Field Guide to the Butterflies of Britain and Europe*. 5th edition. London: Collins.

HIGGS, A. J., 1981. Island biogeography theory and nature reserve design. *Journal of Biogeography*, 8: 117–124.

HIGGS, A. J. and USHER, M. B., 1980. Should nature reserves be large or small? *Nature*, 285: 568–569.

HILL. M. O., 1973. Diversity and eveness: a unifying notation and its consequences. *Ecology*, 54: 427–432.

HINTON, H. E., 1951. Myrmecophilous Lycaenidae and other Lepidoptera – a summary. *Proceedings of the South London Entomological and Natural History Society*, 1949–50: 111–175.

HINTON, H. E., 1970. Insect eggshells. *Scientific American*, 223: 84–91.

HINTON, H. E., 1981. *The Biology of Insect Eggs*. (3 volumes). Oxford: Pergamon Press.

HOCKIN, D. C., 1980. The biogeography of the butterflies of the Mediterranean islands. *Nota Lepidopterologica*, 3: 119–125.

HOCKIN, D. C., 1981. The environmental determinants of the insular butterfly faunas of the British Isles. *Biological Journal of the Linnean Society*, 16: 63–70.

HØEGH-GULDBERG, O., 1966. North European groups of *Aricia allous*; their variability and relationship to *Aricia agestis* (Denis and Schiff.). *Natura Jutlandica*, 13: 1–184.

HØEGH-GULDBERG, O., 1968. Evolutionary trends in the genus *Aricia* (Lep.). *Natura Jutlandica*, 14: 3–76.

HØEGH-GULDBERG, O., 1971a. *Polyommatus icarus* undersidevariabilitet. Kuldeforsog med pupper. *Lepidoptera, KBH.*, 2: 2–5.

HØEGH-GULDBERG, O., 1971b. Pupal sound production of some Lycaenidae. *Journal of Research on the Lepidoptera*, 10: 127–147.

HØEGH-GULDBERG, O., 1974. Natural pattern variation and the effects of cold in the genus *Aricia* R. L. (Lep., Lycaenidae) (*Aricia* Study No. 14). *Proceedings of the British Entomological and Natural History Society*, 7: 37–44.

HØEGH-GULDBERG, O., 1979. The relationship of *Aricia agestis* (Lycaenidae) and its closest relatives in Europe. *Nota Lepidopterologica*, 2: 35–39.

HØEGH-GULDBERG, O., 1985. Southern Palaearctic *Aricia* from Portugal to Afghanistan (*Aricia* studies No. 19). *Proceedings of the 3rd European Congress of Lepidoptera, Cambridge 1982*: 87–96.

HØEGH-GULDBERG, O. and HANSEN, A. L., 1977. Phenotypic wing pattern modification by very brief periods of chilling of pupating *Aricia artaxerxes* (Lycaenidae). *Aricia* studies No. 16. *Journal of the Lepidopterist's Society*, 31: 223–231.

HØEGH-GULDBERG, O. and JARVIS, F. V. L., 1970. Central and north European *Aricia* (Lep.). Relationships, heredity, evolution. *Natura Jutlandica*, 15: 1–119.

HØEGH-GULDBERG, O. and JARVIS, F. V. L., 1974. Polymorphism in *Ariciae* (Lep. Rhop.) in the field and laboratory. The biological relationship between two subspecies of *Aricia artaxerxes* (F.) and temperature experiments on an F2 generation and on *A.artaxerxes* ssp. *salmacis*. *Natura Jutlandica*, 17: 99–120; 121–129.

HÖLLDOBLER, B. and WILSON, E. O., 1990. *The Ants*, Berlin: Springer-Verlag.

HOOPER, M. D., 1980. (Ed.) *Proceedings of Symposium on Area and Isolation, held at Monks Wood Experimental Station, October 1980*. Cambridge: Institute of Terrestrial Ecology.

HOVANITZ, W., 1948. Differences in the field activity of two color phases of *Colias* butterflies at various times of day. *Contributions from the Laboratory of Vertebrate Biology of the University of Michigan*, 41: 1–37.

HOWARTH, T. G., 1971. Descriptions of a new British subspecies of *Pararge aegeria* L., (Lep., Satyridae) and an aberration of *Cupido minimus* Fuessly., (Lep., Lycaenidae). *Entomologist's Gazette*, 22: 117–118.

HOWARTH, T. G., 1973. *South's British Butterflies*. London: F. Warne.

HUNTLEY, B., 1988. Europe. In B. Huntley and T. Webb (Eds.), *Vegetation History*: 341–383. Dordrecht, The Netherlands: Kluwer Academic Publishers.

HUNTLEY, B., 1990. European vegetation history: palaeovegetation map from pollen data—13,000 years BP to present. *Journal of Quaternary Science*, 5: 103–122.

HURLBERT, S. H., 1971. The non-concept of species diversity: a critique and alternative parameters. *Ecology*, 52: 577–586.

HUTCHINSON, G. E., 1957. Concluding remarks. *Cold*

Spring Harbour Symposium on Quantitative Biology, 22: 415–427.

ILSE, D., 1928. Über den Farbensinn der Tagfalter. *Zeitschrift fuer Vergleichende Physiologie*, 8: 658–692.

ILSE, D., 1937. New observations on responses to colours in egg-laying butterlies. *Nature*, 140: 544–545.

INOUYE, D. W., 1978. Resource partitioning in bumblebees: experimental studies of foraging behaviour. *Ecology*, 59: 672–678.

IVERSEN, J., 1958. The bearing of glacial and interglacial epochs on the formation and extinction of plant taxa. In O. Hedberg (Ed.), *Systematics of Today; Proceedings of a Symposium held at the University of Uppsala in Commemoration of the 250th Anniversary of the Birth of Carolus Linnaeus*: 210–215. Uppsala Universitets Arsskrift 1958. Uppsala: Lundequistska Bokhandeln.

JAKŠIĆ, P., 1988. *Privremene Karte Rasrostranjenosti Dnevnih Leptira Jugoslavije (Lepidoptera: Rhopalocera)*. Zagreb: Societas Entomologia Jugoslavica.

JANZEN, D. H., 1983. No park is an island: increase in interference from outside as park size decreases. *Oikos*, 41: 402–410.

JARDINE, W. G., 1975. Chronology of Holocene marine transgression and regression in south-west Scotland. *Boreas*, 4: 173–196.

JARMAN, R., 1984–86. *Exmoor Heath Fritillary Survey 1984, 1985 and 1986*. Bridgewater: Unpublished reports of Somerset Trust for Nature Conservation.

JÄRVI, T., SILLEN-TULLBERG, B., and WIKLUND, C., 1981. The cost of being aposematic. An experimental study of the predation on larvae of *Papilio machaon* by the great tit *Parus major*. *Oikos*, 36: 267–272.

JARVIS, F. V. L., 1962. The genetic relationships between *Aricia agestis* Denis and Schiff., and its subspecies *artaxerxes*. *Proceedings and Transactions of the South London Entomological and Natural History Society*, 106–122.

JARVIS, F. V. L., 1966. The genus *Aricia* (Lep., Rhopalocera) in Britain. *Proceedings and Transactions of the South London Entomological and Natural History Society*, 37–60.

JARVIS, F. V. L., 1968. The status of *Polyommatus salmacis* Stephens (Lep., Lycaenidae). *The Entomologist*, 101: 21–22.

JARVIS, F. V. L., 1969. A biological study of *Aricia artaxerxes* ssp. *salmacis* (Stephens). *Proceedings of the British Entomological and Natural History Society*, 2: 107–117.

JARVIS, F. V. L., 1976. Further research into the relationship of *Aricia agestis* (D. and S.) and *Aricia artaxerxes* (F.) in Britain (Lep., Lycaenidae). *Proceedings of the British Entomological and Natural History Society*, 9: 85–94.

JARZEMBOWSKI, E. A., 1980. Fossil insects from the Bembridge marls, palaeogene of the Isle of Wight, southern England. *Bulletin of the British Museum (Natural History), Geology Series*, 33: 237–293.

JELGERSMA, S., 1979. Sea-level changes in the North Sea Basin. In W. Oele, R. T. E. Schuttenhelme and A. J. Wiggers (Eds.), *The Quaternary History of the North Sea*: 233–248. Symposia Universitatis Annum Quingentesium Celebrantis (Volume 2) Uppsala: Almquist and Wiksell.

JELNES, J. E., 1975a. Isoenzyme heterogeneity in Danish populations of *Aricia artaxerxes* (Lepidoptera, Rhopalocera). *Hereditas*, 79: 47–52.

JELNES, J. E., 1975b. Isoenzyme heterogeneity in Danish populations of *Aricia agestis* (Lepidoptera, Rhopalocera). *Hereditas*, 79: 53–60.

JELNES, J. E., 1975c. A comparative electrophoretic study on Danish species of *Aricia* (Lepidoptera, Rhopalocera). *Hereditas*, 79: 61–66.

JENNERSTEN, O., 1984. Flower visitation and pollination efficiency of some North European butterflies. *Oecologia*, 63: 80–89.

JENSEN, S. R., NEILSON, B. J., and DAHLGREN, R., 1975. Iridoid compounds, their occurrence and systematics: occurrence in the angiosperms. *Botaniska Notiser*, 128: 148–180.

JERMY, T., 1976. Insect–hostplant relationships: Coevolution or sequential evolution? In T. Jermy (Ed.), *The Hostplant in Relation to Insect Behaviour and Reproduction*: 109–113. New York: Plenum Press.

JERMY, T., 1984. Evolution of insect/hostplant relationships. *American Naturalist*, 124: 609–630.

JOHNSTON, G. and MEDICIELO, M., 1979. Female butterfly guarding eggs. *Antenna*, 3: 94.

JOLLY, G. M., 1965. Explicit estimates from capture-recapture data with both death and immigration – stochastic model. *Biometrika*, 52: 225–247.

JONES, C., 1973. *The Conservation of Chalk Downland in Dorset*. Dorchester: Dorset County Council.

JONES, R. E. 1977. Movement patterns and egg distribution in cabbage butterflies. *Journal of Animal Ecology*, 46: 195–212.

JONES, R. E., GILBERT, N., GUPPY, M., and NEALIS, V., 1980. Long-distance movement of *Pieris rapae*. *Journal of Animal Ecology*, 49: 629–642.

JONES, R. E., HART, J. R., and BULL, G. D., 1982. Temperature, size, and egg production in the cabbage butterfly *Pieris rapae*. *Australian Journal of Zoology*, 30: 223–232.

JONES, K. N., ODENDAAL, F. J., and EHRLICH, P. R., 1986. Evidence against the spermataphore as paternal investment in checkerspot butterflies (*Euphydryas*: Nymphalidae). *American Midland Naturalist*, 116: 1–5.

JONES, M. J. and LACE. L. A. (in press). Distribution and behaviour of European and Madeiran Speckled wood butterflies. *Biological Journal of the Linnean Society*.

JONES, M. J., LACE, L. A., HOUNSOME, M. V., and HAMER, K., 1987. The butterflies and birds of Madeira and La Gomera: taxon cycles and human influence. *Biological Journal of the Linnean Society*, 31: 95–111.

KANE, S., 1982. Notes on the acoustic signals of a Neotropical satyrid butterfly. *Journal of the Lepidopterists' Society*, 36: 200–206.

KANZ, J. E., 1977. The orientation of migrant and non-migrant butterflies *Danaus plexipppus*. *Psyche*, 84: 120–141.

KARLSSON, B., 1987. Variation in egg weight, oviposition rate and reproductive reserves with female age in a natural population of the speckled wood butterfly, *Pararge aegeria*. *Ecological Entomology*, 12: 473–476.

KARLSSON, B. and WIKLUND, C. 1985. Egg weight variation in relation to egg mortality and starvation endurance of newly hatched larvae in some satyrid butterflies. *Ecological Entomology*, 10: 205–211.

KENNEDY, C. E. J. and SOUTHWOOD, T. R. E., 1984. The number of species of insects associated with British trees: a reanalysis. *Journal of Animal Ecology*, 53: 455–478.

KERNER, VON MARILAUN, A., 1878. *Flowers and their Unbidden Guests*. London: Kegan Paul.

KETTLEWELL, H. B. D., 1973. *The Evolution of Melanism*. Oxford: Clarendon Press.

KEVAN, P. G. and BAKER, H. G. 1983. Insects as flower visitors and pollinators. *Annual Review of Entomology*, 28: 407–453.

KILLINGBECK, J. P., 1985. *Creating and Maintaining a Garden to Attract Butterflies*. Walsall: National Association for Environmental Conservation.

KILTIE, R. A., 1988. Countershading: universally deceptive or deceptively universal? *Trends in Ecology and Evolution*, 3: 21–23.

KINGSOLVER, J. G., 1981. *Thermoregulatory Strategies in* Colias *Butterflies*. Ph.D. Thesis, Stanford University, California.

KINGSOLVER, J. G., 1983. Thermoregulation and flight in *Colias* butterflies: elevational patterns and mechanistic limitations. *Ecology*, 64: 534–545.

KINGSOLVER, J. G., 1985a. Thermal ecology of *Pieris* butterflies (Lepidoptera: Pieridae): a new mechanism of behavioural thermoregulation. *Oecologia*, 66: 540–545.

KINGSOLVER, J. G., 1985b. Thermoregulatory significance of wing melanization in *Pieris* butterflies (Lepidoptera: Pieridae): physics, posture, and pattern. *Oecologia*, 66: 546–553.

KINGSOLVER, J. G., 1987. Evolution and coadaptation of thermoregulatory behavior and wing pigmentation pattern in Pierid butterflies. *Evolution*, 41: 472–490.

KINGSOLVER, J. G. and DANIEL, T. L., 1979. On the mechanics and energetics of nectar feeding in butterflies. *Journal of Theoretical Biology*, 76: 167–179.

KINGSOLVER, J. G. and MOFFAT, R. J., 1983. Thermoregulation and the determinants of heat transfer in *Colias* butterflies. *Oecologia*, 53: 27–33.

KINGSOLVER, J. G. and WATT, W. B., 1983a. Thermoregulatory strategies in *Colias* butterflies: thermal stress and the limits to adaptation in temporally varying environments. *American Naturalist*, 121: 32–55.

KINGSOLVER, J. G. and WATT, W. B., 1983b. Mechanistic constraints and optimality models: thermoregulatory strategies in *Colias* butterflies. *Ecology*, 65: 1835–1839.

KINGSOLVER, J. G. and WIERNASZ, D. C., 1987. Dissecting correlated characters: adaptive aspects of phenotypic covariation in melanization pattern of *Pieris* butterflies. *Evolution*, 41: 491–503.

KITCHING, R. L., 1981. Egg clustering and the southern hemisphere lycaenids: comments on a paper by N. E. Stamp. *American Naturalist*, 118: 423–425.

KITCHING, R. L. and LUKE, B., 1985. The Myrmecophilous organs of the larvae of some British Lycaenidae (Lepidoptera): a comparative study. *Journal of Natural History*, 19: 259–276.

KNUTH, P., 1906. *Handbook of Flower Pollination* Vol. 1. Translated by J. R. Ainsworth Davis. Oxford: Clarendon Press.

KÖHLER, W. and FELDETTO, W., 1935. Experimentelle untersuchungen über die modifikabilität der flügelzeichnung, ihrer systeme und elemente in den sensiblen perioden von *Vanessa urticae*, nebst einigen beobachtungen an *Vanessa io*. *Archiv der Julius Klaus-Stiftung fur Vererbungsforschung, Sozialanthropologie und Rassenhygiene*, 10: 315–453.

KOLP, O., 1976. Submarine Uferterrasen der südlichen Ost-und Nordsee als Marken des Holozänen Meeransteigs und der Überflutungsphasen der Ostsee. *Petermanns Geographische Mitteilungen*, 120: 1–23.

KOREY, K. A., 1981. Species number, generation length and the molecular clock. *Evolution*, 35: 139–147.

KOSTROWICKI, A. S., 1969. *Geography of the Palearctic Papilionidae (Lep.)*. Krakow: Pantswowe Wydawnictowe Naukowe.

KUDRNA, O., 1986. Aspects of the Conservation of Butterflies in Europe. In *Butterflies of Europe*, Volume 8. Wiesbaden: Aula-Verlag.

KÜHN, A. 1926. Über die änderung des Ziechnungsmusters von Schmetterlingen durch Temperaturreize und das Grundschema der Nymphalidenzeichnung. *Nachrichten von der Gesselschaft der Wissenschaften zu Göttingen Mathematisch-physikalische Klasse*, 1926: 120–141.

LABINE, P. A., 1964. Population biology of the butterfly *Euphydryas editha*. I. Barriers to multiple insemination. *Evolution*, 18: 335–336.

LABINE, P. A., 1966. Population biology of the butterfly *Euphydryas editha*. IV. Sperm precedence. *Evolution*, 20: 580–586.

LABINE, P. A., 1968. The population biology of the butterfly *Euphydryas editha*. VIII. Oviposition and its relation to patterns of oviposition in other butterflies. *Evolution*, 22: 799–805.

LACE, L. A. and JONES, M. J., 1984. Habitat preferences and status of the Madeiran butterfly fauna. *Boletim do Museu Municipal do Funchal*, 36: 162–176.

LAMB, H. H., 1977. *Climate Present, Past and Future*. Volume 2: *Climatic History and the Future*. London: Methuen.

LANE, C., 1957. Preliminary note on insects eaten and rejected by a tame shama (*Kittacincla malabarica*) with the suggestion that in certain species of butterflies and moths females are less palatable than males. *Entomologist's Monthly Magazine*, 93: 172–179.

LANE, R. P., 1984. Host specificity of ectoparasitic midges on butterflies. In R. I. Vane-Wright and P. R. Ackery (Eds.), *The Biology of Butterflies*. Symposium of the Royal Entomological Society, Number 11: 105–108. London: Academic Press.

LARSEN, T. B. 1974. *Butterflies of the Lebanon*. Beirut: National Council for Scientific Research.

LAVERY, T. A., 1989. The heath fritillary (*Mellicta athalia* Rott.): did it really occur in Ireland? *Bulletin of the Amateur Entomologists' Society*, 48: 158–159.

LEATHER, S. R., WATT, A. D., and FORREST, G. I., 1987. Insect induced changes in lodgepole pine (*Pinus contorta*): the effect of previous defoliation on oviposition, growth and survival of the pine beauty moth *Panolis flammea*. *Ecological Entomology*, 12: 275–281.

LEDERHOUSE, R. C., CODELLA, S. G., and COWELL, P. J., 1987. Diurnal predation on roosting butterflies during inclement weather: A substantial source of mortality in the black swallowtail, *Papilio polyxenes* (Lepidoptera: Papilionidae). *Journal of the New York Entomological Society*, 95: 310–319.

LEES, E., 1962. On the voltinism of *Coenonympha pamphilus* (L.) (Lep., Satyridae). *The Entomologist*, 95: 5–6.

LEES, E., 1965. Further observations on the voltinism of *Coenonympha pamphilus* (L.) (Lep., Satyridae). *The Entomologist*, 98: 43–45.

LEES, E., 1969. Voltinism of *Polyommatus icarus* Rott. (Lep., Lycaenidae) in Britain. *The Entomologist*, 102: 194.

LEES, E. and ARCHER, D. M., 1974. Ecology of *Pieris napi* (L.) (Lep., Pieridae) in Britain. *Entomologist's Gazette*, 25: 231–237.

LEES, E. and ARCHER, D. M., 1980. Diapause in various populations of *Pieris napi* L. from different parts of the British Isles. *Journal of Research on the Lepidoptera*, 19: 96–100.

LEES, E. and TILLEY, R. J. D., 1980. Influence of photoperiod and temperature on larval development in *Pararge aegeria* (L.) (Lepidoptera: Satyridae). *Entomologist's Gazette*, 31: 3–6.

LEESTMANS, R., 1983. Les Lepidoptères fossiles trouvés en France. *Linneana Belgica*, IX: 64–89.

LESSE, H. de, 1951. Divisions generiques et subgenriques des anciens genres *Satyrus* et *Eumenis*. *Revue Français de Lepidopterologie*, 13: 39–43.

LESSE, H. de, 1960. Speciation et variation chromosomique chez les Lepidoptères Rhopaloceres. *Annales des Sciences Naturelles Zoologie* 2: 1–223.

LESSE, H. de, 1969. Les nombres de chromosomes dans le groupe de *Lysandra coridon* (Lep. Lycaenidae). *Annales de la Société Entomologique de France (Nouvelle Serie)*, 5: 469–518.

LESSE, H. de, 1970. Les nombres de chromosomes a l'appui d'une systematique du groupe de *Lysandra coridon* (Lycaenidae). *Alexanor*, 6: 203–224.

LEVINS, R., 1970. Extinction. In M. Gerstenhaber (Ed.), *Some Mathematical Questions in Biology*: 77–107. Providence: American Mathematical Society.

LEWONTIN, R. C., 1974. *The genetic basis of evolutionary change*. New York: Columbia University Press.

LEY, C. and WATT, W. B., 1989. Testing the 'mimicry' explanation for the *Colias 'alba'* polymorphism: palatability of *Colias* and other butterflies to wild bird predators. *Functional Ecology*, 3: 183–192.

LIEBERT, T. G. and BRAKEFIELD, P. M., 1990. The genetics of colour polymorphism in the aposematic Jersey tiger moth *Callimorpha quadripunctata*. *Heredity*, 64: 87–92.

LLOYD, H. G., 1970. Past myxomatosis rabbit populations in England and Wales. *European Plant Protection Organisation Publications, Series A*, 58: 197–215.

LLOYD, H. G., 1981. Biological observations on post-myxomatosis wild rabbit populations in Britain 1955–79. In K. Myers and C. D. Macinnes (Eds.), *Proceedings of the World Lagomorph Conference*: 623–68. Ontario: University of Guelph.

LONG, A. G., 1979a. The orange-tip in Midlothian. *Entomologist's Record and Journal of Variation*, 91: 219.

LONG, A. G., 1979b. The return of the orange-tip. *Entomologist's Record and Journal of Variation*, 91: 16–17, 161–158.

LONG, A. G., 1979c. The orange-tip in E. Lothian. *Entomologist's Record and Journal of Variation*, 91: 246.

LONG, A. G., 1980. The mystery of the orange-tip butterfly. *Entomologist's Record and Journal of Variation*, 92: 292–293.

LONG, A. G., 1981. Orange-tips in Peebleshire. *Entomologist's Record and Journal of Variation*, 93: 141.

LONG, R., 1970. Rhopalocera (Lepidoptera) of the Channel Islands. *Entomologist's Gazette*, 21: 241–251.

LONGSTAFF, G. B., 1906. Some rest-attitudes of butterflies. *Transactions of the Entomological Society of London*, Part 1: 97–118.

LORIMER, R. I., 1983. *The Lepidoptera of the Orkney Islands*. Faringdon, Oxon.: E. W. Classey.

LORKOVIĆ, Z., 1938. Studien ueber den Spezies-begriff. II. Artberechtigung von *Everes argiades* Pall., *E.alcetas* Hffgg. und *E.decolorata* Stgr. *Mitteilungen der Munchener Entomologischen Gesellschaften*, 28: 215–246.

LORKOVIĆ Z., 1943. Modifikationen und Rassen von *Everes argiades* Pall. und ihre Beziehungen zu den klimatischen Faktoren ihrer Verbreitungsgebiete. *Mitteilungen der Munchener Entomologischen Gesellschaften*, 33: 431–478.

LORKOVIĆ, Z., 1958. Some peculiarities of spatially and sexually restricted gene exchange in the *Erebia tyndarus* group. *Cold Spring Harbour Symposium of Quantitative Biology*, 23: 319–325.

LORKOVIĆ, Z., 1962. The genetics and reproductive isolating mechanisms of the *Pieris napi-bryoniae* group. *Journal of the Lepidopterists' Society*, 16: 5–19, 105–127.

LORKOVIĆ, Z., 1986. Enzyme electrophoresis and interspecific hybridization in Pieridae (Lep.). *Journal of Research on the Lepidoptera*, 24: 334–358.

LORKOVIĆ, Z. and LESSE, H. de, 1954. Experiences de croisements dans le genre *Erebia*. *Bulletin de la Societé Entomologique de France*, 79: 31–39.

LOYA, Y., 1972. Community structure and species diversity of hermatypic corals at Eilat, Red Sea. *Marine Biology*, 13: 100–123.

LUNDGREN, L., 1977. The role of intra- and inter-specific male: male interactions in *Polyommatus icarus* Rott. and some other species of blues. *Journal of Research on the Lepidoptera*, 16: 249–264.

MABEY, R., 1980. *The Common Ground*. London: Hutchinson.

MACARTHUR, R. H., 1972. *Geographical Ecology: Patterns in the Distribution of Species*. Princeton, New Jersey: Princeton University Press.

MACARTHUR, R. H. and WILSON, E. D., 1967. *The Theory of Island Biogeography*. Princeton: Princeton University Press.

MAGNUS, D. B. E., 1950. Beobachtungen zur Balz und Eiablage des Kaisermantels *Argynnis paphia*. *Zietschrift fur Tierpsychologie*, 7: 435–449.

MAGNUS, D. B. E., 1958. Experimental analysis of some overoptimal sign-stimuli in the mating behaviour of the fritillary butterfly *Argynnis paphia*. *Proceedings of the 10th International Congress of Entomology*, 2: 405–418.

MAISCH, A. and BUCKMANN, D., 1987. The control of cuticular melanin and lutein incorporation in the morphological colour adaptation of a nymphalid pupa, *Inachis io* L. *Journal of Insect Physiology*, 33: 393–402.

MAJERUS, M. E. N., 1979. The status of *Anthocharis cardamines hibernica* Williams (Lepidoptera: Pieridae), with special reference to the Isle of Man. *Entomologist's Gazette*, 30: 245–248.

MALICKY, H., 1969 Versuch einer analyse der okologischen beziehungen zwischen Lycaeniden (Lepidoptera) und Formiciden (Hymenoptera). *Tijdschrift voor Entomologie*, 112: 213–298.

MANLEY, B. F. J. and PARR, M. J., 1968. A new method of estimating population size, survivorship, and birth rate from capture-recapture data. *Transactions of the Society of British Entomology*, 18: 81–89.

MANLEY, G. L., 1974. Central England temperatures: monthly means 1659–1973. *Quarterly Journal of the Royal Meteorological Society*, 100: 389–405.

MARGULES, C., HIGGS, A. J., and RAFE, R. W., 1982. Modern biogeographic theory: are there any lessons for nature reserve design? *Biological Conservation*, 24: 115–128.

MARSH, N. and ROTHSCHILD, M., 1974. Aposematic and cryptic Lepidoptera tested on the mouse. *Journal of Zoology*, 174: 89–122.

MARSHALL, L. D., 1980. *Paternal Investment in* Colias philodice-eurytheme *Butterflies (Lepidoptera: Pieridae)*. M. Sc. Thesis. Arizona State University.

MARSHALL, L. D., 1982. Male nutrient investment in the lepidoptera: what nutrients should males invest? *American Naturalist*, 120: 273–279.

MASCHWITZ, U., WÜST, M., and SCHURIAN, K., 1975. Bläulingsraupen als Zuckerlieferanten für Ameisen. *Oecologia*, 18: 17–21.

MASETTI, M. and SCALI, V. 1972 Ecological adjustments of the reproductive biology in *Maniola jurtina* from Tuscany. *Atti dell'Accademia nazionale dei Lincei. Rendiconti*, 53: 460–470.

MASON, C. F., SHREEVE, T. G., and MCCOY, E. D., 1984. Butterflies and woodlands in Britain. *Biological Conservation*, 28: 377–379.

MASON, L. G., EHRLICH, P. R., and EMMEL, T. C., 1967. The population biology of the butterfly, *Euphydryas editha*. V. Character clusters and asymmetry. *Evolution*, 21: 85–91.

MASON, L. G., EHRLICH, P. R., and EMMEL, T. C., 1968. The population biology of the butterfly, *Euphydryas editha*. VI. Phenetics of the Jasper Ridge colony, 1965–66. *Evolution*, 22: 46–54.

MAY, P. G., 1985. Nectar uptake rates and optimal nectar concentrations of two butterfly species. *Oecologia*, 66: 381–386.

MAYR, E., 1963. *Animal Species and Evolution*. Cambridge, Massachusetts: Harvard University Press.

MAZEL, R., 1986. Contactes parapatriques entre *Melanargia galathea* L. et *M.lachesis* (Lep., Satyridae). *Nota Lepidopterologica*, 9: 81–91.

MAZOKHIN-PORSHNYAKOV, G. A., 1969. *Insect vision*. New York: Plenum Press.

MCKAY, H. V., 1988. *Plant Variability and Egg-Laying by Butterflies*. Ph.D. Thesis (CNAA), Oxford Polytechnic.

MCLEAN, J. A. and BENNETT, R. B., 1978. Characterization of two *Gnathotrichus sulcatus* populations by X-ray energy spectrometry. *Environmental Entomology*, 7: 93–96.

MCNEIL, J. N. and DUCHESNE, R. M., 1977. Transport of hay and its importance in the passive dispersal of the European skipper *Thymelicus lineola* (Lepidoptera, Hesperiidae). *Canadian Entomology*, 109: 1253–1256.

MCWHIRTER, K. G., 1957. A further analysis of variability in *Maniola jurtina*. *Heredity*, 11: 359–371.

MCWHIRTER, K. G., and CREED, E. R., 1971. An analysis of spot placing in the meadow brown butterfly *Maniola jurtina*. In E. R. Creed (Ed.), *Ecological Genetics and Evolution: Essays in Honour of E.B.Ford*. Oxford: Blackwell Scientific Publications.

MELLANBY, K., 1970. *Pesticides and Pollution*. 2nd edition. London: Collins.

MELLANBY, K., 1981. *Farming and Wildlife*. London: Collins.

MELLING, T., 1984a. Discovery of the larvae of the large heath in the wild. *Entomologist's Record and Journal of Variation*, 96: 236.

MELLING, T., 1984b. Extraordinary abilities of larvae of the large heath. *Entomologist's Record and Journal of Variation*, 96: 236.

MELLING, T., 1987. *The Ecology and Population Structure of a Butterfly Cline*. Ph.D. Thesis, University of Newcastle.

MENDEL, H. and PIOTROWSKI, S. H., 1986. *The Butterflies of Suffolk, an Atlas and History*. Ipswich: Suffolk Naturalists' Society.

METEOROLOGICAL OFFICE, 1975a. *Climatological Memorandum No 72. Maps of Average Bright Sunshine over the United Kingdom 1941–1970*. Bracknell, Berks.: Climatological Services.

METEOROLOGICAL OFFICE, 1975b. *Climatological Memorandum No 73. Maps of Mean and Extreme Temperature over the United Kingdom 1941–1970*. Bracknell, Berks.: Climatological Services.

MICHAELIS, H. N., 1963. Lepidoptera. In H. Nelson (Ed.), *The Insects of the Malham Tarn Area. Proceedings of the Leeds Philosophical and Literary Society*, IX: 36–44.

MIDDLETON, J. and MERRIAM, G., 1983. Distribution of woodland species in farmland woods. *Journal of Applied Ecology*, 20: 625–644.

MILLER, K., 1977. *Simple Surveying Techniques for Small Expeditions*. Glasgow: R. Maclehose.

MILLER, L. D., 1968. The higher classification, phylogeny and zoogeography of the Satyridae. *Memoirs of the American Entomological Society*, 24: 1–174.

MILLER, S. E., 1984. Butterflies of the Californian Channel Islands. *Journal of Research into the Lepidoptera*, 23: 282–296.

MILLIMAN, J. D. and EMERY, K. O., 1968. Sea-levels during the past 35,000 years. *Science*, 162: 1121–1123.

MITCHELL, G. F., 1970. The Quaternary deposits between Fennit and Spa on the north shore of Tralee

Bay, county Kerry. *Proceedings of the Royal Irish Academy*, 70: 141–162.

MITCHELL, G. F., 1977. Periglacial Ireland. *Philosophical Transactions of the Royal Society of London B*, 280: 199–209.

MITCHELL, G. F., and WATTS, W. A., 1970. The history of the Ericaceae in Ireland during the Quaternary epoch. In D. Walker and R. G. West (Eds.), *Studies in the Vegetational History of the British Isles*. Cambridge: Cambridge University Press.

MITCHELL, J. B., 1965. *Historical Geography*. London: English Universities Press.

MITCHELL, N. D., 1978. Differential host selection by *Pieris brassicae* (the large white) on *Brassica oleracea* ssp. *oleracea* (the wild cabbage). *Entomologia Experimentalis et Applicata*, 22: 208–219.

MOORE, N. W., 1983. Ecological effects of pesticides. In A. Warren and F. B. Goldsmith (Eds.), *Conservation in perspective*: 159–175. Chichester: Wiley.

MOORE, S. D., 1987. Male-biased mortality in the butterfly *Euphydryas editha*: a novel cost of mate-acquisition. *American Naturalist*, 130: 306–309.

MORIARTY, F., 1969a. Butterflies and insecticides. *Entomologist's Record and Journal of Variation*, 81: 276–278.

MORIARTY, F., 1969b. The sublethal effects of synthetic insecticides on insects. *Biological Review*, 44: 321–357.

MORRIS, M. G., 1967. The representation of butterflies on British statutory nature reserves. *Entomologist's Gazette*, 18: 57–68.

MORRIS, M. G., 1971. The management of grassland for the conservation of invertebrate animals. In E. Duffey and A. S. Watt (Eds.), *The Scientific Management Animal and Plant Communities for Conservation*: 527–552. Oxford: Blackwell Scientific Publications.

MORSE, D. H., 1977. Resource partitioning in bumblebees: the role of behavioural factors. *Science*, 197: 678–680.

MORSE, D. H., 1978. Interactions among bumble bees on roses. *Insectes Sociaux (Paris)*, 25: 365–371.

MORTON, A. C. G., 1979. Isolation as a factor responsible for the decline of the large blue butterfly *Maculinea arion* L., in Great Britain. *Entomologist's Monthly Magazine*, 114: 247–249.

MORTON, A. C. G., 1982. The effects of marking and capture on recapture frequencies of butterflies. *Oecologia*, 53: 105–110.

MORTON, A. C. G., 1983. Butterfly conservation – the need for a captive breeding institute. *Biological Conservation*, 25: 19–33.

MORTON, A. C. G., 1985. *The Population Biology of an Insect with a Restricted Distribution*: Cupido minimus *Fuessly (Lepidoptera, Lycaenidae)*. Ph.D. Thesis, University of Southampton.

MUGGLETON, J. and BENHAM, B. R., 1975. Isolation and the decline of the large blue butterfly (*Maculinea arion*) in Great Britain. *Biological Conservation*, 7: 119–128.

MÜLLER, H. J., 1955. Die saisonformenbildung von *Araschnia levana*. *Naturwissenschaften*, 42: 134–135.

MUNROE, E., 1948. *The Geographical Distribution of Butterflies in the West Indies*. Ph.D. Thesis, Cornell University.

MURPHY, D. D., 1983. Nectar constraints on the distribution of egg masses by the checkerspot butterfly *Euphydryas chalcedona* (Lep.: Nymphalidae). *Environmental Entomology*, 12: 463–465.

MURPHY, D. D., 1984. Butterflies and their nectar plants: the role of the checkerspot butterfly *Euphydryas editha* as a pollen vector. *Oikos*, 43: 113–117.

MURPHY, D. D., LAUNER, A. E., and EHRLICH, P. R., 1983. The role of adult feeding in egg production and population dynamics of the checkerspot butterfly *Euphydryas editha*. *Oecologia*, 56: 257–263.

MURPHY, D. D., MENNINGER, M. S., and EHRLICH, P. R., 1984. Nectar source distribution as a determinant of oviposition host species in *Euphydryas chalcedona*. *Oecologia*, 62: 269–271.

MUSK, L. F., 1985. Glacial and post-glacial climatic conditions in north-west England. In R. H. Johnson (Ed.), *The Geomorphology of North-West England*: 59–79. Manchester: Manchester University Press.

NASH, R., 1975. The butterflies of Ireland. *Proceedings of the British Entomological and Natural History Society*, 7: 69–73.

NATURE CONSERVANCY COUNCIL, 1981. *The Conservation of Butterflies*. Shrewsbury: Nature Conservancy Council.

NATURE CONSERVANCY COUNCIL, 1984. *Nature Conservation in Great Britain*. Shrewsbury: Nature Conservancy Council.

NATURE CONSERVANCY COUNCIL, 1987a. *Thirteenth Report: 1 April 1986 to 31 March 1987*. Peterborough: Nature Conservancy Council.

NATURE CONSERVANCY COUNCIL, 1987b. *Conversion and Extensification of Production: Implications and Opportunities for Nature Conservation*. Peterborough: Nature Conservancy Council.

NATURE CONSERVANCY COUNCIL, 1990. *Sixteenth Report*. Peterborough: Nature Conservancy Council.

NEI, M., MURUYAMA, T., and CHAKRABORTY, R.

1975. The bottleneck effect and genetic variability in populations. *Evolution*, 29, 1–10.

NEKRUTENKO, Y. P., 1965. Tertiary Nymphalid butterflies and some phylogenetic aspects of systematic lepidopterology. *Journal of Research on the Lepidoptera*, 4: 149–158.

NELSON, J. M., 1971. The invertebrates of an area of Pennine moorland within the Moor House Nature Reserve in Northern England. *Transactions of the Society for British Entomology*, 19: 173–235.

NICOLAI, V., 1986. The bark of trees: thermal properties, microclimate and fauna. *Oecologia* 69: 148–160.

NIJHOUT, H. F., 1980. Pattern formation on lepidopteran wings: determination of an eyespot. *Developmental Biology*, 80: 267–274.

NIJHOUT, H. F., 1985. The developmental physiology of color patterns in Lepidoptera. *Advances in Insect Physiology*, 18: 181–247.

NIJHOUT, H. F., 1986 Pattern and pattern diversity on Lepidopteran wings. *Bioscience*, 36: 527–533.

NIJHOUT, H. F., 1991. *The Development and Evolution of Butterfly Wing Patterns*. Washington: Smithsonian Institute Press.

NIJHOUT, H. F. and WRAY, G. A., 1986. Homologies in the colour patterns of the genus *Charaxes* (Lepidoptera: Nymphalidae). *Biological Journal of the Linnean Society*, 28: 387–410.

NORDSTRÖM, F., 1955. *The Distribution of Fennoscandian Butterflies*. Nordisk Stockholm: Familjeboks AB.

NOSS, R. F., 1987. Corridors in real landscapes: a reply to Simberloff and Cox. *Conservation Biology*, 1: 159–164.

NYLIN, S., 1988. Hostplant specialization and seasonality in a polyphagous butterfly, *Polygonia c-album* (Nymphalidae). *Oikos*, 53: 381–386.

NYLIN, S., WICKMAN, P. O., and WIKLUND, C., 1989. Seasonal plasticity in growth and development of the speckled wood butterfly, *Pararge aegeria* (Satyrinae). *Biological Journal of the Linnean Society*, 38: 155–171.

OATES, M. R., 1985. *Garden Plants for Butterflies*. Farnham, Surrey: Brian Marsterton.

OATES, M. R. and WARREN, M. S., 1990. *A Review of Butterfly Introductions in Britain and Ireland*. Research and survey in conservation series. Peterborough: Nature Conservancy Council.

ODENDAAL, F. J., EHRLICH, P. R., and THOMAS, F. C., 1985. Structure and function of the antennae of *Euphydryas editha* (Lepidoptera: Nymphalidae). *Journal of Morphology*, 184: 3–22.

OEHMIG, S., 1980. *Pararge aegeria* L. auf Madeira (Satyridae). *Nota Lepidopterologica*, 6: 60.

O'HARE, G. P., 1974. Lichens and bark acidification as indicators of air pollution in west central Scotland. *Journal of Biogeography*, 1: 135–146.

OHSAKI, N., 1979. Comparative population studies of three Pierid butterflies, *P.rapae*, *P.melete* and *P.napi* living in the same area. I. Ecological requirements for habitat resources in the adults. *Researches on Population Ecology*, 20: 278–296.

OHSAKI, N., 1980. Comparative population studies of three *Pieris* butterflies, *P.rapae*, *P.melete* and *P.napi*, living in the same area. II. Utilization of patchy habitats by adults through migratory and non-migratory movements. *Researches in Population Ecology*, 22: 163–183.

OLIVER, C. G., 1972a. Genetic differentiation between English and French populations of the satyrid butterfly *Pararge megera*. *Heredity*, 29: 307–313.

OLIVER, C. G., 1972b. Genetic and phenotypic differentiation and geographic distance in four species of Lepidoptera. *Evolution*, 26: 221–241.

OLIVER, C. G., 1977. Genetic incompatability between populations of the Nymphalid butterfly *Boloria selene* from England and the United States. *Heredity*, 39: 279–285.

OLIVER, C. G., 1979. Genetic differentiation and hybrid viability between some Lepidoptera species. *American Naturalist*, 114: 681–694.

OWEN, D. F., 1953. Habitat selection in *Argynnis aglaia* (L.) and *A.cydippe* (L.) in the British Isles. *The Entomologist*, 86: 128–129.

OWEN, D. F., 1959. Ecological segregation in butterflies in Britain. *Entomologist's Gazette*, 10: 27–38.

OWEN, D. F., 1975. Estimating the abundance and diversity of butterflies. *Biological Conservation*, 8: 173–183.

OWEN, D. F., 1976. Conservation of butterflies in garden habitats. *Environmental Conservation*, 3: 285–290.

OWEN, D. F., SHREEVE, T. G., and SMITH, A. G., 1986. Colonization of Madeira by the speckled wood butterfly *Pararge aegeria* (Lepidoptera: Satyridae), and its impact on the endemic *Pararge xiphia*. *Ecological Entomology*, 11: 349–352.

OWEN, J., 1983. The most neglected wildlife habitat of all. *New Scientist*, 97: 9–11.

PALMER, R. M. and YOUNG, M. R., 1977. The origin and distribution of Aberdeenshire and Kincardine Lepidoptera. *Entomologist's Record and Journal of Variation*, 89: 285–292.

PARKER, W. E. and GATEHOUSE, A. G., 1985. Genetic factors controlling flight performance and migration in the African armyworm moth *Spodoptera exempta* (Lep. Noctuidae). *Bulletin of Entomological Research*, 75: 49–64.

PARKIN, D. T., 1979. *An Introduction to Evolutionary Genetics*. London: Edward Arnold.

PARR, M. J., GASKELL, T. J., and GEORGE, B. J., 1968. Capture-recapture methods of estimating animal numbers. *Journal of Biological Education*, 2: 95–117.

PARSONS, J. A. and ROTHSCHILD, M., 1964. Rhodanese in the larva and pupa of the common blue butterfly *Polyommatus icarus* (Rott; Lepidoptera). *Entomologist's Gazette*, 15: 58–59.

PASTEUR, G., 1982. A classificatory review of mimicry systems. *Annual Review of Ecology and Systematics*, 13: 169–199.

PEACHEY, C. A., 1978. *The Butterflies of Bernwood Forest: A Preliminary Report May–Sept. 1978*. Unpublished report. Oxford Polytechnic.

PEACHEY, C. A., 1980. *The Ecology of the Butterfly Community of Bernwood Forest*. M. Phil. Thesis (CNAA), Oxford Polytechnic.

PEACHEY, C. A., 1982. *National Butterfly Review: The Representation of Butterflies on National Nature Reserves*. Unpublished report. London: Nature Conservancy Council.

PEACOCK, J. D., GRAHAM, D. K., and WILKINSON, I. P., 1978. Late Glacial and Post-Glacial marine environments at Ardyne, Scotland, and their significance in the interpretation of the history of the Clyde sea area. *Report of the Institute of Geological Sciences* No. 78/17.

PEARSON, G. W., PILCHER, J. R., BAILLIE, M. G. L., and HILLAM, J., 1977. Absolute radiocarbon dating using a low altitude European tree ring calibration. *Nature*, 270: 25–28.

PEET, R. K., 1974. The measurement of species diversity. *Annual Review of Ecology and Systematics*, 5: 285–307.

PENNINGTON, W., 1969. *The History of British Vegetation*. Oxford: Pergamon Press.

PERCIVAL, M. S., 1961. Types of nectar in Angiosperms. *New Phytologist*, 60: 235–281.

PETERKEN, G. F., 1981. *Woodland Conservation and Management*. London: Chapman and Hall.

PETERKEN, G. F. and GAME, M., 1984. Historical factors affecting the number and distribution of vascular plant species in the woodlands of central Lincolnshire. *Journal of Ecology*, 72: 155–182.

PETERSEN, B. 1947. Die geographische variation einiger fennoskanischer Lepidopteran. *Zoologischen Bidrag*, 26: 329–531.

PETERSEN, B. 1954. Egg-laying and habitat selection in some Pieris species. *Entomologisk Tidskrift*, 74: 194–203.

PICARD, F. 1922. Contribution a l'étude des parasites de *Pieris brassicae* L. *Bulletin Biologique*, 56: 54–130.

PIELOU, E. C., 1975. *Ecological Diversity*. New York: Wiley.

PIERCE, N. E., 1984. Amplified species diversity: a case study of an Australian lycaenid butterfly and its attendant ants. In R. I. Vane-Wright and P. R. Ackery (Eds.), *The Biology of Butterflies*. Symposium of the Royal Entomological Society, Number 11: 197–200. London: Academic Press.

PIERCE, N. E. and EASTEAL, S., 1986. The selective advantage of attendant ants for the larvae of a lycaenid butterfly, *Glaucopsyche lygdamus*. *Journal of Animal Ecology*, 55: 451–462.

PIERCE, N. E. and ELGAR, M. A., 1985. The influence of ants on hostplant selection by *Jalmenus evagorus*, a myrmecophilous lycaenid butterfly. *Behavioural Ecology and Sociobiology*, 16: 209–222.

PIVNICK, K. A. and MCNEIL, J. N. 1985. Effects of nectar concentration on butterfly feeding: measured feeding rates for *Thymelicus lineola* (Lepidoptera: Hesperiidae) and a general feeding model for adult Lepidoptera. *Oecologia*, 66: 226–237.

PLANT, C. W., 1987. *Atlas of Butterflies of the London Area*. London: The London Natural History Society, Passmore Edwards Museum.

PLOUVIER, C., 1964. Sur la presence de logonoside dans les escorces de quelques Lonicera (Caprifoliacées) et Hydrangea (Saxifragacées). *Comptes Rendus de l'Academie des Sciences Serie III Sciences de la Vie*, 258: 3919–3922.

PLOWRIGHT, R. C. and OWEN, D. F., 1980. The evolutionary significance of bumble-bee color patterns: a mimetic interpretation. *Evolution*, 34: 622–637.

POLLARD, A. J. and BRIGGS, D., 1984. Genecological studies of *Urtica dioica* L. III. Stinging hairs and plant-herbivore interactions. *New Phytologist*, 97: 507–522.

POLLARD, E., 1977. A method for assessing changes in the abundance of butterflies. *Biological Conservation*, 12: 115–134.

POLLARD, E., 1979a. A national scheme for monitoring the abundance of butterflies: the first three years. *Proceedings of the British Entomological and Natural History Society*, 12: 77–90.

POLLARD, E., 1979b. Population ecology and change in range of the white admiral butterfly *Ladoga camilla* (L.) in England. *Ecological Entomology*, 4: 61–74.

POLLARD, E., 1981. Aspects of the ecology of the meadow brown butterfly, *Maniola jurtina*. *Entomologist's Gazette*, 32: 67–74.

POLLARD, E., 1982. Monitoring the abundance of butterflies in relation to the management of a nature reserve. *Biological Conservation*, 24: 317–328.

POLLARD, E., 1985. Larvae of *Celastrina argiolus* (L.) (Lepidoptera: Lycaenidae) on male holly bushes. *Entomologist's Gazette*, 36: 3.

POLLARD, E., 1988. Temperature, rainfall and butterfly numbers. *Journal of Applied Ecology*, 25: 819–828.

POLLARD, E., 1991. Synchrony of population fluctuations: the dominant influence of widespread factors on local butterfly populations. *Oikos*, 60: 7–10.

POLLARD, E. and HALL, M. L., 1980. Possible movement of *Gonepteryx rhamni* (L.) (Lepidoptera: Pieridae) between hibernating and breeding areas. *Entomologist's Gazette*, 31: 217–220.

POLLARD, E., HALL, M. L., and BIBBY, T. J., 1986. Monitoring the abundance of butterflies 1976–1985. *Research and Survey in Nature Conservation Series*. Number 2. Peterborough: Nature Conservancy Council.

PORTER, A. H. and SHAPIRO, A. M., 1990. Lock-and-key hypothesis: lack of mechanical isolation in a butterfly (Lep: Pieridae) hybrid zone. *Annals of the Entomological Society of America*, 83: 107–114.

PORTER, K., 1981. *The Population Dynamics of Small Colonies of the Butterfly* Euphydryas aurinia. D. Phil. Thesis, University of Oxford.

PORTER, K., 1982. Basking behaviour in larvae of the butterfly *Euphydryas aurinia*. *Oikos*, 38: 308–312.

PORTER, K., 1983. Multivoltinism in *Apanteles bignelli* and the influence of weather on synchronization with its host *Euphydryas aurinia*. *Entomologia Experimentalis et Applicata*, 34: 155–162.

PORTER, K., 1984. Sunshine, sex-ratio and behaviour of *Euphydryas aurinia* larvae. In R. I. Vane-Wright and P. R. Ackery (Eds.), *The Biology of Butterflies*. Symposium of the Royal Entomological Society, Number 11: 309–311. London: Academic Press.

PRATT, C., 1983. A modern review of the demise of *Aporia crataegi* L. the black-veined white. *Entomologist's Record and Journal of Variation*, 95: 45–52.

PRATT, C., 1986. A history and investigation into the fluctuations of *Polygonia c-album* L. *Entomologist's Record and Journal of Variation*, 98: 197–203; 244–250 and 99: 21–27; 69–80.

PREECE, R. C., COXON, P., and ROBINSON, J. E., 1986. New biostratigraphical evidence of the Post Glacial colonization of Ireland and for Mesolithic forest disturbance during the Holocene. *Journal of Biogeography*, 13: 487–509.

PRESTON, F. W., 1962. The canonical distribution of commoness and rarity. *Ecology*, 43: 185–215; 410–432.

PRICE, P. W., 1975. *Insect Ecology*. New York: Wiley-Interscience.

PRICE, R. J., 1983. *Scotland's Environment during the last 30,000 years*. Edinburgh: Scottish Academic Press.

PROCTOR, M. and YEO, P., 1979. *The Pollination of Flowers*. London: Collins.

PULLIN, A. S. 1986a. *Life History Strategies of the Butterflies,* Inachis io *and* Aglais urticae, *Feeding on Nettle,* Urtica dioica. Ph.D. Thesis (CNAA), Oxford Polytechnic.

PULLIN, A. S., 1986b. Influence of the food plant, *Urtica dioica*, on larval development, feeding efficiency, and voltinism of a specialist insect, *Inachis io*. *Holarctic Ecology*, 9: 72–78.

PULLIN, A. S., 1986c. Effect of photoperiod and temperature on the life cycle of different populations of the peacock butterfly *Inachis io*. *Entomologia Experimentalis et Applicata*, 41: 237–242.

PULLIN, A. S., 1987. Changes in leaf quality following clipping and regrowth of *Urtica dioica*, and consequences for a specialist insect herbivore, *Aglais urticae*. *Oikos*, 49: 39–45.

PYÖRNILÄ, M., 1976, 1977 Parasitism in *Aglais urticae* (L.) (Lep., Nymphalidae). *Annales Entomologici Fennici*, 42: 26–33, 133–139, 151–161 and 43: 21–27.

RACKHAM, O., 1975. Temperatures of plant communities as measured by pyrometric and other methods. In G. C. Evans, R. Bainbridge, and O. Rackham (Eds.), *Light as an ecological factor II*: 423–449. Oxford: Blackwell Scientific Publications.

RACKHAM, O., 1980. *Ancient Woodland: its History, Vegetation and Uses in England*. London: Edward Arnold.

RACKHAM, O., 1986. *The History of the Countryside*. London: Dent.

RAFE, R. W. and JEFFERSON, R. G., 1983. The status of *Melanargia galathea* (Lep. Satyridae) in the Yorkshire Wolds. *Naturalist*, 108: 3–7.

RANDS, M. R. W. and SOTHERTON, N. W., 1986. Pesticide use on cereal crops and changes in the abundance of butterflies on arable farmland in England. *Biological Conservation*, 36: 71–82.

RATHKE, B. J., 1976. Competition and co-existence

within a guild of herbivorous insects. *Ecology*, 57: 76–87.

RAUSHER, M. D., 1978. Search image for leaf shape in a butterfly. *Science*, 200: 1071–1073.

RAVENSCROFT, N. O. M., 1990. The ecology and conservation of the silver-studded blue *Plebejus argus* L. on the sandlings of East Anglia, England. *Biological Conservation*, 53: 21–36.

RAVENSCROFT, N. O. M. 1991. The chequered skipper in Scotland. *British Wildlife*, 2: 270–275.

RAWLINS, J. E. and LEDERHOUSE, R. C., 1978. The influence of environmental factors on roosting in the black swallowtail, *Papilio polyxenes asterius* Stoll. (Papilionidae). *Journal of the Lepidopterist's Society*, 32: 145–159.

READ, M., 1985. *The Ecology and Conservation of the Silver-Studded Blue Butterfly*. Unpublished report. Joint Committee for the Conservation of British Insects.

READ. M., 1987. Heathland and the silver-studded blue in Devon. *Nature in Devon*, 7: 5–18.

REDWAY, D. B., 1981. Some comments on the reported occurrence of *Erebia epiphron* (Knoch) (Lepidoptera: Satyridae) in Ireland during the nineteenth century. *Entomologist's Gazette*, 32: 157–159.

REED, T. M., 1985. The number of butterfly species on British islands. *Proceedings of the 3rd Congress of European Lepidoptera, Cambridge*, 1982: 146–152.

REICHHOLF, J., 1973. Die Bedeutung nicht bewirtschafteter Wiesen für unsere Tagfalter. *Natur und Landschaft*, 48: 80–81.

RENWICK, J. A. and RADKE, C. D., 1983. Chemical recognition of hostplants for oviposition by the cabbage butterfly, *Pieris rapae* (Lepidoptera: Pieridae). *Environmental Entomology*, 12: 446–450.

RHOADES, D. F., 1983. Responses of alder and willow to attack by caterpillars and web-worms: evidence for pheromonal sensitivity of willows. In P. A. Hedin (Ed.), *Plant Resistence to Insects*. American Chemical Society Symposium, 208: 56–68.

RICHARDS, O. W., 1940. Biology of the small white butterfly (*Pieris rapae*) with special reference to factors controlling its abundance. *Journal of Animal Ecology*, 9: 243–288.

RICKLEFS, R. E. and COX, G. W. 1972. Taxon cycles in the West Indian avifauna. *American Naturalist*, 106: 195–219.

RILEY, A. M. and LOXDALE, H. D., 1988. Possible adaptive significance of 'tail' structure and 'false-head' lycaenid butterflies. *Entomologist's Record and Journal of Variation*, 100: 59–61.

ROBBINS, R. K., 1980. The lycaenid 'false-head' hypothesis: historical review and quantitative analysis. *Journal of the Lepidopterist's Society*, 34: 194–208.

ROBBINS, R. K., 1981. The lycaenid false-head' hypothesis: predation and wing pattern variation of lycaenid butterflies. *American Naturalist*, 118: 770–775.

ROBBINS, R. K. and AIELLO, A., 1982. Foodplant and oviposition records for Panamanian Lycaenidae and Riodinidae. *Journal of the Lepidopterist's Society*, 36: 65–75.

ROBERTS, B. K., 1977. *Rural Settlement in Britain*. London: Hutchinson University Library.

ROBERTSON, T. S., 1980. Seasonal variation in *Pararge aegeria* (Linnaeus) (Lepidoptera: Satyridae): a biometrical study. *Entomologist's Gazette*, 31: 151–156.

ROBINSON, R., 1971. *Lepidoptera genetics*. Oxford: Pergamon Press.

RODWELL, J., 1982–1989. *Keys to the National Vegetation Classification*. Peterborough: Nature Conservancy Council.

ROER, H., 1961. Ergebnisse mehrjähringer Markierungsversuche zür Erforschunge der fluge-und Wandergehwohnheiten europäischer Schmetterlinge. *Zoologische Anzue*, 167: 456–463.

ROER, H., 1962. Experimenetelle Untersuchungen zum Migrationsverhalten des Kleinen Fuchs (*Aglais urticae*). *Bietrischifte für Entomolgische*, 1: 528–554.

ROFF, D. A., 1973a. On the accuracy of some mark-recapture estimators. *Oecologia*, 12: 15–34.

ROFF, D. A., 1973b. An examination of some statistical tests used in the analysis of mark-recapture data. *Oecologia*, 12: 35–54.

ROLAND, J., 1982. Melanism and diel activity of alpine *Colias* (Lepidoptera: Pieridae). *Oecologia*, 53: 214–221.

ROTHSCHILD, M., 1971. Speculations about mimicry with Henry Ford. In E. R. Creed (Ed.), *Ecological Genetics and Evolution*: 202–223. Oxford: Blackwell Scientific Publications.

ROTHSCHILD, M., 1972. Secondary plant substances and warning coloration in insects. In H. F.van Emden (Ed.), *Insect Plant Relationships*. Symposium of the Royal Entomological Society, Number 6: 59–83. Oxford: Blackwell Scientific Publications.

ROTHSCHILD, M., 1984. British aposematic lepidoptera. In M. Emmet and J. Heath (Eds.), *The Moths and Butterflies of Great Britain and Ireland* (Volume 2): 8–6. Colchester: Harley Books.

ROTHSCHILD, M. and SCHOONHOVEN, L. M., 1977. Assessment of egg load by *Pieris brassicae* (Lepidoptera: Pieridae). *Nature*, 266: 352–355.

ROTHSCHILD, M., VALADON, G., and MUMMERY,

R., 1977. Carotenoids of the pupae of the large white butterfly (*Pieris brassicae*) and the small white butterfly (*Pieris rapae*). *Journal of Zoology*, 181: 323–339.

RUDDIMAN, W. F. and MCINTYRE, A., 1976. North Atlantic palaeoclimatic changes over the last 600,000 years. *Geological Society of America Memoir*, 145: 111–146.

RUDDIMAN, W. F. and MCINTYRE, A., 1981. The North Atlantic ocean during the last deglaciation. *Palaeogeography, Palaeoclimatology and Palaeoecology*, 35: 145–214.

RUSSWURM, A. D. A., 1978. *Aberrations of British Butterflies*. Faringdon, Oxon: E. W. Classey.

RUTOWSKI, R. L., 1979. The butterfly as an honest salesman. *Animal Behaviour*, 27: 1269–1270.

RUTOWSKI, R. L., 1982. Mate choice and lepidopteran mating behaviour. *Florida Entomologist*, 65: 72–82.

SARGENT, C., 1984. *Britain's Railway Vegetation*. Cambridge: Institute of Terrestrial Ecology.

SAUER, J. D., 1969. Oceanic islands and Geographical theory: a review. *Geographical Review*, 59: 582–593.

SCALI, V., 1971. Imaginal diapause and gonadal maturation of *Maniola jurtina* from Tuscany. *Journal of Animal Ecology*, 40: 467–472.

SCALI, V., 1972. Spot-distribution in *Maniola jurtina*: Tuscan archipelago, 1968–1970. *Heredity*, 29: 25–36.

SCALI, V. and MASETTI, M., 1975. Variazioni intrastagionali dello spotting e selezione in *Maniola jurtina*. *Atti dell'Academia nazionale dei Lincei. Rendiconti*, 58: 244–257.

SCHLICHTING, C. D., 1986. The evolution of phenotypic plasticity in plants. *Annual Review of Ecology and Systematics*, 17: 667–693.

SCHMIDT, J. M. and SMITH, J. J., 1987. Short interval time measurement by a parasitic wasp. *Science*, 237: 903–905.

SCHROTH, M. and MASCHWITZ, U., 1984. Zur Larvalbiologie und Wirtsfindung von *Maculinea teleius* (Lepidoptera: Lycaenidae), eines Parasiten von *Myrmica laevinodis* (Hymenoptera: Formicidae). *Entomologia Generalis*, 9: 225–230.

SCOBLE, M. J., 1991. Classification of the Lepidoptera. In A. M. Emmet and J. Heath (Eds.), *The Moths and Butterflies of Great Britain and Ireland*. Volume 7, Part 2. *Lasiocampidae to Thyatridae*: 11–45. Colchester, Harley Books.

SCOTT, J. A., 1972. Biogeography of Antillean butterflies. *Biotropica*, 4: 32–45.

SCOTT, J. A., 1974. Mate-locating behaviour in butterflies. *American Midland Naturalist*, 91: 103–117.

SCOTT, J. A., 1977. Competitive exclusion due to mate-searching behaviour, male-female emergence lags and fluctuation in number of progeny in model invertebrate populations. *Journal of Animal Ecology*, 46: 909–924.

SCOTT, J. A., 1986. On the morphology of the Macrolepidoptera, including a reassessment of the relationships to Cossoidea and Castnioidea, and the reassignment of Mimallonidae to Pyraloidea. *Journal of Research on the Lepidoptera*, 25: 30–38

SCOTT, J. A. and WRIGHT, D. A., 1990. Butterfly phylogeny and fossils. In: O. Kudrna (Ed.) *Butterflies of Europe*, Volume 2: 152–208. Wiesbaden: Aula-Verlag.

SCRIBER, J. M., 1984a. Larval foodplant utilization by the world Papilionidae (Lep.). Latitudinal gradients reappraised. *Tokurana Acta Rhopalocera*, 6/7: 1–50.

SCRIBER, J. M., 1984b. Hostplant suitability. In W. J. Bell and R. T. Cardé (Eds.), *Chemical Ecology of Insects*. 159–202. London: Chapman and Hall.

SCRIBER, J. M. and SLANSKY, F., 1981. The nutritional ecology of immature insects. *Annual Review of Entomology*, 26: 183–211.

SCUDDER, S. H., 1875. Fossil butterflies. *Memoire of the American Association for the Advancement of Science* 1: 1–99.

SHANNON, C. E. and WEAVER, W., 1949. *The Mathematical Theory of Communication*. Urbana, Illinois: University of Illinois Press.

SHAPIRO, A. M., 1970. The role of sexual behaviour in density related dispersal of pierid butterflies. *American Naturalist*, 104: 367–372.

SHAPIRO, A. M., 1975a. The temporal component of butterfly species diversity. In M. L. Cody and J. M. Diamond (Eds.), *Ecology and Evolution of Communities*: 181–195. Cambridge, Massachusetts: Belknap Press.

SHAPIRO, A. M., 1975b. Interspecific competition and the history of *Pieris rapae* in North America. *Entomologist's Record and Journal of Variation*, 87: 17–19.

SHAPIRO, A. M., 1976. Seasonal polyphenism. *Evolutionary Biology*, 9: 259–333.

SHAPIRO, A. M., 1978. The assumption of adaptivity in genital morphology. *Journal of Research on the Lepidoptera*, 17: 68–72.

SHAPIRO, A. M., 1981. The Pierid red-egg syndrome. *American Naturalist*, 117: 276–294.

SHAPIRO, A. M. and CARDÉ, R. T., 1970. Habitat selection and competition among sibling species of satyrid butterflies. *Evolution*, 24: 48–54.

SHAW, M. R., 1977. On the distribution of some Satyridae (Lep.) larvae at a coastal site in relation to their Ichneumonid (Hym.) parasite. *Entomologist's Gazette*, 28: 133–134.

SHEAIL, J., 1971. *Rabbits and their History*. Newton Abbot: David and Charles.

SHELLEY, T. E. and LUDWIG, D., 1985. Thermoregulatory behaviour of the butterfly *Calisto nubila* (Satyridae) in a Puerto Rican forest. *Oikos*, 44: 229–233.

SHENNAN, I., 1982. Interpretation of Flandrian sea-level data from the Fenland, England. In J. T. Greensmith and M. J. Tooley (Eds.), Sea-Level Movements during the Last Deglacial Hemicycle. *Proceeding of the Geological Society*, 93: 53–63.

SHEPPARD, P. M., 1969. Evolutionary genetics of animal populations: the study of natural populations. *Proceedings of the International Congress of Genetics*, 3: 261–279.

SHIELDS, O., 1967. Hilltopping. *Journal of Research on the Lepidoptera*, 6: 69–178.

SHIELDS, O., 1976. Fossil butterflies and the evolution of Lepidoptera. *Journal of Research on the Lepidoptera*, 15: 132–143.

SHIRT, D. B. 1987. (Ed.) *British Red Data Books*: 2. *Insects*. Peterborough: Nature Conservancy Council.

SHOARD, M., 1980. *The Theft of the Countryside*. London: Temple Smith.

SHORROCKS, B., 1978. *The Genesis of Diversity*. London: Hodder and Stoughton.

SHREEVE, T. G., 1984. Habitat selection, mate-location, and microclimatic constraints on the activity of the speckled wood butterfly *Pararge aegeria*. *Oikos*, 42: 371–377.

SHREEVE, T. G., 1985. *The Population Biology of the Speckled Wood Butterfly* Pararge aegeria (*L.*) (*Lepidoptera*: *Satyridae*). Ph.D. Thesis (CNAA), Oxford Polytechnic.

SHREEVE, T. G., 1986a. Egg-laying by the speckled wood butterfly (*Pararge aegeria*): the role of female behaviour, hostplant abundance and temperature. *Ecological Entomology*, 11: 229–236.

SHREEVE, T. G., 1986b. The effect of weather on the life cycle of the speckled wood butterfly *Pararge aegeria*. *Ecological Entomology*, 11: 325–332.

SHREEVE, T. G., 1987. The mate-location behaviour of the male speckled wood butterfly, *Pararge aegeria*, and the effect of phenotypic differences in hindwing spotting. *Animal Behaviour*, 35: 682–690.

SHREEVE, T. G., 1989. The extended flight period of *Maniola jurtina* (Lepidoptera, Satyridae) on chalk downland: seasonal changes of the adult phenotype and evidence for a population of mixed origins. *The Entomologist*, 108: 202–215.

SHREEVE, T. G., 1990. Microhabitat use and hindwing phenotype in *Hipparchia semele* (Lepidoptera, Satyrinae): thermoregulation and background matching. *Ecological Entomology*, 15: 201–213.

SHREEVE, T. G. and DENNIS, R. L. H., 1991. How mobile is the British butterfly fauna? *British Butterfly Conservation Society News*, 48: 20–23.

SHREEVE, T. G. and DENNIS, R. L. H., 1992. The development of butterfly settling posture; the role of predators, climate, hostplant-habitat and phylogeny. *Biological Journal of the Linnean Society*, 45: 57–69.

SHREEVE, T. G. and MASON, C. F., 1980. The number of butterfly species in woodlands. *Oecologia*, 45: 414–418.

SHREEVE, T. G. and SMITH, A. G. (in press). The role of weather related habitat use on the impact of the European speckled wood butterfly *Pararge aegeria* on the endemic *Pararge xiphia* on the island of Madeira. *Biological Journal of the Linnean Society*.

SHUEY, J. A., 1986. Comments on Clench's temporal sequencing of Hesperiid communities. *Journal of Research on the Lepidoptera*, 25: 202–206.

SIEGEL, S., 1956. *Non-parametric Statistics for the Behavioural Sciences*. London: McGraw-Hill Kogakusha.

SILBERGLIED, R. E., 1977. Communication in the Lepidoptera. In T. A. Sebeok (Ed.), *How Animals Communicate*: 362–402. Bloomington, Indiana: Indiana State University.

SILBERGLIED, R. E., 1984 Visual communication and sexual selection among butterflies. In R. I. Vane-Wright and P. R. Ackery (Eds.) *The Biology of Butterflies*. Symposium of the Royal Entomological Society, Number 11: 207–223. London: Academic Press.

SILBERGLIED, R. E., AIELLO, A., and WINDSOR, D. M., 1980. Disruptive coloration in butterflies: lack of support in *Anartia fatima*. *Science*, 209: 617–619.

SILBERGLIED, R. E. and TAYLOR, O. R., 1978. Ultraviolet reflection and its behavioral role in the courtship of the sulfur butterflies, *Colias eurytheme* and *C.philodice*. *Behavioral Ecology and Sociobiology*, 3: 203–243.

SIMBERLOFF, D. and ABELE, L. G., 1984. Conservation and obfuscation: subdivision of reserves. *Oikos*, 42: 399–401.

SIMCOX, D. J. and THOMAS, J. A., 1979. *The Glanville Fritillary*. Joint Committee for the Conservation of British Insects Report.

SIMMONS, I. G., 1979. *Biogeography*: *Natural and Cultural*. London: Edward Arnold.

SIMMONS, I. G. and TOOLEY, M. J., 1981. *The Environment in British Prehistory*. London: Duckworth.

SIMPSON, E. H., 1949. Measurement of diversity. *Nature*, 163: 688.

SINGER, M. C., 1972. Complex components of habitat suitability in a butterfly species. *Science*, 176: 75–77.

SINGER, M. C., 1982. Sexual selection for small size in male butterflies. *American Naturalist*, 119: 440–443.

SISSONS, J. B. and BROOKS, C. L., 1971. Dating of early postglacial land and sea-level changes in the western Forth Valley. *Natural and Physical Sciences*, 234: 124–127.

SLANSKY, F., 1974. Relationship of larval foodplants and voltinism patterns in temperate butterflies. *Psyche*, 81: 243–253.

SMART, P., 1976. *The Illustrated Encyclopaedia of the Butterfly World in Colour*. London: Hamlyn.

SMITH, A. G., 1978. Environmental factors influencing pupal colour determination in Lepidoptera. I. Experiments with *Papilio polytes*, *Papilio demoleus* and *Papilio polyxnes*. *Proceedings of the Royal Society of London B*, 200: 295–329.

SMITH, A. G., 1980. Environmental factors influencing pupal colour determination in Lepidoptera. II. Experiments with *Pieris rapae*, *Pieris napi* and *Pieris brassicae*. *Proceedings of the Royal Society of London B*, 207: 163–186.

SMITH, A. G., HURLEY, A. M., and BRIDEN, J. C. 1981. *Phanerozoic palaeocontinental world maps*. Cambridge: Cambridge University Press.

SMITH, C. J., 1980. *Ecology of the English Chalk*. London: Academic Press.

SMITH, D. A. S., 1984. Mate selection in butterflies: competition, coyness, choice and chauvinism. In R. I. Vane-Wright and P. R. Ackery (Eds.), *The Biology of Butterflies*. Symposium of the Royal Entomological Society, Number 11: 225–244. London: Academic Press.

SMITH, F. W., 1949. The distribution of the orange-tip buttefly, *Euchloe cardamines* in Scotland. *Scottish Naturalist*, 61: 32–35.

SMITH, R. and BROWN, D., 1979. *The Lepidoptera of Warwickshire. A Provisional List. Pt.1. Butterflies 1900 – 1977*. Warwick: Warwick Museum.

SMITH, R. H., SIBLY, R. M., and MOLLER, H., 1987. Control of size and fecundity in *Pieris rapae*: towards a theory of butterfly life cycles. *Journal of Animal Ecology*, 56: 341–350.

SMYLLIE, W. J., 1992. The brown argus butterfly in Britain – a range of *Aricia* hybrids. *The Entomologist*, 111: 27–37.

SNEATH, P. H. A. and SOKAL, R. R., 1973. *Numerical Taxonomy. The Principles and Practice of Numerical Classification*. San Francisco: W. H. Freeman.

SOULÉ, M. E., 1980. Thresholds for survival: maintaining fitness and evolutionary potential. In M. E. Soulé and B. A. Wilcox (Eds.), *Conservation Biology: An Evolutionary-Ecological Perspective*: 151–169. Sunderland: Sinauer Association Incorporated.

SOUTHWOOD, T. R. E., 1962. Migration of terrestrial arthropods in relation to habitat. *Biological Reviews*, 37: 171–214.

SOUTHWOOD, T. R. E., 1977. Habitat, the templet for ecological strategies? *Journal of Animal Ecology*, 46: 337–365.

SOUTHWOOD, T. R. E., 1978. *Ecological Methods: with Particular Reference to the Study of Insect Populations*. 2nd edition. London: Methuen.

SPICER, R. A. and CHAPMAN, J. L., 1990. Climate change and the evolution of high-latitude terrestrial vegetation and floras. *Trends in Ecology and Evolution*, 5: 279–284.

SPOONER, G. M., 1963. On the causes of the decline of *Maculinea arion* L. (Lep. Lycaenidae) in Britain. *The Entomologist*, 96: 199–210.

SPRATT, D. A. and SIMMONS, I. G., 1976. Prehistoric activity and environment on the North Yorks Moors. *Journal of the Archaeological Sciences*, 3: 193–210.

STALLWOOD, B. R., 1972–73. A preliminary survey of the food and feeding habits of adult butterflies. *The Bulletin of the Amateur Entomologists Society*, 31: 25–27; 54–46 and 32: 64–72; 108–114; 174–181.

STAMP, N. E., 1980. Egg deposition patterns in butterflies: why do some species cluster their eggs rather than deposit them singly? *American Naturalist*, 115: 367–380.

STAMP, N. E. and BOWERS, M. D., 1988. Direct and indirect effects of predatory wasps (*Polistes* sp.: Vespidae) on gregarious caterpillars (*Hemileuca lucina*: Saturniidae). *Oecologia*, 75: 619–624.

STANTON, M. L., 1982. Searching in a patchy environment: foodplant selection by *Colias p. eriphyle* butterflies. *Ecology*, 63: 839–853.

STARKEL, L., 1977. The palaeogeography of mid and east Europe during the last cold stage, with west European comparisons. *Philosophical Transactions of the Royal Society of London B*, 280: 351–372.

STEEL, C. and KHAN, R., 1987. *Guidelines for the Management of Rides and Open Spaces in Woodlands*. Bristol: Forestry Commission (West England).

STEEL, C. A. and PARSONS, M., 1985. *National Butterfly Review: The Representation of Butterflies on Royal*

Society for the Protection of Birds Reserves. Unpublished Report. Peterborough: Nature Conservancy Council.

STEEL, C. A. and STEEL, D., 1985. *Butterflies of Berkshire, Buckinghamshire and Oxfordshire*. Oxford: Pisces Publications.

STEPHENS, D. E. A., 1988. Flower-rich landscapes: habitat creation for butterflies. *Proceedings of the National Turf Grass Council Conference: Wildflowers '87*. Bingley, West Yorkshire: Turfgrass Council, 14: 46–53.

STEPHENS, D. E. A. and WARREN, M. S. (in press). *The Importance of Garden Habitats to Butterfly Populations*. Joint Committee for the Conservation of British Insects.

STERN, V. G. and SMITH, R. F., 1960. Factors affecting egg production and oviposition in populations of *Colias eurytheme* Boisduval (Lep.:Pieridae). *Hilgardia*, 29: 411–454.

STOKES, J., 1988. Pollination of the wild gladiolus (*Gladiolus illyricis*) in the New Forest, Hampshire, by butterflies. *News of the British Butterfly Conservation Society*, 41: 38–39.

STREET, F. A., 1980. Ice Age environments. In A. Sherrat (Ed.), *The Cambridge Encyclopaedia of Archaeology*: 5–56. Cambridge: Cambridge University Press.

STRONG, D. R., 1982. Harmonious coexistence of hispine beetles on *Heliconia* in experimental and natural communities. *Ecology*, 63: 1039–1049.

STRONG, D. R., LAWTON, J. H., and SOUTHWOOD, T. R. E., 1984. *Insects on Plants. Community Patterns and Mechanisms*. London: Blackwell Scientific Publications.

STUART, A. J., 1974. Pleistocene history of the British vertebrate fauna. *Biological Review*, 49: 225–266.

STUBBS, A. E., 1979. Changes in London's insect fauna in the last 100 years. *Proceedings and Transactions of the British Entomological and Natural History Society*, 12: 48–57.

STUBBS, A. E., 1991. Protected British butterflies; interpretation of section 9 and schedule 5 of the Wildlife and Countryside Act 1981. *The Entomologist*, 110: 100–102.

STUBBS, D., 1988. *Towards an Introduction Policy: Conservation Guidelines for the Introduction and Reintroduction of Living Organisms into the Wild in Great Britain*. London Ecology Centre: Wildlife Link.

SYNGE, F. M., 1977. Records of sea-levels during the Late Devensian. *Philosophical Transactions of the Royal Society of London, B*, 280: 211–228.

TAUBER, M. J., TAUBER, C. A., and MASAKI, S., 1986. *Seasonal Adaptations of Insects*. Oxford: Oxford University Press.

TAX, M. H., 1989. *Atlas van de Nederlandse Dagvlinders*, Vereniging tot Behoud van Natuurmonumenten in Nederland. Wageningen: Gravenland Vlinderstichting.

TAYLOR, J. A., 1975. Chronometers and chronicles, a study of palaeo-environments in west central Wales. *Progress in Geography*, 5: 248–334.

TAYLOR, O. R., 1973. Reproductive isolation in *Colias eurytheme* and *C.philodice*: use of olfaction in mate selection. *Annals of the Entomological Society of America*, 66: 621–626.

THOMAS, C. D., 1983. *The Ecology and Status of* Plebejus argus *L. in North West Britain*. M. Sc. Thesis, University of Wales, Bangor.

THOMAS, C. D., 1984. Oviposition and egg load assessment by *Anthocharis cardamines* (L.) (Lep.: Pieridae). *Entomologist's Gazette*, 35: 145–148.

THOMAS, C. D., 1985a. Specializations and polyphagy of *Plebejus argus* (Lepidoptera: Lycaenidae) in North Wales. *Ecological Entomology*, 10: 325–340.

THOMAS, C. D., 1985b. The status and conservation of the butterfly *Plebejus argus* L. (Lep: Lycaenidae) in North Wales. *Biological Conservation*, 33: 29–51.

THOMAS, C. D., in press. *Plebejus argus* L. In: T. R. New (Ed.), *The Conservation Biology of Lycaenidae*. Cambridge: I. U. C. N.

THOMAS, J. A., 1974. *Ecological Studies of Hairstreak Butterflies*. Ph.D. Thesis, University of Leicester.

THOMAS, J. A., 1975a. *The Black Hairstreak: Conservation Report*. Unpublished report. Institute of Terrestrial Ecology/ Nature Conservancy Council.

THOMAS, J. A., 1975b. Some observations on the early stages of the purple hairstreak butterfly, *Quercusia quercus* (Linnaeus) (Lep., Lycaenidae). *Entomologist's Gazette*, 26: 224–226.

THOMAS, J. A., 1976. *Report on the Ecology and Conservation of the Large Blue Butterfly*. I.T.E. Project 400.

THOMAS, J. A., 1977. *Second Report on the Ecology and Conservation of the Large Blue Butterfly*. I.T.E. Project No. 400.

THOMAS, J. A., 1980. Why did the large blue become extinct in Britain? *Oryx*, 15, 243–247.

THOMAS, J. A., 1983a. The ecology and conservation of *Lysandra bellargus* (Lepidoptera; Lycaenidae) in Britain. *Journal of Applied Ecology*, 20: 59–83.

THOMAS, J. A., 1983b. The ecology and status of *Thymelicus acteon* (Lep., Hesperiidae) in Britain. *Ecological Entomology*, 8: 427–435.

THOMAS, J. A., 1983c. A quick method for estimating

butterfly numbers during surveys. *Biological Conservation*, 27: 195–211.

THOMAS, J. A., 1983d. A 'Watch' census of common British butterflies. *Journal of Biological Education*, 17: 33–338.

THOMAS, J. A., 1983e. *The Ecological Segregation of Two Species of Large Blue Butterfly* (*Maculinea* spp.). I. T. E. Annual report 1982: 82–83. Cambridge: Institute of Terrestrial Ecology.

THOMAS, J. A., 1984a. The conservation of butterflies in temperate countries: past efforts and lessons for the future. In R. I. Vane-Wright and P. R. Ackery (Eds.) *The Biology of Butterflies*. Symposium of the Royal Entomological Society, Number 11: 333–353. London: Academic Press.

THOMAS, J. A., 1984b. Changes in the status of the Adonis blue and Lulworth skipper in Dorset. *Proceedings of the Dorset Natural History and Archeological Society*, 106: 94–96.

THOMAS, J. A., 1984c. The behaviour and habitat requirements of *Maculinea nausithous* (the dusky large blue butterfly) and *M.teleius* (the scarce large blue) in France. *Biological Conservation*, 28: 325–347.

THOMAS, J. A., 1986. *RSNC Guide to Butterflies of the British Isles*. London: Country Life.

THOMAS, J. A., 1989. Ecological lessons from the re-introduction of Lepidoptera. *The Entomologist*, 108: 56–68.

THOMAS, J. A., 1991. Rare species conservation: case studies of European butterflies. In I. F. Spellerburg, F. B. Goldsmith, and M. G. Morris (Eds.), *The Scientific Management of Temperate Communities for Conservation*: 149–197. Oxford: Blackwell Scientific Publications.

THOMAS, J. A., ELMES, G. W., WARDLAW, J. C., and WOYCIECHOWSKI, M., 1989. Hostplant specificity of *Maculinea* butterflies in *Myrmica* ants nests. *Oecologia*, 79: 452–457.

THOMAS, J. A. and LEWINGTON, R., 1991. *The Butterflies of Britain and Ireland*. London: Dorling Kindersley for the National Trust.

THOMAS, J. A. and SIMCOX, D. J., 1982. A quick method for estimating larval populations of *Melitaea cinxia* during surveys. *Biological Conservation*, 22: 315–322.

THOMAS, J. A. and SNAZELL, R. 1989. Declining fritillaries; the next challenge in the conservation of Britain's butterflies. *Annual Report of the Institute of Terrestrial Ecology*, 1989: 54–56.

THOMAS, J. A., THOMAS, C. D., SIMCOX, D. J., and CLARKE, R. T., 1986. The ecology and declining status of the silver-spotted skipper butterfly (*Hesperia comma*) in Britain. *Journal of Applied Ecology*, 23: 365–380.

THOMAS, J. A. and WARDLAW, J. C., 1990. The effect of queen ants on the survival of *Maculinea arion* larvae in *Myrmica* ant nests. *Oecologia*, 85: 87–91.

THOMAS, J. A. and WARDLAW, J. C. (in press). The capacity of a *Myrmica* nest to support a predacious species of *Maculinea* butterfly. *Oecologia*.

THOMAS, J. A. and WEBB, N. R. 1984. *Butterflies of Dorset*. Dorchester: Dorset Natural History and Archaeological Society.

THOMPSON, J. A., 1952. Butterflies in the coastal region of North Wales. *Entomologist's Record and Journal of Variation*, 64: 161–164.

THOMPSON, J. N., 1989. Concepts of coevolution. *Trends in Ecology and Evolution*, 4: 179–183.

THOMSON, G., 1970. The distribution and nature of *Pieris napi thomsoni* Warren (Lep., Pieridae). *Entomologist's Record and Journal of Variation*, 82: 255–261.

THOMSON, G., 1973. Geographical variation of *Maniola jurtina* L., (Lep., Satyridae). *Tijdschrift voor Entomologie*, 116: 185–226.

THOMSON, G., 1975. Les Races de *Maniola jurtina* L. en France et pays voisins. *Alexanor*, 9: 23–32, 53–66.

THOMSON, G., 1976. Le genre *Maniola* Schrank (Lepidoptera, Satyridae): notes sur les genitalia mâles et femelles. *Linneana Belgica* 6: 126–142.

THOMSON, G. 1980. *The Butterflies of Scotland. A Natural History*. London: Croom Helm.

THOMSON, G. 1987. *Enzyme Variation at Morphological Boundaries in Maniola and Related Genera (Lepidoptera: Nymphalidae: Satyridae)*. Ph.D. Thesis, University of Stirling.

THORPE, J. P., 1982. The molecular clock hypothesis: biochemical evolution, genetic differentiation and systematics. *Annual Review of Ecology and Systematics*, 13: 139–168.

THORSTEINSON, A. J., 1960. Host selection in phytophagous insects. *Annual Review of Entomology*, 5: 193–218.

THROCKMORTON, L. H., 1978. Molecular phylogenetics. In J. A. Rhomberger, R. H. Foote, L. Knutson, and P. L. Lentz (Eds.), *Biosystematics in Agriculture*: New York: Wiley.

TILLEY, R. J. D., 1986. *Melanargia galathea* (L.) and *M.galathea lachesis* (Hübner) in the south of France (Lep., Satyridae). *Entomologist's Gazette*, 37: 1–5.

TINBERGEN, N., 1941. Ethologischebeobachtungen am Samtfalter, *Satyrus semele*. *Journal für Ornithologie*, 89: 132–144.

TINBERGEN, N., 1972. The courtship of the Grayling *Eumenis* (=*Satyrus*) *semele* (L.). In N. Tinbergen (Ed.), *The Animal in its World: Explorations of an Ethologist*. Volume 1, *Field Studies* 1932–1972: 197–249. London: Allen and Unwin.

TINBERGEN, N., 1958. *Curious Naturalists*. London: Country Life.

TINBERGEN, N., MEEUSE, B. J. D., BOEREMA, L. K., and VAROSSIEAU, W. W., 1942. Die Balz des Samtfalters, *Eumenis* (= *Satyrus*) *semele*. *Zeitschrift für Tierpsychologie*, 5: 182–226.

TINDALE, N. B., 1985. A butterfly-moth (Lep. Castriidae) from the Oligocene shales of Florissant, Colorado. *Journal of Research on the Lepidoptera*, 24: 31–40.

TOOLEY, M. J., 1974. Sea-level changes during the last 9000 years in north-west England. *Geographical Journal*, 140: 18–42.

TOOLEY, M. J., 1981. Methods of reconstruction. In G. Simmons and M. J. Tooley (Eds.), *The Environment in British Prehistory*: 1–48. London: Duckworth.

TOWNSEND, C. R., 1987. Bioenergetics: linking ecology, biochemistry and evolutionary theory. *Trends in Ecology and Evolution*, 2: 3–4.

TSUJI, J. S., KINGSOLVER, J. G., and WATT, W. B., 1986. Thermal physiological ecology of *Colias* butterflies in flight. *Oecologia*, 69: 161–170.

TUBBS, R., 1978. The breeding of butterflies, with special reference to the genetics of aberrational forms. *Proceedings of the British Entomological and Natural History Society*, 11: 77–87.

TUDOR, O. and PARKIN, D. T., 1979. Studies on phenotypic variation in *Maniola jurtina* (Lepidoptera: Satyridae) in the Wyre Forest, England. *Heredity*, 42: 91–104.

TURNER, J. R. G., 1963. A quantitative study of a Welsh colony of the large heath butterfly, *Coenonympha tullia* Müller. *Proceedings of the Royal Entomological Society of London A*, 38: 101–112.

TURNER, J. R. G., 1984. Mimicry: the palatability spectrum and its consequences. In R. I. Vane-Wright and P. R. Ackery (Eds.), *The Biology of Butterflies*. Symposium of the Royal Entomological Society, Number 11: 141–161. London: Academic Press.

TURNER, J. R. G., 1986. Why are there so few butterflies in Liverpool? Homage to Alfred Russell Wallace. *Antenna*, 10: 18–24.

TURNER, J. R. G., 1987. The evolutionary dynamics of Batesian and Müllerian mimicry: similarities and differences. *Ecological Entomology*, 12: 81–95.

TURNER, J. R. G., GATEHOUSE, C. M., and COREY, C. A., 1987. Does solar energy control organic diversity? Butterflies, moths and the British climate. *Oikos*, 48: 195–205.

URQUHART, F. A., 1960. *The Monarch Butterfly*. Toronto: University of Toronto Press.

USHER, M. B., 1983. Species diversity: a comment on a paper by W. B. Yapp. *Field Studies*, 5: 825–832.

VAN EMDEN, F. I., 1954. *Handbooks for the Identification of British Insects*. Volume X, Part 4a. *Diptera*: *Cyclorrhapha*. London: Royal Entomological Society.

VAN NOORDWIJK, A. J., VAN BALEN, J. H., and SCHARLOO, W., 1980. Heretability of ecologically important traits in the great tit. *Ardea*, 68: 193–203.

VANE-WRIGHT, R. I., 1975. An integrated classification for polymorphism and sexual dimorphism in butterflies. *Journal of Zoology*, 177: 329–337.

VANE-WRIGHT, R. I., 1976. A unified classification of mimetic resmblances. *Biological Journal of the Linnean Society*, 8: 25–56.

VANE-WRIGHT, R. I., 1978. Ecological and behavioural origins of diversity in butterflies. In L. A. Mound and N. Waloff (Eds.), *Diversity of Insect Faunas*. Symposium of the Royal Entomological Society, Number 9: 56–71. Oxford: Blackwell Scientific Piblications.

VANE-WRIGHT, R. I., 1986. The snake hiss of hibernating peacocks – audio-Batesian mimicry. *Antenna*, 10: 5–6.

VERITY, R., 1916. The British races of butterflies: their relationships and nomenclature. *Entomologist's Record and Journal of Variation*, 28: 73–80, 97–102, 128–133, 165–174.

WADDINGTON, C. H. 1957. *The Strategy of the Genes*. London: Allen and Unwin.

WAGENER, P. S., 1984. *Melanargia lachesis*, est elle un espece differente de *M.galathea*, oui ou non? *Nota Lepidopterologica*, 7: 375–386.

WAGENER, P. S., 1988. What are the valid names for the two genetically different taxa currently included within *Pontia daplidice* (Linneus, 1758) (Lep: Pieridae)? *Nota Lepidopterologica*, 11: 21–38.

WALDBAUER, G. P. and STERNBURG, J. G., 1983. A pitfall in using painted insects in studies of protective coloration. *Evolution*, 37: 1085–1086.

WALES-SMITH, B. G., 1971. Monthly and annual totals of rainfall representative of Kew, Surrey, from 1697 to 1970. *Meteorological Magazine*, 100: 45–60.

WALKER, M. J. C. and LOWE, J. J., 1990. Reconstructing the environmental history of the last glacial-interglacial transition: evidence from the Isle of Skye, Inner Hebrides, Scotland. *Quaternary Science Reviews*, 9: 15–49.

WARING, P., 1984. *A survey of the Butterflies and Moths of Bentley Wood, Wiltshire*. Unpublished report. Nature Conservancy Council, South Region.

WARNECKE, G., 1958. Origin and history of the insect fauna of the nothern Palaearctic. *Proceedings of the 10th International Congress of Entomology* (1956), 1: 719–730.

WARREN, M. S., 1981. *The Ecology of the Wood White Butterfly*, Leptidea sinapis (*Lepidoptera, Pieridae*). Ph.D. Thesis, University of Cambridge.

WARREN, M. S., 1984. The biology and status of the wood white butterfly, *Leptidea sinapis* L. (Lepidoptera, Pieridae) in the British Isles. *Entomologist's Gazette*, 35: 207–223.

WARREN, M. S., 1985a. The influence of shade on butterfly numbers in woodland rides, with special reference to the wood white *Leptidea sinapis*. *Biological Conservation*, 33: 147–164.

WARREN, M. S., 1985b. Habitat utilisation by larvae of the heath fritillary *Mellicta athalia* and other related species in S. E. France. *Bulletin of the British Ecological Society*, 16: 24–26.

WARREN, M. S., 1986. Notes on habitat selection and larval hostplants of the brown argus, *Aricia agestis* (Denis and Schiffermüller), marsh fritillary *Euphydryas aurinia* (Rott.) and painted lady, *Vanessa cardui* (L.) in 1985. *Entomologist's Gazette*, 37: 65–67.

WARREN, M. S., 1987a. The ecology and conservation of the heath fritillary butterfly, *Mellicta athalia*. I. Host selection and phenology. *Journal of Applied Ecology*, 24: 467–482.

WARREN, M. S., 1987b. The ecology and conservation of the heath fritillary butterfly, *Mellicta athalia*. II. Adult population structure and mobility. *Journal of Applied Ecology*, 24: 483–498.

WARREN, M. S., 1987c. The ecology and conservation of the heath fritillary butterfly, *Mellicta athalia*. III. Population dynamics and the effect of habitat management. *Journal of Applied Ecology*, 24: 499–513.

WARREN, M. S., 1989. Pheasants and fritillaries: is there really any evidence that pheasant rearing may have caused butterfly declines? *British Journal of Entomology and Natural History*, 2: 169–175.

WARREN, M. S., 1990. The conservation of *Eurodryas aurinia* in the United Kingdom. In, *Colloquy on the Berne Convention Invertebrates and their Conservation: Conclusions and Summaries*: 71–74. Strasbourg: Council of Europe.

WARREN, M. S., 1991a. The successful conservation of an endangered species, the heath fritillary butterfly

Mellicta athalia, in Britain. *Biological Conservation*, 55: 37–56.

WARREN, M. S., 1991b. Bringing the chequered skipper back to England: a study of its habitats in northern Europe. *British Butterfly Conservation Society News*, 48: 47–51.

WARREN, M. S. (in press). A review of butterfly conservation in central southern Britain. *Biological Conservation*.

WARREN, M. S. and FULLER, M., 1990. *The Management of Woodland Rides and Glades for Wildlife*. Peterborough: Nature Conservancy Council.

WARREN, M. S. and KEY, R. S., 1991. Woodlands: past, present and potential for insects. In N. M. Collins and J. A. Thomas (Eds.), *The Conservation of Insects and their Habitats*: 155–212. London: Academic Press.

WARREN, M. S., POLLARD, E., and BIBBY, T. J., 1986. Annual and long-term changes in a population of the wood white butterfly *Leptidea sinapis*. *Journal of Animal Ecology*, 55: 707–719.

WARREN, M. S. and STEPHENS, D. E. A., 1989. Habitat design and management for butterflies. *The Entomologist*, 108: 123–134.

WARREN, M. S. and THOMAS, J. A., 1992. Butterfly responses to coppicing. In G. P. Buckley (Ed.), *The Ecological Effects of Coppicing*: 249–70. London: Chapman and Hall.

WARREN, M. S., THOMAS, C. D. and THOMAS, J. A., 1981. *The Heath Fritillary: Survey and Conservation Report*. Joint Committee for the Conservation of British Insects: unpublished report.

WARREN, M. S., THOMAS, C. D. and THOMAS, J. A., 1984. The status of the Heath Fritillary butterfly *Mellicta athalia* Rott. in Britain. *Biological Conservation*, 29: 287–305.

WASSERTHAL, L. T., 1975. The role of butterfly wings in regulation of body temperature. *Journal of Insect Physiology*, 21: 1921–1930.

WASSERTHAL, L. T., 1983. Haemolymph flows in the wings of Pierid butterflies visualised by vital staining (Insecta, Lepidoptera). *Zoomorphology*, 103: 177–92.

WATSON, E., 1977. The periglacial environment of Great Britain during the Devensian. *Philosophical Transactions of the Royal Society of London B*, 280: 183–197.

WATT, W. B., 1968. Adaptive significance of pigment polymorphisms in *Colias* butterflies. I. Variation of melanin pigment in relation to thermoregulation. *Evolution*, 22: 437–458.

WATT, W. B., 1969. Adaptive significance of pigment

polymorphisms in *Colias* butterflies. II. Thermoregulation and photoperiodically controlled melanin variation in *Colias philodice eurytheme*. *Proceedings of the National Academy of Sciences USA*, 63: 767–774.

WATT. W. B., 1973. Adaptive significance of pigment polymorphisms in *Colias* butterflies. III. Progress in the study of the 'alba' variant. *Evolution*, 27: 537–548.

WATT, W. B., 1977. Adaptation at specific loci. I. Natural selection on phosphoglucose isomerase of *Colias* butterflies: biochemical and population aspects. *Genetics*, 87: 177–194.

WATT, W. B., 1983. Adaptation at specific loci. II. Demographic and biochemical elements in the maintenance of the *Colias* PGI polymorphism. *Genetics*, 103: 691–724.

WATT, W. B., 1985. Bioenergetics and evolutionary genetics – opportunities for new synthesis. *American Naturalist*, 125: 118–143.

WATT, W. B., CARTER, P. A., and BLOWER, S. M., 1985. Adaptation at specific loci. IV. Differential mating success among glycolytic allozyme genotypes of *Colias* butterflies. *Genetics*, 109: 157–175.

WATT, W. B., CASSIN, R. C., and SWAN, M. S., 1983. Adaptation at specific loci. III. Field behavior and survivorship differencs among *Colias* PGI genotypes are predictable from *in vitro* biochemistry. *Genetics*, 103: 725–739.

WATT, W. B., CHEW, F. S., SNYDER, L. R. G., WATT, A. G., and ROTHSCHILD, D. E. 1977. Population structure of Pierid butterflies. I. Numbers and movements of some montane *Colias* species. *Oecologia*, 27: 1–22.

WATT, W. B., HOCH, P. C., and MILLS, S. G., 1974. Nectar resource use by *Colias* butterflies. Chemical and visual aspects. *Oecologia*, 14: 353–374.

WATT, W. B., KREMEN, C., and CARTER, P., 1989. Testing the 'mimicry' explanation for the *Colias 'alba'* polymorphism: patterns of co-occurrence of *Colias* and Pierine buterflies. *Functional Ecology*, 3: 193–199.

WAY, J. M., 1977. Roadside verges and conservation in Britain; a review. *Biological Conservation*, 12: 65–74.

WAY, M. J., 1962. Mutualism between ants and honeydew-producing Homoptera. *Proceedings of the South London Entomological and Natural History Society*, 1962: 307–343.

WEIS-FOGH, T., 1972. Energetics of hovering flight in hummingbirds and in *Drosophila*. *Journal of Experimental Biology*, 56: 79–104.

WELLS, T. C. E., BELL, S., and FROST, A., 1981. *Creating Attractive Grasslands Using Native Plant Species*. Shrewsbury: Nature Conservancy Council.

WELLS, T. C. E., BELL, S., and FROST, A., 1986. *Wild Flower Grasslands from Crop-Grown Seed and Hay-Bales*. Focus on Nature Conservation Series No. 15. Peterborough: Nature Conservancy Council.

WEST, D. A. and HAZEL, W. N., 1979. Natural pupation sites of swallowtail butterflies (Lepidoptera: Papilioninae): *Papilio polyxenes* Fabr., *P.galaucus* L. and *Battus philenor* (L.). *Ecological Entomology*, 4: 387–392.

WEST, D. A. and HAZEL, W. N., 1982. An experimental test of natural selection for pupation site in swallowtail butterflies. *Evolution*, 36: 152–159.

WEST, D. A. and HAZEL, W. N., 1985. Pupal colour dimorphism in swallowtail butterflies: timing of the sensitive period and environmental control. *Physiological Entomology*, 10: 113–119.

WEST, R. G., 1969. *Pleistocene Geology and Biology*. London: Longman.

WHALLEY, P., 1986. A review of the current fossil evidence of Lepidoptera in the Mesozoic. *Biological Journal of the Linnean Society*, 28: 253–271.

WHEELER, A. S., 1982. *Erebia epiphron* Knoch (Lep., Satyridae) reared on a two-year life cycle. *Proceedings of the British Entomological and Natural History Society*, 15: 28.

WHITE, R. R. and SINGER. M. C., 1974. Geographical distribution of hostplant choice in *Euphydryas editha*. *Journal of the Lepidopterist's Society*, 28: 103–107.

WHITTEN, A. J., 1991. Recovery and hope for Britain's rare species. *British Wildlife*, 2: 219–229.

WICKMAN, P. O., 1985a. The influence of temperature on the territorial and mate locating behaviour of the small heath butterfly, *Coenonympha pamphilus* (L.) (Lepidoptera: Satyridae). *Behavioural Ecology and Sociobiology*, 16: 233–238.

WICKMAN, P. O., 1985b. Territorial defence and mating success in males of the small heath butterfly, *Coenonympha pamphilus* L. (Lepidoptera: Satyridae). *Animal Behaviour*, 33: 1162–1168.

WICKMAN, P. O., 1987. *Mate Searching Behaviour of Satyrine Butterflies*. Ph.D. Thesis, University of Stockholm.

WICKMAN, P. O., 1988. Dynamics of mate searching behaviour in a hilltopping butterfly *Lasiommata megera* (L.). The effects of weather and male density. *Zoological Journal of the Linnean Society*, 93: 357–377.

WICKMAN, P. O. and KARLSSON, B., 1987. Changes in egg colour, egg weight and oviposition rate with the number of eggs laid by wild females of the small heath butterfly, *Coenonympha pamphilus*. *Ecological Entomology*, 12: 109–114.

WICKMAN, P. O. and WIKLUND, C. 1983. Territorial defence and its seasonal decline in the speckled wood (*Pararge aegeria*). *Animal Behaviour*, 31: 1206–1216.

WIGGLESWORTH, V. B., 1964. *The Life of Insects*. London: Weidenfeld and Nicolson.

WIGGLESWORTH, V. B., 1970. *Insect Hormones*. Edinburgh: Oliver and Boyd.

WIKLUND, C., 1974. Oviposition preferences in *Papilio machaon* in relation to the hostplants of the larvae. *Entomologia Experimentalis et Applicata*, 17: 189–198.

WIKLUND, C., 1975. Pupal colour polymorphism in *Papilio machaon* and the survival in the field of cryptic versus non-cryptic pupae. *Transactions of the Royal Entomological Society of London*, 127: 73–84.

WIKLUND, C., 1977a. Oviposition, feeding and spatial separation of breeding and foraging habitats in a population of *Leptidea sinapis* (Lepidoptera). *Oikos*, 28: 56–68.

WIKLUND, C., 1977b. Courtship behaviour in relation to female monogamy in *Leptidea sinapis* (Lepidoptera). *Oikos*, 29: 275–283.

WIKLUND, C., 1981. Generalist vs. specialist oviposition behaviour in *Papilio machaon* (Lepidoptera) and functional aspects on hierarchy of oviposition preferences. *Oikos*, 35: 163–170.

WIKLUND, C., 1982. Behavioural shift from courtship solicitation to mate avoidance in female ringlet butterflies (*Aphantopus hyperantus*) after copulation. *Animal Behaviour*, 30: 790–793.

WIKLUND, C., 1984. Egg-laying patterns in butterflies in relation to their phenology and the visual apparency and abundance of their host plants. *Oecologia*, 63: 23–29.

WIKLUND, C. and ÅHRBERG, C., 1978. Hostplants, nectar source plants, and habitat selection of males and females of *Anthocharis cardamines* (Lepidoptera). *Oikos*, 31: 169–183.

WIKLUND, C., ERIKSSON, T., and LUNDBERG, H., 1979. The wood white butterfly *Leptidea sinapis* and its nectar plants: a case of mutualism or parasitism? *Oikos*, 33: 358–362.

WIKLUND, C. and FAGERSTRÖM, T., 1977. Why do males emerge before females? *Oecologia*, 31: 153–158.

WIKLUND, C. and JÄRVI, T., 1982. Survival of distasteful insects after being attacked by naive birds: a reappraisal of the theory of aposematic coloration evolving through individual selection. *Evolution*, 36: 998–1002.

WIKLUND, C., KARLSSON, B., and FORSBERG, J., 1987. Adaptive *versus* constraint explanations for egg-to-body size relationships in two butterfly families. *American Naturalist*, 130: 828–838.

WIKLUND, C. and PERSSON, A., 1983. Fecundity, egg weight variation and its relation to offspring fitness in the speckled wood butterfly, *Pararge aegeria*, or why don't butterfly females lay more eggs? *Oikos*, 40: 53–63.

WIKLUND, C., PERSSON, A., and WICKMAN, P. O., 1983. Larval aestivation and direct development as alternative strategies in the speckled wood butterfly, *Pararge aegeria*, in Sweden. *Ecological Entomology*, 8: 233–238.

WILEY, E. O., 1981. *Phylogenetics, The Theory and Practice of Phylogenetic Systematics*. New York: Wiley.

WILLIAMS, C. B., 1958. *Insect Migration*. London: Collins.

WILLIAMS, C. B., 1964. *Patterns in the Balance of Nature*, London: Academic Press.

WILLIAMS, R. B. G., 1975. The British climate during the last glaciation; an interpretation based on periglacial phenomena. In A. E. Wright and F. Moseley (Eds.), *Ice Ages; Ancient and Modern*: 95–117. Geological Journal, Special Issue No. 6. Liverpool: Seel House Press.

WILLMOTT, K. J., 1985. A survey of glanville fritillary roosting sites. *News of the British Butterfly Conservation Society*, 35: 35–36.

WILLMOTT, K. J. 1987. *The Ecology and Conservation of the Purple Emperor Butterfly* (*Apatura iris*). Unpublished Report on Project BSP/2 for World Wildlife Fund, 1984–1986.

WILSON, A. J., 1985. Flavinoid pigments in marbled white butterfly are dependent on flavinoid content of larval diet. *Journal of Chemical Ecology*, 11: 1161–1179.

WILSON, E. O., 1961. The nature of the taxon cycle in the Melanesian ant fauna. *American Naturalist*, 95: 169–193.

WILSON, E. O., 1971. *The Insect Societies*. Cambridge, Massachusetts: Belknap Press.

WOOLHOUSE, M. E. J., 1983. The theory and practice of the species-area effect, applied to the breeding birds of British woods. *Biological Conservation*, 27: 315–332.

WORMS, C. G. M. de, 1949. An account of some of the British forms of *Plebejus argus* L. *Report of the Raven Entomological and Natural History Society for 1949*: 28–30.

WOURMS, M. K. and WASSERMAN, F. E., 1985. Butterfly wing markings are more advantageous during handling than during the initial strike of an avian predator. *Evolution*, 39: 845–851.

WRIGHT, D. H., 1983. Species-energy theory: an extension of the species-area theory. *Oikos*, 41: 496–506.

WYKES, G. R., 1952. An investigation of the sugars present in nectar of flowers of various species. *New Phytologist*, 51: 210–215.

YOUNG, A. M., 1979. The evolution of eye spots in tropical butterflies in response to feeding on rotten fruit: an hypothesis. *Journal of the New York Entomological Society*, 87: 66–77

YOUNG, A. M., 1980. The interaction of predators and 'eyespot butterflies' feeding on rotting fruits and soupy fungi in tropical forests. *Entomologist's Record and Journal of Variation*, 90: 63–69.

ZEUNER, F. E., 1962. Notes on the evolution of Rhopalocera (Lep.). *11th International Congress of Entomology*, 1: 310–313.

Index

UNIVERSITY OF GREENWICH LIBRARY